Self-Organization in Electrochemical Systems II

For further volumes:
http://www.springer.com/series/7386

Monographs in Electrochemistry

Surprisingly, a large number of important topics in electrochemistry is not covered by up-to-date monographs and series on the market, some topics are even not covered at all. The series Monographs in Electrochemistry fills this gap by publishing indepth monographs written by experienced and distinguished electrochemists, covering both theory and applications. The focus is set on existing as well as emerging methods for researchers, engineers, and practitioners active in the many and often interdisciplinary fields, where electrochemistry plays a key role. These fields will range – among others – from analytical and environmental sciences to sensors, materials sciences and biochemical research.

Information about published and forthcoming volumes is available at
http://www.springer.com/series/7386

Series Editor: Fritz Scholz, University of Greifswald, Germany

Marek Orlik

Self-Organization in Electrochemical Systems II

Spatiotemporal Patterns and Control of Chaos

Springer

Marek Orlik
University of Warsaw
Faculty of Chemistry
Warsaw
Poland

ISSN 1865-1836 ISSN 1865-1844 (electronic)
ISBN 978-3-642-27626-2 ISBN 978-3-642-27627-9 (eBook)
DOI 10.1007/978-3-642-27627-9
Springer Heidelberg New York Dordrecht London

Library of Congress Control Number: 2012940712

Printed on acid-free paper

Springer is part of Springer Science+Business Media (www.springer.com)

Preface

Self-organization is one of the most important and most general features of nature, being practically omnipresent in our world, viz., in physical and inorganic systems, in organic and living systems, and even in social systems. Already 200 years ago, self-organization phenomena have been observed in electrochemical experiments and much later a vast number of carefully designed electrochemical experiments have been described where self-organization plays a role. Electrochemistry lends itself for such studies in a very special way, as it allows easy control and measurement of the electrode potential and current. Therefore, it is no surprise that the quantitative data of electrochemical experiments have given a very sound basis for mathematical modelling of self-organization. Professor Dr. Marek Orlik is an experienced electrochemist who now specializes in the physical chemistry of self-organization. His profound knowledge of mathematics, physics, and chemistry, together with his clear-cut thinking and his experimental abilities, enables him to present the theoretical background and the experimental details of self-organization in electrochemistry in a very lucid and appealing way. Professor Orlik is a disciple of the Warsaw electrochemical school. He did his PhD with Zbigniew Galus, and worked as postdoc with Gerhard Gritzner (Linz), and he was an *Alexander von Humboldt Fellow* with Karl Doblhofer and Gerhard Ertl in Berlin.

The publishing house Springer and the editor of the series *Monographs in Electrochemistry* regard it as a big fortune that Marek Orlik accepted the invitation to write this monograph because it is the first comprehensive description of that topic, and it is clearly a very seriously needed monograph. When starting to write this monograph, the author quickly realized that the topic cannot be adequately covered in one volume because the mathematical and physical background needs careful and extended explanations. We are thankful to Springer for agreeing to publish this monograph in two volumes, allowing the author to present both the theoretical and the experimental side in detail. Writing such 2-volume monograph is a task which absorbs all energy for several years, and it is not only an intellectual achievement, but also physically very demanding, especially when considering that the author has all the duties of a professor at a chemistry department of a major university! I am sure that the appreciation of the readers will give Marek Orlik the deserved reward and I hope that the monograph will stimulate further studies of this important branch of physical chemistry.

Greifswald, May 2012

Fritz Scholz
University of Greifswald, Germany

Introduction

The world is a *dynamical system* and this system is generally *nonlinear*. This simple sentence describes the fundamental feature of phenomena described in these books. Although most of them belong to the world of electrochemistry, it is extremely important to note that their dynamics exhibit striking analogies, when compared with other types of dynamical systems, like chemical, physical, and biological ones, and even with some social processes or phenomena occurring in the capital market. Since each of these categories deals with its specific language, the universalities in the dynamic behaviors manifest themselves clearly only at the level of mathematical description, which in turn is based on the behavior of solutions of *nonlinear differential equations*. Therefore, *nonlinear dynamics* is really an interdisciplinary science. Electrochemical processes described in these books are thus only specific examples of more general features of nonlinear dynamic processes.

Many phenomena which we experience in everyday life, not only in the scientific laboratory, are nonlinear in their dynamics, but at the elementary level of education their mathematical description is often linearized, like in the case of diffusion transport or heat conduction processes. The advantage of such simplification is the relative simplicity of the mathematical form of respective equations and of the way in which the solutions of them can be obtained. When the differential equation (or the set of them) is nonlinear, the analytical solution is often unknown and one has to use the computer to solve the problem numerically. That is why the invention and development of digital computers assisted the progress in nonlinear dynamics. In course of computer calculations not only numerical solutions of various problems were obtained, but also completely unexpected phenomena were discovered. A seminal example is the observation made in 1963 by E.N. Lorenz, who discovered the unpredictable long-term evolution of the solutions of only three differential equations. This invention led to the idea of *deterministic chaos*, which means complex, aperiodic dynamics generated by entirely deterministic dynamical system, without any assumed stochasticity. One should remember that such unpredictability was treated already at the end of nineteenth century by H. Poincaré who analyzed *qualitatively* the chaotic dynamics of a three-body system, the problem of not having explicit solution in terms of I. Newton's analytical methods. In fact, Poincaré was the pioneer of a modern approach to complex systems in which the emphasis is on the *qualitative type* of dynamics rather than on strict quantitative solutions, frequently not existing.

Deterministic chaos is however an extremely complex example of dynamic phenomena, which manifests itself as aperiodic oscillatory variations of the system's state. But the oscillations can be also periodic, like pulsating of our heart. In chemistry, the best known examples are the periodic variations of the color of the solution in which the Belousov–Zhabotinsky redox reaction is running. In electrochemistry, the oldest examples of analogous phenomena include oscillations of electric current upon anodic electrodissolution of some metals. Periodic dissolution of metals was described as early as in 1828 by G.Th. Fechner, who reported repetitive bursts of effervescence (gas bubbles evolution) during the dissolution of iron in nitric acid.

One should note an extremely important, common feature of those oscillations: they can set in absolutely spontaneously, i.e., without any external periodic perturbation of the system. Therefore, we consider such behavior a *dynamic self-organization*, i.e., the *spontaneous*, coherent dynamic behavior of the system's components, leading, e.g., to periodic (or more complex) variations of its entire characteristics. Furthermore, under appropriate conditions also the spatial symmetry of the dynamical system can be broken and then various types of spatial or spatiotemporal patterns can also *spontaneously* develop. A famous example of stationary patterns of that type was theoretically predicted by A. Turing in his seminal work published in the year 1952. The experimental confirmation of his ideas happened only in 1990s for chemical systems and at the beginning of twenty-first century for electrochemical processes. Thus, self-organization may occur both in time and in space.

One should emphasize that such self-organization phenomena can occur only beyond the state of equilibrium. This is justified by the second law of thermodynamics which forbids the decrease of entropy of Universe. Since creation of any order, including dynamic self-organization, decreases the entropy, there must occur simultaneously an irreversible, dissipative process, in which the production of entropy at least compensates its decrease. This is the thermodynamic condition. In turn, the important kinetic requirement is that the differential equations defining the dynamical systems must be nonlinear. However, it is still not a sufficient condition for the occurrence of self-organization. Analysis of the origin of such phenomena clearly indicates also the necessity of positive and negative *feedback loops* in the steps composing the entire mechanism of a given process. Therefore, understanding of given manifestation of self-organization requires, among others, identification of appropriate feedback loops. In fact, even the simplest case of temporal oscillatory behavior can be understood as the interplay of such feedback loops: during the operation of the positive feedback, the concentration of given species increases quickly in time until the negative feedback loop takes over the control and leads to a decrease in this concentration, creating the conditions for which positive feedback eventually sets in again, etc. In other words, the oscillatory behavior requires the presence of a fast positive and a slow negative feedback loops in the system's dynamics.

A typical sequence of events is that even relatively uncomplicated system, described in terms of deterministic evolution equations, may exhibit sudden change to a completely new, qualitatively different behavior upon smooth variations of the control parameter. A (too) simple deterministic approach, assuming the smooth response of the system's behavior upon increasing distance from equilibrium, must be replaced by a far more sophisticated view, accepting the existence of sudden transitions, mathematically called the *bifurcations*. It is like for the tourist who left the flatlands and entered the mountain area—he has to consider the presence of chasms which can suddenly and dramatically change his situation. This is more than just a simple conclusion; this is a new view of the way in which dynamic processes may develop in nature.

Now it is high time to define the place of electrochemistry in the area of such phenomena. Chemical reactions can exhibit linear or nonlinear dynamics, but electrochemical processes are always inherently nonlinear. This is clearly evidenced even in the simplest case of electric current dependent exponentially on the voltage applied between the electrodes. Furthermore, compared to chemical systems, in electrochemical practice it is extremely easy to drive the system smoothly away from the equilibrium state, by appropriate increase of the voltage applied or the current density imposed. Thus, two fundamental conditions for dynamic self-organization are met, but of course not in every process the appropriate, destabilizing feedback loops can operate, so not every electrochemical process is automatically a source of self-organization under any conditions. Such additional conditions will be shown in these books, based on both numerous experimental examples and theoretical considerations.

In these books thus various kinds of *temporal*, *spatial*, and *spatiotemporal self-organization* in electrochemical systems are described. In spite of specific features of such systems, like the presence and the structure of the electrode–electrolyte interface, the reader will notice analogies in the bifurcation schemes between electrochemical and chemical systems, i.e., those universalities that are so striking at the level of mathematical description. The reader will also be able to notice the evolution of explanations of such phenomena which took place over recent decades. While early works were usually purely electrochemical, i.e., the authors looked for the source of instabilities only in the properties of the electrode–electrolyte interface, later works reflect increasing interest in a treatment based on the concepts of nonlinear dynamics. In particular, as the source of dynamic instabilities, the characteristics of not only the single interface, but of the entire electric circuit could be considered. The books contain numerous examples of both approaches and the reader will be able to choose which of them appears to him/her more convincing. The books' content is organized so that the description of particular electrochemical systems is preceded by an introduction to basic concepts of nonlinear dynamics, in order to help the reader unfamiliar with this discipline to understand at least the fundamental concepts and methods of stability analysis. Also for this purpose this introductory part utilizes selected chemical processes for illustration of self-organization and extends their description for basic stability analysis of electrochemical systems. The next chapters include the description of electrochemical dynamical systems, according to the author's personal selection of papers, including also very recent works. Noteworthy, the presentation of the systems is not always limited to their laboratory construction, but indicates, if possible, their relevance to realistic objects and processes, including systems of biological importance, like neurons in living matter. In fact, in order to understand better the conduction of nerve impulse, it is necessary to update earlier knowledge and models of this process for the recently made progress in self-organization in electrochemical, spatially extended systems.

It is the author's hope that these monographs will trigger and increase the interest of electrochemists, and hopefully also of the students and researchers working in other areas of science, in the modern, interdisciplinary and fascinating subject of dynamic self-organization.

Warsaw, May 2012 Marek Orlik

Contents

Contents for Volume I

Chapter 1
Theoretical Background of Spatial and Spatiotemporal Patterns in Dynamical Systems

1.1 Chemical Reaction–Diffusion Systems

1.1.1 Basic Characteristics of Spatiotemporal Instabilities

In volume I we described the dynamical systems, the *entire* state of which was the same at a given time: the whole system either remained in the steady state or was oscillating with the same period and amplitude and the oscillation shape, irrespective of the location. Such situation corresponds to perfect mixing of homogeneous solutions or, in the case of heterogeneous processes, to perfectly uniform catalyst or electrode surfaces. However, when the solution is not perfectly mixed, or is not mixed at all, or the solid catalyst or electrode surface of realistic structure is in contact with reacting solution, local differences (gradients) of concentrations of species and/or of the electric field are no longer homogenized and give rise to the diffusion and/or migration transport components directed parallel to such surfaces. In this way different sites of the system can "communicate" with each other, exchanging information about their present states, i.e., there appear certain *couplings* between them, giving rise to various dissipative patterns. At a first glance, this conclusion may appear surprising, as the diffusion is usually considered a process that homogenizes a system, and does not assist in keeping or, all the more, developing any inhomogeneity. However, under nonequilibrium conditions, the structure-forming role of diffusion (and other kinds of transport), in cooperation with the chemical or electrochemical process of appropriate characteristics, becomes evident, both experimentally and theoretically. Another type of patterns can emerge if the spatial symmetry of the system is broken by the convective transport which self-organizes into various patterns; the pattern of convective flows determines then the spatial distribution of the system components. Due to their specific characteristics, patterns of convective origin will be described in a separate Chap. 5.

M. Orlik, *Self-Organization in Electrochemical Systems II*,
Monographs in Electrochemistry, DOI 10.1007/978-3-642-27627-9_1,

One should emphasize that the mechanisms and conditions for the formation of spatial and spatiotemporal patterns, the latter ones meaning the concentration fronts traveling through the system and called also "chemical waves," are of tremendous importance for understanding of such phenomena in chemical and biological systems. For example, the biological morphogenesis, when initially identical stem cells differentiate into the specialized tissues may be imagined as the effect of the onset and the progress of the traveling wave of chemical species (like Ca^{2+} ions) which carries information on further different development of initially identical cells. The onset and development of concentric blue wave fronts of the oxidized (Fe^{3+}) form of the ferroin catalyst in the thin layer of the initially red, homogeneous solution with the Belousov–Zhabotinsky reaction running is the simplest and visually attractive chemical model of morphogenesis (Fig. 1.1) [1]. Among natural electrochemical systems a good example of the traveling front is the onset and conduction of the nerve impulse along the neuron.

Mathematical description of such phenomena involves *partial differential equations* (PDE), since dynamic variables, like concentrations, become now the function of both time and spatial coordinate(s). Every such equation includes thus kinetic terms stemming from chemical reaction mechanism and terms describing the spatial transport of the species. In the case of diffusion as the only transport involved, such system of PDE, for N dynamical variables (reagent concentrations) involved in the process, has a following a general form:

$$\frac{\partial c_i(t,x,y,z)}{\partial t} = f_i[\mathbf{c}(t,x,y,z)] \equiv R[\mathbf{c}] + D_i \nabla^2 c_i \quad \text{for } i = 1, \ldots, N \tag{1.1}$$

where $R[\mathbf{c}]$ describes the chemically driven kinetics of the variation of a given species of concentration c_i, while $D_i \nabla^2 c_i$ describes the diffusion of the ith reactant, according to the Fick's law. Generally, the ith species may be produced inside the reactor due to chemical reaction and/or diffuse into/out of that reactor. This is the fundamental equation which will be further considered for various boundary conditions and characteristics of the reaction medium. Later, when electrochemical processes will be discussed, it will be shown how the PDE of the type seen in Eq. (1.1) can be conveniently transformed to the ordinary one (ODE), thus significantly simplifying all calculations and the stability analysis.

The basic type of spatiotemporal patterns that can emerge due to the coupling of chemical kinetics and diffusion transport is the *wave fronts* (also called *trigger waves*) which can appear in excitable media (to which patterns shown in Fig. 1.1 belong). Also the *phase waves* can be distinguished in which no transport is involved, since, e.g., two separate parts of the system can simply reach the same state in different times. If the kinetics of the oscillatory reaction is different in various parts of the system, the emerging pattern is termed a *kinematic wave*. In turn, spatial (stationary) dissipative structures can be the classical *Turing patterns* if particular conditions of their formation are met. For understanding of basic mechanisms underlying such instabilities, it is useful to characterize those simplest cases in terms of appropriate models, keeping in mind, that both in chemical and electrochemical systems also more complicated patterns can develop.

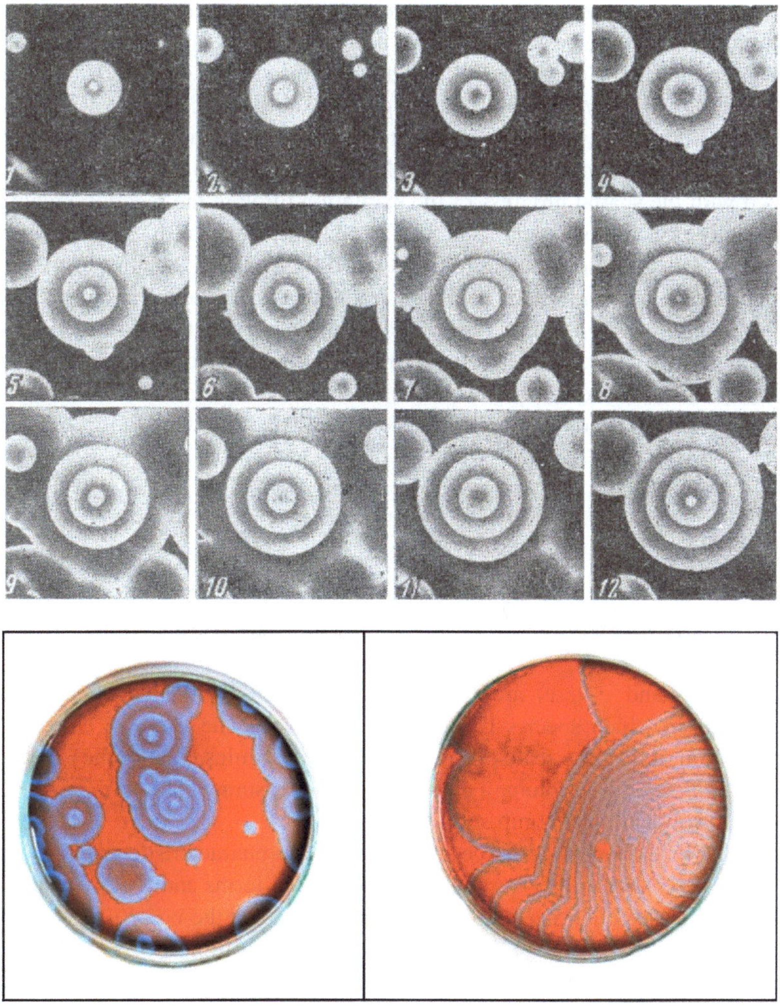

Fig. 1.1 (*Top*) Monochromatic photographic images of concentric traveling fronts in the thin layer of the solution with the Belousov–Zhabotinsky process catalyzed by ferroin. *Blue (bright)* concentric fronts develop in the initially *red (dark)* solution. Intervals between images 1–12: 30 s. Initial concentrations of reactants (M): bromomalonic acid: 0.16; $NaBrO_3$: 0.23; $Fe(phen)_3^{3+}$: 0.003; H_2SO_4: 0.27; temperature 20 °C. Period of oscillations of leading center $T_{LC} \approx 55$ s; front velocity $v = 0.01$ cm s^{-1}, wavelength of leading center $\lambda_{LC} \approx 0.55$ cm (reproduced from [1]); (*Bottom*) Colorful images of two of possible schemes of development of chemical waves in the BZ reaction in the Petri dish

1.1.2 Spatiotemporal Patterns in Excitable Chemical Media

The concept of *excitability* was already mentioned in Sect. 1.6 of volume I. Generally excitability means such a dynamic feature of the system that small perturbations are quickly damped, while sufficiently large ones are enhanced and result in an extensive trip of the system through the phase space before the return to the initial state. In other words, only a sufficiently large perturbation gives rise to a large amplitude single oscillation, through which the system goes back to the initial state, called then *excitable*. Both above-given examples—chemical waves in the Belousov–Zhabotinsky reaction and the impulse conduction along the neuron (axon)— are associated with such excitable characteristics of the nonlinear process, assisted with the spatial progress of the excitation through the reaction medium.

One may explain this crucial feature of excitability in terms of either the thermodynamic or kinetic characteristics of at least two-dimensional dynamical system [2]. We shall choose here the kinetic approach, invoking two dynamic variables. In Chap. 1 of volume I, we used mainly symbols x, y for such variables but now we shall denote them as u and v, respectively, in order to avoid conflict with typical convention of notation of Cartesian spatial coordinates: (x, y, z). Thus, the two-variable (u, v) model reaction–diffusion system is now defined by the following PDE:

$$\frac{\partial u}{\partial \tau} = f(u,v) + D_u \Delta u = f(u,v) + D_u\left[\frac{\partial^2 u}{\partial x^2} + \frac{\partial^2 u}{\partial y^2} + \frac{\partial^2 u}{\partial z^2}\right] \tag{1.2}$$

$$\frac{\partial v}{\partial \tau} = \varepsilon g(u,v) + D_v \Delta v = \varepsilon g(u,v) + D_v\left[\frac{\partial^2 v}{\partial x^2} + \frac{\partial^2 v}{\partial y^2} + \frac{\partial^2 v}{\partial z^2}\right] \tag{1.3}$$

where $f(u, v)$ and $g(u, v)$ are functions determined by the chemical reaction mechanisms in which species U and V are involved. The value of ε determines the separation of the time scale of u (τ_u) and v (τ_v) dynamics, like for the van der Pol equation in Sect. 1.6 of volume I. We shall further assume that $\varepsilon = \tau_u/\tau_v \ll 1$ and thus v is a slow variable, compared to u.

In order to explain the source of the excitable characteristics of this model system, first we shall forget for a while the diffusion terms and focus on the u and v nullclines of the equations of chemical kinetics, i.e., on the $v = f(u)$ dependences coming from Eqs. (1.2) and (1.3) upon applying the conditions: $f(u, v) = 0$ and $g(u, v) = 0$. We shall assume that control parameters were chosen so that those nullclines have a course shown in Fig. 1.2.

Such a shape of nullclines corresponds to the characteristics of two-dimensional OREGONATOR, [cf. Eqs. (1.100) and (1.101), volume I], with u and v replacing x and y, respectively:

$$v = -\frac{qu + (1-q)u^2 - u^3}{f(q-u)} \tag{1.4}$$

$$v = u \tag{1.5}$$

However, for the case shown in Fig. 1.2, the values of f and q parameters were chosen so that the course of nullclines is different from that in Fig. 1.30 of volume I.

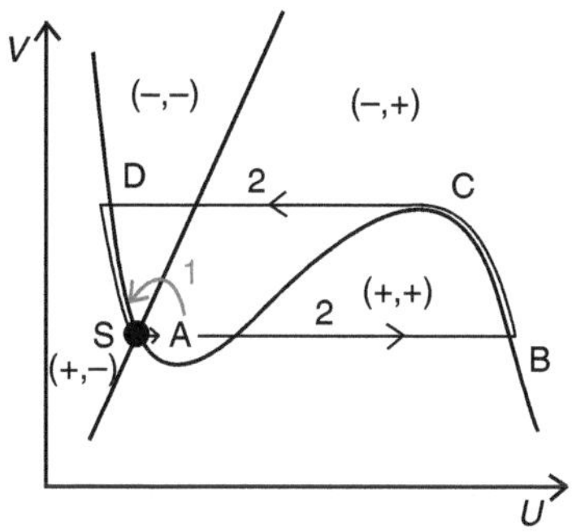

Fig. 1.2 The shape of the *u* (*curved line*) and *v* (*straight line*) nullclines (1.4), (1.5) of dynamical system: two-dimensional OREGONATOR, explaining the excitability of the (locally) stable steady state S. *Symbols in brackets* denote the pairs of signs of the (d*u*/d*t*, d*v*/d*t*) derivatives in the respective sections of the phase space. If the perturbation $\delta u > 0$ is relatively small, the state S is retained through a small amplitude single oscillation [trajectory (1)]; if δu exceeds the AC section of the *u*-nullcline, the perturbation is enhanced to point B and the state S is retained through a large amplitude oscillation, along the trajectory (2) (after [2] and [3])

The stable homogeneous steady state S, defined by the intersection of nullclines in Fig. 1.2 is *excitable*. For such system's characteristics, relatively *small* deviations of variable *u* from its steady-state value at point S, *not exceeding the AC section* of *u*-nullcline, are asymptotically damped, but larger perturbations induce the trip of the system along the A–B–C–D–S pathway in the phase space, until the return to the original state S. In other words, there is a certain threshold, exceeding of which causes a long-distance trip of the system's state, with the amplitude of *u* and *v* variations much larger than in the case of small, subcritical perturbations and now also independent of the magnitude of supercritical perturbations. Such an evolution of the system's state can immediately be associated with the above-mentioned principal scheme of activation of nerve that is enhanced only above certain minimum value of perturbation (of appropriate polarization).

The signs of derivatives (d*u*/d*t*, d*v*/d*t*) in Fig. 1.2 explain the course of respective trajectories. As long as the (positive) perturbation of S toward point A leaves the system on the left-hand side of the AC section of the *u*-nullcline, the sign of d*u*/d*t* remains negative, so the positive perturbation is immediately damped. However, if the initial positive perturbation drives the system on the opposite side of the AC section, the sign of d*u*/d*t* changes to positive, so the increase of *u* proceeds further, now relatively quickly, until the point B is crossed, when the d*u*/d*t* sign returns to a negative value. Further slow trip along the BC section, fast along CD and slow return along DS may be understood analogously as in the case of the van der Pol model, described in Sect. 1.6 of volume I.

It is now useful to extend specific terminology referred to three different states which the system experiences upon its passage along the S–A–B–C–D–S loop: *excitable*, *excited*, and *refractory* states. The state S was already defined as the excitable one. Upon sufficiently large perturbation the system quickly switches to the *excited* states BC, in which it lasts for some time, before it quickly goes to state D and more slowly returns to S. Along the BC path system remains in the so-called *refractory phase*, during which the system is not excitable, i.e., cannot be excited again, and in principle this refractory state persists effectively up to the beginning of the DS path, since then the middle branch of u-nullcline is still far away from the actual states [4]. Only after the return to state S the system recovers the feature of excitability.

In terms of excitability it is now possible to describe a class of spatiotemporal patterns, like those in the Belousov–Zhabotinsky (BZ) reaction (Fig. 1.1). In excitable media such patterns appear when the locally stable steady state, like that denoted with S in Fig. 1.2, but now for the *spatially distributed reaction–diffusion system*, undergoes a sufficiently large local perturbation. Further explanation requires invoking the kinetic characteristics of the BZ reaction. This process contains an autocatalytic step of the production of $HBrO_2$ which can be kinetically important if the concentration of Br^- ions is low enough [see, e.g., the FKN mechanism of this process or the corresponding OREGONATOR model (1.89–1.93, volume I)]. Initially the whole solution is red colored due to predominant presence of Fe(II) ions in ferroin complex. The local fluctuation of $[Br^-]$ ions occurs spontaneously or can be induced by adding a small portion of $AgNO_3$ solution which causes the local precipitation of AgBr. In such a place the local concentration of $HBrO_2$ can quickly (autocatalytically) increase and this situation, in terms of Fig. 1.2, can be interpreted as the switching of the system to the excited state B in which also the ferroin catalyst occurs predominantly in an oxidized form, i.e., containing Fe^{3+} ions—the solution is then blue. Since this excitation occurred only locally, a steep concentration gradient of $HBrO_2$ appears now between the excited area and the *excitable surroundings*, which induces fast $HBrO_2$ diffusion there. The substantial inflow of $HBrO_2$ into these surroundings switches it into the blue-colored excited state, while in the places that were excited before, the system passes the BCDS trajectory, entering the refractory phase, when the excitation is not possible for some time. The return to initial state S, after passing certain distance by the front, means local consumption of $HBrO_2$, reduction of Fe^{3+} ions back to Fe^{2+} (return of red color of ferroin complex), and production of Br^- ions. Figure 1.3 correlates these processes with the spatial progress of a $[HBrO_2]$ front along one dimension that corresponds to the BZ process running in a long tube of small diameter (capillary) reactor. The contribution from diffusion causes that the system does not stop in the state S, but once set into "swinging motion" generates the repeating wave fronts.

With respect to the rate of this front propagation it is intuitively understandable that it should be the faster, the greater is the rate constant of the autocatalytic production of $HBrO_2$ and its diffusion coefficient. In conjunction with the FKN mechanism of the BZ process, the following dependence was derived [5, 6]:

$$v_{\text{front}} = 2(Dk'[BrO_3^-][H^+])^{1/2} \tag{1.6}$$

The square-root dependence of v_{front} on the $[BrO_3^-]$ and $[H^+]$ concentrations was confirmed experimentally. The conclusion is also that if the concentrations of these reactants are constant, the front travels through the excitable medium with

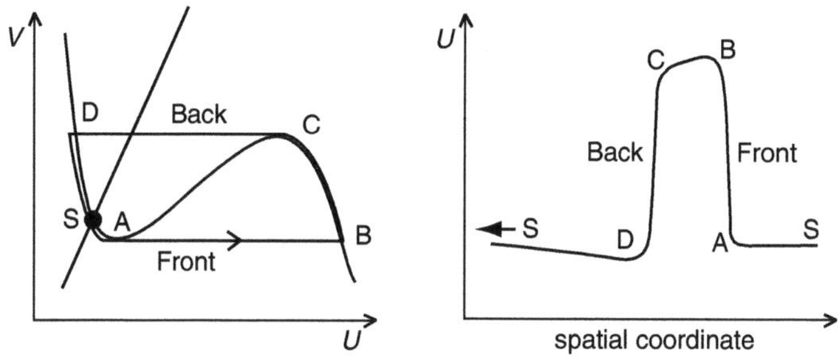

Fig. 1.3 Correlation of the excitable characteristics of the chemical reaction (*left*) with the formation and progress of the pulse wave in the one-dimensional spatially distributed reaction–diffusion system. The pulse passes through every point of the system's spatial coordinate from the left to the right. Every point on the space coordinate axis follows the cycle S–A–B–C–D–S on the *u*–*v* plane, but at different times. Reproduced from [2] with kind permission from Springer Science + Business Media

constant velocity. Under typical conditions and concentrations this rate is of the order of a few mm min^{-1}. It is useful to compare this value immediately with the rate of propagation of excitation along the squid axon, which is as high as up to 10^3 cm s^{-1} [2]. This difference is understandable if one realizes that in the latter case not only diffusion, but also migration of ions in the electric field contributes to the propagation of wave front. This point will be elaborated in more detail in further parts of this chapter, describing recent achievements in understanding of spatio-temporal patterns in electrochemical systems.

Taking the pulse in a BZ system as a representative example of the front in excitable media with coupled chemical and diffusion processes, one can indicate some its features which may be to some extent generalized for other systems [2]. Obviously, if the front and the back of the wave move with the same velocity, the pulse is stable and vanishes only at the boundary of the reactor. Otherwise, if the back moves faster than the front, the pulse undergoes gradual destruction. If the front moves faster than the back, the excited zone widens as a function of time. If the initial perturbation is imposed somewhere in the middle of the capillary, the resulting wave front splits into two ones, traveling in the opposite directions with the same velocity. On the other hand, if two pulses initiated in two different places of the capillary move toward each other, after collision they annihilate, since the surroundings of those wave fronts immediately after collision can only be in excited or refractory, but in excitable states.

Finally, let us note an important condition of instability, specific for spatially distributed systems. Since the pulse in Fig. 1.3 has certain finite width, it is obvious that this kind of instability must be affected by the available space of the system. Such a pulse will develop and travel through the system only if it has sufficient space for that. In other words, if initial fluctuation is spatially too large, compared to the system's size, it may decay, instead of giving rise to macroscopically developed concentration front.

The spatial progress of the concentration fronts in more than one-dimensional system is of course more complicated. In two-dimensional systems, well approximated by a thin layer of the solution in the Petri dish, circular or spiral waves (called *target patterns*) develop. In the case of the BZ reaction, the generation of concentric waves repeats periodically, giving rise to the patterns of concentric waves of a common origin (Fig. 1.1). There were numerous attempts to understand both the origin and the repeated birth of this set of concentric waves [3]. One of the explanations assumes spontaneous local concentration fluctuation that exceeds the threshold of instability and causes initiation of the wave. Another concept, involving heterogeneous source, assumes that the wave starts from the place which contains the *pacemaker*, e.g., the dust particle that locally changes the excitable into oscillatory dynamics (i.e., in terms of the nullcline representation the system's characteristics change from that shown in Fig. 1.3 to that in Fig. 1.30 of volume I), while away from the pacemaker the medium remains excitable. Then the phase wave born in the oscillatory center turns into trigger wave propagating through an excitable bulk medium, and the next front starts after the oscillatory region has gone through its cycle and the excitable part has returned to its resting state after the refractory period [3]. In our recent modeling of the circular waves in the BZ system, aimed also to explain the spatial distribution of solution temperature (see Fig. 1.4), measured experimentally with infrared camera [7], we defined the initial state of the model system as exhibiting subcritical Hopf bifurcation. Although Fig. 1.4 shows only single wave fronts, prolonged numerical simulations showed the spontaneous periodic birth of the next wave fronts.

In three dimensions, spherical or helical patterns can be observed [2]. In the course of numerous research studies of these phenomena, the initially applied aqueous solutions were later replaced with gel media (cf., e.g., [8]) which hinder undesirable convection and facilitate studies in three dimensions. One should note that the velocity of the circular wave propagation depends on the local front curvature K which is defined as the inverse of radius $K = -1/r$. Then the relationship between the speed of a planar wave (v_0) and the speed v of an identical wave with the curvature $1/r$ is defined by the *eikonal equation* [4]:

$$v = v_0 - D/r \tag{1.7}$$

where D is the diffusion coefficient of the autocatalytic species. An important consequence of this dependence is the stability of the circular wave front spreading in the homogeneous excitable medium: deformation into more planar form means acceleration of the front, while its local enhanced curvature retards the front, so it always will return to the circular shape. Furthermore, at least theoretically, if the curvature reaches the critically high value $1/r_{crit} = v_0/D$, the circular front stops to move ($v = 0$). Hence, spatial fluctuations of the size smaller than the critical radius r_{crit} (in the case of BZ reaction in highly excitable medium this is ca. 20–50 μm [2, 4]) will be *damped*. So, the fluctuation must be large enough to give rise to macroscopically observable circular wave fronts. The physical reason for the existence of this critical radius is that in the case of high curvature, the dilution of

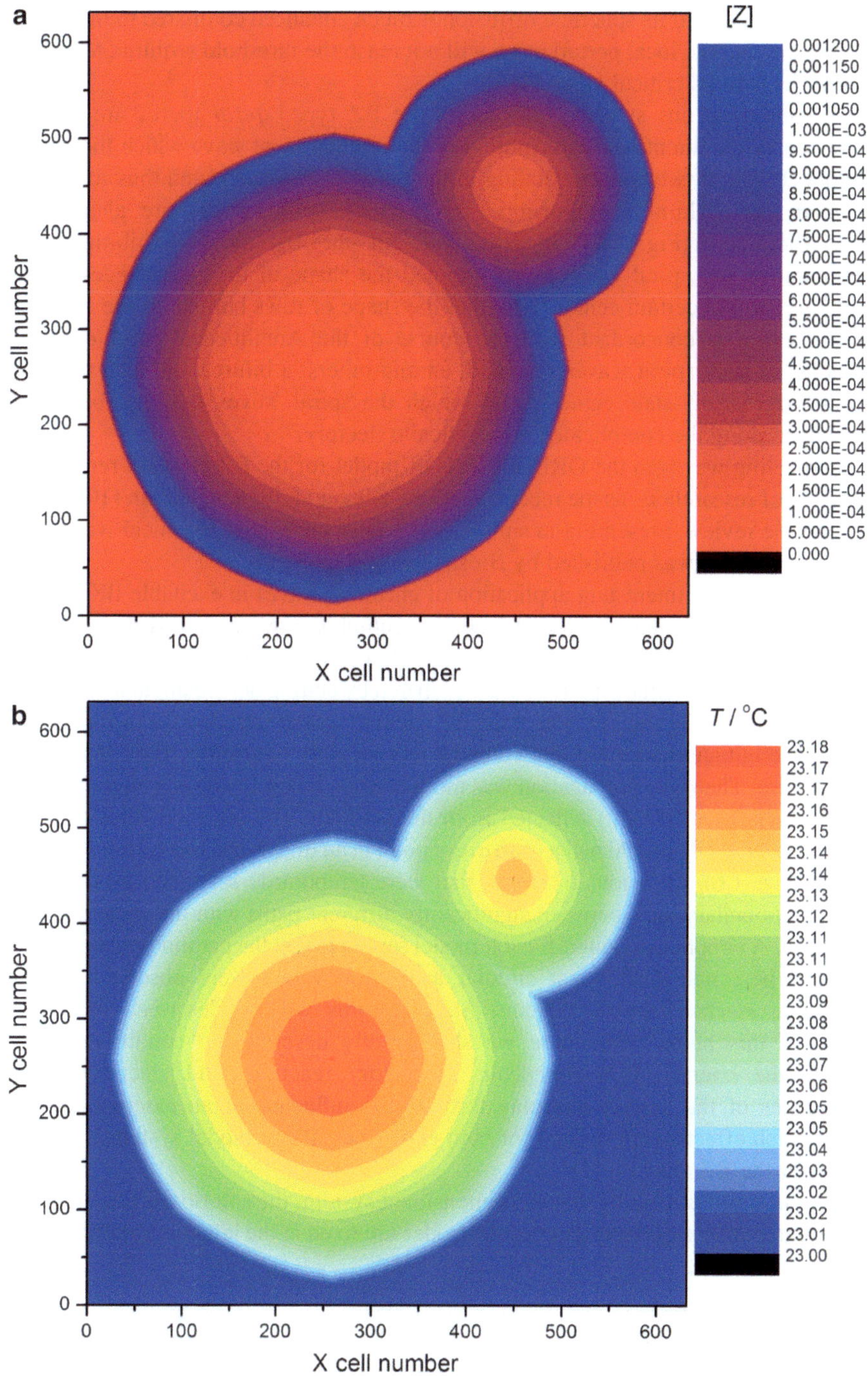

Fig. 1.4 Modeled two-dimensional maps of (**a**) the concentration distribution of species Z in the OREGONATOR and (**b**) the corresponding temperature distribution. Model perturbations

autocatalyst due to its spherical diffusion is much enhanced compared to the planar diffusion, and the local perturbation will not reach the threshold required to reduce local [Br^-] to its critical value [4].

In homogeneous systems, like again the BZ type, *spiral waves* are usually induced by the intentional interruption of the circular front wave which then splits into a couple of two spirals, rotating in the opposite directions and thus forming a pair of two "chiral" components [2, 4]. Such spiral waves are also called "reverberators" (Fig. 1.5). In particular, for the BZ media exhibiting high excitability, the spiral wave fronts attained the shape of the Archimedes spiral, rotating around certain center (note that the shape of rows burned on the compact disc surface is concordant with the course of the Archimedes spiral $r = a\varphi$). Studies of such spiral waves revealed, among others, a bifurcation, as a result of which the steady-state center, from which the spiral waves develop, begins to meander along the complicated, epicyclical trajectory.

Traveling waves in the OREGONATOR model for the BZ reaction remain the interest of researchers, as the recent works by Kiss et al. [9] and Merkin [10] prove. A concise review, presenting in simple way basic concepts of chemical waves and spatial patterns, was published by Biosa et al. [11].

A particularly interesting application of chemical waves in excitable BZ system, exploited to find minimum-length paths in complex labyrinths, was described by Steinbock et al. [12]. The vinyl-acrylic membrane was saturated with BZ reaction mixture and impenetrable barriers were prepared by cutting out rectangular regions of the membrane. The composition of the BZ solution was prepared so that no spontaneous wave initiation occurred, but could be reproducibly initiated by the contact with silver wire. The wave propagation was monitored by image analysis with monochromatic light ($\lambda = 500$ nm) reflected from the medium. The obtained velocity fields provided maps of optimal paths from every point in an image grid to a particular target point (Fig. 1.6). Collisions of waves that were temporarily separated by obstacles marked boundary lines between significantly different paths with the same absolute distance. A simple reaction–diffusion model served to test the pathfinding algorithm.

It is here noteworthy that circular and spiral waves were observed also in excitable *electrochemical* systems. Such patterns will be described in further parts of this book, while here we shall briefly describe the conditions under which the course of *homogeneous* oscillatory reactions and electrochemical properties of the system interplay through the influence of an external electric field. Such effects were studied for both experimental and model systems, particularly for the BZ reaction [2].

In the simplest case of the one-dimensional system, the solution is placed in the horizontal tube (capillary) and the electrodes are fixed at the left and right boundary

---

Fig. 1.4 (continued) were introduced in a form of [Y] = 1×10^{-7} M, being ca. ¼ of the steady-state concentration of Y, imposed into two single spatial cells of the coordinates: $(M_x, M_y) = (259, 259)$ at $t = 0$ s and $(M_x, M_y) = (449, 449)$ at $t = 3.0$ s (see [7] for more details). Reprinted with permission from [7] Copyright 2010 American Chemical Society

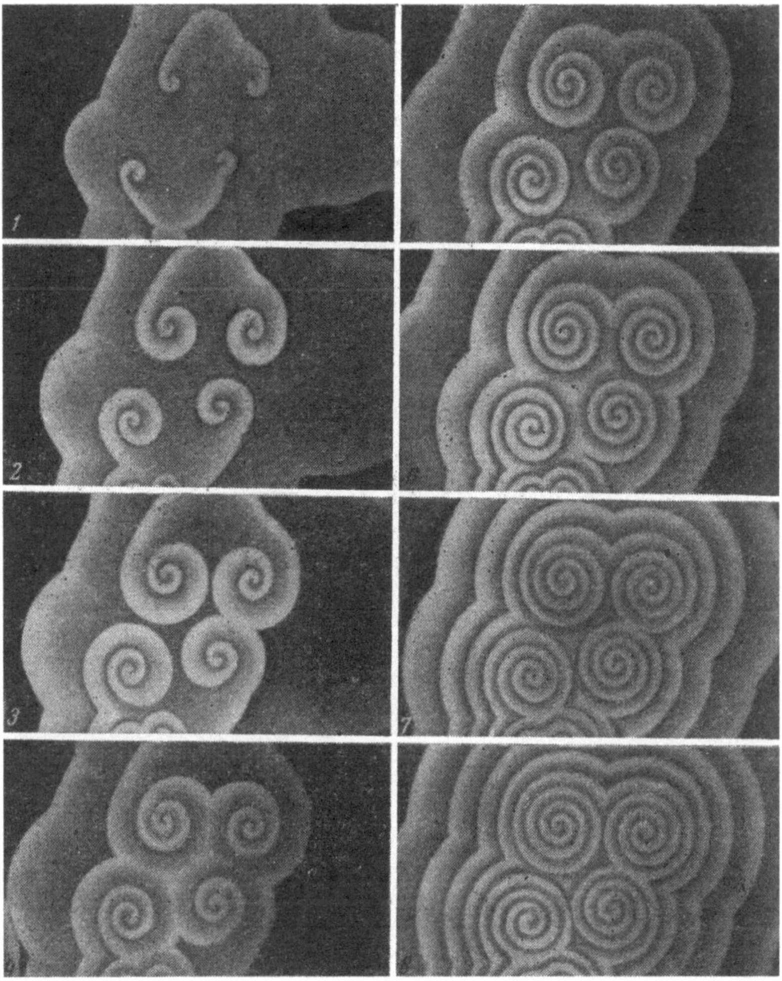

Fig. 1.5 Photographic images of spiral waves (reverberators) in the thin layer of the solution with the Belousov–Zhabotinsky process catalyzed by ferroin. Experimental conditions the same as in Fig. 1.1. Period of evolving spiral $T_R \approx 15$ s; wavelength of pattern $\lambda_R \approx 0.15$ cm. Reproduced from [1]

of this system. Upon sufficient voltage applied between the electrodes the electrolysis begins. We will not consider here the nature of charge-transfer processes at each electrode, since they are not crucial for understanding of the changes that occur in characteristics of the pulse waves developing in the solution bulk. The flow of electric current is associated with migration of all ions, of which we shall consider here the Br^- ions which, as mentioned earlier, serve as a switch between the two groups of reactions. They were denoted as groups A and B in the FKN or the OREGONATOR model, and one of them (B) involves the autocatalytic production of $HBrO_2$, predominating below critical Br^- concentration. The migration of

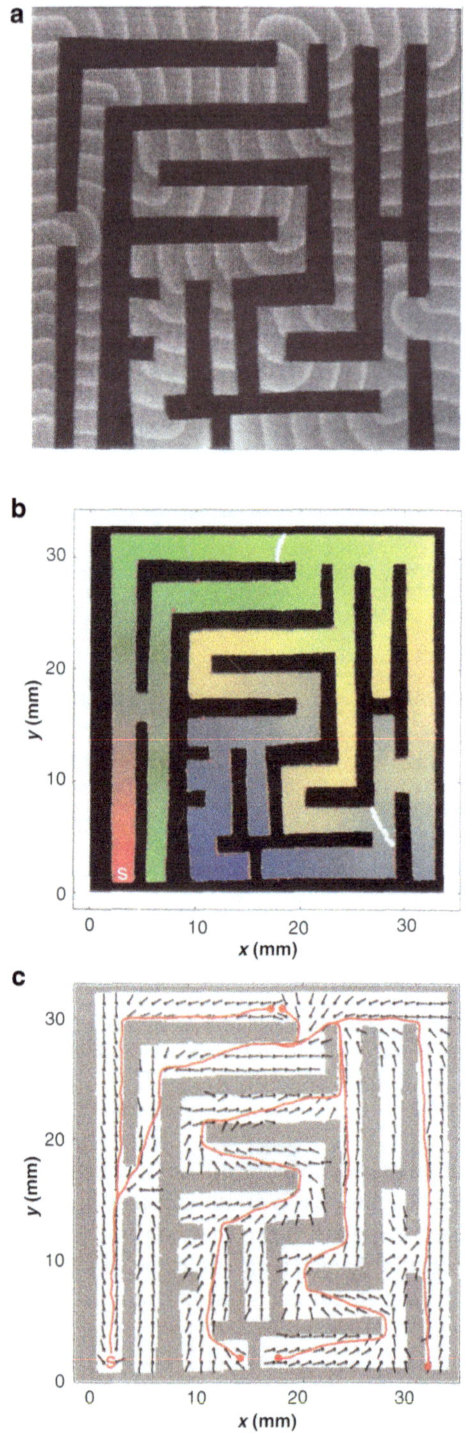

Fig. 1.6 (**a**) Chemical wave propagating through a BZ membrane labyrinth. A sequence of 50 images obtained at 50-s intervals was superimposed to form the composite image. The wave was initiated in the *lower left corner* of the maze. The total area of the maze is 3.2 cm by 3.2 cm, with obstacles appearing as *black rectangular* segments. (**b**) Color map representing the time difference between wave initiation and local excitation for all points in the labyrinth. A sequence of 250 images obtained at 10-s intervals was used to form the time-indexed composite image (yielding a spacing of ~4 pixels between successive front positions). *Red*, *green*, *yellow*, and *blue* correspond to successively longer times over the total elapsed time of 2,500 s. White lines show path boundaries. (**c**) Velocity field describing the local wave propagation direction. *Small black dots* represent the origins of vectors. The shortest paths between five points and the target point S are shown in *red*. From [12]. Reprinted with permission of AAAS

Br^- ions toward the anode modifies locally the excitability of the system and thus the velocity of wave propagation. The following effects can occur. The electric field slows down the pulse wave moving toward the cathode since the moving front meets with the counterflow of Br^- ions which locally decrease the system's excitability. In turn, the wave originally moving toward the anode, like Br^- ions, undergoes then acceleration [2]. For sufficiently high electric field (ca. only 10 V cm^{-1} or higher) even the inversion of the direction of the wave propagation may occur. Also, single pulse wave, traveling toward cathode, may split into two ones traveling in the opposite direction; in the course of this splitting an original front survives and the new one is born at the back of the original one [2].

For the two-dimensional systems, the effect of the electric field (up to 10 V cm^{-1}, corresponding to the current of 50 mA) was particularly intensively studied for a couple of spiral waves. Generally all the effects are again caused by the migration of Br^- ions toward the anode. The most trivial effect of the electric field in this case is a distortion of the spiral waves to an extent that is proportional to the cosine of the angle between the vectors of the electric field and of the velocity of wave propagation. But the effect of the field can be also much more sophisticated. For example, it was found that upon electric field the origin of the spiral migrates toward the anode. Furthermore, an additional component of this motion develops which is oriented perpendicularly to the vector of the electric field. The direction of the vector of this new component depends on the direction ("chirality") of the spiral evolution. This means that a couple of counterrotating spirals is wandering through the solution toward the anode, but their origins become gradually more and more separated. The reversal of polarity of the electric field causes the return of both spiral components to their original configuration and further they can even collide, forming a circular wave, i.e., a typical target pattern in BZ reaction, from which such spiral waves are usually born [2].

The reader interested in more recent achievements in the studies of the electric field effects on traveling waves in the OREGONATOR model is advised to consult the work by Kiss et al. [13] in which also previous important works in this subject are mentioned. Of particular importance are pioneer works of H. Ševčikova, M. Marek, I. Schreiber, S.C. Müller, P. De Kepper et al. (cf., e.g., [14–22]).

Various spatiotemporal patterns, including circular and spiral ones, were reported also in the oscillatory kinetics in heterogeneous catalysis which processes are not described in this book. The reader interested in temporal and spatiotemporal self-organization in such systems is therefore advised to consult the relevant literature, based, e.g., on the review articles by Ertl and Imbihl [23–25].

1.1.3 Linear Stability Analysis of the Reaction–Diffusion Systems

Linear stability analysis, presented in Sect. 1.3 of volume I for homogeneous systems, can be extended for spatially distributed system. It is useful to precede the present analysis with recalling the remark made in the previous section, saying that the chance of spatial fluctuation to be enhanced to macroscopic dimensions is dependent on its size, compared to the size of the spatially distributed system. Fluctuations which are spatially too large may decay, while sufficiently small ones can be enhanced under appropriate conditions. We shall see whether the linear stability analysis is able to reveal this dependence.

Let us invoke again the general form of differential equations for the reaction–diffusion systems in Eq. (1.1) that will be now a subject of linear stability analysis. The solutions of Eq. (1.1) depend also on the boundary conditions, determined by the construction of the given reactor. For example, if the reaction is carried out in a reactor with impermeable walls, the zero diffusion fluxes at the system's boundary have to be applied (this means the appropriate von Neumann boundary conditions). If the system is practically one dimensional, i.e., if the patterns can develop only along the z-axis of the thin tube (capillary) of length L, then the condition $\partial c_i/\partial z = 0$ must hold both at $z = 0$ and $z = L$.

The principle of linear stability analysis of the system of Eq. (1.1) for spatially distributed system is generally the same as for homogeneous systems and includes (1) the determination of a steady state(s) and (2) analysis of evolution of the infinitesimal perturbation of a given state.

1.1.3.1 The Principle of Stability Analysis

We shall consider first the steady state which is *uniform in space*, i.e., its parameters are calculated from the condition $\partial c_i/\partial t = 0$ referred to only "chemical" terms $R[\mathbf{c}]$ in Eq. (1.1), with diffusion terms omitted. This state will be further perturbed with the spatiotemporal fluctuation assumed in the form:

$$\delta_i = \alpha_i \exp(\lambda_n t) u_n(x, y, z) \tag{1.8}$$

which means the quantity proportional to the product of the eigenfunctions of the time derivative and of Laplace operator (diffusion terms). In this expression there appears an integer $n = 1, 2, 3, \ldots$; in the case of $n = 0$ the problem reduces to the homogeneous perturbation. It is useful to continue further analysis for the simple case of one-dimensional system, being the reactor of length L. The von Neumann boundary conditions $dc_i/dz = 0$ determine the form of u_n[1]:

[1] for the Dirichlet boundary conditions, defined as zero perturbations at $z = 0$ and $z = L$ borders, $u_n = \sin(n\pi z/L)$; $n = 1,2,3, \ldots$

$$u_n(z) = \cos\left(\frac{n\pi z}{L}\right) \tag{1.9}$$

in which $n = 0,1,2,\ldots$ specifies the wavelength *of the perturbation*. The value of n is extremely important for stability analysis, since the *system may appear stable with respect to perturbation of certain wavelengths, but unstable with respect to those of different wavelengths* [3]. In other words, the boundary conditions determine available (discrete) values of wave number. The *spatiotemporal* evolution of perturbation of steady-state concentrations $[(c_{i,\text{ss}});\ i = 1, \ldots, N]$ is thus described by the dependence:

$$c_i(t,z) = c_{i,\text{ss}} + \alpha_i \exp(\lambda_n t)\cos(n\pi z/L) \tag{1.10}$$

where $\alpha_i \exp(\lambda_n t)$, denoted further as $A_{i,n}(t)$ is the time-only dependent amplitude of the respective spatiotemporal perturbation. Now one can introduce Eq. (1.10) to Eq. (1.1) and obtain dependences suitable for stability analysis. This analysis will be now represented in terms of general dependences, illustrated later by the application to the Brusselator model.

1.1.3.2 Generalized Approach

According to principles of linear stability analysis, applied to system defined by Eq. (1.1), the temporal evolution of spatiotemporal perturbations δ of the steady state is described by [2]:

$$\frac{\partial \boldsymbol{\delta}}{\partial t} = \mathbf{J}_0\boldsymbol{\delta} + \mathbf{D}\nabla^2\boldsymbol{\delta} \tag{1.11}$$

in which $\mathbf{J}_0$ means the Jacobian matrix for the homogeneous chemical steady state and $\mathbf{D}$ is the matrix of diffusion coefficients which in this general form takes into account possible mutual interactions of individual diffusing species (then $D_{ij} \neq 0$ for $i \neq j$). If the species diffuse independently of each other, then matrix $\mathbf{D}$ becomes diagonal, with the diffusion coefficient of individual species $D_i \equiv D_{ii}$. Applying the *spatial Fourier transformation* allows to expand the vectors of perturbations into series involving all possible wavenumbers:

$$\delta_i = \sum_{n=0}^{\infty} A_{i,n}(t)\cos(n\pi z/L) \quad i = 1,\ldots,N \tag{1.12}$$

In Eq. (1.12), $A_{i,n}$ mean the time-dependent amplitudes of perturbations. In terms of this transformation, PDE (1.11) can be replaced with the N ordinary differential equations, for given wavenumber:

$$\frac{\mathrm{d}A_{i,n}}{\mathrm{d}t} = \left(\mathbf{J}_0 - \frac{n^2\pi^2}{L^2}\mathbf{D}\right)A_{i,n} = \mathbf{J}_0^{\mathbf{D}}A_{i,n} \quad i = 1,\ldots,N \tag{1.13}$$

Note that in this notation, a new matrix $\mathbf{J}_0^{\mathbf{D}}$, being a "chemical" Jacobian matrix $\mathbf{J}_0$ extended for the diffusion terms, was defined. The general solution of Eq. (1.13) is given by:

$$A_{i,n}(t) = A_{i,n}^0 \exp\left[\left(\mathbf{J}_0 - \frac{n^2\pi^2}{L^2}\mathbf{D}\right)t\right] = A_{i,n}^0 \exp\left[\mathbf{J}_0^{\mathbf{D}}\, t\right] \tag{1.14}$$

Thus, the temporal evolution of every Fourier component of the perturbations of a steady state, $\boldsymbol{\delta} = \mathbf{c} - \mathbf{c}_{ss}$, depends on the (generally complex) eigenvalues of the matrix $\mathbf{J}_0^{\mathbf{D}}$. As for homogeneous systems, if for given wavelength of the perturbation, all eigenvalues have negative real parts, then this perturbation will asymptotically decay as a function of time and the steady state is stable with respect to this perturbation. However, if though one of the real parts of eigenvalues is positive, for any wavelength, the steady state appears to be unstable, i.e., the corresponding Fourier component $A_{i,n}$ will increase in time. Quite formally, in terms of Eq. (1.14), one can also interpret the stability of a spatially distributed system with respect to inhomogeneous fluctuations as *formally* equivalent to the stability of a fictitious homogeneous system of a kinetics $\partial c_i/\partial t = R_i - (n^2\pi^2 D_i/L^2)c_i$ with respect to homogeneous fluctuations.

Also analogously to the homogeneous case, the eigenvalues λ of matrix $\mathbf{J}_0^{\mathbf{D}}$ are the solutions of the characteristic equation:

$$\mathrm{Det}\left[\lambda\mathbf{I} - \mathbf{J}_0 + \left(\frac{n^2\pi^2}{L^2}\right)\mathbf{D}\right] = 0 \tag{1.15}$$

which is equivalent to:

$$\lambda^N + c_1(n)\lambda^{N-1} + \ldots + c_N(n) = 0 \tag{1.16}$$

where $c_i(n)$ are polynomials in n^2. Analytical solution of this problem is possible for $N \leq 3$.

Below we shall concentrate in more detail on the solution for $N = 2$ [2]. Let us assume the following notation, in which a_{ij} mean the elements of the "chemical" (2×2) Jacobian matrix $\mathbf{J}_0$:

$$\mathrm{Tr}(\mathbf{J}_0) = a_{11} + a_{22} \tag{1.17}$$

$$\mathrm{Tr}(\mathbf{D}) = D_{11} + D_{22} \tag{1.18}$$

$$\mathrm{Tr}(\mathbf{J}_0\mathbf{D}) = a_{11}D_{11} + a_{12}D_{21} + a_{21}D_{12} + a_{22}D_{22} \tag{1.19}$$

$$\mathrm{Det}(\mathbf{J}_0) = a_{11}a_{22} - a_{12}a_{21} \tag{1.20}$$

$$\mathrm{Det}(\mathbf{D}) = D_{11}D_{22} - D_{12}D_{21} \tag{1.21}$$

The characteristic equation is:

$$\lambda^2 + c_1(n)\lambda + c_2(n) = 0 \tag{1.22}$$

with coefficients:

$$c_1(n) = \frac{n^2\pi^2}{L^2}\mathrm{Tr}(\mathbf{D}) - \mathrm{Tr}(\mathbf{J}_0) \tag{1.23}$$

$$c_2(n) = \left[\frac{n^2\pi^2}{L^2}\right]^2 \mathrm{Det}(\mathbf{D}) + \frac{n^2\pi^2}{L^2}\{\mathrm{Tr}(\mathbf{J}_0\mathbf{D}) - \mathrm{Tr}(\mathbf{J}_0)\mathrm{Tr}(\mathbf{D})\} + \mathrm{Det}(\mathbf{J}_0) \tag{1.24}$$

The solution of Eq. (1.22) is:

$$\lambda_{1,2}(n) = \frac{-(n^2\pi^2/L^2)\mathrm{Tr}(\mathbf{D}) + \mathrm{Tr}(\mathbf{J}_0) \pm \sqrt{\Gamma}}{2} \tag{1.25}$$

with

$$\begin{aligned}\Gamma = {} & \left(\frac{n^2\pi^2}{L^2}\right)^2 \left\{[\mathrm{Tr}(\mathbf{D})]^2 - 4\mathrm{Det}(\mathbf{D})\right\} \\ & + \left(\frac{n^2\pi^2}{L^2}\right)\{2\mathrm{Tr}(\mathbf{J}_0)\mathrm{Tr}(\mathbf{D}) - 4\mathrm{Tr}(\mathbf{J}_0\mathbf{D})\} + \left\{[\mathrm{Tr}(\mathbf{J}_0)]^2 - 4\mathrm{Det}(\mathbf{J}_0)\right\}\end{aligned} \tag{1.26}$$

It is important to emphasize that the steady state reacts particularly strongly on the perturbations of a wavelength that corresponds to high positive eigenvalue λ_n. Furthermore, preferred is then the formation of such pattern which corresponds to the highest value of the real part of the eigenvalue $\mathrm{Re}(\lambda_n)$ [2].

It is instructive to consider two limiting cases of wavelengths, corresponding to $n \to 0$ and $n \to \infty$ in Eq. (1.25) which are expressed by Eqs. (1.27) and (1.28), respectively:

$$\lim_{n\to 0} \lambda_{1,2} = \frac{\mathrm{Tr}(\mathbf{J}_0) \pm \sqrt{[\mathrm{Tr}(\mathbf{J}_0)]^2 - 4\mathrm{Det}(\mathbf{J}_0)}}{2} \tag{1.27}$$

$$\lim_{n\to \infty} \lambda_{1,2} = \frac{-(n^2\pi^2/L^2)\mathrm{Tr}(\mathbf{D}) \pm \sqrt{[\mathrm{Tr}(\mathbf{D})]^2 - 4\mathrm{Det}(\mathbf{D})}}{2} \tag{1.28}$$

Equation (1.27) is obviously identical with the solution for the homogeneous system seen in Eq. (1.28) (vol. I), i.e., in when kinetics of chemical reactions decide the system's behavior, whereas Eq. (1.28) (vol. I), not including any kinetic term, describes the purely diffusion control of the system's dynamics. Mathematically, the bifurcations leading to qualitatively new solutions in spatially distributed systems occur, when the real part of at least one λ value changes its sign, i.e., when $\mathrm{Re}(\lambda) = 0$. One can distinguish two different cases corresponding to this condition: the imaginary part of λ can be simultaneously either zero: $\mathrm{Im}(\lambda) = 0$ (this means the so-called "real bifurcation") or nonzero: $\mathrm{Im}(\lambda) \neq 0$ (this means the "complex bifurcation") [2]. In terms of the considerations for the homogeneous system (Sect. 1.3, volume I) it becomes clear that the real bifurcation leads to spatial patterns, while complex bifurcation leads to oscillatory solutions (spatiotemporal oscillations). In other words, the complex bifurcation is an analog of the Hopf bifurcation applied now to the spatially distributed, reaction–diffusion systems.

1.1.3.3 Stability Analysis of the Brusselator Model with Diffusion

The above general dependences are often illustrated with the stability analysis of the Brusselator model [2, 3, 26, 27]:

$$\begin{gathered} \mathrm{A} \rightarrow \mathrm{X} \\ \mathrm{B} + \mathrm{X} \rightarrow \mathrm{Y} + \mathrm{D} \\ 2\mathrm{X} + 3\mathrm{Y} \rightarrow 3\mathrm{X} \\ \mathrm{X} \rightarrow \mathrm{E} \end{gathered} \tag{1.29}$$

with incorporated one-dimensional diffusion transport:

$$\frac{\partial[\mathrm{X}](t,z)}{\partial t} = [\mathrm{A}] + [\mathrm{X}]^2[\mathrm{Y}] - ([\mathrm{B}]+1)[\mathrm{X}] + D_{\mathrm{X}}\frac{\partial^2[\mathrm{X}]}{\partial z^2} \tag{1.30}$$

$$\frac{\partial[\mathrm{Y}](t,z)}{\partial t} = [\mathrm{B}][\mathrm{X}] - [\mathrm{X}]^2[\mathrm{Y}] + D_{\mathrm{Y}}\frac{\partial^2[\mathrm{Y}]}{\partial z^2} \tag{1.31}$$

where [A], [B], [C], [D], [X], and [Y] mean the concentrations of respective species. D_{X} and D_{Y} mean the diffusion coefficients of the respective species. In the following, we assume that X and Y diffuse independently, so the matrix of diffusion coefficients **D** is diagonal. The (exemplary) von Neumann boundary conditions of the reaction running in the long capillary are:

$$\left[\frac{\mathrm{d}[\mathrm{X}]}{\mathrm{d}z}\right]_{z=0} = \left[\frac{\mathrm{d}[\mathrm{Y}]}{\mathrm{d}z}\right]_{z=0} = \left[\frac{\mathrm{d}[\mathrm{X}]}{\mathrm{d}z}\right]_{z=L} = \left[\frac{\mathrm{d}[\mathrm{Y}]}{\mathrm{d}z}\right]_{z=L} = 0 \tag{1.32}$$

For fixed concentrations of A and B species, the *homogeneous* steady state is defined as $([X]_{ss}, [Y]_{ss}) = ([A], [B]/[A])$. Linear stability analysis of this reaction–diffusion problem, done either in terms of substitution of trial solution from Eqs. (1.10) to (1.30) and (1.31), or based on general principles summarized earlier, leads to the extended Jacobian matrix:

$$\mathbf{J}_0 - (n\pi/L)^2\mathbf{D} = \begin{bmatrix} [B] - 1 - D_X\left(\frac{n\pi}{L}\right)^2 & [A]^2 \\ -[B] & -[A]^2 - D_Y\left(\frac{n\pi}{L}\right)^2 \end{bmatrix} \tag{1.33}$$

The roots of the characteristic equation are thus given by:

$$\lambda_{1,2}(n) = \frac{-(n\pi/L)^2(D_X + D_Y) + [B] - 1 - [A]^2 \pm \sqrt{\Gamma}}{2} \tag{1.34}$$

with

$$\Gamma = \left\{1 - [B] - [A]^2 + \left(\frac{n\pi}{L}\right)^2(D_X - D_Y)\right\}^2 - 4[A]^2[B] \tag{1.35}$$

Let us take as an example of instability the complex Hopf bifurcation. First, in order to meet the condition for oscillatory instability, $\mathrm{Im}(\lambda_{1,2}) \neq 0$, i.e., $\Gamma < 0$, so the following inequality has to be fulfilled:

$$\left\{1 - [B] - [A]^2 + \left(\frac{n\pi}{L}\right)^2(D_X - D_Y)\right\}^2 < 4[A]^2[B] \tag{1.36}$$

Next, the condition $\mathrm{Re}(\lambda_{1,2}) = 0$ is equivalent to:

$$[B]_H = 1 + [A]^2 + \left(\frac{n\pi}{L}\right)^2(D_X + D_Y) \tag{1.37}$$

i.e., this instability manifests itself for $[B] > [B]_H$. Thus, compared to the homogeneous Brusselator, for which the Hopf bifurcation occurs for $[B]_{H,hom} = 1 + [A]^2$, the present bifurcation point is shifted to higher value, due to the diffusion term.

These inequalities allow one to conclude that spatiotemporal instabilities in the Brusselator with diffusion occurring in one-dimensional space are favored, if (1) [B] is high enough, (2) the diffusion coefficients D_X and D_Y are low enough, (3) the size of the system (L) is large enough (there should be enough space for at least one full wavelength to develop a pattern); for $L \to 0$ the respective sides of both inequalities tend to infinity and meeting the instability condition is practically unrealistic, (4) the value of n is, on one hand, greater than 0, since otherwise the system would be homogeneous, where spatiotemporal patterns do not develop, but on the other hand, should not exceed certain maximum value, dependent on [A], D_X, D_Y, and L, i.e., for given parameters, there is an interval of n, or of the wavelengths of perturbations, that give rise to oscillatory patterns over the entire system.

Other dynamical regimes in Brusselator with diffusion the interested reader can find, e.g., in [26, 28, 29] and references cited therein. Here we shall only briefly describe the case of the above-mentioned real (or Turing) bifurcation in Brusselator [2] (for more detailed characteristics of Turing patterns, see next section). For this bifurcation to occur, eigenvalues have to be real, and a positive λ occurs if:

$$\left\{[\mathrm{B}] - 1 - \frac{n^2\pi^2}{L^2}D_x\right\}\left\{[\mathrm{A}]^2 + \frac{n^2\pi^2}{L^2}D_y\right\} - [\mathrm{A}]^2[\mathrm{B}] > 0 \tag{1.38}$$

For the case when (1.38) is equal to zero, one derives the relationship for the bifurcation curve:

$$[\mathrm{B}] = 1 + D_\mathrm{X}\left(\frac{n\pi}{L}\right)^2 + \left(\frac{D_\mathrm{X}}{D_\mathrm{Y}}\right)[\mathrm{A}]^2 + \frac{[\mathrm{A}]^2}{D_\mathrm{Y}(n\pi/L)^2} \tag{1.39}$$

which constitutes the dependence of B on the wavelength of the perturbation, expressed by the integer value of n (see Fig. 1.7).

This diagram shows that upon increasing [B] the first Turing instability sets in for a critical value $[\mathrm{B}]_\mathrm{T}$, closest to the minimum of the stability curve:

$$[\mathrm{B}]_\mathrm{T} = \left[1 + \sqrt{\frac{D_\mathrm{X}}{D_\mathrm{Y}}}[\mathrm{A}]\right]^2 \tag{1.40}$$

As an example, for $[\mathrm{A}] = 1, D_\mathrm{X} = 1 \times 10^{-6}$ and $D_\mathrm{Y} = 1 \times 10^{-5}$ the following critical (bifurcation) values of [B] can be calculated and, in consequence, respective dynamical regimes can be diagnosed:

1. In the *homogeneous* system without diffusion, the onset of X and Y oscillations due to the Hopf bifurcation, leading to temporal oscillations, occurs for $[\mathrm{B}] = [\mathrm{A}]^2 + 1 = 2$.
2. In the system with diffusion, the Hopf bifurcation, leading to spatiotemporal oscillations, is increased for the diffusion factor: $[\mathrm{B}]_\mathrm{H} = 2 + 1.1 \times 10^{-5}(n\pi/L)^2$.
3. In the system with diffusion, the Turing bifurcation, leading to spatial patterns, occurs near $[\mathrm{B}]_\mathrm{T} = 1.732$, for integer value of n, closest to this minimum value of B. For this value of [B] the homogeneous system would still remain stable. Thus, thanks to diffusion, the homogeneous system loses its stability and patterns are formed before the onset of homogeneous oscillations. This is one of the key features of the Turing bifurcation: this is diffusion which destabilizes the initially homogeneous stable steady state. But in order to achieve the onset of the Turing bifurcation before the Hopf point, the diffusion coefficient of X had to be smaller than that of Y and, as we shall see later, such a difference in diffusion coefficients is one of the necessary conditions for the formation of the Turing patterns. In turn, similar diffusion coefficients favor the oscillatory instabilities.

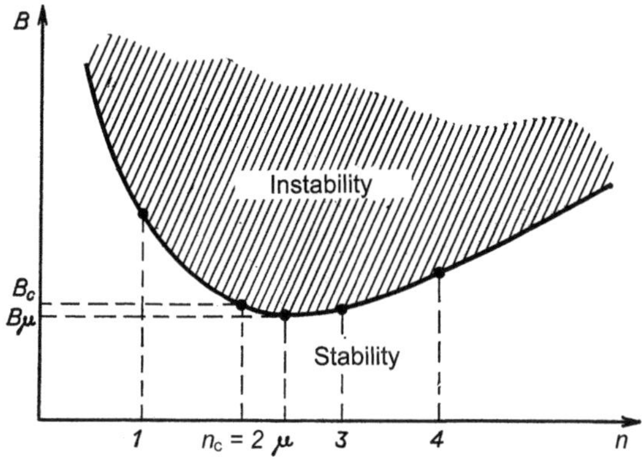

Fig. 1.7 The dependence of critical value of parameter B, leading to Turing pattern formation in the Brusselator model, for constant A, D_X, and D_Y parameters. Points corresponding to *integer* values of n are the points of bifurcation of stationary solutions. The first bifurcation point is located in close vicinity of the minimum of $B = f(n)$ dependence (after [30], with kind permission of J. Wiley & Sons Inc., Copyright 1977 J. Wiley & Sons Inc.)

4. If the sequence of the diffusion coefficients is reversed: $D_X = 1 \times 10^{-5}$, $D_Y = 1 \times 10^{-6}$, the Turing bifurcation would formally occur close to $[B]_T = 17.3$, but earlier the system will switch into homogeneous oscillations at $[B] = 2$. Also, B will reach earlier the critical value $[B]_H = 2 + 1.1(n\pi/L)^2$, for which spatiotemporal oscillations set in. So, the Turing bifurcation will then be "covered" by earlier bifurcations occurring upon increase in [B]. However, note that a detailed discussion of the onset of particular instabilities requires also the definition of the boundary conditions.

1.1.3.4 The Interaction of the Hopf and the Turing Bifurcations

As the discussed examples of the Brusselator dynamics showed, in the same system the Hopf and the Turing bifurcations may occur, in a sequence dependent on particular model parameters. If upon increasing [B], the condition of the Turing bifurcation [Eq. (1.40)] is met first, from the homogenous stable steady state the stable spatial (Turing) structures emerge, otherwise, if the condition of the Hopf bifurcation [Eq. (1.37)] is met first, the spatiotemporal oscillations set in and this regime will prevent further formation of Turing patterns.

There is also possible a particularly interesting case, when the conditions of the Hopf and the Turing bifurcation coincide, i.e., when the *Turing–Hopf bifurcation* occurs. In this case the critical values of [B] defined by Eqs. (1.37) and (1.40) are equal (for the case of Dirichlet boundary conditions) [2]:

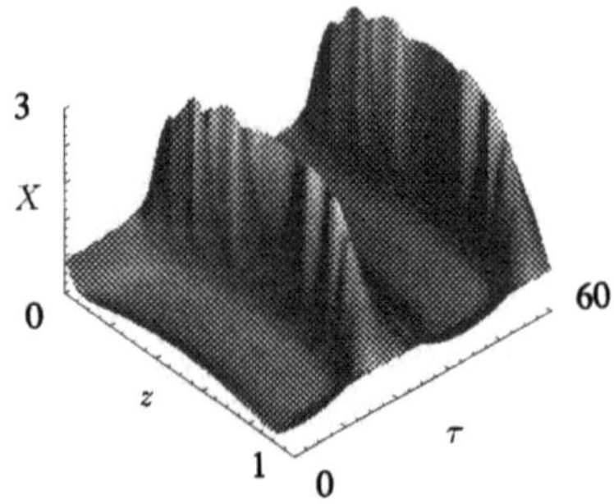

Fig. 1.8 Spatiotemporal oscillations (*phase waves*) in one-dimensional Brusselator with diffusion obtained for [A] = 2, [B] = 5.2 and diffusion coefficients $D_X = D_Y = 4 \times 10^{-3}$; Hopf bifurcation occurred at $[B]_H = 5{,}008$, while Turing bifurcation would occur at $[B]_T = 9.0$, but is masked by earlier oscillatory regime. Dirichlet boundary conditions were assumed. Reproduced with kind permission from Springer Science + Business Media: [2]

$$1 + [A]^2 + \left(\frac{n\pi}{L}\right)^2 (D_X + D_Y) = \left\{1 + \sqrt{\frac{D_X}{D_Y}}[A]\right\}^2 \tag{1.41}$$

so it is the *codimension-2* bifurcation.[2]

The interaction of Hopf and Turing bifurcations causes new spatial and spatiotemporal dynamics, including complex oscillations or multistability. New interesting dynamical regimes appear already close to the collision of the Hopf and Turing bifurcations. Studies of bifurcation interaction for three model oscillatory systems, including Brusselator, were described a.o. by Rovinsky and Metzinger [31].

1.1.3.5 Examples of Model Turing Patterns

It is useful to compare the above theoretical analysis with results of exemplary, representative numerical calculations of concentration profiles in time and space. Figure 1.8 shows the examples of the spatiotemporal oscillations (phase waves), while Fig. 1.9 illustrates stationary concentration profiles corresponding to Turing patterns in one-dimensional Brusselator with diffusion.

Extension of the model space to two dimensions allowed Erneux and Herschkowitz-Kaufman [32] to calculate periodical wave fronts that can resemble those reported experimentally in thin layer of solutions placed, e.g., in Petri dish.

For the Brusselator with diffusion occurring along two spatial dimensions also Turing patterns can be obtained (Fig. 1.10), and they exhibit striking similarity to the experimentally reported, quite generic Turing hexagonal or stripe patterns.

---

[2] Codimension of bifurcation means the number of system's control parameters which have to be tuned to achieve the bifurcation. The Hopf and saddle-node bifurcations are codimension-1 bifurcations. Besides the Turing-Hopf bifurcation, another codimension-2 bifurcation occurs at a *cusp point* in which two lines of saddle-node bifurcations meet tangentially.

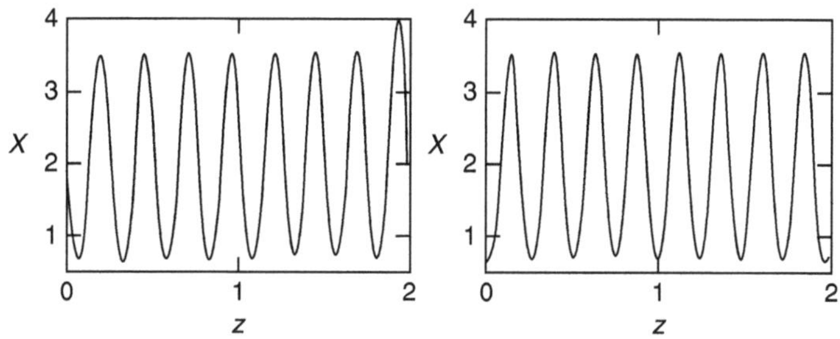

Fig. 1.9 Stationary oscillatory concentration profiles (Turing patterns) in one-dimensional Brusselator with diffusion for $D_X = 1.6 \times 10^{-3}$, $D_Y = 6.0 \times 10^{-3}$, assuming Dirichlet (left) and von Neumann (right) boundary conditions. Reproduced with kind permission from Springer Science + Business Media: [2]

Fig. 1.10 Hexagonal and striped Turing patterns visualizing stationary spatial distribution of X in the Brusselator model with two-dimensional diffusion. Reproduced from [2] with kind permission from Springer Science + Business Media

Finally, we shall come back to Brusselator with one-dimensional diffusion, in order to indicate that introduction of even one spatial dimension to this two-variable chemical dynamical system can generate chaotic solutions which in the absence of diffusion would be possible only for at least three ordinary differential equations (Fig. 1.11).

For the electrochemists it can be of certain interest to learn about the effect of the external electric field on the development of model Turing patterns generated by Brusselator. The aim of such studies was to build up the simple models of the influence of the electric field in membranes, mimicking analogous processes in biological membranes. Typically, X and Y forms acquire the positive signs: X^+, Y^- and the transport of such univalent cations includes, besides diffusion, also migration in the electric field. In such exemplary calculations, A, B, D_X, and D_Y was maintained constant, and the electric field intensity was the only parameter tuned. In this way switching from one type of the stationary Turing patterns to another, e.g., from the point like to the stripes was obtained (Fig. 1.12).

In the case of electrode processes, one should take into account that the formation of pattern on the electrode surface means that it is no longer equipotential and it can be necessary to consider also the spatial distribution of *local currents* which summarize to the total current. Emergence of nonuniform local currents means

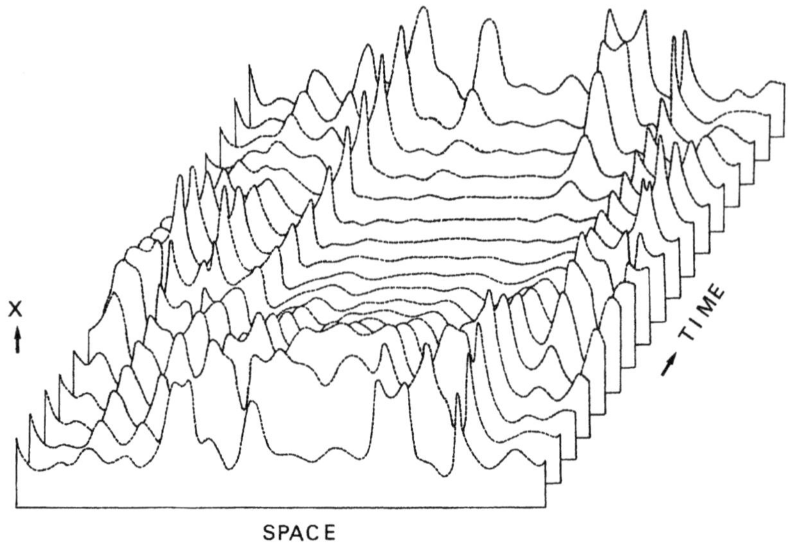

Fig. 1.11 Chaotic (aperiodic) propagation of the X concentration wave in the Brusselator with one-dimensional diffusion for $A = 2, B = 6.5, D_X = 1, D_Y = 0$. Reprinted from [33], Copyright 1981, with permission from Elsevier

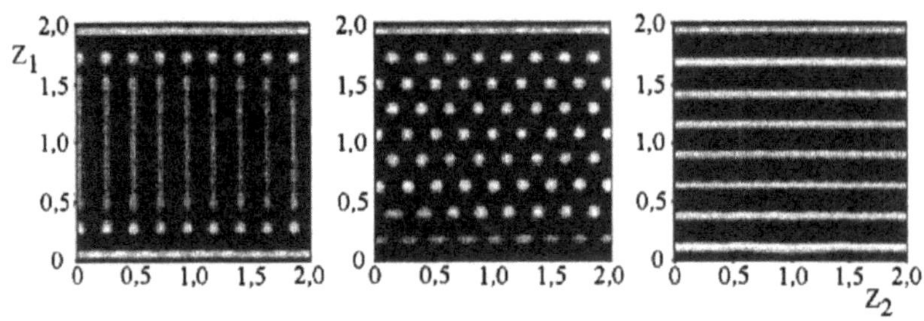

Fig. 1.12 Turing patterns in the ionic (X^+) version of Brusselator model. Along the coordinate Z_1 the von Neumann boundary conditions were assumed, along Z_2—the Dirichlet boundary conditions. Physically it means that system's borders parallel to Z_2 are in contact with the reservoirs ensuring constant boundary concentrations of species and contain also the electrodes. The electric field, parallel to coordinate Z_1, causes switching between striped and point patterns. Reproduced from [2] with kind permission from Springer Science + Business Media

further the existence of concentration, conductivity, and electric potential gradients in the solution, parallel to the electrode surface. Thus, a fluctuation of even single local current may spread over the entire surface due to the *spatial coupling*. Under appropriate conditions, such fluctuation may give rise to patterns on a macroscopic level. Such instabilities will be analyzed further in this chapter and following chapters.

At the end of this section we shall mention the same limitation of the analysis as for homogeneous systems: all the above predictions on the spatial or spatiotemporal instability were based on the *linearized* models which are originally nonlinear, as the Brusselator. Numerical integrations of nonlinear differential equations may reveal dynamic additional dynamic regimes, not predicted by the linear stability analysis.

1.1.4 The Turing Bifurcation

In the previous section, taking Brusselator as an example, we introduced the basic properties and conditions of formation of stationary dissipative patterns, which were predicted by A.M. Turing in his seminal paper of 1952 [34]. In fact, Brusselator is a simplified reaction scheme proposed by Turing, in which all reaction steps were assumed irreversible and rate constants equal to unity. In this section a more general theoretical description of stationary Turing patterns is presented and is followed by brief description of their experimental realization in chemical systems.

Turing patterns are of tremendous importance due to their role in understanding the biological morphogenesis at both microscopic (e.g., patterns in bacterial colonies) and macroscopic levels (e.g., patterns on the skins of animals, like zebras or tigers) which can be considered the structures of such type [35–37]. Murray in his *melanogenesis model* [37, 38] has described the spots on the animals skins as the places in which chemical reaction led to the local accumulation of the pigment melanin, synthesized from tyrosine with the participation of the enzyme tyrosinase. The aim of this model was also to explain the dependence of the pattern on the size of the animal, as an example of a more general relation between the characteristic size (wavelength) of the pattern and the space available for its development.

Let us now specify general conditions under which stationary Turing patterns can emerge. First, the chemical dynamical system is assumed to be stable in the absence of diffusion. In other words, the spatially homogeneous steady state is stable with respect to *spatially uniform* perturbations. However, if diffusion transport is possible in this system, its state can become unstable with respect to the *spatially nonuniform, infinitesimal* perturbations and then *Turing patterns* can develop as a result of the *Turing bifurcation.*[3] Typically they take a form of stripes or spots distributed uniformly in the system (cf. Fig. 1.10). The distance between the stripes or the spots, i.e., the intrinsic wavelength of the pattern, depends on both the kinetics of chemical reactions (rate constants) and diffusion coefficients of the reagents. The size of the system does not determine this characteristic wavelength, as long as this size is much larger than the wavelength, along all spatial coordinates. Thus, the pattern requires enough space to develop.

In order to formulate quantitative conditions for the formation of the Turing patterns in simple system [3], we shall consider again the dynamics of the local

[3] Although usually Turing patterns are considered to be stationary, it is also possible to consider nonstationary Turing patterns (see further in this section).

concentrations due to coupling of chemical kinetics and diffusion [Eq. (1.1)]. We further assume that two species X and Y are involved in the system's dynamics and that their diffusion develops only along a single coordinate r:

$$\frac{\partial x}{\partial t} = f(x,y) + D_{\mathrm{X}}\left(\frac{\partial^2 x}{\partial r^2}\right) \tag{1.42}$$

$$\frac{\partial y}{\partial t} = g(x,y) + D_{\mathrm{Y}}\left(\frac{\partial^2 y}{\partial r^2}\right) \tag{1.43}$$

with x and y denoting the respective concentrations: $x \equiv [\mathrm{X}]$, $y \equiv [\mathrm{Y}]$.

The *spatially uniform* (homogeneous) steady state (x_{ss}, y_{ss}) is defined by:

$$f(x,y) = 0 \tag{1.44}$$

$$g(x,y) = 0 \tag{1.45}$$

According to the first Turing condition, this state should be *stable* with respect to *spatially uniform fluctuations*, i.e., fluctuations not inducing concentration gradients and thereby do not triggering the diffusion. In terms of linear analysis, this stability requires that the trace and the determinant of the Jacobi matrix $\mathbf{J}_0$ of the linearized chemical system [Eqs. (1.42) and (1.43)]:

$$\mathbf{J}_0 = \begin{bmatrix} a_{11} & a_{12} \\ a_{21} & a_{22} \end{bmatrix} = \begin{bmatrix} \frac{\partial f}{\partial x} & \frac{\partial f}{\partial y} \\ \frac{\partial g}{\partial x} & \frac{\partial g}{\partial y} \end{bmatrix} \tag{1.46}$$

meet the conditions:

$$\mathrm{Tr}(\mathbf{J}_0) = a_{11} + a_{22} < 0 \tag{1.47}$$

$$\mathrm{Det}(\mathbf{J}_0) = a_{11}a_{22} - a_{12}a_{21} > 0 \tag{1.48}$$

since only then the real parts of eigenvalues λ in terms $\exp(\lambda t)$, describing the temporal change of perturbations, are all negative, i.e., all perturbations are asymptotically damped (cf. Sect. 1.3, volume I).

According to the procedure described in the previous section, we construct the Jacobian matrix for the system with diffusion:

$$\mathbf{J}_0^{\mathbf{D}} = \begin{bmatrix} a'_{11} & a'_{12} \\ a'_{21} & a'_{22} \end{bmatrix} = \begin{bmatrix} a_{11} - q^2 D_{\mathrm{X}} & a_{12} \\ a_{21} & a_{22} - q^2 D_{\mathrm{Y}} \end{bmatrix} \tag{1.49}$$

with $q \equiv n\pi/L$ (thus, q^{-1} [m] defining the spatial scale of the perturbation). Since now the inhomogeneous perturbation should *destabilize* the steady state of chemical-diffusion system, the properties of matrix $\mathbf{J}_0^\mathbf{D}$ should meet the following *instability* conditions:

$$\mathrm{Tr}\left(\mathbf{J}_0^\mathbf{D}\right) = (a_{11} + a_{22}) - q^2(D_\mathrm{X} + D_\mathrm{Y}) > 0 \tag{1.50}$$

or:

$$\mathrm{Det}\left(\mathbf{J}_0^\mathbf{D}\right) = \left(a_{11} - q^2 D_\mathrm{X}\right)\left(a_{22} - q^2 D_\mathrm{Y}\right) - a_{12}a_{21} < 0 \tag{1.51}$$

Let us note that in view of Eq. (1.47) the condition in Eq. (1.50) cannot be met. Thus, the diffusion-induced instability is possible only if the condition in Eq. (1.51) is fulfilled. It is convenient to rearrange this expression to a quadratic function of $\theta \equiv q^2$ [3]:

$$\begin{aligned} F(\theta) &= D_\mathrm{X} D_\mathrm{Y} \theta^2 - (a_{11} D_\mathrm{Y} + a_{22} D_\mathrm{X})\theta + a_{11}a_{22} - a_{12}a_{21} \\ &= D_\mathrm{X} D_\mathrm{Y} \theta^2 - (a_{11} D_\mathrm{Y} + a_{22} D_\mathrm{X})\theta + \mathrm{Det}(\mathbf{J}_\mathrm{O}) < 0 \end{aligned} \tag{1.52}$$

The $F(\theta)$ dependence is a parabola directed upwards, since D_X and D_Y are positive. Also, in view of Eq. (1.48), $\mathrm{Det}(\mathbf{J}_\mathrm{O})$ is positive. Thus, the only possibility for the $\mathrm{Det}\left(\mathbf{J}_0^\mathbf{D}\right)$ to become negative is a sufficiently positive value of the $(a_{11}D_\mathrm{Y} + a_{11}D_\mathrm{X})\theta$ term:

$$(a_{11}D_\mathrm{Y} + a_{22}D_\mathrm{X}) > \frac{D_\mathrm{X} D_\mathrm{Y} \theta^2 + \mathrm{Det}(\mathbf{J}_\mathrm{O})}{\theta} \tag{1.53}$$

This negative value will be first achieved by the minimum of the parabola that occurs at:

$$\theta_\mathrm{min} = \frac{1}{2}\left(\frac{a_{11}}{D_\mathrm{X}} + \frac{a_{22}}{D_\mathrm{Y}}\right) \tag{1.54}$$

By combination of Eqs. (1.53) and (1.54) one obtains the equivalent instability condition:

$$D_\mathrm{Y} a_{11} + D_\mathrm{X} a_{22} > 2\sqrt{D_\mathrm{X} D_\mathrm{Y} (a_{11}a_{22} - a_{12}a_{21})} > 0 \tag{1.55}$$

or:

$$D_\mathrm{Y} a_{11} + D_\mathrm{X} a_{22} > 2\sqrt{D_\mathrm{X} D_\mathrm{Y} \mathrm{Det}(\mathbf{J}_\mathrm{O})} > 0 \tag{1.56}$$

Alternative derivation of condition in Eq. (1.55) or Eq. (1.56) involves the search of existence of two real roots of the quadratic equation (1.52) through condition of the positive sign of its discriminant.

Concluding, the so far derived criteria for the Turing instability consist of conditions as in Eqs. (1.47), (1.48) and (1.55) or (1.56). It is thus sufficient to know the Jacobian matrix for the *chemical system* and the diffusion coefficients of X and Y species.

The above listed conditions are worthy of further analysis which will reveal their deeper kinetic sense.

(a) Condition in Eq. (1.47) means that the matrix elements a_{11} and a_{22} cannot be both positive, while condition in Eq. (1.55) or (1.56) means that they cannot be also both negative. Thus, one of them, e.g., a_{11} must be positive, and then a_{22} must be negative, with $|a_{22}| > a_{11}$.
(b) Positive value of $a_{11} \equiv \partial/\partial x[\partial x/\mathrm{d}t]$ means that an increase in [X] enhances production of X—such a species is a self-activator or simply its own activator. In turn, negative value of $a_{22} \equiv \partial/\partial y[\partial y/\mathrm{d}t]$ means that an increase in [Y] slows down its own production, so the species Y is a self-inhibitor or its own inhibitor (see Sect. 1.8, volume I). That is why the system [Eqs. (1.42) and 1.43)] in which Turing bifurcation occurs belongs to the class of the activator–inhibitor model.
(c) From conditions in Eqs. (1.47) and (1.48) it follows that a_{12} and a_{21} must also have opposite signs, with $|a_{12}a_{21}| > |a_{11}a_{22}|$. The conclusions collected in points (a)–(c) are the premise for the signs of the elements of Jacobian matrix $\mathbf{J}_0$ (1.46) (see below).
(d) From the condition in Eq. (1.55) or Eq. (1.56) it follows that the diffusion coefficients D_X and D_Y must be different, otherwise $D_X a_{22} + D_Y a_{11}$ would always be negative. Furthermore, since in our case $a_{22} < 0$ and $|a_{22}| > a_{11}$, the diffusion coefficient of X (the activator) must be lower than the diffusion coefficient of Y (the inhibitor):

$$\frac{D_X}{D_Y} < \frac{a_{11}}{|a_{22}|} < 1 \tag{1.57}$$

This condition for the Turing bifurcation was indicated already in the previous section, for Brusselator as the model discussed there. In fact, faster diffusion of inhibitor Y, compared to the activator X, is crucial for the formation of stationary Turing patterns. The latter condition is convincingly illustrated by schematic development of the Turing patterns, shown in Fig. 1.13 [39]. It also explains why the formation of Turing patterns can be understood as an interaction between short-range activation and long-range inhibition [35].

Based on all above criteria and conditions for the Turing bifurcation, one can understand the distribution of signs of the elements of the Jacobian of chemical system which corresponds to the pure activator–inhibitor system [cf. also (1.114), volume I]:

$$\begin{bmatrix} + & - \\ + & - \end{bmatrix} \tag{1.58}$$

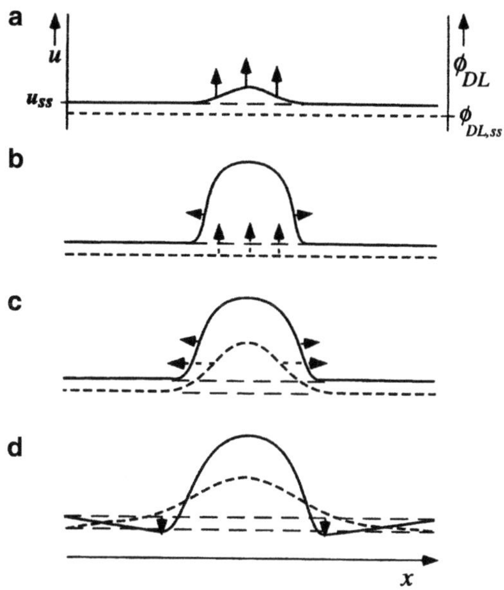

Fig. 1.13 Illustration of the emergence of a Turing structure. (**a**) The activator (*u*, *solid curve*) is locally slightly disturbed from its steady-state value u_{ss}. (**b**) This triggers the self-enhanced production of activator that *slowly* diffuses also in neighboring regions. (**c**) An increased activator concentration leads also to the production of the inhibitor (*short-dashed curve*), which diffuses much *more quickly* out of this "activator spot." (**d**) Due to the fast diffusion of the inhibitor, its concentration increases faster outside the spot than the activator concentration. As the inhibitor consumes the activator, the concentration of the latter is always kept low outside the "spot." In this way a stationary profile can develop with regions of high activator concentration that are surrounded by clouds of increased inhibitor concentration. This particular scheme refers to an electrochemical system with the electrode (double layer) potential ϕ_{dl} as a fast inhibitor and chemical species *u* as a slow activator, respectively. Reprinted from [39], Copyright 2001, with permission from Elsevier

In fact, $a_{11} > 0$ means that X is a self-activator; $a_{21} > 0$ means that X is an activator also for Y which is a self-inhibitor ($a_{22} < 0$); finally, $a_{12} < 0$ reflects the inhibitory effect of Y on the production of X. Note that if $a_{12} > 0$ and $a_{21} < 0$, the signs in the Jacobian matrix correspond to cross-activation-inhibition system (Fig. 1.33, vol. I), like for the Brusselator, [B] >1. Formal transformation $z = -y$ will convert this system to that of pure activator-inhibitor type [35].

Although Turing has published his seminal paper in 1952, the first experimentally obtained stationary chemical patterns of this kind were reported only in 1990s by I. Epstein group, for the so-called CIMA (chlorite–iodine–malonic acid) or CDIMA (chlorine dioxide–iodine–malonic acid) oscillators, with the soluble starch used as an indicator of triiodide ions [3]. The history of the route toward this success shows how many experimental problems had to be resolved and how smart ideas had to come to the mind of researchers. First, the unambiguous detection of the steady-state patterns required an open system, constantly fed with reactants, as in any closed arrangement all patterns would have only a transient character. Furthermore, the diffusion coefficients of the activator and the inhibitor had to differ

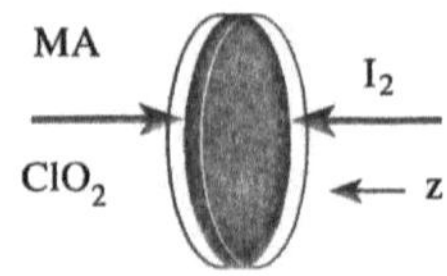

Fig. 1.14 Schematic diagram of reactor for studying Turing pattern formation. Reproduced from [3] by permission of Oxford University Press, Inc.

markedly (not less than ca. for one order of magnitude), which condition is not easy to meet in the same, aqueous phase, since the diffusion coefficients of many species in the same solution are quite similar to each other, with the exception of H^+ or OH^- ions in water which diffuse approximately ten times faster. The turning point in the construction of the experimental arrangement was the use of a slab of gel which both faces remained in contact with circulating solutions, as in the CSTR reactor. The reactants diffused into gel and met near its middle, when the pattern formation occurs [3, 40] (see Fig. 1.14).

The problem of significantly different diffusion coefficients of the activator and the inhibitor was solved in a following way. A key role was ascribed to the starch indicator (amylose) which was added to acrylamide monomer prior to its polymerization. In the gel phase, the diffusivity of starch (polymeric amylose chains) was practically negligible. When iodine molecules and iodide ions (iodine species are the activator) diffuse into the gel phase and form blue complex, the motion of these particles becomes severely slowed down, compared to chlorine species (chlorite ions, chlorine dioxide) being an inhibitor. In order to achieve sufficiently large differences between the rate of transport of the activator and the inhibitor, the concentration of the starch in the gel phase must also be sufficiently high. Also the temperature had to be maintained at 4 °C. Later it was found that not only starch, but also polyvinyl alcohol (PVA) and polyacrylamide gel were able to bind iodine and iodide to an extent allowing to form the Turing patterns. The blue color of the iodine–iodide–starch complex was, however, the most convenient way of detection of patterns formed in the gel phase. One should also note that experimental observations of the Turing patterns were supported by successful theoretical modeling of the CIMA/CDIMA oscillators (Fig. 1.15) [41].

Concerning ***electrochemical systems***, only in the year 2000 Mazouz and Krischer have found out that Turing patterns should emerge from the electrochemical steady state, if the system characterizes with the region of the **S-shaped** negative differential resistance (S-NDR) [42]. This theoretical prediction was soon after confirmed experimentally by the same group [43]. More detailed description of these and other types of electrochemical patterns are presented later in this book.

At the end of this section one should note that the above stationary patterns, ascribed to the name of A. Turing, were not the only ones predicted in his paper [34]. Less popular, but becoming recently a subject of increasing interest, is the predicted oscillatory instability with finite wavelengths that results in "genuine traveling waves" [44]. The corresponding transition to this regime is termed the "wave bifurcation" (WB) or "wave instability," or "oscillatory Turing bifurcation."

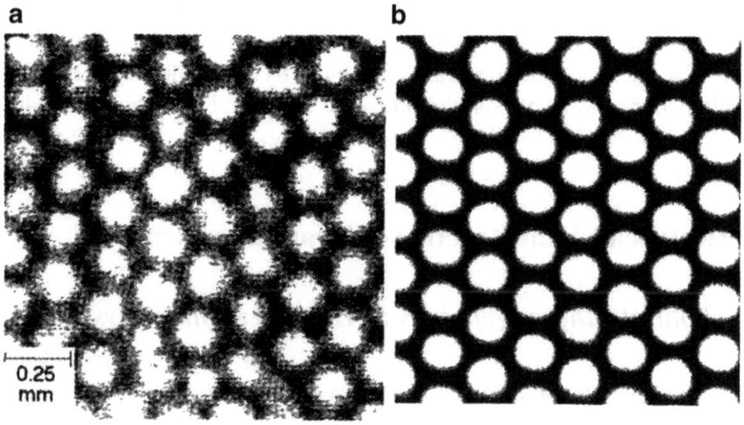

Fig. 1.15 (**a**) Experimental and (**b**) calculated Turing patterns in the CDIMA reaction. Bright pixels represent low [I^-], (**a**) $[ClO_2]_0 = 1 \times 10^{-3}$ M, $[I_2]_0 = 8 \times 10^{-4}$ M, $[MA]_0 = 1 \times 10^{-2}$ M, [S] = 3% w/v; (**b**) $[ClO_2]_0 = 1.5 \times 10^{-4}$ M, $[I_2]_0 = 7 \times 10^{-4}$ M, $[MA]_0 = 2 \times 10^{-3}$ M, $K[S] = [SI_3^-]/[I^-][I_2] = 1.83 \times 10^{-4}$ M^{-1} [3, 41]. Reproduced with permission from [41] http://link.aps.org/abstract/PRL/v69/p2729. Copyright 1992 by The American Physical Society

Turing has shown that at least three species (dynamical variables) are required to obtain this instability in a reaction–diffusion system with a diagonal diffusion matrix. For the case of three-species and spatially one-dimensional (coordinate r) system, if the Fourier mode, written now in complex notation:

$$u = U\exp(\lambda t \pm ikr) \tag{1.59}$$

is substituted into the corresponding linearized PDE, the characteristic equation takes a form (cf. Sect. 1.10, volume I):

$$\lambda^3 - \mathrm{Tr}(\mathbf{J})\lambda^2 + F\lambda - \mathrm{Det}(\mathbf{J}) = 0 \tag{1.60}$$

where Tr(**J**) is the trace of the corresponding Jacobian matrix **J**, F is the sum of all the second-order determinants of the Jacobian matrix, and Det(**J**) is the determinant. Of interest here is the case with one real negative and a complex conjugate pair of eigenvalues:

$$\lambda_1 < 0; \quad \lambda_{2,3} = \alpha \pm \mathrm{i}\omega \tag{1.61}$$

Then the wave instability occurs when $\alpha > 0$ at $k > 0$, and $\alpha < 0$ at $k = 0$, with k being here a wave number. In the vicinity of supercritical wave bifurcation, traveling waves (TW), standing waves (SW), and target patterns (TP) can emerge in reaction–diffusion systems. Also interactions between stationary Turing and wave (oscillatory Turing) bifurcations can give rise to large variety of spatiotemporal patterns, including modulated Turing structures, modulated standing waves, and combinations of Turing structures and spiral waves [45]. Such phenomena occurring also in electrochemical systems will be described in further chapter of this book.

The reader interested more in waves and patterns in chemical systems is advised to consult also the classical monograph edited by Kapral and Showalter [46].

1.2 Electrochemical Reaction-Migration Systems

1.2.1 Spatial Inhomogeneities in Electrochemical Systems

Although the general mathematical conditions for the formation of spatiotemporal patterns, presented in Sect. 1.1 for chemical systems, remain valid also for electrochemical processes, the latter ones have specific characteristics which require separate treatment of self-organization in such systems. One should note that the progress in understanding of formation of spatiotemporal or spatial dissipative patterns on electrode surfaces occurred only within recent two decades [39, 47–49].

The formation of such patterns means that the electrode|solution interface cannot be considered equipotential, so the local currents differ from one place to another. Emergence of nonuniform local currents means further the existence of nonzero, parallel to the electrode surface, gradients of concentration, conductivity, and electric potential in the solution. Evolution of such nonuniformities on a macroscopic scale means that a fluctuation of even single local current spreads over the entire electrode surface. This occurs due to *spatial couplings*. Under appropriate conditions, such couplings may give rise to various kinds of patterns on the electrode surface. One of the crucial factors deciding the occurrence of spatiotemporal/spatial patterns on the electrode surface is a spatial arrangement of the working (WE), reference (RE), and counter (CE) electrodes, as well as their own geometry, which determine the distribution of the electric potential in the solution.

As a simple introductory example, let us consider the electrode arrangement consisting of the large flat disk counterelectrode and the parallel disk working electrode, of smaller radius (ρ_0) due to embedding in an insulator. The counterelectrode will be assumed as a distant equipotential plane. Then one can calculate the current distribution and equipotential lines, indicating that the current density is infinite at radial coordinate equal to the radius of the working electrode, meaning that at the edge of this electrode the effective resistance of the electrolyte vanishes and increases when approaching the center of a disk [48]. Let us assume further that this electrode undergoes electrochemical dissolution which process, due to, e.g., passivation, exhibits the N-NDR region. Typical bifurcation diagram of such dynamical system *without* spatial inhomogeneities, for constant serial resistance R_s, was already shown in Fig. 2.8 of volume I, but now we shall reinterpret its shape in terms of varying resistance, dependent on the radial distance ρ from the center of the working electrode. Then, it may happen that, e.g., the edge of the electrode is in the passive state, the

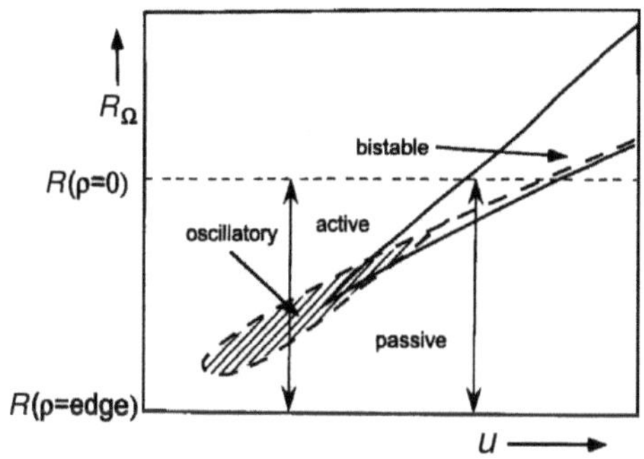

Fig. 1.16 Skeleton bifurcation diagram of an N-NDR oscillator for the case of spatial distribution of resistance R_Ω along the radius (ρ coordinate) of the disk electrode. $R(\rho = 0)$: effective resistance at the center of a disk electrode; $R(\rho = \text{edge}) = 0$: effective resistance at the edge of the disk electrode [48]. Copyright Wiley-VCH Verlag GmbH & Co. KGaA. Reproduced with permission

center is in active state, while the intermediate range would be oscillatory (Fig. 1.16). The inhomogeneity of the dynamics developing on the circular Fe electrode was experimentally proved [50, 51] and it was shown that the current oscillations were accompanied by the growth of a zone with higher reflectance, emerging from the outer rim of the electrode and propagating toward its center.

In turn, let us consider the case in which the electrode is in bistable regime, and initially *both* stable steady states are present in different parts of the electrodes: one part of the electrode is in active (high-current) and the other part is in the passive (low-current) state. The corresponding profile of the double layer potential (ϕ_{dl} symbol is used here instead of alternative E[4]) is sketched with solid line in Fig. 1.17 [39, 47]. Note that the term "local dynamics" in the caption (and further henceforth) means the dynamics of the electrode process at this interface which in this case is characterized with the N-NDR region.

Such a situation means that both parts of the electrode differ with the current flowing and with the electric potential. The gradient of the electric field parallel to the electrode surface induces migration currents which tend to homogenize the electric potential distribution. However, in the course of this process some states corresponding to smoothed potential profile cross the horizontal separatrix, composed of unstable saddle points. This means that they enter the basin of attraction of the opposite attractor: the perturbed states lying above this separatrix tend to the upper state, while those lying now below are attracted by the lower state. Figure 1.17 shows how in this way the original potential profile is restored but in different position, so the potential front has moved to the left. One can say that this front moves due to the interplay of the electrochemical process exhibiting N-NDR region and *spatial coupling* through side migration currents. Further analysis can include

[4] strictly speaking, electrode potential $E = \phi_{\text{dl}} - \phi_{\text{dl}}(\text{RE})$, but since $\phi_{\text{dl}}(\text{RE}) = \text{const}$, the dynamics of $dE/dt = d\phi_{\text{dl}}/dt$

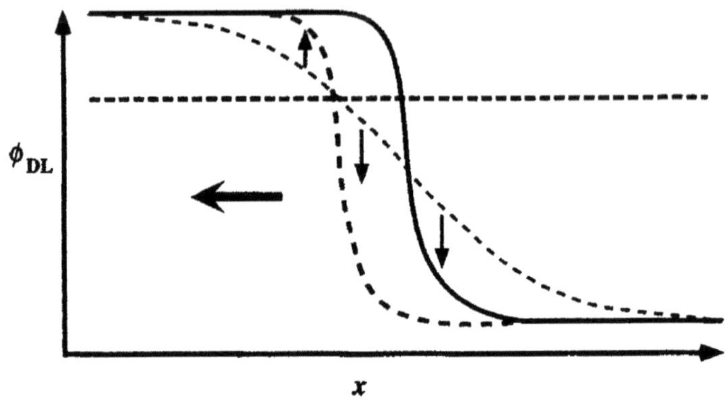

Fig. 1.17 Illustration of front motion in a bistable system due to the interplay of homogeneous dynamics and spatial coupling. The *solid line* represents an initial situation where the electrode is at small and large values of x in a different stable steady state. The *horizontal dashed curve* parallel to the x-axis indicates the ϕ_{dl} value of the saddle point. The potential difference induces cross-migration currents which tend to smooth out the front (*dashed thin curve*) thereby driving a region adjacent to the front interface on the other side of separatrix (*saddle-point*). The homogeneous dynamics try to drive the potential values back to the stationary state that is on the same side of the separatrix (*arrows*). In this way a sharp potential profile is re-established (*dashed thick curve*), however at different position; the front has moved to the *left*. Reprinted from [39], Copyright 2001, with permission from Elsevier

the spatial dynamics of the system exhibiting not only bistability, but also oscillatory behavior, caused by the destabilization due to NDR characteristics. A simplified, introductory model of such situation was elaborated by Mazouz et al. [52].

The above examples will be developed into a more systematic description of the mechanism of formation of spatiotemporal patterns, in which various kinds of spatial coupling play a crucial role.

1.2.2 Types of Spatial Coupling in Spatially Extended Electrochemical Systems

One of the first attempts to model the formation of spatial or spatiotemporal structures on the electrode surfaces was taken up by Jorne [53] who, using linear stability analysis, indicated the conditions under which the thin-layer electrochemical system with diffusion, in the three-dimensional space, could produce various (e.g., hexagonal) patterns of concentration of electroactive species. Later, Koper and Sluyters [54] have suggested a more realistic approach, in which the electrode process exhibited the negative differential resistance, the migration of ions was considered and the inhomogeneous distribution of local currents at the electrode surface and in the solution was taken into account.

The significant progress in understanding of the formation of electrochemical patterns was made, starting from 1990s, when the Berlin groups from the Fritz-

Haber Institute, led by G. Ertl, have elaborated a modern theory of such phenomena. These works proved that the problem of spatial or spatiotemporal patterns was by no means trivial and involved various detailed sources of such instabilities. Theoretical treatment of this problem is mathematically rather advanced, so the reader interested in all its details will be sometimes advised to consult the original papers or relevant reviews. Here we shall focus on the essential features of the models and their conclusions, in order to indicate various types of mechanisms leading to spatiotemporal or spatial self-organization. Shortly speaking, the way toward understanding of possible source of pattern formation goes through recognizing of possible spatial couplings that can arise in electrochemical systems and their interplay with the unstable systems of the N-NDR and S-NDR characteristics.

At the basic level of understanding, it is clear that if the small, local perturbation of the chemical or an electrochemical system develops into patterns of macroscopic size, there must exist a communication between this initial perturbation place and its surroundings. This communication occurs through the transport phenomena, like diffusion (caused by concentration gradients) and, for ionic species, through migration (caused by electric potential gradient). In consequence, the general basis for the description of the spatial or spatiotemporal evolution of any pattern is given by the equation:

$$\frac{\partial c_i}{\partial t} = g_i(c_j, k_m) - \text{div } \mathbf{f}_i \tag{1.62}$$

where $g_i(c_j, k_m)$ is the term describing the kinetics of the chemical (or electrochemical) process, called also the "local dynamics," whereas div $\mathbf{f}_i = (\partial f_{i,x}/\partial x) + (\partial f_{i,y}/\partial y) + (\partial f_{i,z}/\partial z)$ is the divergence of the vector of the flux ($\mathbf{f}_i$) of ith species, transported generally through diffusion, migration, and convection: $\mathbf{f}_i = \mathbf{f}_{i,\text{dif}} + \mathbf{f}_{i,\text{migr}} + \mathbf{f}_{i,\text{conv}}$. Since in this chapter we do not consider the convection patterns, we shall further neglect the last term.

For the explanation of stationary or nonstationary spatial patterns in *chemical* systems (cf. previous section) it was diffusion which caused the communication of various places of the system, differing with the concentration of the given species. In other words, different parts of the systems were *coupled* through diffusion, due to which the information about different states was exchanged. The divergence of the vector of the diffusion flux $\mathbf{f}_{i,\text{dif}} = -D_i \nabla c_i$ can be expressed as the product of the diffusion coefficient and the Laplacian of the concentration: div $\mathbf{f}_{i,\text{dif}} = -D_i \nabla^2 c_i$.

In electrochemical systems one has to include also the migration transport. In the description of nonlinear spatiotemporal dissipative patterns, the relative contribution from diffusion often can be even neglected. Hence the spatial inhomogeneities occur largely due to inhomogeneous distribution of the electric potential in the solution and at the electrode|solution interface. The gradients of the electrical potential parallel to the electrode surface cause a flow of corresponding local migration currents which are a way of communication between the local perturbation and its surroundings, i.e., are responsible for the spatial coupling. Due to differences in mathematical description of migration and diffusion, the detailed form of equations describing patterns in electrochemical systems is different from those for purely chemical ones. Certain analogy can appear only if in the expression of the form [Eq. (1.62)] it is possible to separate its right-hand side into the "homogeneous" part, describing the system's dynamics in the absence of any inhomogeneities, and the "heterogeneous" part, in which transport processes responsible for the coupling between different parts of the system locations are included. In the following, such couplings, specific of electrochemical systems, will be described in an introductory manner, while the details of those theoretical constructions the interested reader can find in original source, including reviews, like, e.g., [39, 47–49].

We shall further consider the process, for which the electrode potential (interfacial potential drop denoted as E or ϕ_{dl}) is an *essential variable*, according to terminology outlined already in Sect. 2.3 of volume I. This means that fixing this quantity (so reducing its role from the variable to the adjustable parameter) would suppress such manifestation of self-organization, like the oscillations.

In the *homogeneous* case, the entire electrode surface can be characterized with the standard equivalent circuit, shown in Fig. 2.6 of volume I. The entire potential drop (external voltage) U between the working and reference electrode is separated into the potential drop within the double layer (ϕ_{dl}), determining the rate of the electron-transfer process at this interface and the U–ϕ_{dl} part in the electrolyte solution extending up to the equipotential (by assumption) plane of the counterelectrode. The charge conservation principle leads then to the single equation for the temporal dynamics, identical to (2.29) or (2.65) of volume I in which symbol E is actually replaced with ϕ_{dl}:

$$C_{\mathrm{d}}\frac{\mathrm{d}\phi_{\mathrm{dl}}}{\mathrm{d}t} = \frac{U-\phi_{\mathrm{dl}}}{R_{\mathrm{s}}} - j_{\mathrm{f}}(c,\phi_{\mathrm{dl}}) \tag{1.63}$$

In this dependence, C_{d} is the differential double layer capacitance per unit area, j_{f} means the current density of the faradaic process(es) and R_{s} means the solution (serial) resistance. The $(U-\phi_{\mathrm{dl}})/R_{\mathrm{s}}$ term describes the total current which is the entire migration current flowing through the solution. We shall now define the geometry of the system: let the coordinate z be perpendicular to the working electrode surface, while coordinates x and y be parallel to it. At coordinate $z=0$ there is located an *equipotential plane*, while the reaction site at the working electrode, where the reacting particle experiences the potential drop occurring within the double layer (ϕ_{dl}), is located at $z=-w$. If this distance is normalized to w, the reaction site is equivalently placed at $z=-1$. The thickness of the double layer will be considered negligible compared to the distance w. The reader of original papers should note that the location of the working electrode can be different in various versions of the model discussed. The equipotential plane, located here at $z=0$, can be identified either with the location of the reference electrode, positioned far enough from the working electrode, so all potential inhomogeneities at the latter one fade, or with the end of the double layer at the

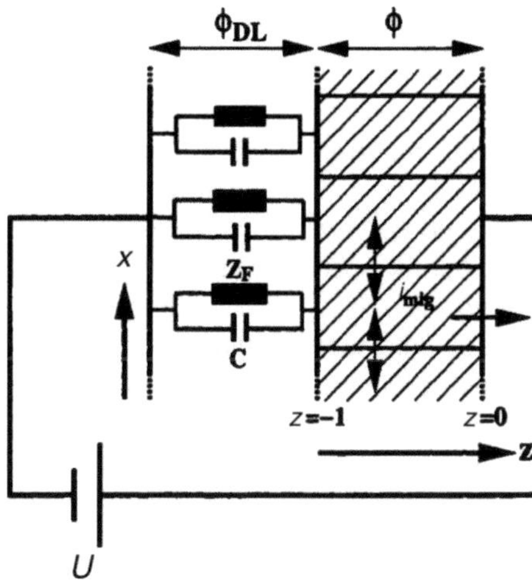

Fig. 1.18 Equivalent circuit of the basic spatially extended system; x—direction parallel to the electrode, z—direction perpendicular to the electrode (reproduced from [47] with kind permission from Springer Science + Business Media B.V.). Note that the electrolyte end of the double layer ($z = -1$) is in formulas below denoted as $z = -w$ (i.e., z is then not normalized to w)

counterelectrode. In the simplest model case, the reference electrode is assumed to be at the height of the counterelectrode. Of course, this latter assumption does *not* correspond to the situation created by the Luggin-Haber capillary, the tip of which closely and locally approaches the working electrode surface. Finally, in the homogeneous case the linear course of the electric potential course between the reaction site and the equipotential plane is assumed. Principal types of couplings, crucial for understanding of pattern formation in electrochemical systems will now be introduced.

1.2.2.1 The Migration (Nonlocal) Coupling

The spatial coupling, indicated in the examples described in Sect. 1.2.1, is a *migration coupling*. In the simplest case (i.e., when only x-coordinate parallel to, and z-coordinate normal to the electrode surface are considered), the inhomogeneity of the working electrode|solution interface can be represented by the set of equivalent circuits, as in Fig. 1.18. This interface is thus not an equipotential surface, but is composed of sectors which can differ with local electric potential and thus migration currents. In consequence, nonzero migration current components, parallel to the electrode surface, appear and they are also visualized in this scheme.

We shall here outline the mathematical description of this situation. In terms of scheme from Fig. 1.18, Eq. (1.63) has to be written separately for every spatial sector, with partial derivatives replacing the ordinary ones, due to the actual dependence of the electric potential on both the temporal and spatial variables:

$$C_{\mathrm{d}}\frac{\partial \phi_{\mathrm{dl}}}{\partial t} = -\sigma \frac{\partial \phi}{\partial z}\bigg|_{z=-w} - j_{\mathrm{f}}(c, \phi_{\mathrm{dl}}) \tag{1.64}$$

This expression describes the local capacitive current as a difference between the migration current density flowing into the double layer at position $z = -w$ (or, alternatively, at -1, depending on normalization) and a local faradaic current density j_{f}. In Eq. (1.64), σ is the specific conductivity of the electrolyte. In order to calculate the derivative $(\partial\phi/\partial z)|_{z=-w}$ one should know the spatial potential distribution in the solution. Outside the double layer usually the electroneutrality condition is accepted for the solution bulk, so the spatial distribution of electric potential ϕ is assumed to meet the Laplace equation: $\nabla^2\phi = 0$. Solving this equation requires also appropriate boundary conditions which are dependent on the geometric arrangement of the electrodes, and in this way this geometric setup affects the formation of spatiotemporal patterns. Also, Eq. (1.64) is a boundary condition for the Laplace equation. Thereby, any local change in the ϕ_{dl} value causes the redistribution of the electric potential in the *whole* electrolyte, including the spatial derivative $(\partial\phi/\partial z)|_{z=-w}$ and thus the local migration current. In other words, due to tendency of the solution to stay electroneutral, the migration currents will change at the whole interface. In this way different sites of the electrode become *coupled through migration.* Note significant quantitative characteristics of that coupling: compared to diffusion, which causes only a local coupling (i.e., limited to close neighborhood), the migration coupling is relatively *long range*, i.e., covers all other positions, although with different strength. That is why the migration coupling is called a ***nonlocal coupling***. Such nonlocal spatial couplings occur in *all* inhomogeneous electrochemical systems.

Crucial for further description of electrochemical inhomogeneities is further development of $(\partial\phi/\partial z)|_{z=-w}$ term in Eq. (1.64). In the *homogeneous* case, the $(\partial\phi/\partial z)|_{z=-w}$ derivative, due to assumed linearity of the potential profile between the working electrode and the common equipotential plane of both reference and counterelectrodes, can be estimated as $\phi(z = -w)/w$ [if the potential at the equipotential counterelectrode plane is taken as the reference value, i.e., $\phi(z = 0) = 0$]:

$$-\sigma\frac{\partial \phi}{\partial z}\bigg|_{z=-w} = \frac{\sigma}{w}\left(\phi|_{z=0} - \phi|_{z=-w}\right) = -\frac{\sigma}{w}\phi|_{z=-w} = \sigma\frac{(U - \phi_{\mathrm{dl}})}{w} \tag{1.65}$$

where w/σ is equivalent to the solution resistance [Eq. (1.63)]. By subtracting this term from the local migration current density for the *inhomogeneous* case, one obtains the contribution from the migration coupling currents originating from lateral differences of the double layer potential. In view of that it is convenient to express Eq. (1.64) in an equivalent form, Eq. (1.66), involving the expression $(U - \phi_{\mathrm{dl}})/w$:

$$C_{\mathrm{d}}\frac{\partial \phi_{\mathrm{dl}}}{\partial t} = \sigma\frac{(U-\phi_{\mathrm{dl}})}{w} - \sigma\left[\left.\frac{\partial \phi}{\partial z}\right|_{z=-w} + \left(\frac{U-\phi_{\mathrm{dl}}}{w}\right)\right] - j_{\mathrm{f}}(c,\phi_{\mathrm{dl}}) \tag{1.66}$$

The term in square brackets in Eq. (1.66), which describes the spatial migration coupling, vanishes in the homogeneous case. The two remaining terms it is thus convenient to group into a single term $f(c,\phi_{\mathrm{dl}})$, which describes the homogeneous solution that does not vanish in the absence of spatial inhomogeneities:

$$f(c,\phi_{\mathrm{dl}}) = \sigma\frac{(U-\phi_{\mathrm{dl}})}{w} - j_{\mathrm{f}}(c,\phi_{\mathrm{dl}}) \tag{1.67}$$

In such notation, Eq. (1.66) is conveniently composed of the "homogeneous" $f(c, \phi_{\mathrm{dl}})$ and "spatial coupling" terms:

$$C_{\mathrm{d}}\frac{\partial \phi_{\mathrm{dl}}}{\partial t} = f(c,\phi_{\mathrm{dl}}) - \sigma\left[\left.\frac{\partial \phi}{\partial z}\right|_{z=-w} + \left(\frac{U-\phi_{\mathrm{dl}}}{w}\right)\right] \tag{1.68}$$

the latter one being mathematically an "migration analogue" to the diffusion term in Eq. (1.62).

As it follows from Eq. (1.68), the migration coupling depends on two parameters: conductivity σ and geometrical parameter w. The conductivity plays a role analogous to that of diffusion coefficient in diffusion coupling, i.e., it determines how fast an inhomogeneous potential distribution is evened out with time (in the absence of electrochemical process causing that inhomogeneity). Thus, σ is a measure of the *strength* of migration coupling. In turn, distance w determines the *range* of coupling, i.e., is a measure of the characteristic distance over which a change in the state at a particular position instantaneously affects neighboring parts [47, 48]. The exact definition of a coupling range depends on the cell geometry, e.g., for the two-dimensional cylinder arrangement of the working and counter electrodes, the range of coupling is appropriately characterized with the quantity $\beta = (2\pi w)/L$, an aspect ratio of the cell, where L is the circumference of the electrode [55]. Then the ratio $\sigma/\beta = 1/R$, when R is the electrolyte resistance between the working electrode and the equipotential plane.

Several modelings of the development of spatiotemporal fronts due to such nonlocal coupling were performed by the group of K. Krischer [47–49]. As an example, we here show results for the process exhibiting bistable characteristics [47, 55]. In these calculations, the electrode was initially defined so that only its small portion was set in the high-current (active) steady state, while the remaining area was set in the low-current (passive) steady state. Figure 1.19 shows the gray scale maps of the front propagation for various combinations of σ and $w(\beta)$. The range of coupling (w) increases from left to right, while the strength of coupling (σ) decreases from top to bottom.

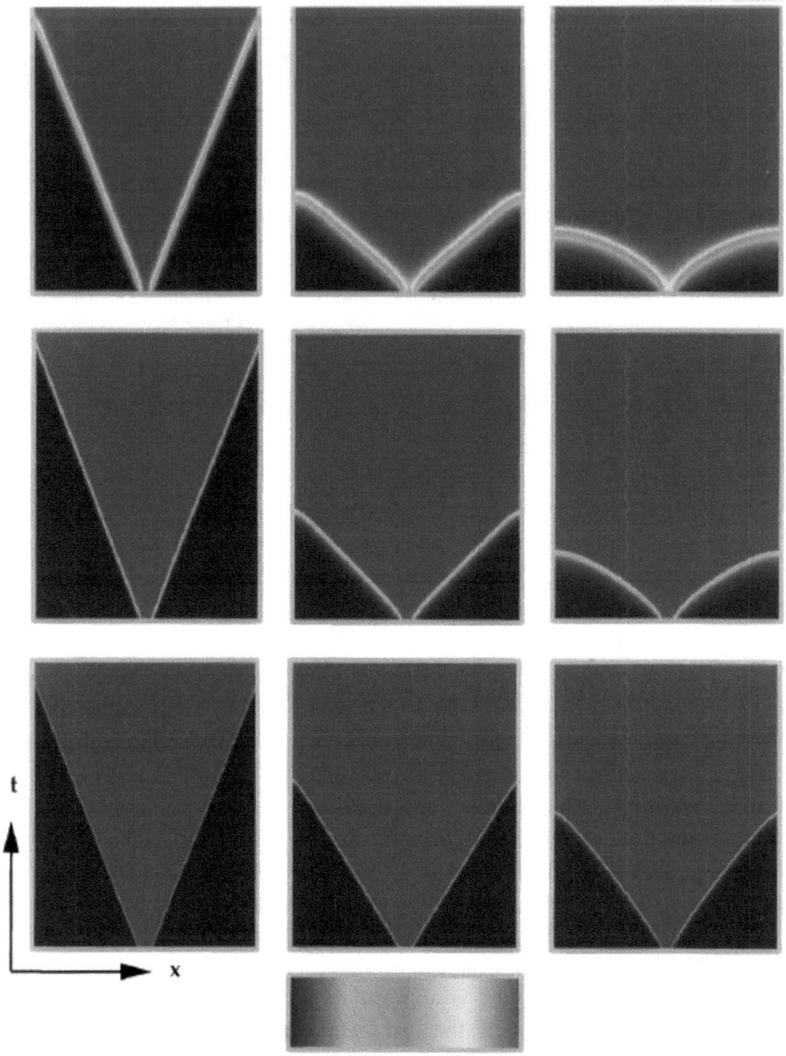

Fig. 1.19 *Gray scale* representation of the spatiotemporal evolution of the potential at the working electrode for different values of coupling strength σ and coupling range β. The coupling strength decreases from top to bottom (top row $\sigma = 0.565$, middle row $\sigma = 0.0565$, and bottom row, $\sigma = 0.00565$). The coupling range increases from left to right (left column, $\beta = 0.1$, middle column, $\beta = 1$, and right column, $\beta = 10$). The *gray scale* given at the bottom was chosen nonmonotonically for clarity (after [47, 55]). Reproduced with permission from [55] http://link.aps.org/abstract/PRE/v55/p2260. Copyright 1997 American Physical Society

This modeling shows that for increasing coupling strength σ the transition of the electrode to another state is faster and covers broader area. On the other hand, if coupling range is small (left column), i.e., is of nearly diffusion nature, the fronts move with constant velocity, while the increase in the coupling range (corresponding to increasing role of migration) causes acceleration in the movements of the front. Experimentally it is very easy to modify the strength of coupling by varying the solution conductivity. It is remarkable that such accelerating fronts were indeed observed by Flätgen and Krischer for the electroreduction of peroxodisulfate to sulfate at silver ring electrodes [56].

Alternative approach to the formulation of the nonlocal coupling term, involving Green function, was also elaborated by Christoph, Hudson et al. [57, 58] (see also Chap. 2). The mathematical formalism of that approach is applicable to any geometry of the electrochemical system and it is possible to generate various patterns, both for the bistable and oscillatory regime of the process, for example, spiral waves and spiral turbulence in the model of the nickel dissolution on a rectangular electrode [58]. In the opinion of the authors, their generalized theoretical approach would be useful, among others, for the description of traveling of excitation pulses along unmyelinated nerve axons, better than the classical Hodgkin–Huxley approach [59], based on the idea of the reaction–diffusion system.

To summarize this subsection, the migration (nonlocal) spatial coupling is *always synchronizing*, i.e., in the absence of the electrochemically generated instability, it always smoothes out any potential gradient tangential to the working electrode, analogously to diffusion smoothing out concentration gradients, but exhibits much longer range.

1.2.2.2 The Global Couplings

Besides the migration coupling, which occurs in all spatially inhomogeneous electrochemical systems, also another way of communication between different sites of the system can occur which is termed the *global coupling*. Such type of coupling occurs with the participation of the electronic device—galvanostat or potentiostat, and these two modes of operation produce different particular schemes of *positive* or *negative* couplings, respectively. Mazouz, Krischer et al. [60] considered the works by Lev et al. [61, 62], the first premise that pattern formation can be affected by the control mode of the electrochemical experiment. In these works, the spatiotemporal behavior of the anodic dissolution of Ni wire was studied. Under galvanostatic conditions, two halves of the wire oscillated out of phase (antiphase oscillations). In turn, under potentiostatic conditions and in the presence of external series resistor in the circuit, traveling waves were observed, with spatial self-organization decaying upon decreasing resistance up to completely homogeneously oscillating state of the wire for zero resistance (see also Chap. 2).

In theoretical analysis of this kind of coupling, it is convenient to consider the galvanostatic mode a limiting case of the potentiostatic mode, when the external resistance R_s (together with applied voltage U) tends to infinity, with the U/R_s ratio defining the imposed current. Thereby, by continuous increase of the external resistance one can smoothly change the control mode from potentiostatic to galvanostatic one. Accordingly, the mathematical analysis of this coupling can be conveniently referred to the corresponding equivalent circuit containing the external resistor which can be appropriately tuned, together with the external voltage. As above, for the explanation of pattern formation, the series of equivalent circuits was ascribed to the electrode/electrolyte interface [63].

It is further of crucial importance to distinguish between the cases of the above-mentioned positive and negative global couplings which occur under galvanostatic and potentiostatic conditions, respectively.

Galvanostatic Case: Positive Global Coupling

The essential scheme of the origin of global coupling is in this case quite simple: the galvanostat forces the flow of constant current which can be considered a sum of partial, local currents: $I = \sum I_k$. Let us assume that one of these partial currents I_k is slightly increased (e.g., due to fluctuation). The total current exceeds then the programmed value and the galvanostat reacts, producing the charging current compensating this difference. In other words, the potential of the *entire* working electrode changes appropriately, so—in spite of only local initial fluctuation of the current, eventually *all* reaction sites at the electrode become affected. That is why this mode of coupling is named *global*. Furthermore, this coupling in the galvanostatic mode exhibits an important dynamical feature: it leads to smoothing of the differences of interfacial potential drops between different electrode sites, i. e., it has a *synchronizing* effect on them; this of course stabilizes the electrode state and does *not* favor the formation of spatial or spatiotemporal patterns.

In order to describe mathematically this kind of coupling it is useful to refer to the above construction of an equivalent circuit under potentiostatic control, with external serial resistor R_{ext} increasing its resistance to infinity, when the potentiostatic control switches to galvanostatic one. The cell resistance R_{cell} between the WE and CE is a sum of the uncompensated resistance R_{uncomp} between the working and reference electrodes and of the compensated resistance R_{comp} between the reference and counter electrodes. The external voltage U distributes between the double layer potential drop (ϕ_{dl}) and the ohmic drop occurring due to all uncompensated resistances. The following expression for the migration term $(\partial\phi/\partial z)|_{z=-w}$ is then obtained [48]:

$$-\sigma\frac{\partial\phi}{\partial z}\bigg|_{-w} = \frac{1}{A(R_{\text{ext}}+R_{\text{cell}})}\phi|_{-w} - \sigma\left[\frac{\partial\phi}{\partial z}-\frac{\phi}{w}\right]_{z=-w} + \frac{1}{AR_{\text{cell}}}\frac{R_{\text{ext}}/R_{\text{cell}}}{(1+R_{\text{ext}}/R_{\text{cell}})}(\langle\phi_{\text{dl}}\rangle-\phi_{\text{dl}}) \tag{1.69}$$

Then the charge conservation principle, after incorporation of the "homogeneous" term into $f(c, \phi_{\text{dl}})$, attains a form:

$$C_{\text{d}}\frac{\partial\phi_{\text{dl}}}{\partial t} = f(c,\phi_{\text{dl}}) - \sigma\left[\frac{\partial\phi}{\partial z}\bigg|_{z=-w} - \left(\frac{\phi}{w}\right)_{z=-w}\right] + \gamma(\langle\phi_{\text{dl}}\rangle-\phi_{\text{dl}}) \tag{1.70}$$

The crucial parameter γ, defined as:

$$\gamma = \frac{1}{AR_{\text{cell}}} \frac{R_{\text{ext}}/R_{\text{cell}}}{(1 + R_{\text{ext}}/R_{\text{cell}})} \tag{1.71}$$

determines the strength of this kind of coupling. Now, since the galvanostatic control is achieved through the increase of R_{ext} to infinity, one realizes that γ increases then from zero to $1/AR_{\text{cell}}$ which is equal to the ratio σ/w. This allows to determine experimental conditions which favor this kind of coupling: *high solution conductivity* or *close distance between the working and the counter electrodes* (the reference electrode is positioned far away from working electrode, in order to avoid negative coupling, described further in this section). Such dependencies emphasize the role of the cell geometry in the occurrence and relative proportions of migration and global couplings [48]. Finally, Eq. (1.71) says that positive global coupling will exist for any nonzero external resistance.

The mathematical form of Eq. (1.70), with the term $+\gamma(\langle\phi_{\text{dl}}\rangle - \phi_{\text{dl}})$ is crucial for the qualification of this kind of coupling as *positive* one. This nomenclature is related to the positive sign of the coefficient: $+\gamma$, preceding the expression in the bracket. Note further that the dependence of the system's reaction on $\langle\phi_{\text{dl}}\rangle$, expressed by Eq. (1.70) illustrates the *global* nature of this coupling. In fact, a global coupling is characterized with the dependence of the temporal evolution of a *local* state variable on the *spatial mean* of this variable calculated for the entire (i.e., globally considered) system. If the interfacial potential drop ϕ_{dl} is locally enhanced (positive local perturbation), its actual average value calculated for the entire surface $\langle\varphi_{\text{dl}}\rangle$ also increases. Thus, in all *other* places, where the average value became higher than local values: $\langle\phi_{\text{dl}}\rangle > \phi_{\text{dl}}$, local ϕ_{dl} will *also* increase, until the expression $+\gamma(\langle\phi_{\text{dl}}\rangle - \phi_{\text{dl}})$ reaches zero at every location. Thus, under galvanostatic conditions, the positive global coupling acts globally on the system in a *synchronizing* manner, similarly as migration coupling, described earlier. The physical sense of the word "positive" is thus that the local change and the global change of the double layer potential (interfacial potential drop) have the same sign.

One can illustrate the manifestation of this kind of coupling in the evolution of bistable electrochemical system that can exist either in low-potential state (1) or high-potential state (2) [49]. If the system remains initially in the homogeneous state (1) and its *small* part undergoes a transition to state (2), the average interfacial potential drop $\langle\phi_{\text{dl}}\rangle$ slightly increases. In consequence, according to Eq. (1.70), all other locations originally in state (1) undergo a change to state (2).

As a complement to this description, one should note that this transition from state (1) to state (2) occurs in the form of an *accelerated front*, as a result of concerted action of migration coupling and global positive coupling. Only for very big values of γ this transition is practically homogeneous [49].

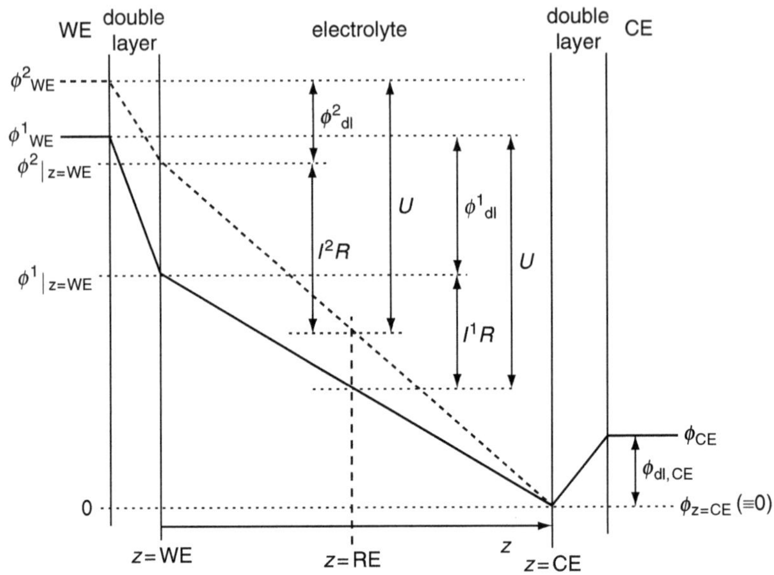

Fig. 1.20 Potential profiles perpendicular to the WE with corresponding potential drops for two different current densities (I^1 and I^2) at the same value of the applied external voltage U, demonstrating the origin of the global coupling. (The potential parallel to the cell is assumed to be uniform) [48]. Copyright Wiley-VCH Verlag GmbH & Co. KGaA. Reproduced with permission

Potentiostatic Case: Negative Global Coupling

In the potentiostatic mode the constant external voltage U is distributed between the potential drop in the double layer (ϕ_{dl}) and the voltage drop in the solution between the working electrode and the location of the reference electrode [48]:

$$U = \phi_{dl}(x,y) + [\phi(x,y)|_{z=WE} - \phi|_{z=RE;\ x,y=RE}] \tag{1.72}$$

When the working electrode potential is locally perturbed, this affects the electric potential distribution in the entire electrolyte, so also at the reference electrode, if it is placed between the working and counterelectrodes. The potentiostat feels this alteration as the deviation of the potential of the working electrode from the programmed value and charges this electrode to a new value, which process modifies the electric potential distribution in the entire electrolyte, until Eq. (1.72) is again fulfilled. The relevant change in the distribution of the potential is shown in Fig. 1.20.

If working and reference electrodes are of the same size and parallel to each other, the entire WE surface experiences the change of the potential, meaning the global nature of this effect. In practice, an equivalent situation can be realized more conveniently by using the working electrode in the form of thin annulus, with the reference electrode located on the axis of the ring.

The specific dynamic feature of the global coupling *under potentiostatic conditions* is its *negative* nature. Thus, opposite to synchronizing effect of the positive global coupling under galvanostatic conditions, in the present case the inhomogeneities at the working electrode increase, i.e., the negative global coupling causes *desynchronization* of different sites of the system. This effect is particularly strongly pronounced if the Haber-Luggin capillary is used for minimization of the ohmic drop between WE and RE. Of course, the amplification of spatial fluctuations means that this kind of global coupling favors formation of various spatial patterns.

In order to describe this type of coupling in more detail, one should distinguish again between the three resistances present in the system: the total cell resistance R_{cell} composed of the uncompensated resistance between the working and the reference electrodes (R_{uncomp}), and the compensated resistance R_{comp} between the reference and counter electrode. The total cell resistance: R_{cell} is also equal to $w/(A\sigma)$, where A is the surface area of the working electrode and w has the same geometrical meaning as above. Then, one derives the following expression for the migration term [48]:

$$-\sigma\left.\frac{\partial\phi}{\partial z}\right|_{z=-w} = \frac{1}{AR_{\text{uncomp}}}\phi|_{z=-w} - \sigma\left[\frac{\partial\phi}{\partial z} - \frac{\phi}{w}\right]_{z=-w} - \frac{1}{AR_{\text{cell}}} \times \frac{R_{\text{comp}}/R_{\text{cell}}}{(1 - R_{\text{comp}}/R_{\text{cell}})}(\langle\phi_{\text{dl}}\rangle - \phi_{\text{dl}}) \tag{1.73}$$

The charge conservation principle, after incorporation of "homogeneous" term into $f(c, \phi_{\text{dl}})$, leads now to the following dependence for the dynamics of the interfacial potential drop, with the last term on the right-hand side describing the contribution from the global feedback:

$$C_{\text{d}}\frac{\partial\phi_{\text{dl}}}{\partial t} = f(c, \phi_{\text{dl}}) - \sigma\left[\frac{\partial\phi}{\partial z} - \frac{\phi}{w}\right]_{z=-w} + \alpha(\langle\phi_{\text{dl}}\rangle - \phi_{\text{dl}}) \tag{1.74}$$

The crucial coefficient α is defined as:

$$\alpha = -\frac{1}{AR_{\text{cell}}}\frac{R_{\text{comp}}/R_{\text{cell}}}{(1 - R_{\text{comp}}/R_{\text{cell}})} \tag{1.75}$$

The global coupling is nonexisting, if $\alpha = 0$ which is the case for $R_{\text{comp}} = 0$. Physically it corresponds to the reference electrode placed at the height or behind the counter electrode, so if one wishes to avoid this kind of coupling, the RE should be placed possibly far from WE. The potential at the RE is then not influenced by any change of ϕ_{dl} at the WE. On the contrary, the global coupling becomes significant when α attains relatively high values, which is the case if the cell resistance is largely compensated ($R_{comp} \rightarrow R_{cell}$, $R_{uncomp} \rightarrow 0$) due to close approach of the working electrode by the reference electrode, e.g., by using a Haber-Luggin capillary [47].

Analogously to the case of galvanostatic conditions, the nature of coupling is indicated by the mathematical form of the term $\alpha(\langle\phi_{dl}\rangle - \phi_{dl})$, with the crucial role of *negative sign* of coefficient α. Therefore, the effect of global coupling *under potentiostatic conditions* has a *reversed* direction, compared to galvanostatic control. Considering the global nature of the coupling we further apply an analogous way of thinking, as above. If ϕ_{dl} is locally enhanced, it causes an increase in $\langle\phi_{dl}\rangle$. Then, due to negative α, in all *other* locations of the electrode, characterized now with the relation: $\phi_{dl} < \langle\phi_{dl}\rangle$, the potential will decrease, so quite opposite to the effect of positive global coupling described earlier. Thus, the present case of a *negative global coupling* has a *desynchronizing* effect on the electrode state and thus favors the inhomogeneities or the pattern formation. The physical sense of the word "negative" means now that the *local* change and the *global* change of the double layer potential have the *opposite* signs.

In terms of the evolution of bistable electrochemical system, one can now describe the manifestation of the negative global coupling in the following way. If, as above, the system remains initially in a low-potential state (1) and becomes locally perturbed to high-potential state (2), the other positions remaining in state (1) are driven to lower potentials, i.e., farther away from state (2). Furthermore, the propagation speed of the front is reduced and, beyond a certain threshold of the coupling strength $|\alpha|$, can become zero, meaning the formation of a stationary dissipative structure, composed of two regions of the interfacial potential, coexisting at the electrode [39].

Although the global couplings under galvanostatic and potentiostatic conditions were described as acting in opposite directions, it is possible to make interesting generalizations [48]. Mathematically, both kinds of coupling differ with the definition of the coupling coefficient [Eqs. (1.71) and (1.75)]. If in definition of γ [Eq. (1.71)], one replaces the external resistance with the *negative* compensated resistance: $R_{ext} = -R_{comp}$, the obtained expression appears to be identical with the definition of coefficient α. Thus, one can interpret this result as the equivalent (desynchronizing) role played by the compensated part of electrolyte resistance under potentiostatic conditions and a negative external resistance under galvanostatic conditions.

Concluding, in this section, three fundamental kinds of coupling occurring in electrochemical systems were described: (1) migration (nonlocal) coupling, existing in every electrochemical system and exhibiting the synchronizing effect on the working electrode state; (2) global synchronizing coupling (occurring under galvanostatic conditions); and (3) global desynchronizing coupling (occurring under potentiostatic conditions). There were also given examples of spatiotemporal patterns modeled for the process, characterized with NDR characteristics, leading to bistability. In the next step, one should consider the interplay of all these three types of coupling with the N-NDR and S-NDR systems, which yields six different combinations. They will be now briefly characterized [39].

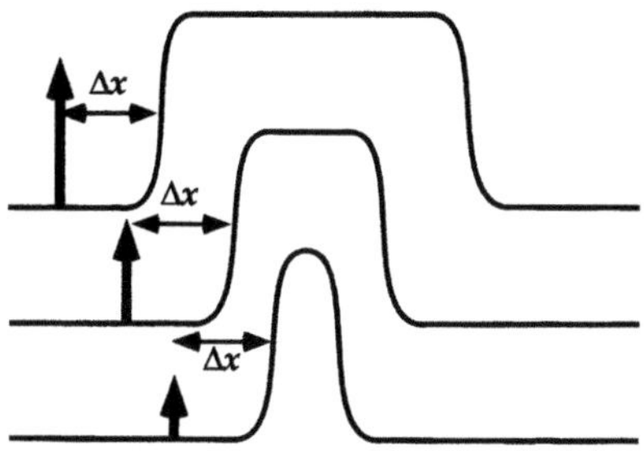

Fig. 1.21 Illustration showing why nonlocal coupling leads to accelerated front motion. The effect of the cross migration current on the temporal evolution of the double layer at certain distance from the interface (e.g., the inflection point) increases with the area that is in a different state. Reprinted from [39], Copyright 2001, with permission from Elsevier

1.2.3 The Interaction of Spatial Couplings with the NDR Systems

1.2.3.1 N-NDR and Migration (Nonlocal) Coupling

Systems with N-NDR (as well as with HN-NDR) characteristics belong to the so-called activator–inhibitor systems, in which the electrode potential is an activator (autocatalytic variable) and the chemical variable (concentration of the electroactive species) is an inhibitor, engaged in the slow negative feedback loop. In principle, migration coupling alone, similarly as the diffusion in chemical systems, in the absence of any electro(chemical) reaction should lead to homogenization of the system. But as interplay of chemical reactions and diffusion can give rise to spatial or spatiotemporal self-organization, also interaction of migration coupling with the activator–inhibitor systems may result in analogous self-organization. In particular, if the coupling acts on the activator variable, the pulses or spiral waves, propagating over long distances without attenuation, are generated [39]. The propagation of the front in an electrochemical bistable system, shown in Fig. 1.17, was a simple model example of such a situation in which the process with the N-NDR characteristics interplays with the migration coupling: the activator—the electrode potential flew along the electrode surface due to migration.

An interesting feature of such interplay is the above-mentioned acceleration of the propagation of the potential wave across the electrode, unlike the reaction–diffusion systems in which concentration fronts move with constant velocity. It's now time to outline the understanding of that effect. This acceleration is explained by the specific properties of the nonlocal coupling, illustrated schematically in Fig. 1.21 which shows the accelerated front motion in the bistable system, starting from the low-potential reference state.

Due to nonlocal coupling, every point along the interface contributes to the spatial coupling term, with its magnitude becoming the larger, the more points are in a state with a double layer potential more positive than the reference state.

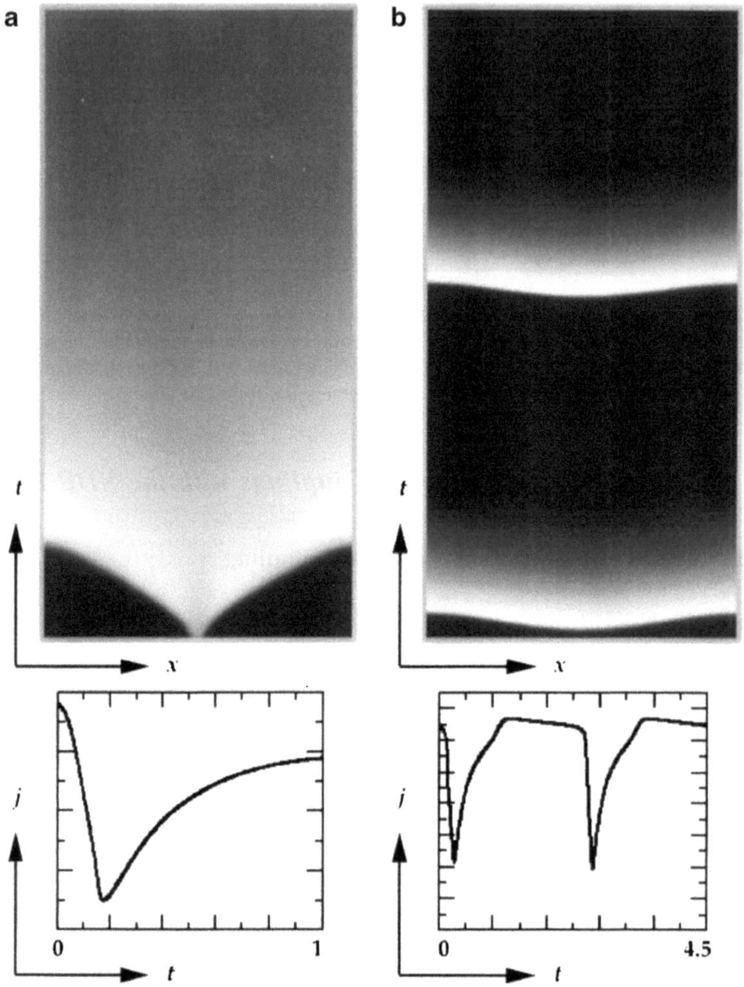

Fig. 1.22 *Gray scale* representation of the calculated spatiotemporal evolution of the potential at the electrode (*top*) and time series of the global current (*bottom*) in the (**a**) bistable and (**b**) oscillatory regime. See [52] for parameter values. Reprinted with permission from [52], Copyright 2001 American Chemical Society

In other words, the front velocity increases due to an increase of the portion of the electrode switched to an alternative (high-potential) state. Such acceleration of the fronts can be observed also as a phenomenon accompanying the relaxation oscillations of current for the N-NDR oscillators [39]. Representative comparisons of the evolution of the bistable and oscillatory model N-NDR systems are shown in Fig. 1.22. When such correlations were first realized, they became an important premise that explanations of temporal bistable and oscillatory dynamics of electrochemical systems, based on assumption of only homogeneous variations of the electrode state, can be a serious simplification. Selected experimental examples of such spatiotemporal dynamics, involving, e.g., anodic dissolution of Fe or Co electrode, and cathodic reduction of $S_2O_8^{2-}$, will be described in Chap. 2.

1.2.3.2 N-NDR and Positive Global coupling

In the case of positive global coupling, observed under galvanostatic conditions (or in the presence of external resistor), one should expect again the synchronizing effect, i.e., the tendency to homogenize the electrode state, thus preventing the development of stable patterns. Let us refer again to the bistable characteristics of the process, which causes that in the initial state a majority of the electrode surface is on the "low-potential" state, while a small part of the electrode surface acquires the "high-potential" state (Fig. 1.21) [39]. In consequence, the actual average double layer potential $<\phi_{dl}>$ is enhanced, compared to the homogeneous case. In terms of the evolution equation (1.70) it means that the local ϕ_{dl} potential will increase in all places where it is smaller than $<\phi_{dl}>$; hence, this positive coupling will cause the switch of the "low-potential" electrode states, through the unstable steady states of the folded diagram of bistability, toward the "high-potential" states. The autocatalytic kinetics of the ϕ_{dl} dynamics makes this transition appropriately fast. The process of the front propagation is accelerated in time since with gradually expanding "high-potential" region, the average double layer potential becomes more positive. Thus, the front acceleration occurs due to *both* nonlocal and positive global coupling, according to similar mechanism. When bistability is replaced with the oscillations due to N-NDR region, their interaction with positive global coupling is a more complex problem and the observed dynamics may depend on the strength of the coupling. For example, for galvanostatic Ni electrodissolution in H_2SO_4 medium, for small and large strengths of global coupling, the relaxation oscillations were associated with homogeneous changes of the electrode state, while for intermediate strengths the formation of various domains was observed which, among others, formed stationary clusters with a certain phase difference between them [39, 65].

1.2.3.3 N-NDR and Negative Global Coupling

The general conclusion drawn above was that the negative global coupling favors the pattern formation. The negative character of this coupling means that if the double layer potential is locally enhanced, and thus the average potential increases, this results in a decrease of the potential at all other electrode sites [cf. Eq. (1.74)]. However, the migration coupling is then also induced and, as acting in the opposite direction, tends to broaden the potential profile. It is convenient to illustrate these effects again for the electrode state in the bistable regime (Fig. 1.21). Let the potential be again initially locally increased, but now the negative global coupling causes the *decrease* of the potential of remaining electrode parts.

In consequence of the competition between these two spatial couplings, the spatial expansion of the high-potential zone is slowed down. Initially, the migration coupling predominates, slightly opposing global coupling, but as long as the high potential state expands (meaning constant increase in $\langle\phi_{\mathrm{dl}}\rangle$), accordingly also the strength of negative coupling rises. If it happens that eventually both migration and negative global couplings balance each other, stationary patterns consisting of high and low potential domains develop [39]. Such phenomena were in fact observed for the electroreduction of $S_2O_8{}^{2-}$ ions at a Ag-ring electrode, close to the bistable region, when the ohmic potential drop was minimized by using Haber-Luggin capillary [66]. In the case of oscillatory processes, the negative global coupling may induce a variety of spatiotemporal patterns, as standing waves and pulses [39]. Examples of such phenomena are described in Chap. 2.

1.2.3.4 S-NDR and Migration (Nonlocal) Coupling

The most important consequence of the interaction between the electrochemical process with S-NDR characteristics and the long-range, migration coupling is the formation of stationary, *Turing patterns*. The principles of formation of such patterns in chemical systems, due to the coupling of reaction kinetics with diffusion of reagents exhibiting various diffusion coefficients, were described in Sect. 1.1.4. In terms of the concept of the activator–inhibitor system, the Turing patterns emerge when the inhibitor, the production of which is accelerated by the high concentration of an activator, diffuses faster than the activator and in this way the spatial progress of the activator becomes hindered (cf. Fig. 1.13, referring explicitly to electrochemical systems). In the kinetic scheme considered here X is an activator since it enhances the production of an inhibitor Y and also is a self-activator, since it autocatalytically produces itself. In turn, Y is both a self-inhibitor and an inhibitor for X.

For electrochemical systems, the theory [42] and the experimental realization [43] of Turing patterns were published only in the years 2000–2001 by the Berlin group from the Fritz-Haber Institute of the Max Planck Society. These works indicated not only the possibility of formation of Turing patterns for the electrochemical systems of the S-NDR type, but it was also suggested that, for such systems, the Turing patterns should form in a huge range of control parameters. The basis for such conclusion is the fact that in electrochemical systems not only diffusion, but also migration transport is present. While in chemical systems the situation in which diffusion coefficients of the activator and the inhibitor differ significantly is rather as an exception than as a rule, in electrochemical systems the characteristic rate of migration is usually much higher than that of diffusion. In S-NDR systems the electrode potential is an inhibitor (and typically a fast variable), while the chemical species, involved in the autocatalytic loop, is an activator which is transported by (relatively slow) diffusion. This is a fundamental premise for the possibility of Turing pattern formation in electrochemical systems of that type.

The concise, simplified description of the theory of formation of electrochemical Turing patterns under potentiostatic conditions is presented below, while the reader

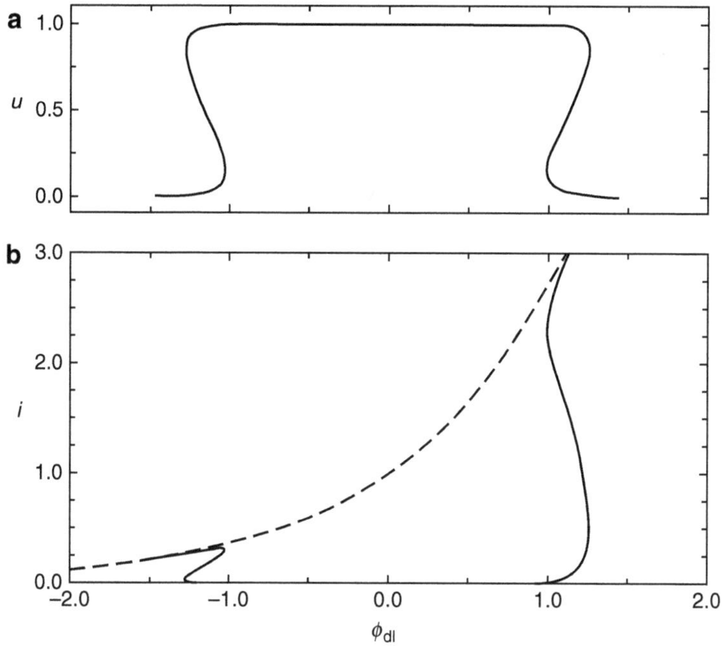

Fig. 1.23 (**a**) Coverage of a neutral species, u, as a function of the electrode potential. The multivalued region results from the attractive interactions between the adsorbed molecules. (**b**) Current–potential characteristics in the absence (*dashed line*) and in the presence (*solid line*) of the adsorbate according to Eq. (1.76). The characteristics possess an S-shaped and Z-shaped region. Reprinted with permission [42]. Copyright 2000 American Chemical Society

interested in the full model construction and all mathematical derivations is advised to consult the original reference [42]. Since the analysis involves migration, and not only diffusion considered originally by Turing [34], the following analysis can be considered a way toward the diagnosis of conditions for the *Turing-like* bifurcations. The S-shaped characteristics of the *I–E* curve, meaning that the current is not a unique function of the electrode potential, require a definition of the autocatalytic kinetics that explains the multivalued *I–E* region. As the model of the relevant electrochemical process there was chosen the electron-transfer process obeying the Butler–Volmer kinetics, but inhibited by the electrochemically inactive adsorbate. This adsorbate is assumed to undergo first-order phase transitions, with strong attractive lateral interactions between the adsorbed particles, leading to multivalued isotherm of the Frumkin type. The corresponding dependence of the electrode coverage u on the electrode potential is depicted in Fig. 1.23a. Under further assumption that within the overpotential region considered, the backward process can be neglected, the faradaic current of the irreversible electron transfer is described with Eq. (1.76):

$$I_f = (1 - u)nFc_r k_{reac} \exp[\alpha nF(\phi_{dl} - V)/RT] \qquad (1.76)$$

where n is the number of electrons transferred per molecule, α is the transfer coefficient, c_r is the bulk concentration of the electroactive species, k_{reac} is the rate constant of the electron transfer, and V denotes here the equilibrium potential of the reaction. This *I*–*E* dependence is depicted in Fig. 1.23b.

The final equations describing the dynamics of the model system, after transformation to dimensionless form, are the following:

$$\frac{\partial u}{\partial t} = \mu f(u, \phi_{dl}) + \frac{\partial^2 u}{\partial x^2} \tag{1.77}$$

$$\frac{\partial \phi_{dl}}{\partial t} = \kappa g(u, \phi_{dl}) + \frac{d}{\beta}(U - \phi_{dl}) - \frac{d}{\beta}\left(\left.\frac{\partial \phi}{\partial z}\right|_{z=-w} + \phi|_{z=-w}\right) \tag{1.78}$$

Equation (1.77) describes the autocatalytic kinetics of variable u (electrode coverage with an electrochemically inactive adsorbate), included in the functional term $f(u, \phi_{dl})$ and the (relatively slow) diffusion of that species, treated as an activator. Equation (1.78) describes the dynamics of the electrode potential—the inhibitor, which, through the last term on the right-hand side, is affected by the cross-migration currents, creating a spatial coupling through the electrolyte. The particular form of the parameters is determined by the accepted geometrical construction of the model system, in which the working electrode has a shape of a thin ring, of a diameter large compared to the thickness of the ring. The electrolyte is modeled on a cylindrical surface with a constant potential at some distance from the electrode, at which the reference electrode is located. Then, x is the direction parallel to the electrode, L is the length of the electrode, coordinate $z = 0$ corresponds to the equipotential surface, identical with the position of the reference electrode where $\phi = 0$, while the working electrode is placed at $z = -w$ (after normalization to w, the working electrode is placed at $z = -1$). The respective parameters are defined in a following way:

$$\beta = \frac{2\pi w}{L} \tag{1.79}$$

$$\mu = \frac{L^2}{4\pi^2 D} k_u \tag{1.80}$$

$$\kappa = \frac{\alpha n^2 F^2 L^2 k_r e^{-V}}{RTC_0 4\pi D} \tag{1.81}$$

$$d = \frac{L\sigma}{2\pi D C_0} \tag{1.82}$$

where D is the diffusion coefficient of the electroactive species, k_u [s^{-1}] and k_r [mol cm^{-2} s^{-1}] are characteristic rates involved in the reaction kinetics of species u and the electron transfer kinetics, respectively, σ—the specific conductivity, C_0 is the specific double layer capacitance, and n is the wavenumber. As above, parameter β can be understood as an aspect ratio of the electrochemical cell, which determines the nonlocality of the migration coupling.

The stability of the steady state defined by Eqs. (1.77) and (1.78) set to zero: $(\partial u/\partial t = \partial \phi_{\mathrm{dl}}/\partial t = 0)$ was analyzed by means of the linear stability analysis. For the Turing bifurcation to occur, the homogeneous steady state alone has to be asymptotically stable [$\mathrm{Tr}(\mathbf{J}_0) < 0$, $\mathrm{Det}(\mathbf{J}_0) > 0$] where $\mathbf{J}_0$ is the Jacobian matrix for the homogeneous system, evaluated at the steady state. Further analysis indicates the following necessary condition for the Turing-like instability [64]:

$$d > \frac{n^2}{k(n)} = \frac{n^2}{n\coth(n\beta) - (1/\beta)} \tag{1.83}$$

This inequality is the electrochemical equivalent of the condition for the Turing bifurcation in a chemical system, requiring the ratio of the diffusion coefficients of the inhibitor and the activator greater than one. Now, dimensionless quantity d, always greater than one for any $n > 1$, irrespective of the value of β, describes the ratio of the rate of migration and diffusion. In order to satisfy condition seen in Eq. (1.83), two parameters: wave number n and aspect ratio of the cell β which is a measure of the nonlocality of the coupling have to be taken into account. Taking value of $D = 10^{-5}\ \mathrm{cm}^2\ \mathrm{s}^{-1}$ and typical values of other parameters, one obtains d of the order 10^3–10^4, meaning the spatial spreading of the potential profile much faster (due to migration) than the spreading of the surface concentration of electroactive species.

Numerical calculations led to the two-parameter bifurcation diagrams, showing the locations of the saddle-node bifurcations for homogeneous systems and Turing bifurcations for the system with reagent transport. Such diagram is shown in Figure 1.24 [39, 42].

Analysis of this diagram indicates an important conclusion, that the formation of the Turing patterns requires nonzero (nonvanishing) electrolyte resistance. In fact, in order for the electrochemical Turing bifurcation to occur, the ohmic potential drops are necessary which transform the S-shaped, multivalued I–E dependence into the single-valued I–U dependence of a positive slope (cf. Fig. 2.10a, b, volume I). Note further that the Turing bifurcations occur within certain interval of the solution conductivity σ which describes the rate of migration. Since Turing patterns require sufficiently fast migration across the electrode, compared to diffusion, the respective bifurcation will occur above certain critical value $\sigma_{\mathrm{crit},1}$. However, and this is a specific feature of electrochemical systems, there is also a second critical value $\sigma_{\mathrm{crit},2}$, above which the Turing patterns vanish, since then the conductivity becomes so high that the homogeneous state becomes asymptotically stable again.

The exemplary stationary Turing patterns, visualized as spatial distributions of the electrode potential, together with two homogeneous distributions, occurring just outside the region of spatial instabilities, are shown in Fig. 1.25.

Further analysis showed the possibility of coexistence of patterned and homogeneous potential distributions, as well as of the coexistence of the patterns with the

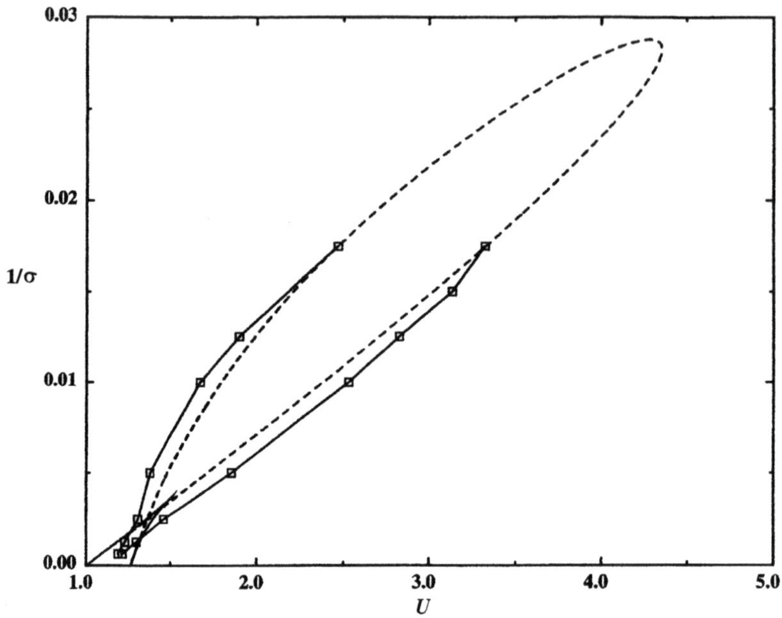

Fig. 1.24 Two-parameter bifurcation diagram showing the locations of the saddle-node (*solid curves* in the *lower left corner*) and Turing bifurcations (*dashed curve*) in the S-NDR system. The curves connecting the *square symbols* give the approximate borders of a coexistence region of a stationary structure and the homogeneous solutions. The diffusion constant of the activator is scaled to 1 [39]. Parameter values: $\beta = 6.25$, $\mu = 10$, $\kappa = 10$. See [42] for other parameters. Reprinted from [39], Copyright 2001, with permission from Elsevier

oscillatory variations of state due to Hopf bifurcation. In the latter case, corresponding to appropriately increased time scale of the activator, the asymptotic system's behavior depends on initial conditions, as the exemplary simulations, presented in Fig. 1.26, illustrate. For other parameter values there were found also nonstationary, but oscillating Turing patterns (having characteristics of mixed-mode oscillations) [42].

At this point it is useful to realize that stationary character of the patterns observed in the given experiment does not mean that the *true* Turing patterns were formed, thus not every observed stationary pattern is automatically a Turing pattern. As mentioned earlier in description of N-NDR system with negative global coupling, stationary patterns were observed during the reduction of $S_2O_8{}^{2-}$ [65]. Since this process is of the N-NDR, and not of the S-NDR type, these patterns cannot be classified as Turing ones. Instead, the formation of two spatial domains on the working electrode surface was there due to negative global coupling, caused by the external circuit, and not only by the migration of ions in the electrolyte which is one of the major factors deciding the emergence of Turing patterns. Also, one of the essential features of the Turing patterns is that their wavelength is independent of the size of the system, but depends only on the rate constants of the system. This condition was not met for $S_2O_8{}^{2-}$ reduction when the wavelength of patterns detected the size of the electrode. Thus,

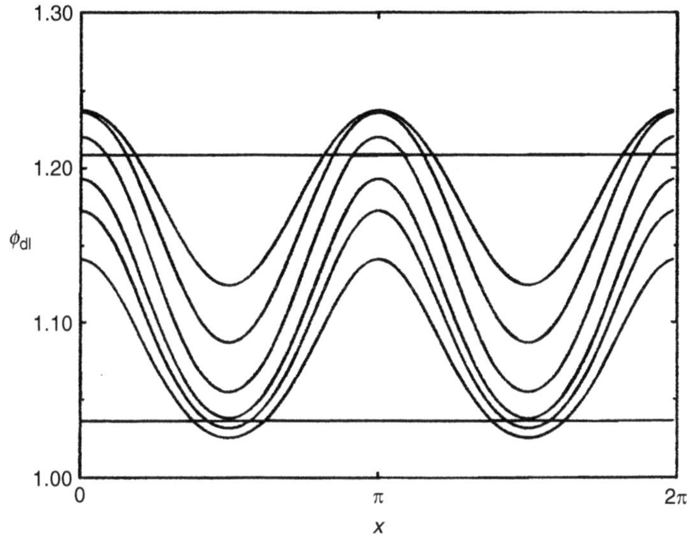

Fig. 1.25 Stationary potential patterns for $d = 50$ and $U =$ (from top to bottom) 2.7 (homogeneous solution on the cathodic side), 2.8, 2.9, 3.1, 3.3, 3.4, 3.5, 3.6 (homogeneous solution on the anodic side) (see [42] for other parameters). Reprinted with permission from [42]. Copyright 2000 American Chemical Society

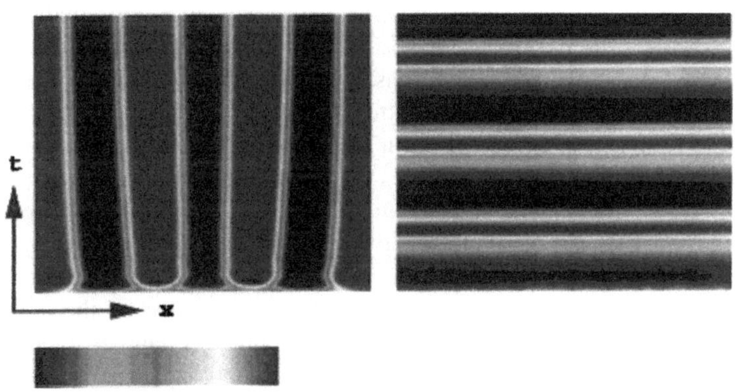

Fig. 1.26 *Gray scale* representation of the evolution of the coverage in a Turing- and Hopf-unstable region for two different initial conditions: *left*, the homogeneous steady state was slightly perturbed in the first three modes; *right*, random initial conditions. The nonlinear *gray scale* is displayed at the bottom of the figure. See [42] for parameter values. Reprinted with permission from [42]. Copyright 2000 American Chemical Society

patterns associated with $S_2O_8{}^{2-}$ reduction appeared to be due to mechanism essentially different than that responsible for the Turing patterns (see also Chap. 2). Their formation was, however, found for the periodate ($IO_4{}^-$) reduction on Au(111), in the presence of camphor (see Chap. 2) [43].

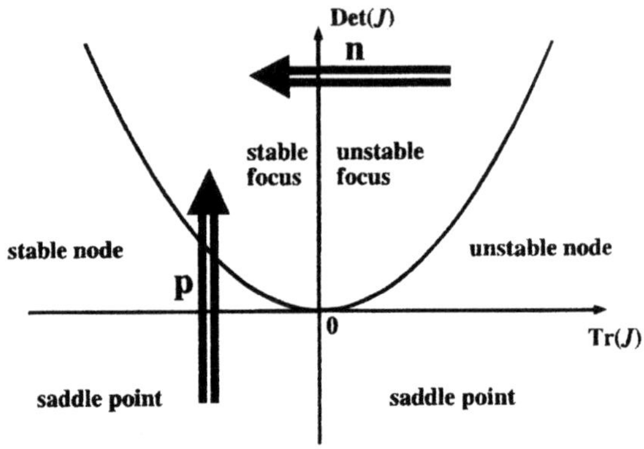

Fig. 1.27 Stability diagram showing how Tr(**J**) and Det(**J**), where **J** is the Jacobian matrix, determine the type and stability of stationary homogeneous solutions of equations used for modeling of the effect of global coupling on the S-NDR system. The *arrow marked* "n" indicates that the homogeneous component of an unstable focus can be stabilized by a negative global coupling; the *arrow marked* "p" indicates that the homogeneous component of a saddle point with Tr(**J**) < 0 can be stabilized by a positive global coupling. Reprinted with permission from [66]. Copyright 2000 American Chemical Society

1.2.3.5 S-NDR and Global Coupling

For S-NDR systems, global coupling favors the occurrence of inhomogeneous solutions—not only the negative coupling, under potentiostatic conditions (like for the N-NDR systems), but also positive one, under galvanostatic conditions [47, 48, 66]. Theoretical considerations, with the global coupling introduced in terms of the external resistance and external voltage, led to the model equations which were analyzed by means of linear stability analysis. This led to the stability diagram shown in Fig. 1.27. It indicates the effect of the positive (arrow "p") and negative (arrow "n") global couplings on the type of solutions of the system of two differential equations, describing the dynamics of the surface coverage of the electrochemically inactive adsorbate (u) and the dynamics of the double layer potential (ϕ_{dl}), corresponding to the S-NDR characteristics: if the steady-state solution is of a saddle type, the positive global coupling can stabilize it, but negative coupling—not. Instead, only negative coupling can stabilize unstable focus.

Concerning further positive global coupling, numerical calculations led to the conclusion that it can cause two essential effects for the pattern formation of the S-NDR systems: (i) the stabilization of Turing-type patterns; (ii) due to stabilization of the saddle point, the creation of a pattern composed of exactly two domains (analogous to the above-mentioned stationary domains that occur for the N-NDR system with negative global coupling). For the case (ii), representative results of calculations are collected in Fig. 1.28, where the inhomogeneous distribution of the steady-state electrode coverage is realized for the intermediate imposed current densities. For relatively low current density, the entire electrode is in the state characterized with high-coverage of u. Upon increasing current density, a low

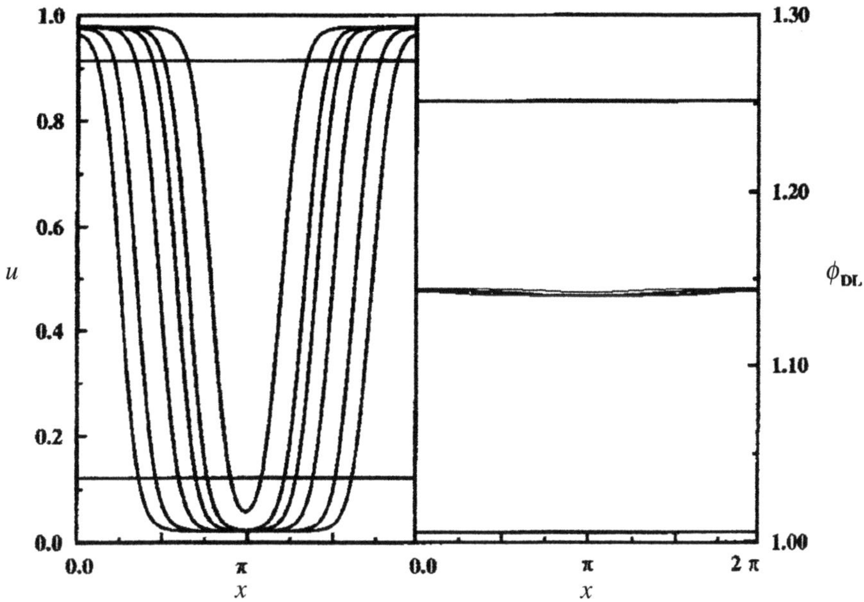

Fig. 1.28 Stationary coverage u (*left*) and potential ϕ_{dl} (*right*) profiles for different values of the galvanostatically set current density calculated for an S-NDR system. The two homogeneous high and low coverage states exist at the upper and lower border of the current interval in which the current density was varied. The middle curves of the ϕ_{dl} profiles correspond to parameter values where u is inhomogeneous [39, 66]. Reprinted with permission from [66], Copyright 2000 American Chemical Society

coverage domain is formed on this homogeneous surface and this domain expands, when the current density increases further. Eventually, at sufficiently high current density, the entire electrode attains a low-coverage state. One should note, according to Fig. 1.28, that the coverages corresponding to coexistence of two domains are above and below of the corresponding homogeneous cases, respectively, and do not change with current, as long as the patterned state persists. Simultaneous spatial variations of the double layer potential are rather modest, compared to u variations. Finally, no oscillatory behaviors were predicted for the interaction of the S-NDR characteristics with the positive global coupling.

In turn, concerning negative global coupling acting on the inhibitor variable, achieved under potentiostatic conditions with reference electrode close to working electrode, model considerations have shown that in this case, contrary to positive global coupling, oscillatory behaviors can be predicted. Antiphase oscillations, pulses, and even more complex spatiotemporal behaviors were found in numerical calculations. The predicted behaviors are collected in the bifurcation diagram, shown in Fig. 1.29 which also includes the Hopf bifurcation for the homogeneous mode.

All the bifurcations are located close to each other. One can note that when the Hopf bifurcation takes place, the homogeneous steady state is already unstable with respect to inhomogeneous perturbations with the wave number $n = 1$ (the *wave*

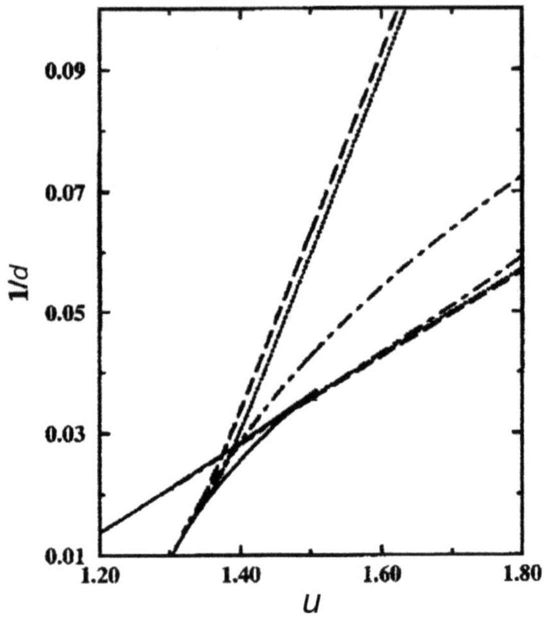

Fig. 1.29 Two-parameter bifurcation diagram in the d^{-1}/U plane (d being proportional to the cell resistance, expresses the ratio between the characteristic rates of diffusion and migration, U is the external voltage across the cell and series resistor). The locations of the following bifurcations are depicted: the saddle-node bifurcation (*solid curve*), the wave bifurcation for the wavenumber $n = 1$ (*long dashed curve*), the Hopf bifurcation of the homogeneous mode (*dotted curve*), and the Turing bifurcation (*dot-dashed curve*). See [66] for the underlying equations and parameters. Reprinted with permission from [66]. Copyright 2000 American Chemical Society

bifurcation then occurs, cf. Sect. 1.1.4). In turn, the Turing instability occurs, for most parameters, within the Hopf-unstable region [66]. Figure 1.30 shows representative simulations of spatiotemporal patterns corresponding to various control parameters.

Figure 1.30a shows the standing wave with wavenumber $n = 1$, Fig. 1.30b: the pulse solution, both emerging as a result of a wave bifurcation, caused by a negative global coupling. In turn, Fig. 1.30c shows the pattern that morphologically resembles the sophisticated case of a codimension-2 Turing–Hopf bifurcation [29], although it does not meet its detailed conditions.

One should note that the experimental verification of the above theoretical predictions is far more difficult than in the case of all other combinations of the NDR characteristics and global couplings. It is because the probability of finding the oscillations for the S-NDR systems is low, since it is difficult to find the system with the activator, characterized with sufficiently fast time scale of its dynamics, compared to the dynamics of the inhibitor. The time scale of the double layer potential (activator in the N-NDR systems) is typically much shorter than the time scale of the chemical variable (inhibitor in the N-NDR systems) and that is why the oscillations in the N-NDR systems are far more likely than in the S-NDR systems [39].

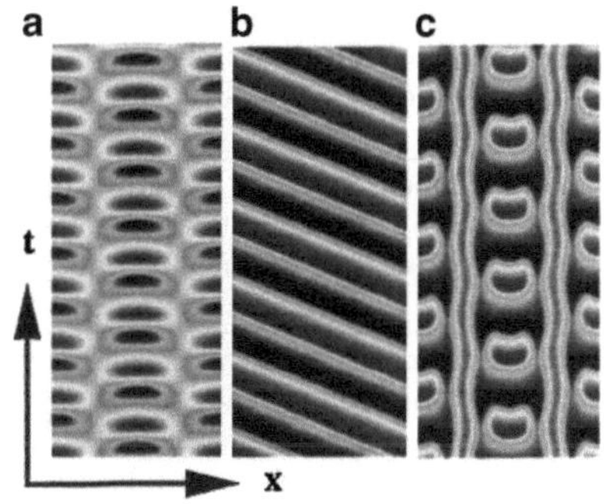

Fig. 1.30 Spatiotemporal evolution of the coverage for three different sets of parameter values: (**a**) $d^{-1} = 0.09, U = 1.6$; (**b**) $d^{-1} = 0.042, U = 1.5$; (**c**) $d^{-1} = 0.0342, U = 1.48$. Reprinted with permission from [66]. Copyright 2000 American Chemical Society

For the sake of generalization, it is noteworthy that analogous effects of global coupling can be found in the processes with S-NDR characteristics occurring at semiconductor electrodes (formation of current filaments) (cf., e.g., [67, 68]), and even for the nonelectrochemical processes, like the heterogeneous catalytic reactions, where the global coupling is caused by the control of the catalyst temperature, and then the patterns consist of hot and cold domains (cf., e.g., [69, 70]).

Summary and Generalization

The content of Sect. 1.2 shows how recent studies significantly enhance the understanding of the pattern formation in different electrochemical systems, characterized with the N-NDR (including HN-NDR) and S-NDR characteristics, for various kinds of emerging couplings. All discussed above typical spatiotemporal behaviors are collected in Table 1.1.

Relevant experimental examples of pattern formation are collected in Chap. 2. Further progress in this area can find potential applications also in the following disciplines [39]:

1. In *bioelectrochemistry*: including the role of the gradient of the electric field, and thus migration, together with the concepts of various couplings, should lead to much better understanding of the formation and progress of activation waves in nerve axons and coupled neurons. In this context one should remember again that the classical Hodgkin–Huxley approach for modeling of such phenomena includes only diffusion transport, i.e., only it takes into account only a local spatial coupling.
2. In *industrial science*: applying of the nonstationary or nonhomogenous conditions can optimize the yield in some processes. The representative example refers to fuel cells based on oxidation of small organic molecules. Usually carbon oxide is then formed as the intermediate species which poisons the Pt

Table 1.1 Summary of typical spatiotemporal behaviors in N-NDR (HN-NDR) and S-NDR systems under different experimental conditions corresponding to different types of spatial coupling (n is the wave number of the resulting pattern) [48].

	Nonlocal migration coupling through the electrolyte (always present)	Negative global coupling: potentiostatic control with Haber-Luggin capillary	Positive global coupling: galvanostatic control
N-NDR (or HN-NDR) systems; electrode potential as an activator	Accelerated waves, inhomogeneous oscillations	Stationary domains, standing waves, antiphase oscillations with $n = 1$; pulses, target patterns, clusters	Enhanced acceleration of potential fronts, mainly homogenizing effect; cluster formation in the oscillatory region
S-NDR systems; electrode potential as an inhibitor	Turing-like structures ($n \geq 1$)	Standing waves, antiphase oscillations with $n = 1$ or mixed-mode structures with $n \geq 1$; pulses	Stationary domains ($n = 1$ or $n \geq 1$)

electrode surface (cf. Chap. 5, volume I). If the course of poisoning and depoisoning of the electrode becomes an oscillatory cycle, the overall reaction rate can exceed the value observed under trivial, stationary conditions, as Schell et al. first reported [71]. During galvanostatic conditions of formic acid, under condition of periodically varying control parameters, the average surface potential was found to decrease considerably, presumably due to sustained decrease in the electrode coverage of CO.

3. In *reaction control* under condition of broken spatial symmetry: according to Strasser et al. [72], in the oxidation of formic acid on the ring Pt electrode (see also Chap. 5, volume I), due to negative global coupling, partial poisoning of the electrode in some region enhances the catalytic activity in a distant region. In this way, the homogeneous potential distribution becomes unstable and various spatiotemporal patterns (including standing wave oscillations) emerge and under such conditions the overall catalytic activity can increase.

References

1. Zhabotinsky AM (1974) Concentration autooscillations. Nauka, Moscow (in Russian)
2. Schneider FW, Münster AF (1996) Nichtlineare Dynamik in der Chemie. Spektrum Akademischer Verlag, Heidelberg
3. Epstein IR, Pojman JA (1998) An introduction to nonlinear chemical dynamics. Oscillations, waves, patterns, and chaos. Oxford University Press, Oxford
4. Scott SK (1994) Oscillations, waves and chaos in chemical kinetics. Oxford Science Publications, Oxford
5. Gray P, Scott SK (1990) Chemical oscillations and instabilities. Non-linear chemical kinetics. Clarendon Press, Oxford

6. Field RJ, Burger M (eds) (1985) Oscillations and travelling waves in chemical systems. Wiley, New York
7. Pekala K, Wiśniewski A, Jurczakowski R, Wiśniewski T, Wojdyga M, Orlik M (2010) Monitoring of spatiotemporal patterns in the oscillatory chemical reactions with the infrared camera: experiments and model interpretation. J Phys Chem A 114:7903–7911
8. Pojman JA, Epstein IR (1990) Convective effects on chemical waves. 1. Mechanisms and stability criteria. J Phys Chem 94:4966–4972
9. Kiss I, Merkin JH, Scott SK, Simon PL (2003) Travelling waves in the Oregonator model for the BZ reaction. Phys Chem Chem Phys 5:5448–5453
10. Merkin JH (2009) Travelling waves in the Oregonator model for the BZ reaction. IMA J Appl Math 74:622–643 doi:10.1093/imamat/hxp009
11. Biosa G, Bastianoni S, Rustici M (2006) Chemical waves. Chem Eur J 12:3430–3437. doi:10.1002/chem.200500929
12. Steinbock O, Tóth Á, Showalter K (1995) Navigating complex labirynths: optimal paths from chemical waves. Science 267:868–871 doi:10.1126/science.267.5199.868
13. Kiss I, Merkin JH, Scott SK, Simon PL (2004) Electric field effects on travelling waves in the Oregonator model for the Belousov-Zhabotinsky reaction. Q J Mech Appl Math 57:467–494. doi:10.1093/qjmam/57.4.467
14. Ševčikova H, Marek M (1983) Chemical waves in an electric field. Physica D 9:140–156
15. Ševčikova H, Marek M, Müller SC (1992) The reversal and splitting of waves in an excitable medium caused by an electric field. Science 257:951–954
16. Kosek J, Ševčikova H, Marek M (1995) Splitting of excitable pulse waves. J Phys Chem 99:6889–6896
17. Ševčikova H, Schreiber I, Marek M (1996) Dynamics of oxidation Belousov-Zhabotinsky waves in an electric field. J Phys Chem 100:19153–19164
18. Ševčikova H, Kosek J, Marek M (1996) Splitting of 2D waves of excitation in a direct current electric field. J Phys Chem 100:1666–1675
19. Müller SC, Steinbock O, Schütze J (1992) Autonomous pacemaker of chemical waves created by spiral annihilation. Physica A 188:47–54
20. Agladze KI, DeKepper P (1992) Influence of electric field on rotating spiral waves in the Belousov-Zhabotinsky reaction. J Phys Chem 96:5239–5242
21. Steinbock O, Schütze J, Müller SC (1992) Electric-field-induced drift and deformation of spiral waves in an excitable medium. Phys Rev Lett 68:248–251
22. Perez-Munuzuri V, Aliev R, Vasiev B, Krinsky VI (1992) Electric current control of spiral wave dynamics. Physica D 56:229–234
23. Ertl G (1990) Oscillatory catalytic reactions at single-crystal surfaces. Adv Catal 37:213–277
24. Imbihl R (1993) Oscillatory reactions on single crystal surfaces. Prog Surf Sci 44:185–343
25. Imbihl R, Ertl G (1995) Oscillatory kinetics in heterogeneous catalysis. Chem Rev 95:697–733
26. Babloyantz A (1986) Molecules, dynamics and life. An introduction to self-organization of matter. Wiley-Interscience, New York
27. Nicolis G (1995) Introduction to nonlinear science. Cambridge University Press, New York
28. De Wit A (1993) Brisure de symétrie spatial et dynamique spatio-temporelle dans les systems reaction-diffusion (in French), Ph.D. thesis. Université Libre de Bruxelles
29. De Wit A (1999) Spatial patterns and spatiotemporal dynamics in chemical systems. Adv Chem Phys 109:435–513
30. Nicolis G, Prigogine I (1977) Self-organization in nonequilibrium systems. From dissipative structures to order through fluctuations. Wiley-Interscience, New York
31. Rovinsky A, Metzinger M (1992) Interaction of Turing and Hopf bifurcations in chemical systems. Phys Rev A 46:6315–6322
32. Erneux T, Herschkowitz-Kaufman M (1977) Rotating waves as asymptotic solutions of a model chemical reaction. J Chem Phys 66:248–250
33. Kuramoto Y (1981) Rhythms and turbulence in populations of chemical oscillators. Physica A 106:128–143

34. Turing AM (1952) The chemical basis for morphogenesis. Philos Trans Soc Lond Ser B 327:37–72
35. Britton NF (2003) Essential mathematical biology. Springer, London
36. Murray JD (2002) Mathematical biology. I. An introduction, 3rd edn. Springer, New York
37. Murray JD (2003) Mathematical biology. II. Spatial models and biomedical applications, 3rd edn. Springer, New York
38. Foryś U (2005) Mathematics in biology. WNT, Warsaw (in Polish)
39. Krischer K (2001) Spontaneous formation of spatiotemporal patterns at the electrode/electrolyte interface. J Electroanal Chem 501:1–21
40. Tam WY, Horsthemke W, Noszticzius Z, Swinney HL (1988) Sustained spiral waves in a continuously fed unstirred chemical reactor. J Chem Phys 88:3395–3396
41. Lengyel I, Kádár S, Epstein IR (1992) Quasi-two dimensional Turing patterns in an imposed gradient. Phys Rev Lett 69:2729–2732
42. Mazouz N, Krischer K (2000) A theoretical study on Turing patterns in electrochemical systems. J Phys Chem 104:6081–6090
43. Li YJ, Oslonovitch J, Mazouz N, Plenge F, Krischer K, Ertl G (2001) Turing-type patterns on electrode surfaces. Science 291:2395–2398
44. Zhabotinsky AM, Dolnik M, Epstein IR (1995) Pattern formation arising from wave instability in a simple reaction-diffusion system. J Chem Phys 103:10306–10314
45. Yang L, Dolnik M, Zhabotinsky AM, Epstein IR (2002) Pattern formation arising from interactions between Turing and wave instabilities. J Chem Phys 117:7259–7265
46. Kapral R, Showalter K (eds) (1995) Chemical waves and patterns. Kluwer, Dordrecht
47. Krischer K (1999) Principles of temporal and spatial pattern formation in electrochemical systems. In: Conway BE, Bockris JO'M, White R (eds) Modern aspects of electrochemistry. Plenum, New York
48. Krischer K (2003) Nonlinear dynamics in electrochemical systems. In: Alkire RC, Kolb DM (eds) Advances in electrochemical science and engineering. Wiley-VCH, Weinheim
49. Krischer K, Mazouz N, Grauel P (2001) Fronts, waves, and stationary patterns in electrochemical systems. Angew Chem Int Ed 40:850–869
50. Pigeaud A, Kirkpatrick HB (1969) A correlated potentiostatic microscopic study of iron passivation in sulfuric acid. Corrosion 25:209
51. Russell PP, Newman J (1983) Experimental determination of the passive-active transition for iron in 1M sulfuric acid. J Electrochem Soc 130:547–553
52. Mazouz N, Krischer K, Flätgen G, Ertl G (1997) Synchronization and pattern formation in electrochemical oscillators: model calculations. J Phys Chem 101:2403–2410
53. Jorne J (1983) Oscillations and concentration patterns in electrochemical systems. Electrochim Acta 28:1713–1717
54. Koper MTM, Sluyters JH (1993) A simplified approach to the modeling of wave propagation at electrode/electrolyte interfaces. Electrochim Acta 38:1535–1544
55. Mazouz N, Flätgen G, Krischer K (1997) Tuning the range of spatial coupling in electrochemical systems: from local via nonlocal to global coupling. Phys Rev E 55:2260–2266
56. Flätgen G, Krischer K (1995) Accelerating fronts in an electrochemical system due to global coupling. Phys Rev E 51:3997–4004
57. Christoph J, Otterstedt RD, Eiswirth M, Jaeger NI, Hudson JL (1999) Negative coupling during oscillatory pattern formation on a ring electrode. J Chem Phys 110:8614–8621
58. Christoph J, Eiswirth M (2002) Theory of electrochemical pattern formation. Chaos 12:215–230
59. Hodgkin AL Huxley AF (1952) A quantitative description of membrane current and its application to conduction and excitation in nerve. J Physiol 117:500–544
60. Mazouz N, Flätgen G, Krischer K, Kevrekidis IG (1998) The impact of the operation mode on pattern formation in electrode reactions. J Electrochem Soc 145:2404–2411
61. Lev O, Sheintuch M, Pismen LM, Yarnitzky C (1998) Standing and propagating wave oscillations in the anodic dissolution of nickel. Nature 336:458–459

62. Lev O, Sheintuch M, Yarnitzky M, Pismen LM (1990) Spatial current distribution during nickel anodic dissolution in sulfuric acid. Chem Eng Sci 45:839–847
63. Flätgen G, Krischer K (1995) A general model for pattern formation in electrode reactions. J Chem Phys 103:5428–5436
64. Kiss IZ, Wang W, Hudson JL (1999) Experiments on arrays of globally coupled periodic electrochemical oscillators. J Phys Chem B 103:11433–11444
65. Grauel P, Christoph J, Flätgen G, Krischer K (1998) Stationary potential patterns during the reduction of peroxodisulfate at Ag ring electrodes. J Phys Chem B 102:10264–10271
66. Krischer K, Mazouz N, Flätgen G (2000) Pattern formation in globally coupled electrochemical systems with an S-shaped current-potential curve. J Phys Chem B 104:7545–7553
67. Niederostheide JJ (ed) (1995) Nonlinear dynamics and pattern formation in semiconductors and devices. Springer, Berlin
68. Meixner M, Rodin P, Schöll E (1998) Fronts in a bistable medium with two global constraints: oscillatory instability and large-amplitude limit-cycle motion. Phys Rev E 58:5586–5591
69. Annalai J, Liauw M, Luss D (1999) Temperature patterns on a hollow cylindrical catalytic pellet. Chaos 9:36–42
70. Mikhailov AS (1994) Foundations of synergetics I, 2nd edn. Springer, Berlin
71. Schell M, Albahadily FN, Safar J (1993) Decreases in the inhibition of the electrocatalyzed oxidation of formic acid by carbon monoxide. J Electroanal Chem 353:303–313
72. Strasser P, Christoph J, Lin WF, Eiswirth M, Hudson JL (2000) Standing wave oscillations in an electrocatalytic reaction. J Phys Chem A 104:1854–1860

Chapter 2
Experimental and Model Spatiotemporal and Spatial Patterns in Electrochemical Systems

2.1 Simple Examples of Dissipative Pattern Formation

In Chaps. 4–6 of volume I, we analyzed the experimental manifestations and the relevant models of temporal dynamic instabilities, i.e., oscillations and multistability characterized by the total (average) current or the potential of the working electrode. In the present chapter, we describe the spatiotemporal and spatial patterns that develop due to nonlocal and global couplings, according to theoretical background outlined in Sect. 1.2. This means, among others, that in the formation of such patterns the inhomogeneous distribution of the electric field at the interface, causing *migration* currents flowing parallel to the electrode surface, will be taken into account. Also, the role of *diffusion* in generating a particular type of patterns will be described. Concerning another type of transport, the *convection* can also be a source of spatial self-organization in electrochemical systems due to its possible self-organized nature under appropriate conditions. However, the patterns of convective origin will be described separately in Chap. 5, as the mechanism of their formation is different from those operating in the case of diffusion–migration systems. Thus, for the purposes of the present chapter we shall assume that the convection does not exist or at least does not play any significant role in the pattern formation discussed. As indicated in Sect. 1.2, the substantial progress in understanding the mechanisms of the formation of such nonconvective dissipative patterns took place only in recent decades, when this challenging task was undertaken by the Berlin group, gathered around Ertl, Krischer, and coworkers in the Fritz-Haber Institute of Max-Planck Society. The success of these investigations was possible not only because of the development of appropriate theory, but also due to employing the modern techniques of analysis of the state of surfaces, as the surface plasmon spectroscopy [1].

Before description of recent achievements in this area, let us mention and comment selected historical examples of spatiotemporal patterns in electrode processes. In 1948, Bonhoeffer has published the paper [2]: "Activation of passive iron as a model for the excitation of nerve." The title referred to the experimental fact that if a piece of passivated iron immersed in concentrated nitric acid is touched

M. Orlik, *Self-Organization in Electrochemical Systems II*,
Monographs in Electrochemistry, DOI 10.1007/978-3-642-27627-9_2,

momentarily with a zinc rod, the iron may become active and the zone of activation will spread over the whole piece of iron. The final state of iron, i.e., active or passive, depends on the concentration of nitric acid (repassivation occurs for intermediate concentrations). In a more sophisticated version of this experiment, when the iron wire was used, and for that intermediate acid concentration, one might observe a spreading wave front of activation, followed by a spreading wave front of repassivation. But in fact, as Bonhoeffer remembers in his paper, this impressive phenomenon was first reported already by Wilhelm Ostwald around 1900 who also suggested that its course resembles the progress of excitation of nerve. Inspired by Ostwald, Heathcote has studied this phenomenon more thoroughly [3, 4] and has found that the iron could not be reactivated if one touched it with zinc rod just after repassivation, but after certain delay, just like the nerve after excitation exhibits certain “dead,” or refractory period, typical of *excitable* characteristics (see Sect. 1.1.2). These analogies were later analyzed by Lillie (cf. e.g., [5–7]) who has reported, among others, the dependence of the velocity of the active zone propagation on the thickness of the surrounding layer of electrolyte, indicating in this way the role of local currents.

Today, we of course know that properties of the Fe/HNO_3 system and the principle of the nerves conductivity are based on very different detailed mechanisms. Concerning the latter process, its most famous four-variable theoretical model, formulated by two Nobel Prize winners A. L. Hodgkin and A. F. Huxley (1952), was based on the various permeability of K^+ and Na^+ ions through the membrane of the neural cell. Due to complexity of the original approach, the mathematically more tractable version, reduced to two dynamic variables, was later elaborated by R. FitzHugh and J. S. Nagumo (cf. e.g., [8, 9]). For the case when the neural cell is not perturbed by external current, but the diffusion associated with axial current in the axon is considered, the following system of equation was derived [9]:

$$\frac{\partial u}{\partial t} = f(u) - v + D\frac{\partial^2 u}{\partial x^2} \tag{2.1}$$

$$\frac{\partial v}{\partial t} = bu - \gamma v \tag{2.2}$$

where:

$$f(u) = u(a - u)(u - 1) \tag{2.3}$$

u is directly related to the membrane potential, $0 < a < 1$ and b and γ are positive constants. The course of nullclines of (2.1) and (2.2) reveals typical excitable characteristics (cf. Sect. 1.1.2).

It is appropriate here to indicate the basic weakness of this approach which assumes only diffusion as the cause of the propagation of neural excitation, while in this process there are involved ions placed in the inhomogeneous electric field. Therefore, the coefficient D in (2.1) is rather an *effective* transport coefficient, named also sometimes the "diffusion" coefficient [9]. Neglecting the obvious contribution from migration in a strict way relating its rate to the gradient of electric field does not allow us to estimate correctly the rate of the impulse propagation along the neuron. In this context, the recent developments of understanding the pattern formation in electrochemical systems, when migration is considered an essential transport, offer a chance to a progress in a more quantitative model of neural systems.

But today we also realize that completely different systems, if defined in terms of appropriately chosen mathematical equations that reflect the "core" of their nonlinear dynamics, exhibit striking qualitative similarities or universalities. The above-mentioned excitability, observed for both the onset of traveling zone of activity on a passivated iron wire and the excitation of propagating nerve impulse, is a good example of that. Later, Bonhoeffer confirmed the excitable characteristics of the passivated iron system: the assembled electrochemical system consisted of Fe wire (cathode) and Pt wire (anode) dipped in nitric acid and connected to the external power source producing current pulse of desired magnitude and duration. For nitric acid of sufficiently high concentration, the Fe wire became strongly passivated and then its activation (followed by repassivation) was induced by the current pulse, but only if the electric charge injected exceeded certain threshold value. The refractory state after the activation was also observed and explained as due to the formation of nitrous acid which removal from the electrode environment required some time, and also due to the formation of oxide layer on the iron surface after its repassivation, which disappeared gradually during the period of recovery of the ability to the next activation. The next important step in developing the studies of this phenomenon was described by Franck [10]. In the experimental arrangement shown in Fig. 2.1, it was possible to monitor the propagation of the activation zone on the surface of Fe wire immersed in the solution of either HNO_3 or H_3PO_4. In fact, these works can be considered the precursors of later serious studies of spatiotemporal patterns on the electrode surfaces.

More recently, studies of spatiotemporal propagation of the electrochemical reaction along an iron electrode in sulfuric acid solution have been undertaken by Nakabayashi et al. [11, 12] in order to correlate the current oscillations with the development of a spatiotemporal pattern on the electrode surface. The propagation of the reaction front of an oxide formation and reduction along the iron wire was monitored by means of a 16-channel platinum potential probe array. A pulse of the current oscillation was found to consist of both a sequential propagation of the oxide reduction, which makes the rise of the current, and a random formation of the oxide at the electrode surface for current decay. The traveling velocity of the oxide reduction increased as the potential of iron electrode became more cathodic. An important observation was that the oscillation waveform and the propagation speed of the reduction front were changed if serial resistance was inserted in the circuit. Moreover, if the propagation was blocked, e.g., by an insulator coating of the electrode, the rising current was disturbed and modulated, having independent occurrence of the reduction propagation at each separated branch of the electrode. These results suggested a coupling between the ohmic potential drop in the solution and the reaction propagation along the electrode surface [12] and are further developed in

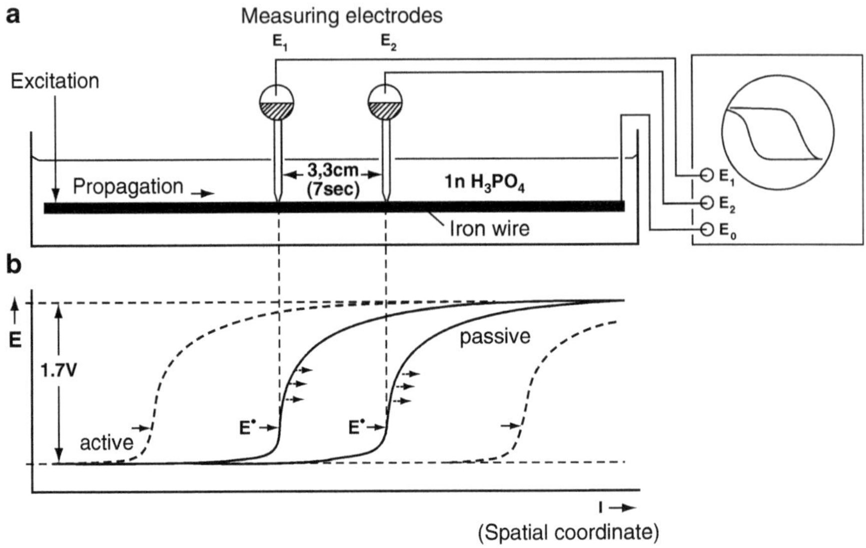

Fig. 2.1 An example of the instability-driven propagation of activation zone along the passivated iron (**a**) experimental setup and (**b**) local profiles of the electric potential associated with the passage of the activation wave through both sensing electrodes. Reprinted with translation into English from [10], with kind permission of Deutsche Akademie der Naturforscher Leopoldina—Nationale Akademie der Wissenschaften

Sect. 3.1.4, devoted to coupled Fe/H_2SO_4 oscillators. The works cited earlier include also instructive spatiotemporal maps of the propagation of the oxide reduction along the iron electrode, correlated with the shape of the oscillatory peak.

For a chemically similar system, consisting of low-carbon steel plates, covered with thin (0.5 mm) layer of oxygen-containing nitric acid, inducing the corrosion, Agladze, Steinbock et al. [13, 14] have described the development of two-dimensional patterns in the form of pulses, rotating spirals, and serpentines, propagating with the velocity of the order of millimeters per second. The corrosion could be initiated locally by breaking the passivation film or, as in classical Bonhoeffer experiments, by touching the surface of the plate with a zinc rod for ca. 1 s. These patterns could be observed visually as traveling brownish bands on the passivated grayish steel. Generally, the pattern formation in the Fe–HNO_3 reaction system depended significantly on the concentration of nitric acid. The morphology of spiral waves resembled analogous phenomena for the excitable Belousov–Zhabotinsky system, however with a striking quantitative difference: in the Fe–HNO_3 system, the recovery times were substantially shorter than excitation times.

The simplest chemical mechanism underlying dissolution of propagating waves in the above Fe–HNO_3 system is the following [13]. In contact with concentrated HNO_3, the dissolution of Fe is hindered due to formation of passivating surface film which contains considerable amount of water of hydration. The local damage of this film causes the Fe oxidation to Fe^{2+} and this triggers the flow of Fe^{2+} ions into the nearby solution layer. In this way, the local electrochemical cell is formed in which

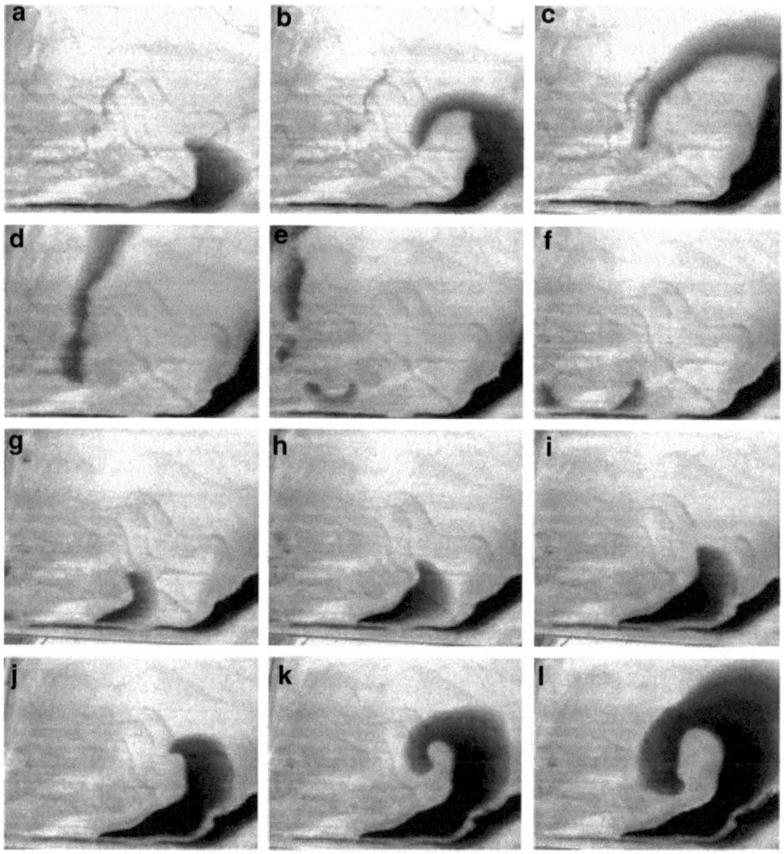

Fig. 2.2 Propagation of waves and rotation of spirals in the iron–nitric acid system. The sequence shows 12 consecutive snapshots of the steel surface obtained at 0.7 s intervals. The wave is initiated at the lower left corner of the horizontal steel plate a few seconds prior to the snapshot (**a**). Under the given experimental conditions, the spiral waves have a rotation period of approximately 6 s. Experimental parameters: 11.9 M nitric acid; temperature of the solution and steel plate, 0 °C; thickness of the solution layer 0.5 mm; field of view: $5.9 \times 4.6\ cm^2$. All experiments are carried out with low-carbon steel under open circuit conditions. Reprinted with permission from [13]. Copyright 2000 American Chemical Society

the local defect in the passivation film is the anode. In consequence, the H^+ ions migrate away from this area, in order to meet the electroneutrality condition. The local increase in pH destabilizes the adjacent passivation film and additional Fe^{2+} is released into the solution. The spatiotemporal progress of this process means the formation of the traveling wave. Noteworthy, this process is again dynamically similar to the nerve excitation. The difference lies in the recovery process. In the Fe/HNO_3 system, this recovery is due to pH equilibration between the bulk solution and a thin layer adjacent to the dissolving surface. Also, in order to explain the formation of visually observed rust, one has to take into account the key role of dissolved oxygen in the recovery process:

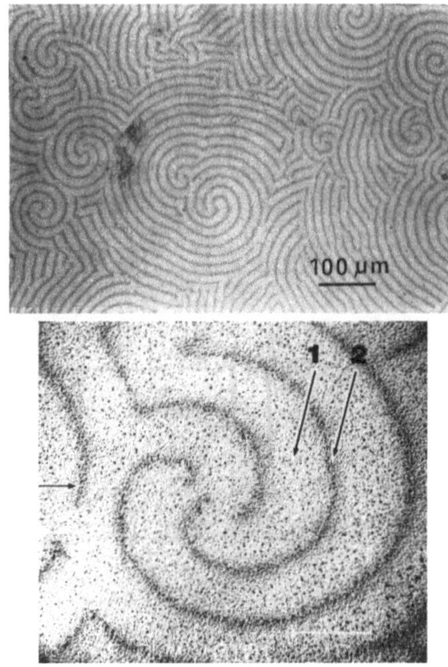

Fig. 2.3 *Top*: The surface of a silver-antimony alloy coating deposited at current density 1.3 A dm^{-2}; *bottom*: Surface region of the alloy coating deposited at current density 1.3 A dm^{-2}. Secondary electron image. Concentration of antimony in the electrolyte: 0.04 M. Reprinted from [17], Copyright 1989, with permission from Elsevier

$$4Fe^{2+} + O_2 + 4H_2O + 2xH_2O \rightarrow 2Fe_2O_3 \cdot xH_2O + 8H^+ \tag{2.4}$$

Further examples of patterned Fe electrodissolution include, among others, the spatiotemporal period doubling for a circular Fe electrode in contact with 1 M H_2SO_4 electrolyte [15]. An attempt to model the spiral waves in this type of systems was described by Koper and Sluyters [16]. Spiral patterns were experimentally observed by Kristev et al. [17] for electrodeposition of Ag–In or Ag–Sb alloys within the range of limiting current, under galvanostatic conditions (Fig. 2.3).

In turn, Gladyshev and Kovaleva [18] have described patterns associated with the anodic oxidation of ternary amalgams, composed of mercury, dissolved alkaline metal and its hydride, formed during parallel electroreduction of alkali metal ions and hydrogen ions on mercury cathode. Anodic polarization of such amalgams causes hydrogen evolution with the formation of dispersed bubbles in the mercury phase. Under appropriate conditions of anodic oxidation, the chaotic oscillations ($f = 0.5$–2 Hz) imposed on the basic, monotonously increasing current were observed and, simultaneously, on the mercury surface one observed circular, concentric waves, running from the center to the peripheral zone of the electrode. The circular patterns were surrounded with very fine hydrogen bubbles. It is possible that nucleation and evolution of hydrogen bubbles in the mercury phase is responsible for the formation of these patterns.

As mentioned earlier, the theoretical treatment of the dissipative spatial or spatiotemporal electrochemical patterns was outlined in Sect. 1.2. The main point is that between different sites of the electrode there emerge spatial couplings, obviously dependent on cell geometry, which have to be taken into account, in addition to the criteria of instability existing for spatially homogeneous systems. A concise summary of possible couplings is given below. There are two principal modes of the spatial coupling. A first type is the *migration coupling*. If the distribution of the electric potential was ideally homogeneous, the vector of migration of charge carriers would be oriented only normal to the electrode surface (this normal direction will further be assumed as concordant with the direction of z axis of the Cartesian coordinate system). However, if different sites of the electrode differ with the local electric potential, the local migration currents exhibit also corresponding spatial distribution. The vector of migration of ions acquires now the components parallel to the electrode surface (x, y directions), since the current flows in those directions due to the local difference in electric potentials. In this way different sites of the electrode do not remain independent, but become coupled through the migration. Such coupling has substantial consequences: if the working electrode potential (interfacial potential drop) changes at a certain position, it will affect the distribution of the potential across the entire electrolyte. This, in turn, will change the local migration current densities at the working electrode. One should emphasize that migration coupling is termed "nonlocal" one, in order to emphasize its greater extent, compared to only local nature of diffusion coupling. On the other hand, the closer the counter electrode (CE) to the working electrode (WE), the more localized becomes this coupling.

The second type of coupling, named the *global coupling* between different sites of the electrode surface, is caused by the feedback in the electronic circuit, working either in the potentiostatic or galvanostatic mode, determining the different particular feedback mechanism. For the *galvanostatic* operation mode, any local fluctuation of the faradaic current at the electrode surface changes the imposed fixed total current. In order to compensate for this difference, the galvanostat provides the additional, charging (capacitive) current which causes the change of the working electrode potential. In this way, all the electrode sites become affected with the initial small fluctuation of the current, therefore this mode of coupling is indeed global in nature. In the *potentiostatic* operation mode, the local fluctuation of the working electrode potential affects the distribution of the electric potential in the entire electrolyte, so also at the position of the reference electrode (RE). The potentiostat feels this alteration as the deviation of the electrode potential from the programmed fixed value and charges the WE electrode to a new value, changing the entire distribution of the electric potential throughout the solution. If the WE and RE electrodes are of the same size and parallel to each other, the entire WE surface experiences the change of the potential, meaning the global nature of the effect. In a more realistic case, the analogous situation is obtained in this way that the working electrode is a thin annulus, and the reference electrode is located on the axis of the ring. In spite of certain similarity in the mechanism of the creation of the global coupling under both operation modes, there is between them a very important difference from the dynamical point of view. In the galvanostatic mode, due to global coupling the interfacial potential drop of various electrode sites becomes equalized. In the potentiostatic mode these differences are amplified. Therefore for the galvanostatic mode, the *synchronization* of various sites occurs and we call that situation an effect of the *positive global coupling*, leading to the stabilization of the electrode state. For the potentiostatic mode, the opposite effect—desynchronization means *negative global coupling*, destabilizing the electrode state. The latter effect is particularly strongly pronounced if WE and RE are positioned close to each other, as in the case of application of a Haber-Luggin capillary, minimizing the ohmic potential drop between WE and RE [19–21]. In fact, Otterstedt, Hudson et al. [27] were the first who reported such spatiotemporal phenomena for the anodic dissolution of Co (see Sect. 2.3). The Co electrode had a shape of a disk or a ring and the tip of the Luggin capillary was placed in its center. In conclusion, the dependence of the range of the coupling on the cell geometry means that it can be tuned experimentally by variation of the relative positions and shapes of the working, reference, and counter electrodes.

Finally, the interactions between various couplings and the negative differential characteristics (of N-NDR or S-NDR type) may give rise to different spatiotemporal or spatial patterns. These theoretical backgrounds will now be illustrated with appropriate experimental examples.

2.2 Patterns in $S_2O_8^{2-}$ Electroreduction

From Sect. 4.1 of volume I we know that the electroreduction of $S_2O_8^{2-}$ is an example of the process with the N-NDR characteristics in the current–potential response. In other words, it is a system of the activator–inhibitor type, in which the electrode potential is an activator (autocatalytic variable) and the chemical variable (concentration of the electroactive species) is an inhibitor, engaged in the slow negative feedback loop. Pattern formation occurs if such process is coupled to migration. More precisely, this interplay engages the activator—the electrode potential which due to migration flows along the electrode surface. Under such conditions, the pulses or spiral waves, propagating over long distances without attenuation, are generated. The model case of the propagation of the pulse in an electrochemical bistable system was shown in Fig. 1.17. One should also remember that, contrary to the reaction–diffusion systems (RDS), when the stationary traveling fronts move with constant velocity, in the electrochemical systems of the above characteristics the acceleration of the front was reported. This acceleration was experimentally observed, e.g., for the peroxodisulfate reduction at the rotating ring Ag electrodes, being the first experimental evidence for such phenomenon [22]. The experimental setup is shown in Fig. 2.4.

The measurements were made for the bistable region of the $S_2O_8^{2-}$ electroreduction, i.e., for the coexisting high-current (active) and low-current (passive) states. The potential microprobes measured the potential in front of the electrode as a function of the angle and time. Figures 2.5, 2.6 clearly show the acceleration in

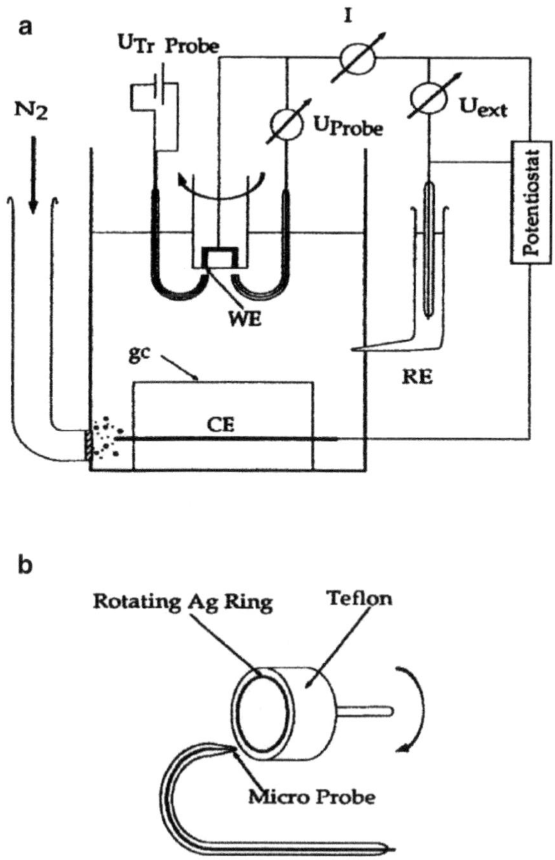

Fig. 2.4 (**a**) Scheme of the experimental setup; WE working electrode, CE counter electrode, RE reference electrode (standard calomel electrode), gc glass cylinder. Between WE and RE the voltage U_{ext} was held constant. The local potential in front of WE was measured by two micropotential probes. In some experiments one of them was exchanged by a trigger probe. (**b**) Enhanced sketch of working electrodes and microprobes. Reprinted with permission from [22] http://link.aps.org/abstract/PRE/v51/p3997. Copyright 1995 by the American Physical Society

the propagation of the front, formed as a result of transition from the passive to the active state.

The role of migration coupling in the acceleration of the front was experimentally proved by the reported increase in the average front velocity with increasing solution conductivity (σ).

The fact that oscillations and transitions in the bistable regime of the $S_2O_8^{2-}$ electroreduction are associated with wave phenomena was further confirmed by the observations made by means of the surface plasmon spectroscopy [23] which also allowed to monitor the fast surface dynamics of this process. Onto one side of the glass prism of high refractive index, a 50-nm-thick Ag film was evaporated.

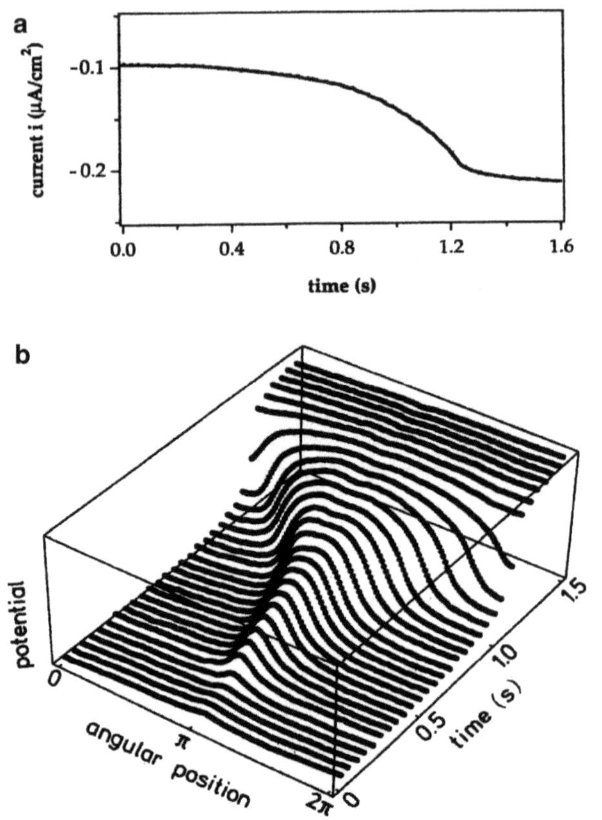

Fig. 2.5 (**a**) Time series of the total current i and (**b**) spatiotemporal plot of the local potential during a transition from the passive (low current density) to the active (high current density) state in the bistable regime at $U_{\mathrm{ext}} = -1.36\,\mathrm{V}, f = 20\,\mathrm{Hz}$, $[Na_2S_2O_8] = 0.1$ mM, $[Na_2SO_4] = 0.1$ mM, and pH 5. The circumference of the electrode is equal to 3.46 cm. Reprinted with permission from [22] http://link.aps.org/abstract/PRE/v51/p3997. Copyright 1995 by the American Physical Society

This Ag film served as a working electrode, whereas the glass prism permitted the excitation of surface plasmons by a p-polarized He–Ne laser beam which was passed through the prism onto the Ag film with a certain angle of incidence. The excitation of surface plasmon (SP) occurs in a narrow interval of the angle of incidence, at which the energy and momentum of the incoming photons and surface plasmons are matched. Experimentally, this excitation manifests itself as a strong decrease in intensity of the reflected laser beam. Furthermore, this phenomenon is potential dependent, so if the potential drop is different at two different locations at the electrode, the intensity of the reflected laser beam at a certain angle of incidence is also different. More precisely, for the system considered here, the intensity in the images is affected by the change in coverage of specifically adsorbing anions (which directly reflect the changes of the electrode potential) and a change in roughness (or a partial oxidation) of the Ag electrode.

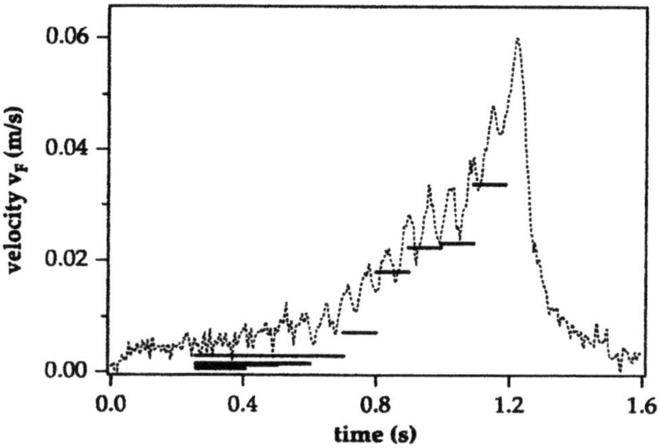

Fig. 2.6 Front velocity versus time for the transition displayed in Fig. 2.5. The *bars* show the mean velocity obtained from the micropotential probes. The *dashed line* displays the velocity obtained from the derivative of the current shown in Fig. 2.5a. Reprinted with permission from [22] http://link.aps.org/abstract/PRE/v51/p3997. Copyright 1995 by the American Physical Society

Figure 2.7 shows exemplary comparison of the temporal variations of the current and the corresponding images, recorded by CCD camera, which illustrate the distribution of the interfacial potential drop. One notices that when the cathodic current of $S_2O_8^{2-}$ reduction rises (at $t = 48$ ms), a new state is formed in the lower right-hand corner of the electrode surface. As a function of time, this zone expands across the electrode and highest intensity of this phenomenon, occurring eventually almost homogeneously on the surface, is obtained some time after the moment at which the current passed through its maximum. Then the intensity slowly decreases. Although the interpretation of intensity images may contain some uncertainties, their analysis (including also studies in the bistable regime) led to conclusion that the characteristic length of the patterns was determined by the temporal dynamics of the potential drop across the double layer at high mass transport, and by the dynamics of the concentration of $S_2O_8^{2-}$ at the electrode at low mass transport. Furthermore, the relatively high velocity of the fronts (from several cm s^{-1} to several m s^{-1}) and their dependence on the uncompensated electrolyte resistance is a strong premise for the important role of migration transport in the spatial coupling as a cause for their spatial progress.

At this point it is useful to invoke an interplay of a negative global coupling, favoring the pattern formation and the, induced by this inhomogeneity, migration coupling that acts in an opposite direction. Then, in the case of bistable system, when one part of the electrode remains in the active state, and the other part in the passive state, the migration coupling initially predominates, being slightly opposed global coupling, but gradually increasing extent of negative global coupling may induce the situation when both couplings balance each other, giving rise to stationary patterns consisting of high and low potential domains [19]. An experimental

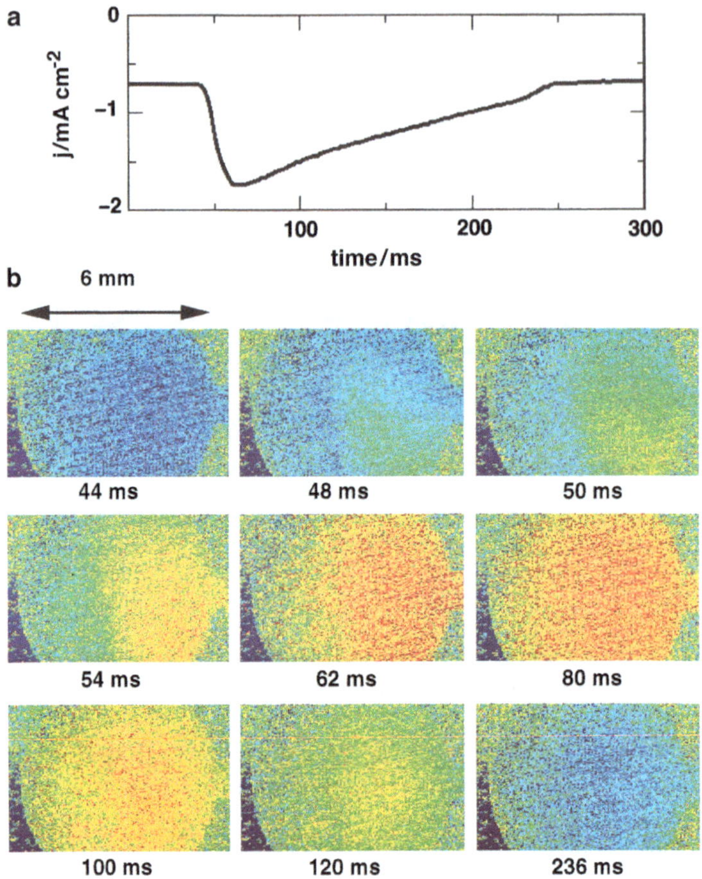

Fig. 2.7 (**a**) Time trace of the global current during the oscillation (the oscillation frequency was about 2 Hz) (**b**) Images of the electrode during the oscillation obtained with the SP microscope. Electrolyte 2 mM $Na_2S_2O_8$, 0.1 mM Na_2SO_4, 0.1 M NaOH, N_2 saturated, impinging jet stirring. Reprinted from [23], Copyright 1996, with permission from Elsevier

illustration of such interplay of couplings was reported by Grauel et al. [24] for the electroreduction of $S_2O_8{}^{2-}$ at Ag ring electrodes, for the parameter values close to or corresponding to the bistable characteristics of that process, and with the Luggin-Haber capillary minimizing the ohmic potential drop. Figure 2.8 shows the distribution of the electrode potential, varying as a function of time, along the ring Ag electrode (i.e., for the angle varying from 0 to 2π).

Comparison of these maps with the *I–E* characteristics of the $S_2O_8{}^{2-}$ electroreduction shows that for the potential values lying within the N-NDR region (Figs. 2.8b, c), the potential distribution along the ring electrode is particularly inhomogeneous and takes a form of patterns, consisting of two domains, the one characterized with the relatively positive potential ("active" domain) and the other one—with the relatively negative potential ("passive" domain), compared to the

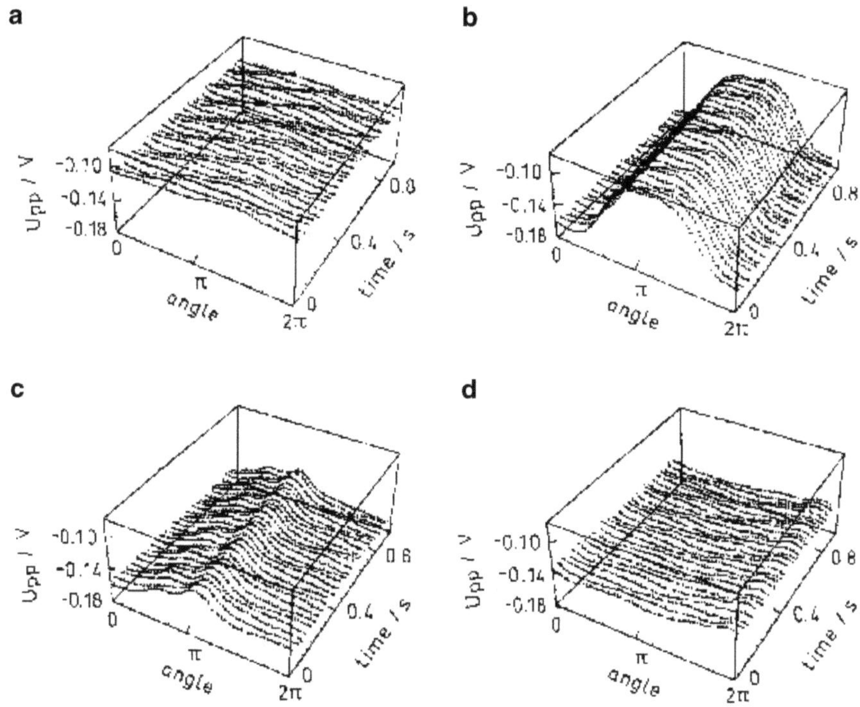

Fig. 2.8 Space–time plot of the local potential measured at four different values of the external potential: (**a**) −1.2 V, (**b**) −1.13 V, (**c**) −1.1 V, (**d**) −1.0 V. Reprinted with permission from [24]. Copyright 1998 American Chemical Society

potential of the NDR region. Furthermore, if the voltage was kept constant, the patterns appeared to be stable, i.e., invariant with time.

It was also found experimentally that insertion of sufficiently large external resistor to the circuit of the working electrodes caused the disappearance of spatial patterns. This is because external serial resistor introduced a positive global coupling which counteracted the negative global coupling and thus the formation of patterns. These experimental results were successfully modeled in terms of the appropriate equivalent circuits, where variation of external resistance tuned the relative contribution of the negative and positive global couplings [24].

2.3 Patterns in Co Electrodissolution

Other experimental examples of the electrochemical accelerated fronts include the dissolution of Co [25, 26] and Fe [11, 12] (see Sect. 2.1) electrodes. We shall briefly describe here the phenomena reported for the dissolution of the ribbon cobalt electrodes in buffered phosphate solutions [26]. The experimental setup is shown in Fig. 2.9.

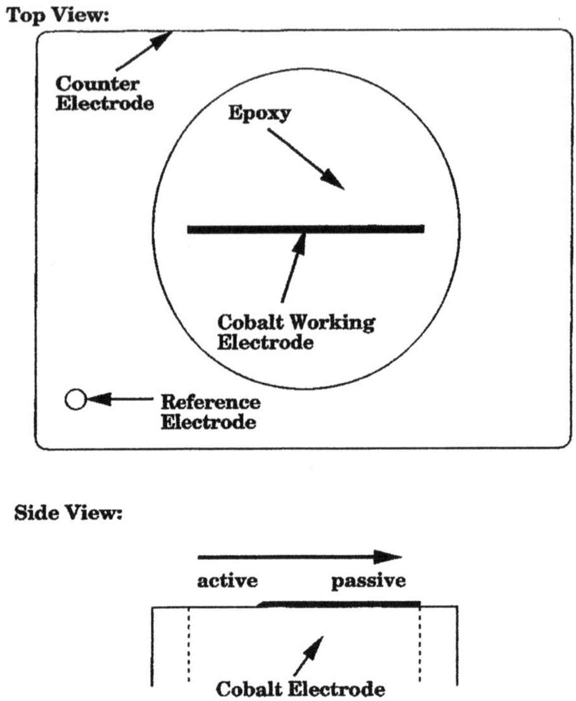

Fig. 2.9 Top and side views of the experimental setup. Reprinted with permission from [26] http://link.aps.org/abstract/PRE/v54/p3744. Copyright 1996 by the American Physical Society

The saturated calomel reference electrode was used and the rectangular copper wire served as a counter electrode. In one of the experiments, the Co electrode was initially passivated by applying sufficiently positive potential which was then moved to the value more negative than the Flade potential. Due to slow rate of the film dissolution, the depassivation of the electrode would take some time, but it was locally accelerated by scratching the surface at one end of the ribbon. The particularly interesting sequence of dynamical events happened if the electrode potential was close to the Flade potential. The resulting small active region immediately began to move toward the center of the ribbon and another active region formed spontaneously at the other end of the electrode, also starting to propagate towards the ribbon center. When both fronts of active zones meet, they annihilate. Then, after a short delay, two new active regions form, one from each end of the ribbon, and they again propagate towards the center.

In one of the experiments corresponding to such conditions, the set of 12 auxiliary reference electrodes was placed along the Co ribbon which was initially passivated by applying sufficiently positive potential which was then stepped to the value slightly more negative than the Flade potential, and the ribbon electrode was locally activated by placing there the drop of acid. The development of spatiotemporal patterns, visualized as maps of equipotential lines, is shown in Fig. 2.10.

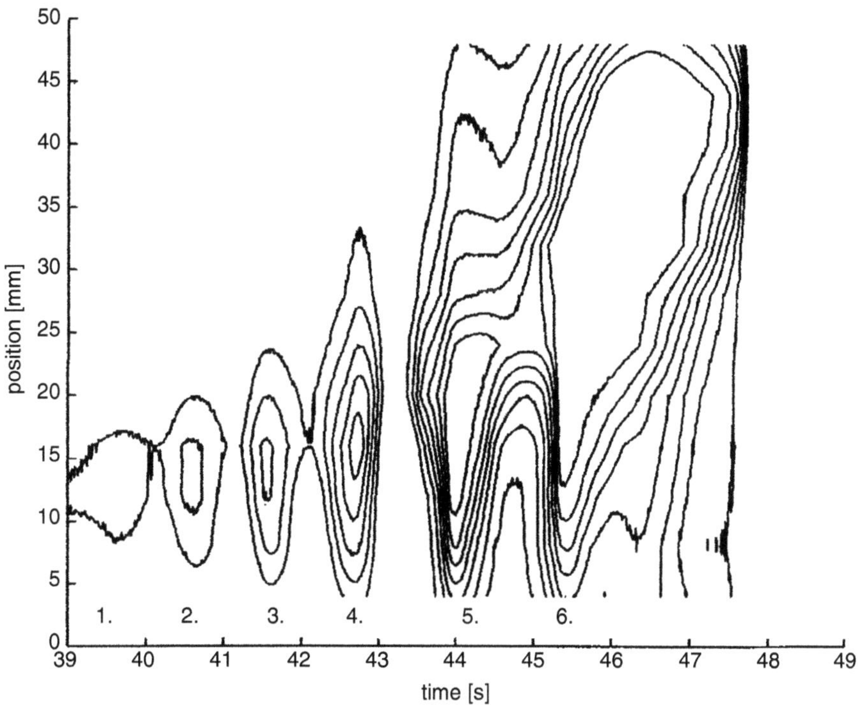

Fig. 2.10 Position–time plot of selected equipotential lines as derived from the signals obtained from 12 auxiliary reference electrodes positioned equidistantly along the ribbon. Time denotes the time elapsed after stepping the potential down from 1,200 to 1,056 mV vs. SCE, and placing the drop of acid on the ribbon at position 0 mm. The numbers denote the maxima in the corresponding current–time curve (see Fig. 2.11) http://link.aps.org/abstract/PRE/v54/p3744. The outside equipotential lines correspond to a potential of 300 mV cathodic with respect to the applied potential of 1,056 mV vs. SCE. Subsequent lines are 50 mV apart. Reprinted with permission from [26] http://link.aps.org/abstract/PRE/v54/p3744. Copyright 1996 by the American Physical Society

It is important to note that in these experiments the leading edge of active zones moved with the velocity increasing in time, so the observed phenomena were one more example of the accelerating fronts.

It is also interesting that if the electrode potential is slightly more positive than the Flade potential, the system exhibits excitable characteristics—not with respect to homogeneous, but *spatially inhomogeneous* fluctuations. According to the basic characteristics of spatiotemporal excitability (cf. Sect. 1.1.2), the too small local fluctuation, which physically means formation of local active area, is damped, while a disturbance exceeding certain threshold expands along the entire electrode. The current, initially enhanced due to local activation, now increases significantly more and when the active area reaches the end(s) of the electrode, this current drops to its original value. In this case, the experimental realization involved Co wire of diameter 1 mm, placed against the Co ribbon, brought into the passive state by sufficiently anodic polarization. Since the Co wire has no protective film, an

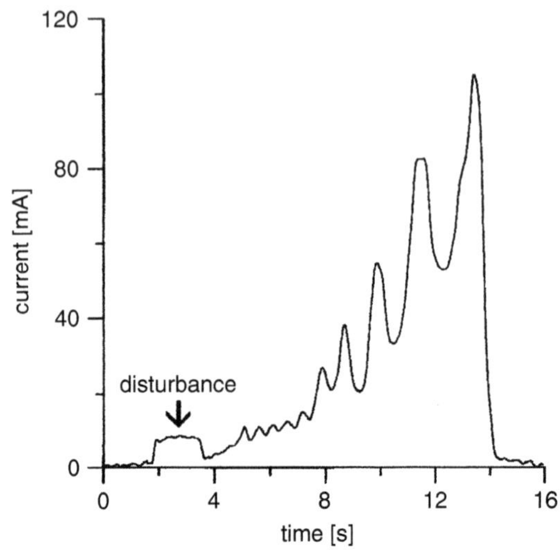

Fig. 2.11 Current–time curve with modulation following initiation of activation in the passive region: 1.25 M phosphate buffer, pH 1.67, applied potential 1,150 mV/SCE. Passivation for 15 s. Disturbance at the end of the ribbon. Reprinted with permission from [26] http://link.aps.org/abstract/PRE/v54/p3744. Copyright 1996 by the American Physical Society

oxidation reaction occurs on it immediately after its potential is brought to that of the ribbon. Oxidation on the wire occurs with a simultaneous local reduction of the passive film on the ribbon Co electrode. The single active region, produced in this way, travels along the ribbon, reaches its end(s), and the system returns to the original passive state. Figure 2.11 shows the corresponding temporal variations of the current observed after activation of Co ribbon at one of its ends.

This *I*–*t* shape can be explained in the following way. The Co ribbon was first passivated for 15 s and then disturbed with the cobalt wire, for ca. 2.5 s, at position close to its end. The initially low current increased then for the time when the disturbance was maintained and after that slightly decreased. As the perturbation was sufficiently large, the active zone produced began to expand as it traveled, since the activation was faster than the passivation; accordingly the current increased significantly. During this travel, the active zone grew and shrank (it “breathed”) and therefore the current exhibited oscillatory variation. Eventually the active zones reached the end of the Co ribbon and the current permanently fell to a very low value, corresponding to the original passive state of the electrode. It is interesting to consider this alternate increase and decrease of the width of the active region when the electrode potential is close to the Flade potential. This phenomenon means that the trailing edge of the active region propagates intermittently in the reverse direction, and that the velocity of this reverse motion is significantly higher than the primary forward velocity of the leading edge. This modulation leads to an additional (intermittent) acceleration of the leading edge and explains an increase in the total current.

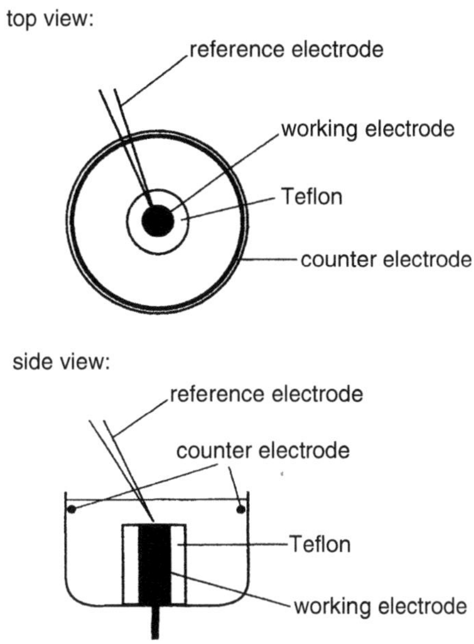

Fig. 2.12 Experimental setup for studies of spatiotemporal patterns on Co electrode. Reproduced from [27] by permission of the Royal Society of Chemistry

When the shape of the cobalt electrode was changed to the disk or the ring geometry [27], the rotating waves were recorded during the anodic oxidation of such electrodes in buffered phosphate solutions media. The experiments were performed under potentiostatic conditions. The three-electrode arrangement, with Co disk of a 12.5 mm diameter, is shown in Fig. 2.12; the ring electrode was obtained by sealing the center part of the disk with adhesive tape leaving an outer ring with a width of 1.25 mm. The reference electrode approached the working electrode with a Haber-Luggin capillary, placed ca. 1 mm away from the surface in the center of Co disk. The copper wire encircling the Co disk at the same height was used as a counter electrode.

Depending on particular experimental conditions, such system can exhibit either bistability, or oscillations, or modulated electrochemical waves. Here we shall briefly describe the formation of spatiotemporal patterns. The experiment starts from stepping the electrode potential from the rest value to a value between 500 and 1,000 mV (i.e., below the Flade potential). Under such conditions, the inhomogeneous distribution of the electrode potential already manifests itself: the center of the Co electrode (close to the tip of the RE) is in active state, while its rest, peripheral area, attains a passive state. Next the potential is stepped to an appropriate, more anodic value, at which most of the Co electrode becomes passivated, while the anodic current significantly decreases. However, further evolution of the electrode surface state is by no means trivial. A small circular active area, starting from the center of the disk, begins to develop, with the diameter growing and moving towards the rim of the electrode; accordingly, anodic current increases. It is possible to observe the border between the passive and active zone visually. At certain moment of the ring development it becomes asymmetric and breaks up into

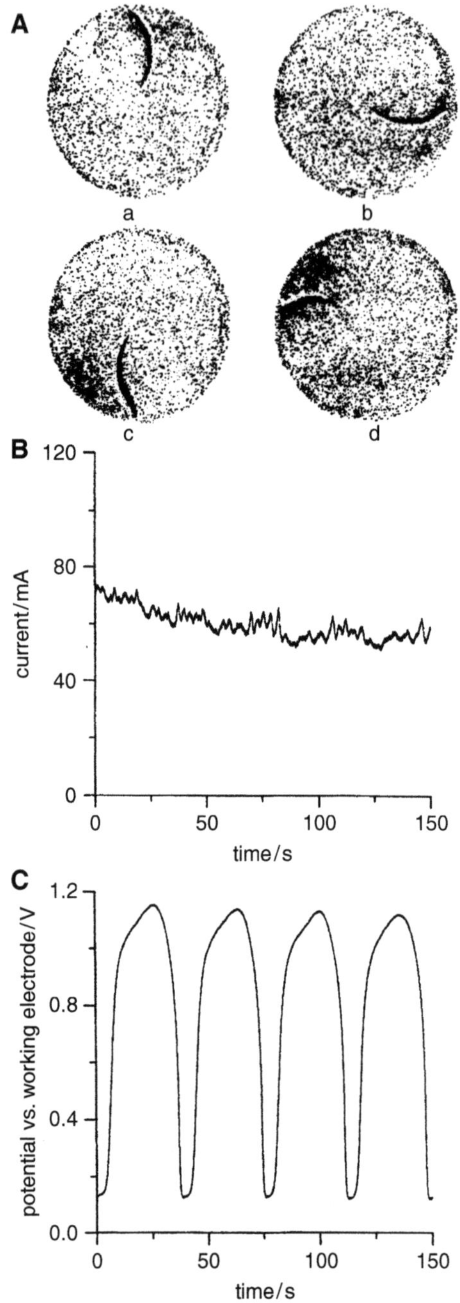

Fig. 2.13 (**A**) Video images of a clockwise rotating wave at 850 mV vs. SCE (first rotation); (a–d): sequence of snapshots for every quarter of the rotation. (**B**) Current–time curve of the rotating wave at 850 mV vs. SCE. (**C**) Potential between the working electrode and an additional calomel reference electrode with a capillary placed at the rim of the working electrode *ca.* 1 mm away from the surface. Reproduced from [27] by permission of the Royal Society of Chemistry

two branches. Of these two branches, one remains stationary and the other one moves in a clockwise or counterclockwise manner around the tip of the reference electrode, with the period of 30–50 s. Figure 2.13 shows the exemplary video

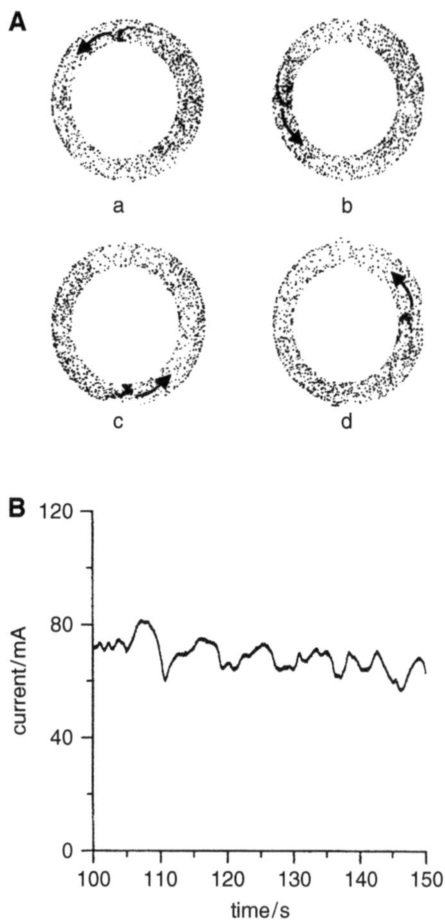

Fig. 2.14 (**A**) Digitized video images of a single counterclockwise rotation wave on the electrode at 700 mV vs. SCE; period of rotation: 9 s; (a–d): sequence of snapshots for every quarter of rotation; the arrow denotes position and direction of propagation of the wave. (**B**) Current–time curve of the single wave on the ring electrode. Reproduced from [27] by permission of The Royal Society of Chemistry

images of the clockwise rotating wave, compared with the temporal dependence of the current and of the potential of the working electrode, measured with an additional reference electrode. The variations of this potential indicate the transitions between the passive and the active state during the passage of the front.

When, in another experiment, the potential was established at more anodic value (+950 mV), the modulated system's behavior is observed, with the activation front moving in the reverse direction. Finally, if the Co disk is modified into the ring electrode, the rotating waves were reported (Fig. 2.14). Also in this case a modulation of the oscillatory characteristics, visualized on the *I–t* dependences, can be observed.

In the preliminary explanation of these phenomena, one should consider the spatial distribution of the potential drop along the electrode radius, caused by the geometrical arrangement of the working and the reference electrodes (Fig. 2.15).

If in certain region of the electrode, the potential is lower than the Flade potential, this part of the electrode remains an active state, otherwise it attains the

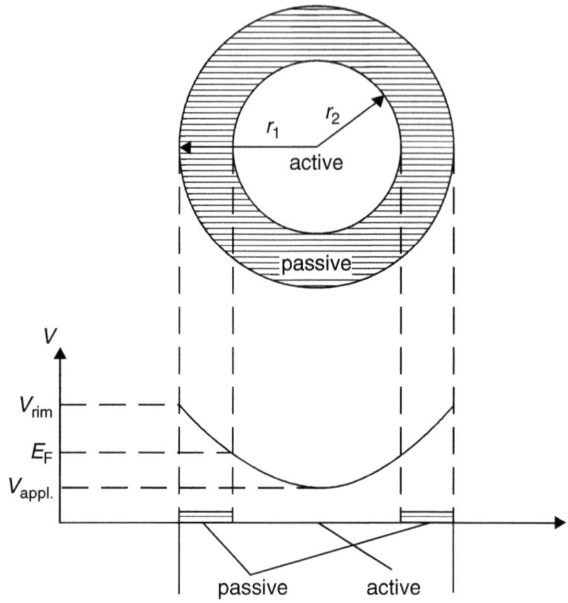

Fig. 2.15 Qualitative dependence of the potential drop on the radial position and hypothetical coverage with a passive layer, V_{rim}—potential at the rim of the working electrode; E_F—Flade potential; V_{appl}—applied potential. Reproduced from [27] by permission of The Royal Society of Chemistry

passive state. In an ideal circular geometry, the boundary between the active and passive regions is circular and for such steady state the current should be constant, determined largely by the active part of the surface (which in turn depends on the external voltage applied). The dynamic situation which develops for the case of rotating waves, shown in Fig. 2.14, corresponds to a quasi-steady state: the active area remains constant as the pulse rotates around the disk. At the border between the active and the passive regions there occurs a significant difference in the local current density and hence in the ohmic potential drop in the electrolyte. Figure 2.15 shows local potential gradients associated with the propagation of rotating waves. Locally, when the electrode potential is shifted to the value more negative than the Flade potential, electrodissolution of passive layer occurs and the advance of the wave is compensated by repassivation in the back of the wave. Evidently, the gradient of electrode potential causes the coupling between different sites of the electrodes and causes the side migration of ions which is involved in the propagation of waves.

2.4 Spatial Patterns in the Ni/H_2SO_4 Oscillator

In Sect. 6.1.4 of volume I, the temporal instabilities and bifurcation analysis of the Ni/H_2SO_4 oscillator, classified as typical HN-NDR system, were described. Here, as the introduction to the coupled Ni/H_2SO_4 oscillators in Sect. 3.1.6, we describe the formation of spatiotemporal patterns during the transpassive nickel dissolution in sulfuric acid medium, reported in the pioneer work by Lev et al. [28].

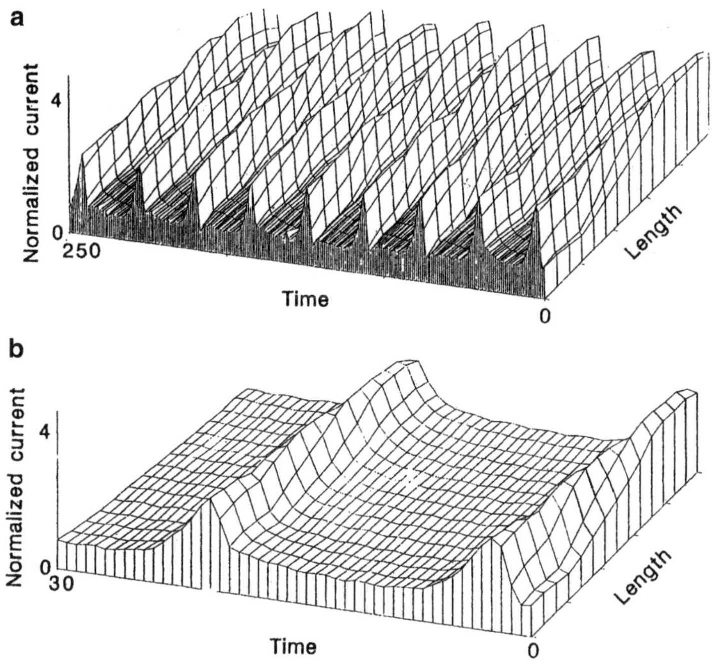

Fig. 2.16 Traveling waves during almost potentiostatic oscillations: (**a**) Three-dimensional representation of local current distribution. (**b**) Enlarged section. Operating conditions: 20-cm nickel wire in 6 N sulfuric acid, 1.67 V relative to saturated calomel electrode. Current normalized relative to 65 mA cm^{-2}, distributed homogeneously. Timescale in scans (50 scans s^{-1}); length scale covers the whole wire. Reproduced by permission from Macmillan Publishers Ltd. [28], Copyright 1988

The electrochemical cell of dimensions 200 × 200 × 40 mm, constructed of plexiglass, held two parallel nickel wires, 0.63 mm in diameter and 200 mm in length, situated 12 mm apart. A calomel reference electrode was situated 40 mm behind the counter electrode at the center of the cell. Like in experiments by Osterwald described in Sect. 6.1.4 of volume I, the operational mode of the electrochemical experiment could have been switched quickly between the potentiostatic and galvanostatic regimes. It is important to note how the operational mode affected the morphology of induced spatiotemporal patterns.

The spatial distribution of local currents was recorded by 15 Ag–AgCl micro-reference electrodes equally spaced along the working electrode, relative to a common calomel electrode situated near the reference electrode [28]. Figure 2.16 shows typical trains of waves (sequence of pulses) that set in, under *almost potentiostatic* conditions, after destabilization of a homogeneous current distribution through the external, serial 1 Ω resistor. This is the presence of that small resistance, and in consequence of ohmic potentials drops, which does not allow to call the experiment to be truly potentiostatic. The pulses started at the right-hand side of the electrode and traveled at nearly constant velocity of ca. 4.5 m s^{-1} to the

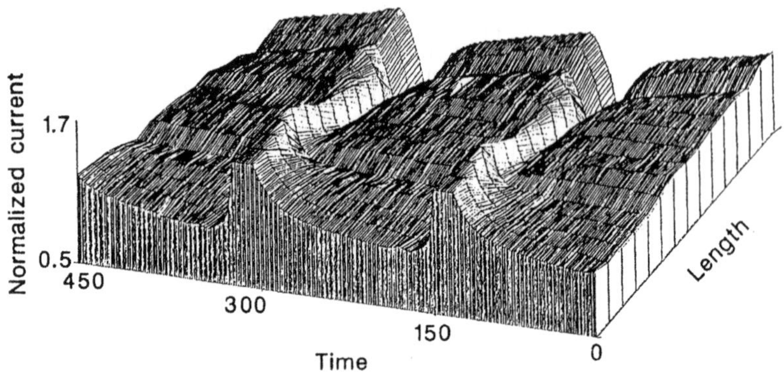

Fig. 2.17 Local current distribution of antiphase oscillations during galvanostatic operation. Operating conditions: 20-cm nickel wire in 2 N sulfuric acid, current is 16.25 mA cm^{-2}; local current normalized relative to the current density at 15.5 mA cm^{-2} as measured under potentiostatic conditions ($I = (v_g - v_0)/(v_p - v_0)$), where v_0, v_g, and v_p are the micro-electrode potential at rest, under galvanostatic and under potentiostatic conditions, respectively. Time and length scales are the same as in Fig. 2.16. Reproduced by permission from Macmillan Publishers Ltd. [28], Copyright 1988

other end of the electrode. This shows good global synchronization among the local oscillators.

Particularly interesting are the antiphase oscillations, shown in Fig. 2.17 which require an analysis in view of more recent understanding of various kinds of spatial couplings operating in electrochemical systems. One should note here that these antiphase oscillations occurred under galvanostatic conditions, where positive global coupling should counteract the pattern formation. This means that in the oscillatory region the effect of positive global coupling is not as straightforward as it could be inferred from the simple theory [19]. More detailed experimental and theoretical studies of these phenomena, the latter ones in terms of appropriately extended Haim et al. model (cf. Sect. 6.1.4, volume I), are reported in [29] and [30], respectively.

2.5 Modeling the Spatiotemporal Patterns in Electrodissolution Processes

2.5.1 The Two-Dimensional Model for Bistable and Oscillatory Process

The progress in understanding of the mechanism of formation of spatiotemporal patterns led to the elaboration of models which allowed to simulate such patterns during metal electrodissolution, for various electrochemical cell geometry, electrolyte conductivity, and double layer capacitance. For example, Bîrzu et al. [31] have

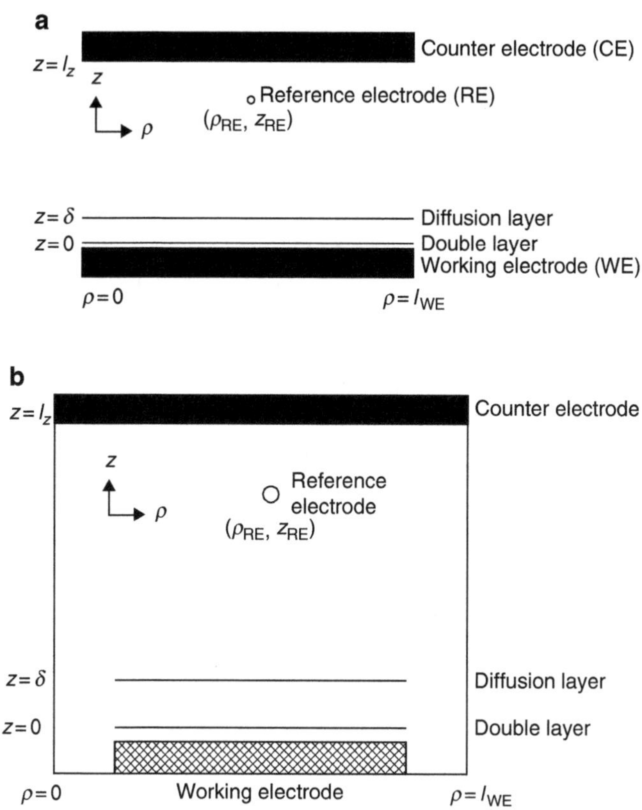

Fig. 2.18 Schematics of the two geometries with working and counter electrodes and a point-like reference electrode: (**a**) Ring geometry, l_{WE} is the circumference of the ring; (**b**) Ribbon electrode embedded in insulator, l_{WE} is the length of the ribbon plus insulating side regions. Reproduced from [31] by permission of the PCCP Owner Societies

developed a numerical model for the bistable and oscillatory process occurring during metal electrodissolution, for assumed two-dimensional geometry, involving ring or ribbon working electrode and a point-like reference electrode located in the electrolyte between the working and the counter electrodes (Fig. 2.18).

Concerning the electrochemical mechanism, this model invokes the principles of the Frank-FitzHugh approach (Sect. 6.1.2.2, volume I) in which the dependence of the Flade potential on pH varying near the electrode surface was one of the important factors deciding the active/passive system's dynamics. The interfacial potential drop (electrode potential) and the concentration of H^+ ions at the electrode surface were assumed as two dynamical variables. The oscillations were then caused by the interplay between diffusion and migration of protons with the pH dependence of the Flade potential. In terms of the Nernst concept of the diffusion layer of thickness δ, the respective equations are:

$$\frac{\partial[\mathrm{H}^+]}{\partial t} = \frac{2D_{\mathrm{H}^+}([\mathrm{H}^+]_{\mathrm{bulk}} - [\mathrm{H}^+]_{\mathrm{WE}})}{\delta^2} + \frac{2\mu[\mathrm{H}^+]_{\mathrm{WE}}}{\delta}\left(\left.\frac{\partial\phi}{\partial z}\right|_{z=0}\right) \tag{2.5}$$

$$C_{\mathrm{d}}\frac{\partial\phi}{\partial t} = \kappa\left(\left.\frac{\partial\phi}{\partial z}\right|_{z=0}\right) + \frac{A}{\exp\left[-\alpha\left(-\phi_{\mathrm{dl}} + V_{\mathrm{f}}([\mathrm{H}^+]_{\mathrm{WE}})\right)\right] + 1} \tag{2.6}$$

where D_{H^+} is the diffusion coefficient of H^+ ions, μ is their migration coefficient, C_{d}—differential double layer capacitance, κ—solution conductivity, ϕ_{dl}—interfacial potential drop at the working electrode, $V_{\mathrm{f}}([\mathrm{H}^+])$—the pH-dependent Flade potential, while A and α are the appropriate constants defining the magnitude of the maximum faradaic reaction current and the slope of the active–passive transition, respectively. These equations served as two of boundary conditions which were used in solving the Laplace equation ($\nabla^2\phi = 0$) for the entire system. The significance of the calculations lies mainly on the indication how important is the localization of the reference electrode on the patterns emerging in the model system, both for the bistable and oscillatory regions of the electrode process. In particular, for the oscillatory regime the direction of the passivation wave could even be reversed by changing the location of the reference electrode to the position relatively apart from the working electrode surface. In this case, the passivation wave starts at the center of the working electrode and moves towards the edges, while the activation waves start at the edges and move toward the center. If the reference electrode is moved closer to the WE, the activation and passivation waves both start at the outer edges and move towards the center.

2.5.2 *The Extension to Three-Dimensional Model*

2.5.2.1 Cell Geometries and Basic Bifurcation Scenarios

A logical development of the two-dimensional approach was the extension of the above model to the three-dimensional geometry, keeping the assumption of the point-like reference electrode (Fig. 2.19) [32].

This modification allowed to model the rotating waves on a ring electrode, antiphase oscillations, spatially homogeneous oscillations, nonuniform stationary state, and the transitions between these regimes. Furthermore, in [33], the accelerating fronts, rotating waves, and homogeneous oscillations were modeled, in relation to experimental studies of electrodissolution of cobalt and iron, in buffered phosphoric acid and sulfuric acid/Na_2SO_4 solutions, respectively, under potentiostatic conditions.

Further development of these model studies involved simulation of patterns in the three-dimensional space, for the *excitable* characteristics of the electrochemical system [34]. These numerical simulations not only showed again how important is the role of the cell geometry (electrodes arrangement) in the type of patterns that

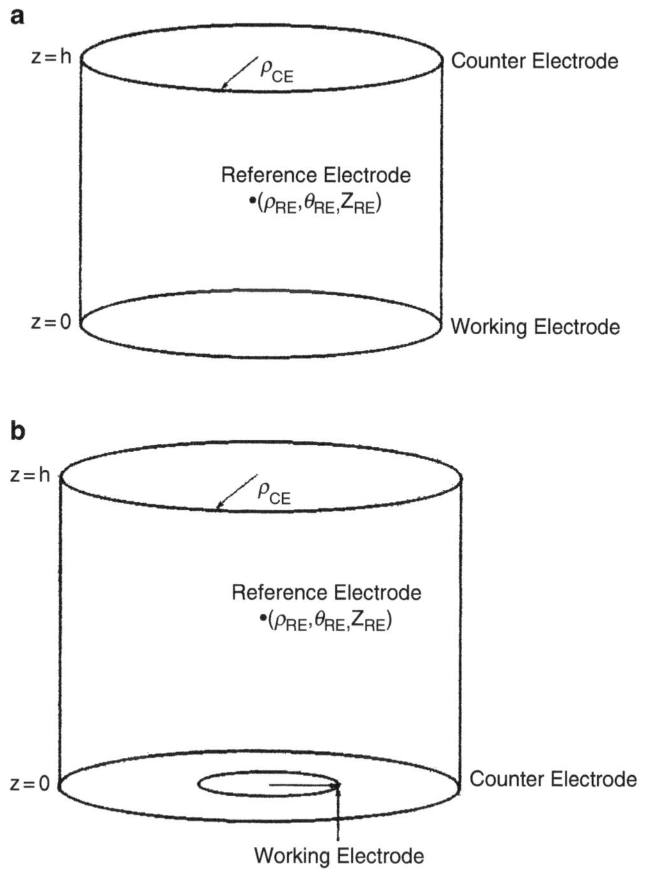

Fig. 2.19 Scheme of electrode arrangement. (**a**) Three-dimensional geometry. The counter electrode in the same radius as the working electrode, ρ_{WE} at a distance h from each other. (**b**) Three-dimensional geometry. The working electrode, with a radius ρ_{WE} and the counter electrode of radius ρ_{CE} are in the same plane. Reprinted from [32], Copyright 2001, with permission from Elsevier

can emerge under given conditions. Not less important are the conclusions indicating that the response of the excitable electrochemical system for the appropriate perturbation can be *remarkably different* from those reported for nonelectrochemical reaction diffusion system. As described in Sect. 1.2.2, the main reason for that is the difference in the range of coupling that exists in such systems: in chemical ones it is only diffusion, i.e., a short-range (or local) coupling, engaging the closest neighborhood of the given site, while in electrochemical systems, the coupling is dominated by migration which, as involving the electric field, is a long-range, or nonlocal type of coupling. The simulations were performed for two cell geometries as shown in Fig. 2.20.

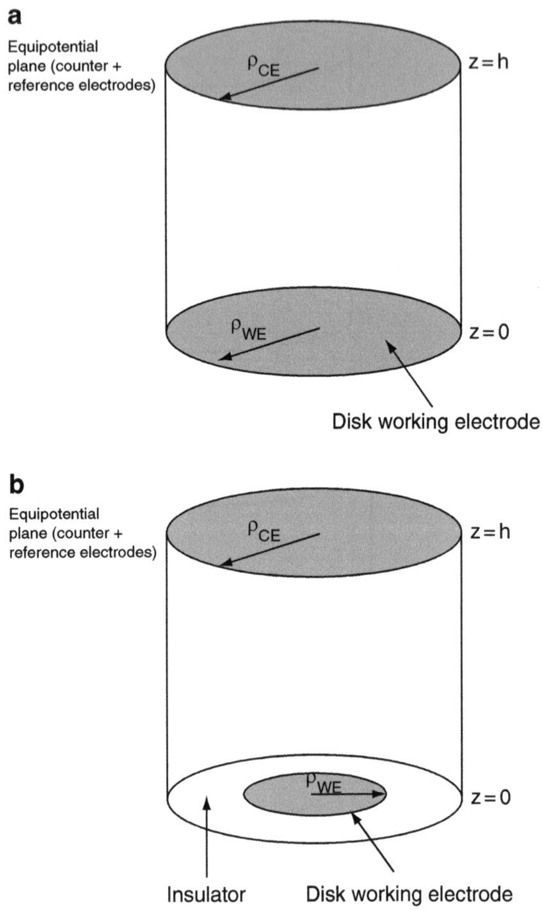

Fig. 2.20 Cell geometries used in the simulations. (**a**) Geometry 1: the disk WE occupies the entire lower plane of the cylindrical cell. (**b**) Geometry 2: the WE is embedded in an insulator. The CE is modeled as an equipotential plane, the RE is assumed to be located in the plane of CE [34]. Reproduced by permission of the PCCP Owner Societies

As in earlier studies, the working electrode symbolizes the metal phase that exhibit active (high anodic current)/passive (low anodic current) transitions under potentiostatic conditions. In the case of geometry (a), every part of the working electrode has the same electric potential, but for geometry (b) it is of course not the case. Since the effective resistance at the rim of the disk is smaller than in the center (see Sect. 1.2), the electrode potential is higher at the rim of the disk than in the center.

Transition from the active to passive state is marked with the region of the N-shaped negative differential resistance (N-NDR). Similarly as in the classical Frank-FitzHugh model, the main carrier of the current in the solution are H^+ ions, so the local pH at the electrode surface is different, depending on whether the electrode is in active or passive state. The destabilization of the system towards the bistability or oscillatory behavior occurs due to the interaction of the NDR characteristics with these changes of local solution pH, meaning the dependence of the Flade potential on pH (being as high as 117 mV pH^{-1}). The calculation details

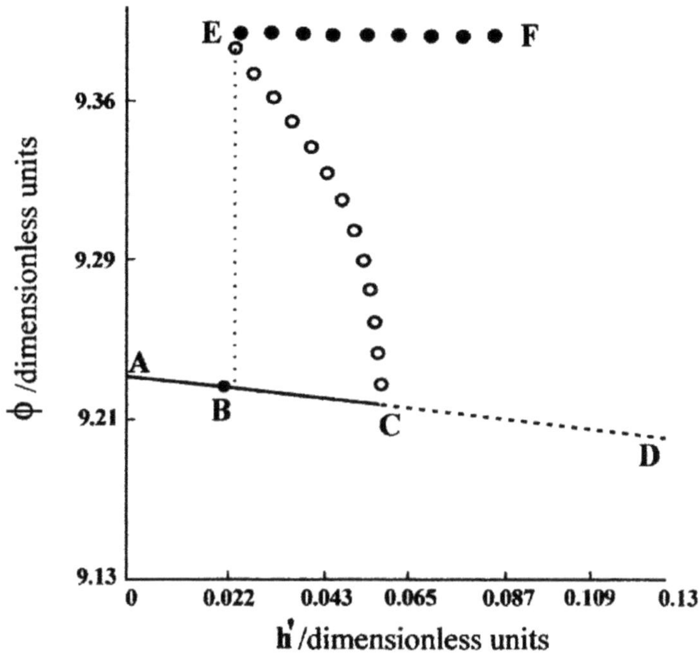

Fig. 2.21 Bifurcation diagram of the uniform system (**a**) as a function of the scaled cell height. Existence regions of the different limit sets: AC—stable (excitable) stationary state; CD—unstable stationary state; CE—unstable limit cycle; EF—stable limit cycle; EC—coexisting range of the stable stationary state and the stable limit cycle. The solid and hollow dots give the maximum of the amplitude of the stable and unstable limit cycles, respectively. C—location of a subcritical Hopf bifurcation. See [34] for underlying equations and parameter values [34]. Reproduced by permission of the PCCP Owner Societies

the interested reader will find in original reference [34], while below we briefly summarize the most representative results of calculations.

Obviously *uniform* steady states exist only for geometry (a) and not for geometry (b) of the cell. Figure 2.21 shows the bifurcation diagram of the uniform system, constructed in terms of scaled height of the cell: $h' = (h-h_0)/h_0$, with $h_0 = 0.023$ cm.

Furthermore, for the uniform system, Fig. 2.22 shows the representative, different phase portraits, that were constructed from the nullclines of the electrode potential and proton concentration dynamics, for two different heights of the cell h, after transformation of the variables to the dimensionless ones. The portrait on Fig. 2.22a, corresponding to h smaller than that for Fig. 2.22b, shows the dynamics typical of excitable behavior—after a small (but exceeding certain threshold) perturbation the system makes a large excursion in the phase space, before it returns to the steady state being a stable focus, thus in the form of damped oscillations. In turn, Fig. 2.22b is a phase portrait for the cell height ca.

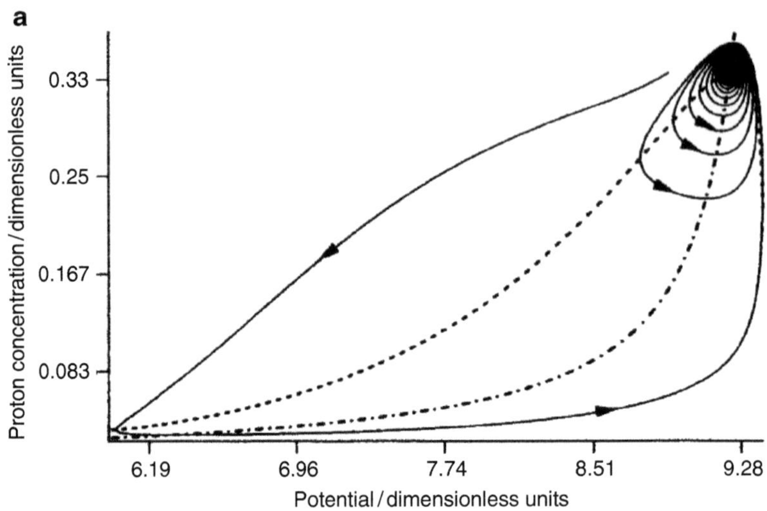

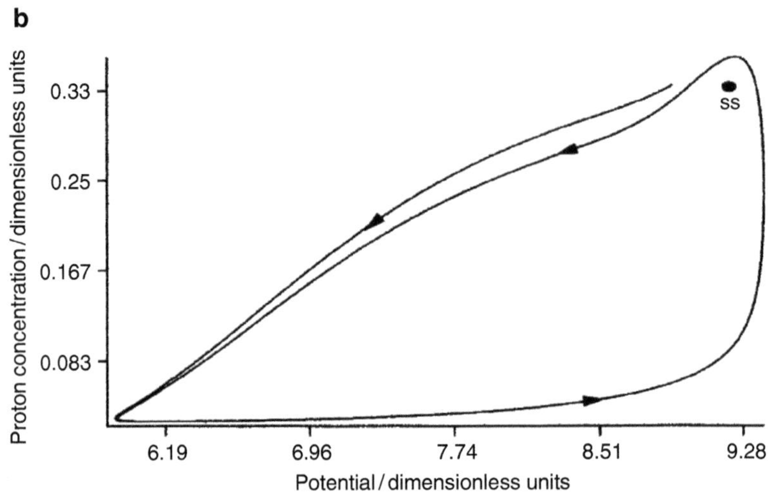

Fig. 2.22 (**a**) Phase portrait of the excitable uniform system. $h' = 0.022$. The *dashed* and *dashed-dotted lines* represent the nullclines for the electrode potential and the proton concentration, respectively. Their intersection is the excitable stationary state: after a small perturbation from the stationary state, the system makes a large excursion before returning in an oscillatory manner to the steady state (*continuous line*). (**b**) Phase portrait for a similar perturbation from the stationary state as in (**a**) but for $h' = 0.043$. At this value of h' the stable stationary state (denoted SS) coexists with a stable limit cycle. See [34] for the parameter values [34]. Reproduced by permission of the PCCP Owner Societies

twice as high as for Fig. 2.22a. In this case, the system is no longer excitable, but the perturbed steady state loses its stability in the subcritical Hopf bifurcation and the resulting asymptotic regime is the continuous trip along the limit cycle.

Applying this bifurcation scenario to the inhomogeneous potential distribution means that upon increasing cell height (h) the subcritical Hopf bifurcation will occur at different values of h for different radial positions—first in the center of the disk and only for sufficiently high values of h it will reach the rim of the disk. Numerical simulations involving the spatial coordinate have shown that the two cell geometries considered yield substantially different spatiotemporal patterns.

2.5.2.2 Patterns in an Inhomogeneous Case

The principal scheme of evolution of spatiotemporal pattern in this case, when the system is excitable, is the pulse generation, the propagation scheme of which depends on the ratio of radii of the working and the counter electrodes: ρ_{WE}/ρ_{CE}. If this ratio is relatively small, the single perturbation of the initially passive (and excitable) state of the entire electrode causes the appearance and expansion of active region, which moves from the excitation point to the opposite side of the disk, where it vanishes and the whole electrode surface returns to the initial steady state (Fig. 2.23a). Simultaneously, the temporal variation of the anodic current indicates a large-amplitude peak followed by small oscillations with a vanishing amplitude, as the asymptotic steady state is a stable focus (Fig. 2.23b).

When the radius of the working electrode is appropriately increased, with the ρ_{WE}/ρ_{CE} ratio kept constant, the system dynamics changes to the coexistence of a stationary nonuniform steady state and a limit cycle. However, in this case, if a single perturbation of the initial passive steady state is not large enough to cause a transition to the limit cycle, it can trigger quite a complex dynamics: a propagation of multiple pulses with pulse reflection (Fig. 2.24). Note, for example, pictures 9–13 which show how, at certain stage of the system's evolution, the traveling pulse reaches the right edge of the disk electrode, then splits and send back a new pulse that travels to the opposite edge where it disappears.

Other modes of perturbations were also studied, leading to different responses of the model system. When the system of the same characteristics as in Fig. 2.24 was perturbed at two locations, symmetric with respect to the center of the disk, a pair of pulses developed and further evolution indicated the onset of periodic oscillations, which constituted the front-like transitions between fairly uniform active and passive states (Fig. 2.25), with total current periodically oscillating as a function of time.

Furthermore, if instead of a single perturbing pulse, an intense, continuous perturbation was applied at one location (again for the same other parameters, as in Fig. 2.24), the onset of asymmetric periodic oscillations was observed. The detailed description of this and other modes of evolution of the system's state in an inhomogeneous case, the reader will find in the original reference [34].

2.5.2.3 Patterns in a Homogeneous Case

Considering the system geometry from Fig. 2.20a, i.e., for the homogeneous case, one should emphasize the substantially different response for the single local perturbation (Fig. 2.26).

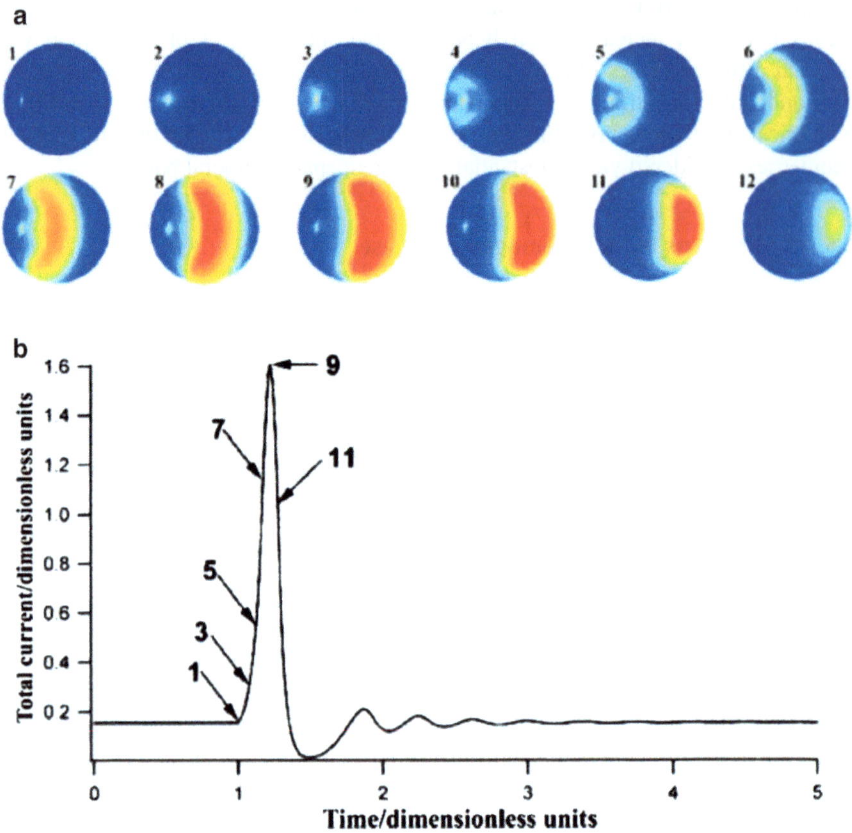

Fig. 2.23 (**a**) Pulse propagation for *geometry 2* and the excitable state between A and B in Fig. 2.21. The excitable stationary state was perturbed by imposing a potential value of 7.74 on a grid element centered at (0.67 ρ_{WE}, $\theta = 0°$), during 0.25 time units. The images show the potential on the disk WE at different times; *blue* represents high potential (passive) and *red* low potential (active). (**b**) Total current. The perturbation starts at approximately 1 time unit. The arrows point to the moments of time corresponding to the frames from (**a**). Parameter values: σ' (proportional to σ/hA) $= 0.262$; γ' (proportional to $\mu\delta/Dh$, μ—mobility of H^+ ions, δ—thickness of the diffusion layer in front of WE, D—diffiusion coefficient of H^+) $= 10.51$; $h' = 0.0696$; $\rho_{\mathrm{WE}}/\rho_{\mathrm{CE}} = 0.8$; $h/\rho_{\mathrm{CE}} = 0.309$; other parameters as in Fig. 2.21 [34]. Reproduced by permission of the PCCP Owner Societies

The active region that appears in this place spreads faster towards the edge of the electrode than towards the center (Frame 2) and, what seems to be most surprising, once the more activated region has touched the edge of the electrode, the activation spreads fast along the entire rim of the disk (Frame 3), the activation of the remaining electrode occurring then from the rim towards the center. This activation is followed by passivation that starts at the electrode rim and spreads over the entire electrode. Subsequently, the system relaxes back to the excitable steady state. It is particularly important to note that this behavior of the model electrochemical system is essentially different from that expected for the nonelectrochemical,

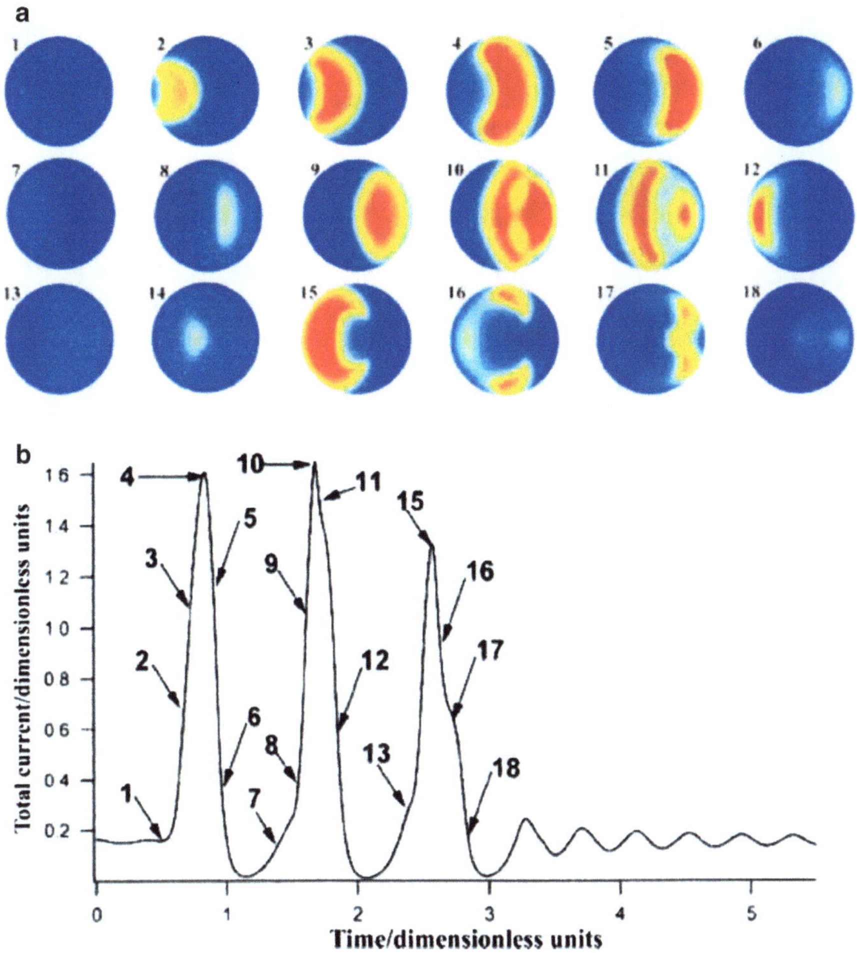

Fig. 2.24 (**a**) Multiple pulse propagation with pulse reflection. With respect to Fig. 2.23, the radius of the cell was increased, $h/\rho_{CE} = 0.1545$, the other parameters are the same as in Fig. 2.23. The images show the potential on the disk at different times. (**b**) Total current corresponding to (**a**). The perturbation starts at approximately 0.5 time units [34]. Reproduced by permission of the PCCP Owner Societies

reaction-diffusion system (RDS), for which a single, circular, propagating pulse would form, locally extinguished upon reaching the system's boundary. This difference illustrates the specific role of the migration (nonlocal) coupling in the generation of the specific system's dynamics, compared to the diffusion (local) coupling.

Furthermore, if larger working electrode is considered, it appears that above critical value of its radius, the system does not relax back to the uniform steady state, but complex spatiotemporal motion prevails, like that shown in Fig. 2.27.

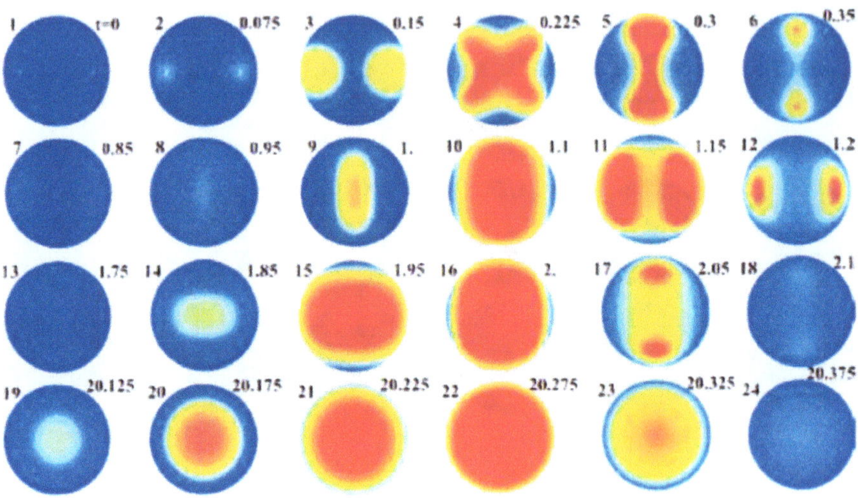

Fig. 2.25 Onset of periodic oscillations. The images show the potential on the disk WE at different times, marked in the upper right side of each frame. Same conditions as in Fig. 2.24, but a second perturbation that is symmetrical with respect to the center is applied [34]. Reproduced by permission of the PCCP Owner Societies

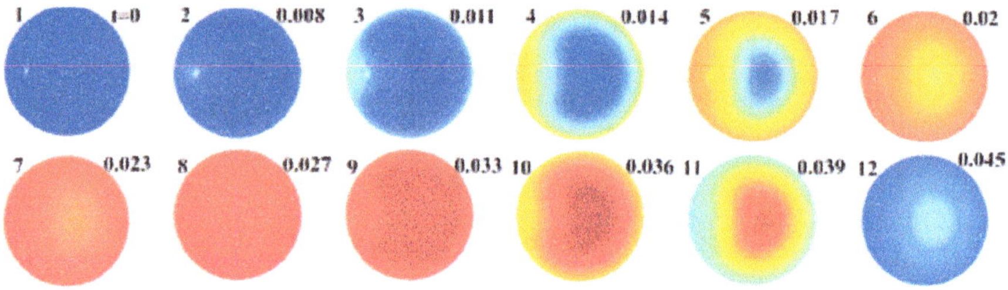

Fig. 2.26 Pulse propagation for *geometry 1* and asymmetric perturbation of the excitable stationary state. The images show the potential on the disk WE at different times, marked in the upper right side of each frame. Parameter values: $\sigma' = 0.275$; $\gamma' = 11$, $h' = 0.022$, $h/\rho_{\mathrm{WE}} = 0.369$. Other parameters as in Fig. 2.21. The perturbation was imposed on a grid element centered at ($0.675\ \rho_{\mathrm{WE}}$, $\theta = 0°$) where ϕ_{dl} was set to 7.738, between 0.25 and 0.35 time units [34]. Reproduced by permission of the PCCP Owner Societies

This is a striking manifestation of the dependence of spatiotemporal dynamics emerging from the excitable steady state on the boundary conditions. Further studies involved, among others, the effect of circular perturbation which resulted in the self-sustained target pattern-like structures, with preserved angular symmetry.

Concluding, the above results show, in terms of appropriate model calculations, the significant role of the cell geometry (in this case—of the way the working electrode is embedded in the electrochemical cell) in the spatiotemporal dynamics of the evolution of the state of the electrode surface in the active/passive region. If the working electrode is surrounded by the insulator, i.e., when the WE radius is

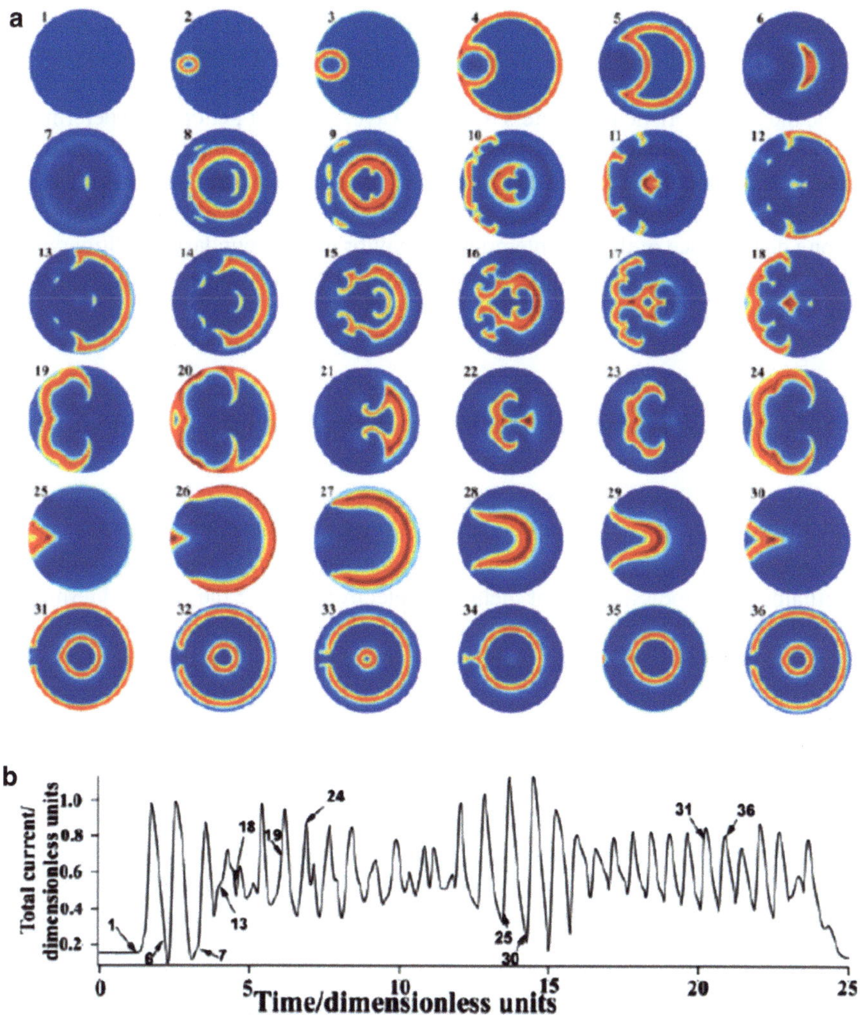

Fig. 2.27 (**a**) Pulse propagation for geometry 1 and asymmetric perturbation of the excitable stable stationary state (the corresponding electrode potential and proton concentration at the stationary state are $\phi^0 = 9.224$; $[H_{WE}{}^+]^0 = 0.338$). The images show the potential on the disk WE at different times. Parameter values: $\sigma' = 0.275$; $\gamma' = 11$, $h' = 0.022$, $h/\rho_{WE} = 0.0369$. Other parameters as in Fig. 2.21. The perturbation was imposed on a grid element centered at (0.686 ρ_{WE}, $\theta = 0°$), where ϕ_{dl} was set to 7.738, between 1.25 and 1.375 time units. (**b**) Total current corresponding to (**a**) [34]. Reproduced by permission of the PCCP Owner Societies

smaller than that of the counter electrode, the rim of the electrode is kept in more passive state than the center. This situation, in cooperation with such specific dynamical characteristics, like the coexistence of the limit cycle with the stationary state and the proximity of the subcritical Hopf bifurcation gives rise to complex spatiotemporal dynamics, dependent also on the magnitude and duration of the perturbation imposed. For the apparently simpler electrode configuration, in which the radii of the WE and CE are equal, the response of the excitable state appears to be fundamentally different from that expected for the chemical, RDS. In the latter ones, the characteristic parameters of spiral waves or target patterns are determined by the kinetic parameters of the process, rather than by the boundary conditions, as long as the domain size is much larger than the characteristic pattern size. In the case of an electrochemical system considered earlier, we have seen the surprising strong dependence of the spatiotemporal behavior on the boundary conditions. This is because the properties of migration coupling depend strongly on the boundary conditions in the plane of the working electrode [34].

2.5.3 Modeling the Electrochemical Turbulence

The same geometries of the electrochemical cell, as shown in Fig. 2.20, were more recently used by Bîrzu and Krischer [35] in the modeling of a two-dimensional electrochemical turbulence (i.e., *spatiotemporal chaotic dynamics*) during the oscillatory electrodissolution of metal disk electrodes. Note the reversed terminology: the terms "Geometry I" and "Geometry II," used in the captions to figures below, will refer now to the "inhomogeneous case" and "homogeneous case," respectively.

We shall briefly describe here the main results. The assumed electrode process exhibited not sinusoidal, but relaxation oscillations which means that the system's parameters were far away from the point of the supercritical Hopf bifurcation. As the bifurcation parameter there was chosen the system's size, defined as the ratio between the cell radius and cell height: $w = (\rho_{WE}/h)$, meaning the "aspect ratio" of the cell. Finally, only the migration (nonlocal) coupling was assumed to exist in the system. The spatiotemporal patterns were developed in two dimensions parallel to the working electrode surface. The set of two dynamical variables consisted of the interfacial potential drop at the working electrode (an activatory variable) ϕ_{dl} and the concentration of H^+ ions at this electrode in the solution, $[H^+]$ (an inhibitory variable).

2.5.3.1 Turbulence in an Inhomogeneous Case

The series of numerical simulations was performed for increasingly larger working electrodes, i.e., for increasing values of bifurcation parameter w. For sufficiently small electrode ($w = 0.764$) the quasi-periodic course of oscillations of total current vs. time was obtained, and no corresponding irregular patterns, but only a simple spatial pattern was generated (Fig. 2.28).

With increasing cell aspect ratio, the spatially coherent patterns became unstable and irregular spatiotemporal states developed, with the typical transient regime depicted in Fig. 2.29.

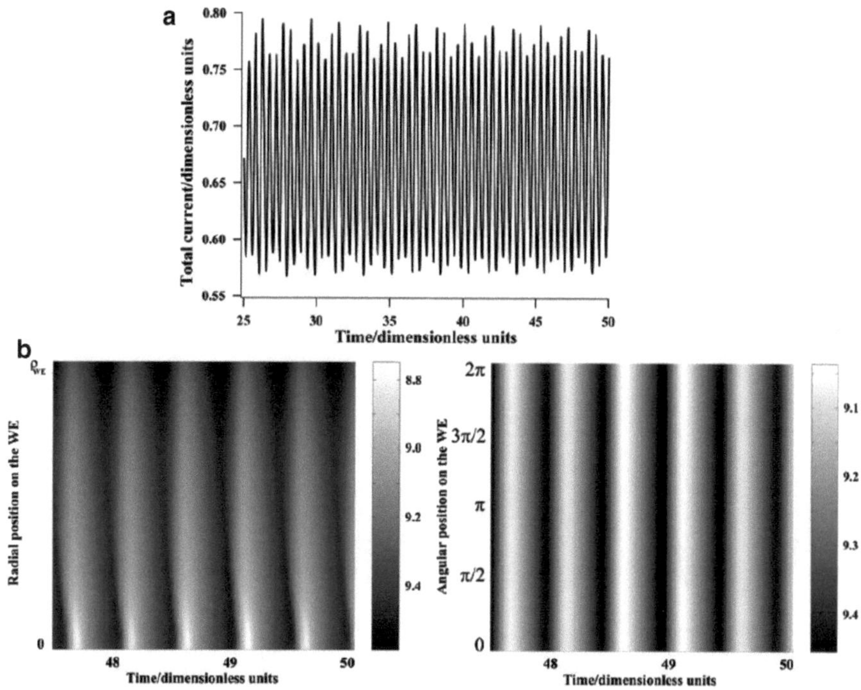

Fig. 2.28 $\rho_{WE}/h = 0.764$. Geometry I: spatially coherent dynamics. (**a**) Asymptotic time series of the total current. (**b**) Asymptotic radial (*left*) and angular (*right*) temporal evolution of ϕ_{dl}. The angular evolution is shown at $\rho = 0.5\rho_{WE}$. (*white* denotes active states and *black* passive ones) [35]. Reproduced by permission of the PCCP Owner Societies

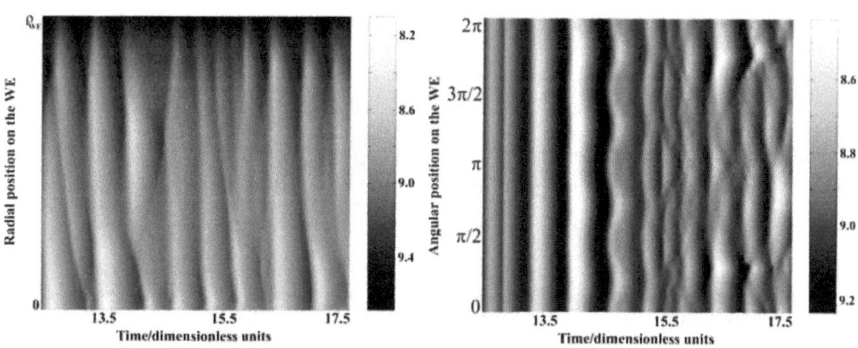

Fig. 2.29 $\rho_{WE}/h = 3.182$. Geometry I: spatiotemporal evolution of the radial and angular dependence of ϕ_{dl} around the angular symmetry breaking. The angular evolution is shown at $\rho = 0.5$ ρ_{WE} [35]. Reproduced by permission of the PCCP Owner Societies

Figure 2.30 shows the snapshots of amplitude and phase for four different, increasing aspect ratios of the cell. These images, which show the symmetric transition states and the fully turbulent (chaotic) regimes, confirm that the system's dynamics becomes more chaotic as the system size increases [35].

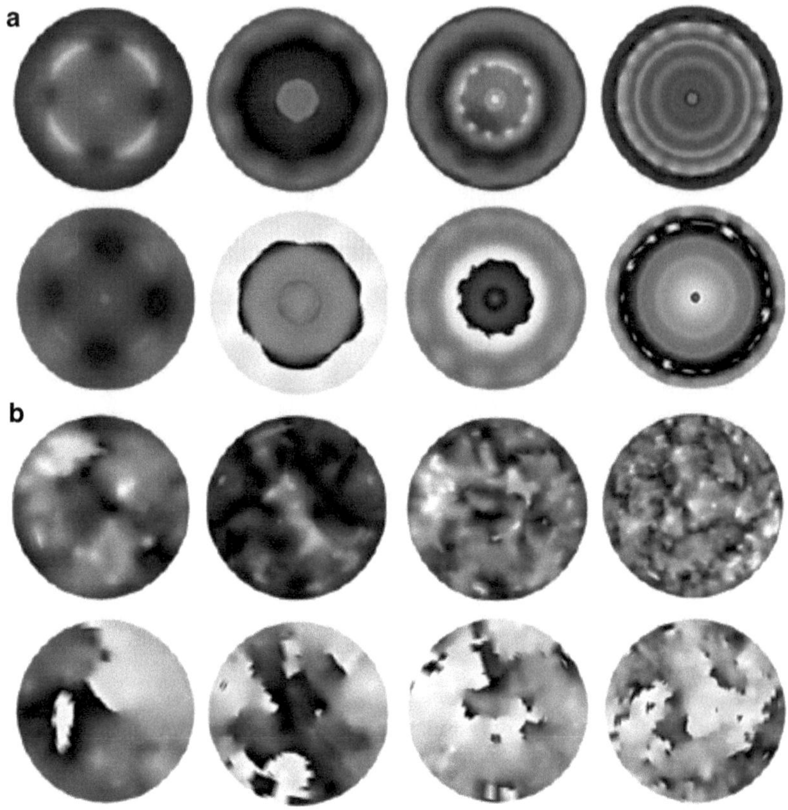

Fig. 2.30 Geometry I: Snapshots of amplitude (*upper row*) and phase (*lower row*) for four different electrode sizes (**a**) during the symmetric transients and (**b**) in the defect turbulent regime. $\rho_{WE}/h = 3.182$ (*first column*), 4.773 (*second column*), 6.366 (*third column*) and 9.549 (*last column*) [35]. Reproduced by permission of the PCCP Owner Societies

The term "defect turbulent regime" in the caption of Fig. 2.31 refers to the idea that the spatial disorder is generated by the so-called defects (meaning the singularities in the field of the oscillation phase).

2.5.3.2 Turbulence in a Homogeneous Case

Model calculations show that even for the homogeneous case ("Geometry II"), although the local oscillatory dynamics does not vary with position, the long-term dynamics may also exhibit irregularity in time and space [35] (see Fig. 2.31) which is presumably caused by the strong spatial coupling along the angular direction close to the rim of the electrode.

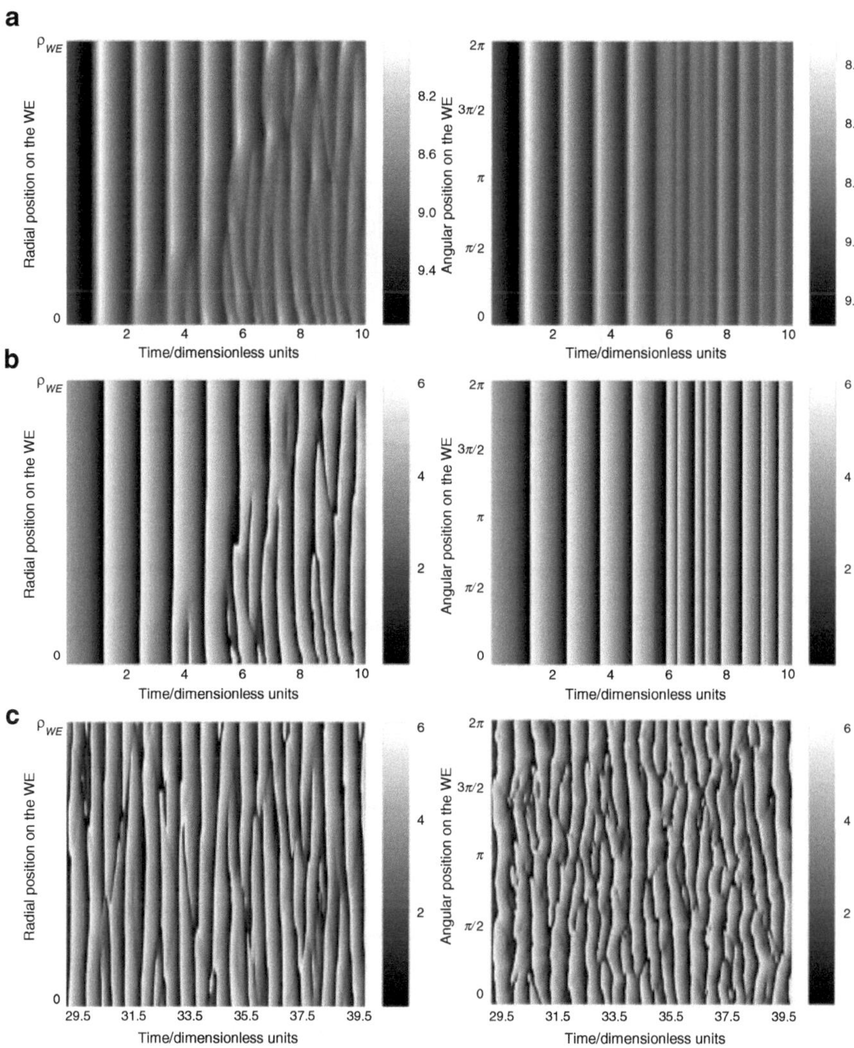

Fig. 2.31 $\rho_{WE}/h = 6.366$, Geometry II: (**a**) radial (*left*) and angular (*right*) cut of ϕ_{dl} for the first ten time units. (**b**) Radial and angular cut of the oscillation phases for the first ten time units. (**c**) Same as (**b**) but for the asymptotic dynamics [35]. Reproduced by permission of the PCCP Owner Societies

The reader interested in more detailed interpretation of these phenomena and their quantitative analysis in terms of such tools of analysis of spatiotemporal data, as the Karhunen–Loève decomposition [36], is advised to consult the original paper [35].

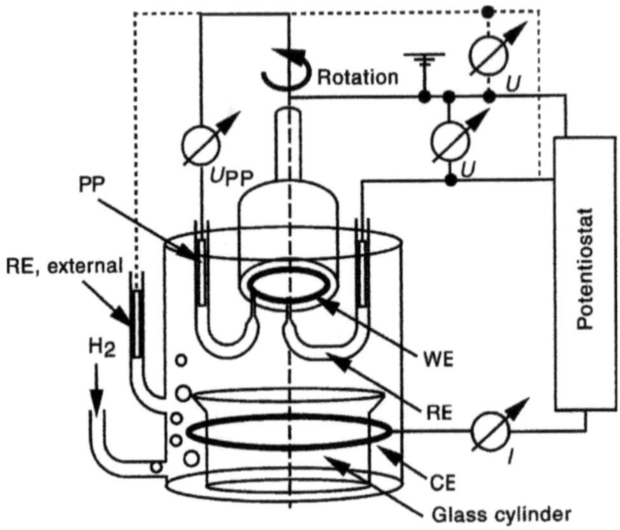

Fig. 2.32 Scheme of the electrochemical cell. WE: working electrode, RE: reference electrode (note that either the RE in the Haber-Luggin capillary with the tip on the axis of the ring or the external RE was used). CE: counter electrode, PP: potential probe [37]. Reproduced by permission of the PCCP Owner Societies

2.6 Patterns in H_2 Oxidation

In complement to temporal instabilities in electrochemical oxidation of H_2 at Pt electrodes (section 5.1, Vol. I), in this section we shall discuss the formation of spatiotemporal patterns in this process, as described by Krischer, Varela et al. in a series of papers [37–41]. In most experiments the specially designed setup was used, which included the ring Pt electrode as the working electrode, at which electrooxidation of H_2 occurred, while the series of microprobes located along this ring allowed to measure the actual spatial distribution of the electric potential (Fig. 2.32).

Essential results of these studies, which confirm theoretical predictions about the conditions of formation of spatiotemporal patterns, can be summarized in the following way.

For the bistable region of the H_2 electrooxidation, when the distance between the WE and RE was relatively large or medium, the transitions between the two homogeneous states occurred through nucleation and growth of the globally stable state from the initially metastable one.

For the bistable region, but for the closest WE–RE distance, when the negative global coupling manifested itself strongly, stationary (nonequilibrium) patterns were observed, consisting of two domains, i.e., with high currents (where Pt surface was bare) and low currents (where Pt–O domains existed on surface), respectively, with the difference in the double layer potentials of two domains reaching more than 1 V! [37].

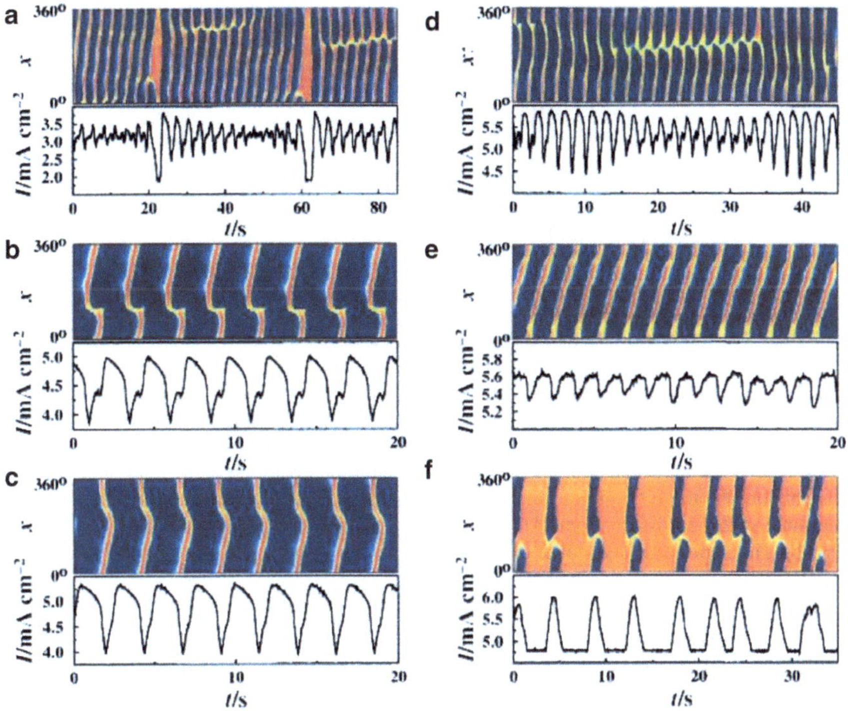

Fig. 2.33 Position–time plots of the local double layer potential and corresponding time series of the global current under potentiostatic conditions with the reference electrode [RE(2)] placed at a distance of 13 mm from the plane of the WE for different values of the external voltage U: (**a**) 1.06, (**b**) 1.37, (**c**) 1.5, (**d**) 1.67, (**e**) 1.76 and (**f**) 1.86 V vs. SHE. Reproduced from [38] by permission of The Royal Society of Chemistry

In the presence of poisoning, electrosorbing ions (i.e., after addition of 0.1 mM HCl and 0.025 M $CuSO_4$ to the 0.5 mM H_2SO_4 solution saturated with H_2), and under potentiostatic conditions, the system's dynamics was studied again for the three positions of the RE with respect to WE, corresponding to zero, intermediate, and maximum global coupling. In the first case, when only migration (nonlocal) coupling could operate, spatially inhomogeneous oscillations were reported and attributed to a spatial bifurcation of the homogeneous limit cycle; the systems appeared to be Benjamin–Feir unstable [42, 43]. In the other extreme case, corresponding to strongest global coupling, the spatiotemporal dynamics was qualitatively different—the quasi-harmonic oscillations of the total current (with the shape dependent on the external potential) were accompanied with the pulse propagating around the ring in a fairly uniform manner. This was the change of the width of the pulse during one rotation around the ring which caused these current oscillations. Finally, at intermediate distance between the RE and WE, a particularly reach dynamics were observed, depending on the external voltage (see Fig. 2.33) [38].

In Fig. 2.33 the following dynamic behaviors are shown: (a) low-voltage, complex spatiotemporal patterns, made up of a sequence of behaviors, that appear

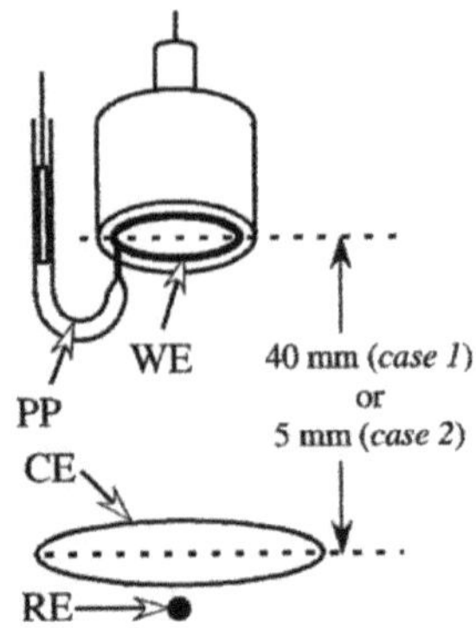

Fig. 2.34 Experimental setup for studies of chaotic patterns in H_2 electrooxidation: WE, working electrode (mean circumference: 85 mm; width: 1 mm), CE: counter electrode, RE: reference electrode (Hg/Hg_2SO_4, sat.), located below the plane of the CE, PP: potential probe. The WE was rotated at 20 Hz and the spatiotemporal evolution of the interfacial potential was obtained by the PP with a resolution of 50 points per rotation. See [41] for further details. Reprinted with permission from [41] http://link.aps.org/abstract/PRL/v94/p174104. Copyright 2005 by The American Physical Society

"isolated" at larger values of U; (b) periodical generation of two pulses, of which only one propagates effectively in one direction, while the other one, which is about to travel into the opposite direction, stops soon and their meeting leads to mutual annihilation; this behavior was considered the formation of a novel pattern type—the "asymmetric target pattern"; (c) similarly to (b), periodical generation, at the position 0°=360°, of two pulses which however now travel in opposite directions and meet in the center of the ring where they annihilate each other (a kind of one-dimensional analog of the "target pattern"); (d) the combination of cases (b) and (c)—the system switches back and forth between the target and the asymmetric target pattern; (e) a pulse traveling around the electrode with slightly changing velocity; (f) a front-like, non-periodic excitation travels once around the ring before it is extinguished again at a certain position [38]. These experimental observations of asymmetric target patterns were supported by perfectly concordant results of numerical simulations in terms of appropriate model [40]. These calculations clearly revealed that the nonlocal migration coupling was crucial for the endogenous persistence of the perturbation that caused the asymmetry.

Furthermore, for the H_2 oxidation, also in the presence of Cu^{2+} and Cl^- poisons, Varela et al. [39] have reported the complex hierarchy of patterns, involving, among others, nested spatial (and temporal) symmetry breakings of the emerging domains, the scenario being presumably a universal route leading to a substructuring of space in systems which experience a global coupling and possess at least two adjacent regions in the parameter space that exhibit dynamical instabilities.

Chaos, i.e., "electrochemical turbulence" was also reported in the H_2 electrooxidation. In these experiments, the distance between WE and the CE was varied, with the tip of RE located always *just below* the CE and such configuration minimized the global coupling, leaving only migration (nonlocal) coupling as the operating one (Fig. 2.34). Experiments were conducted for two different distances

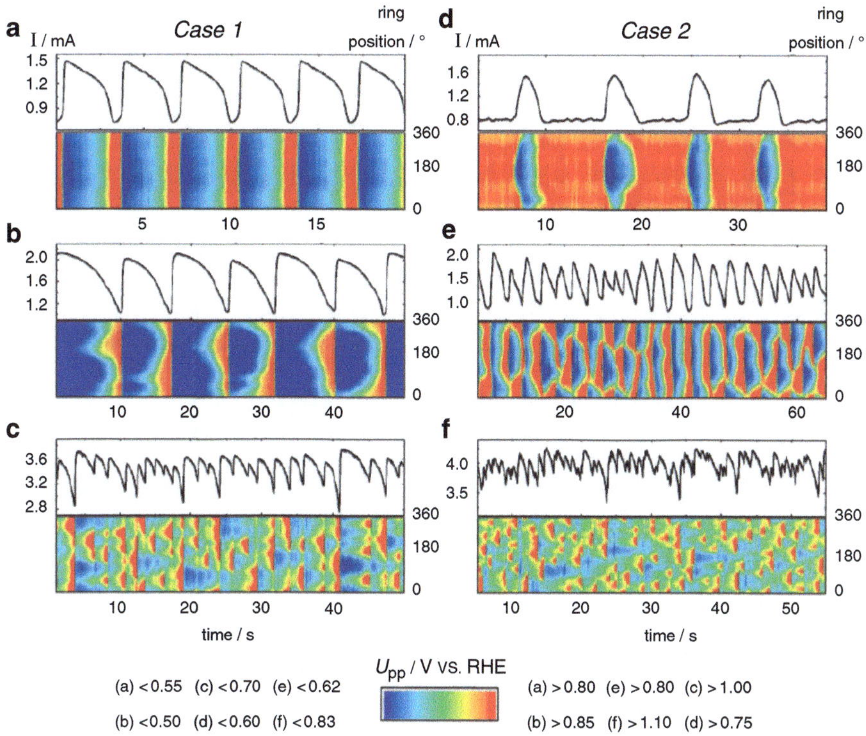

Fig. 2.35 Global current time series and spatiotemporal evolution of the interfacial potential, ϕ_{dl}, as a function of position on the ring and time for different values of the applied voltage U and electrode placement corresponding to case 1 [*plates* (**a**), (**b**), and (**c**)] and case 2 [(**d**), (**e**), and (**f**)]. U: (**a**) 1.06 V, (**b**) 1.19 V, (**c**) 2.14 V, (**d**) 0.82 V, (**e**) 0.89 V, and (**f**) 1.54 V. Electrolyte: H_2-saturated, aqueous 0.5 mM H_2SO_4 solution containing 0.1 mM HCl and 0.01 mM $CuSO_4$. A continuous flow of H_2 was maintained throughout the experiments. Reprinted with permission from [41] http://link.aps.org/abstract/PRL/v94/p174104. Copyright 2005 by The American Physical Society

between the WE and CE: 40 mm (case 1) and 5 mm (case 2) to realize different coupling ranges.

For this experimental arrangement, Fig. 2.35 shows the global current–time series and spatiotemporal evolution of the interfacial potential (denoted by ϕ_{dl}, as usual as an alternative for E) for two WE–CE distances (case 1 and case 2) and different voltages. In particular, Figs. 2.35c, f show irregular oscillations of total current and associated irregularities in pattern formation.

In the analysis of these phenomena, it was advantageous to transform the potential values into amplitude and phase variables. Then, for strongly irregular data there were found the locations, at which the amplitude vanished and the phase changed abruptly. It was possible to determine the density of such space–time defects which was increasing with voltage, starting from zero and indicating a transition into a defect turbulent regime. In turn, the irregular phase

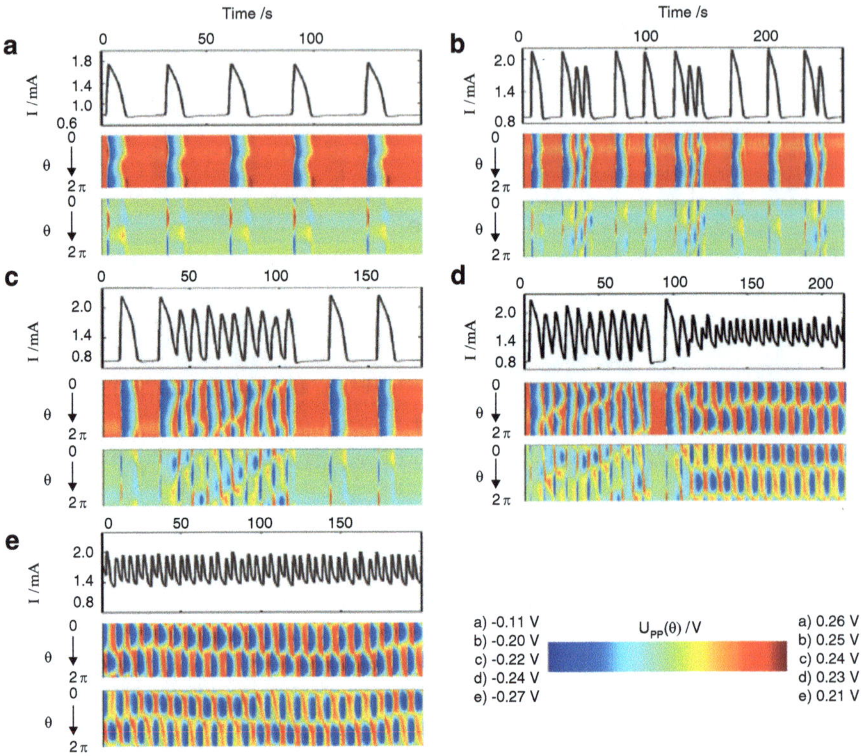

Fig. 2.36 For global coupling strength $|\rho| \approx 0.3$: Time series of the global current (*upper plates*), spatiotemporal evolution of the interfacial potential drop (*middle plates*), and spatiotemporal evolution of the inhomogeneous part of the interfacial potential drop (*lower plates*) for weak global coupling about five different values of the applied voltage U: (**a**) 1.02 V, (**b**) 1.11 V, (**c**) 1.14 V, (**d**) 1.17 V, (**e**) 1.18 V. Cell resistance $R_\Omega = 510 \pm 20\ \Omega$, compensated resistance $R_c = 160 \pm 20\ \Omega$. Electrolyte composition: 0.5 mM H_2SO_4, 0.1 mM HCl, and 0.01 mM $CuSO_4$, H_2—saturated. Rotation rate of the electrode $\omega = 20$ Hz. The transformation of the measured interfacial potential drops (U_{PP}) into the *color code* is given at the *bottom*. The values to the *left* and *right* of the *color bar* denote the minimum and maximum U_{PP} values, respectively [39]. Reproduced by permission of the PCCP Owner Societies

modulation dominated, in the absence of defects, at lower voltages. So, irregular series shown in Fig. 2.36 are the experimental manifestation of a transition from the phase to defect turbulence. For the sake of generalization, it is important to note that quite analogous sequence of events was previously found in theoretical treatment of spatiotemporal chaos, done in the framework of the complex Ginzburg–Landau (CGLE) equation for diffusively coupled oscillators close to a supercritical Hopf bifurcation. In terms of this analysis, the one-dimensional system enters the region of spatiotemporal chaos via the above-mentioned Benjamin–Feir (BF) instability which first leads to phase turbulence followed by defect turbulence (phase dislocations) upon farther driving system away from the BF point [44]. The present analysis proves that similar scenario occurs also for the system with nonlocal (nondiffusive) coupling which is, in addition, far from a supercritical Hopf bifurcation.

Furthermore, the H_2 electrooxidation in the presence of poisons was studied under conditions favoring the existence of a global coupling [39] which was this time introduced either by placing the RE between the WE and CE, on the axis of WE (cf. Fig. 2.21) or by compensating part of the cell resistance with an external negative impedance device (NID). Experimental conditions were chosen so that the migration coupling induced the spatiotemporal chaos (see above). Two parameters were varied systematically—the strength of global coupling and the externally applied voltage. Without global coupling, increasing U led to a transition from a spatially modulated limit cycle, via a phase turbulent regime, to space–time defect turbulence, while the presence of even weak global coupling suppressed the occurrence of defect turbulence, with the formation of regular two-phase clusters (domains). The strength of global coupling substantially affected the transition scenarios and the final dynamical states reached at high values of U. At appropriately high strength of global coupling, a second type of two-phase cluster was observed, as well as two types of irregular cluster patterns, with the irregularly moving boundaries, and accompanied with the emergence and disappearance of clusters through splitting and merging the boundaries, respectively. At even higher coupling strengths, the five-phase clusters were stabilized and finally, when the coupling was the highest, the cluster patterns transformed into strongly modulated pulses. Of a rich set of dynamical modes, shown in [39] we shall choose here, as an example, the simplest case of weak global coupling [expressed quantitatively as the value of parameter $|\rho| = 0.3$ (which varies from 0 to 1)] illustrated by Fig. 2.36. Initially, at low value of U, nearly periodic relaxation oscillations of the global current, synchronized with the interfacial potential of the entire electrode, are observed, as in the case described in the previous section. However, further system evolution following the increase in U is now completely different due to the presence of a negative global coupling. Instead of formation of space–time defects, a transition to a two-phase cluster took place: two domains formed on the electrode, which oscillated with identical frequency but 180° out of phase. Locations within each of the two domains oscillated with approximately identical amplitude and phase. Of other type of clusters found under global coupling conditions, two were found as novel ones.

In conclusion, it is evident that H_2 electrooxidation on Pt, particularly in the presence of Cl^- and Cu^{2+} ions, is a source of variety of spatiotemporal regimes. The reader interested in analysis of rich sets of experimental data as well their theoretical treatment is advised to consult the respective original references.

At the end of this section, we shall mention that the spatiotemporal patterns emerging during H_2 electrooxidation have been used by Krischer et al. to compare two essential methods of inducing the global coupling: either by close position of RE to WE or by insertion of an external resistor in series to the WE, in the potentiostatic experiment. A certain difference between these two ways of creating such coupling consists in that in the latter case it is easy to adjust the coupling strength by varying the external resistance, for the given cell resistance, but in the former case this strength is a more complicated function of RE position, cell geometry, and electrolyte conductivity (see, e.g., the approach presented in [52]). In a more recent paper, Krischer et al. [45] have presented an alternative approach to this problem, in terms of general equivalent circuit of an electrochemical cell and by introducing the subcircuit of the negative impedance device (NID). Theoretical considerations, verified experimentally by the studies of H_2 oxidation on Pt ring electrode, in the presence of Cu^{2+} and Cl^- poisons, proved that the same strength of global feedback can be caused by two equivalent methods of compensation of the cell resistance: either by placing the tip of the reference electrode close to the WE (using Haber-Luggin capillary) or by inserting the serial negative ohmic resistance, of the absolute value equal to the cell resistance. This finding is important, since the use of external negative resistor exerts the same coupling effect independently of the relative arrangement and geometry of the electrodes, which dependence was the main problem in case of using the Haber-Luggin capillary (see Section 1.2.2). In fact, in both cases, the cyclic voltammograms, the (low-amplitude) current oscillations, and spatiotemporal evolution of the electrode potential during H_2 electrooxidation were found almost identical. In other words, the global feedback realized through the external, electronic ohmic drop compensation is always strictly global, i.e., all locations of the electrode are coupled with the same strength. Such application of NID was further used by Plenge et al. [46], this time for the IO_4^- electroreduction on the 50 nm thick, rectangular Ag(111) electrode evaporated onto a glass plate. This process, in neutral supporting electrolyte ($NaClO_4$), is characterized with the N-NDR region. The spatial patterns were monitored with surface plasmon spectroscopy. Upon inserting the appropriate negative impedance, the surface plasmon images indicated correlation between the current oscillations and nonuniform changes of the optical properties of the interface. A striking result was that the two halves of the electrode oscillated alternately in a harmonic manner, the resulting pattern resembling the standing wave born in a wave bifurcation.

Finally, a recent theoretical study by Baba and Krischer [47] of the model of oscillatory electrooxidation of hydrogen in the presence of poisons under galvanostatic conditions led to generation of mixed-mode (MMO) oscillations for low preset currents, in the homogeneous case. For spatially extended systems, there were formed cluster patterns exhibiting interesting features resulting from a combination of the properties of the homogeneous oscillations in the model and spatial coupling (see original reference for details)

2.7 Patterns in CO Oxidation

In Sect. 3.9 of volume I, the effect of spatiotemporal dynamics on the impedance spectra was described, taking as a model system the bulk electrooxidation of CO on a rotating Pt electrode which process is characterized with an S-NDR region in the

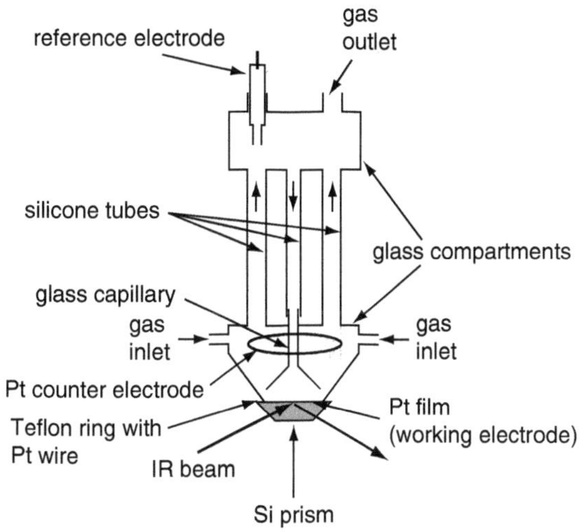

Fig. 2.37 Electrochemical cell used for IR measurements. A chemically deposited Pt film was used as the working electrode. The radiation totally reflected at the metal/liquid interface is analyzed. The geometrical area of Pt film in contact with the electrolyte is approximately 175 mm^2. Reprinted with permission from [50]. Copyright 2008 American Chemical Society

current–potential curve [48]. Such characteristic was found to occur due to the interaction between the competitive Langmuir–Hinshelwood mechanism and the mass transport of CO towards the electrode. It was also found experimentally that direct measurements of the S-shaped polarization curve, under controlled conditions, could not explicitly indicate the NDR region which could be suggested only by current–potential curve at fast galvanodynamic scans [49] (cf. Sect. 5.2, volume I). For the purposes of this section, it is important to note that one reason for this was found the formation of spatial patterns on the electrodes. The emergence of such patterns was recently proved by Morschl et al. [50] by means of spatially resolved infrared spectroscopy used for the detection of electrode areas covered and free of CO. IR measurements were performed during galvanostatic oxidation of CO on Pt film electrode in a setup shown in Fig. 2.37.

Representative results of the experiments are shown in Fig. 2.38. It is shown that the way in which Pt electrode is covered with CO depends on imposed current. For applied current up to 1.3 mA, Pt surface is completely covered with CO (symbolized red color). When the current is increased up to 1.8 mA, the studied area splits into regions still covered with CO (red zone) and (almost) free of CO (blue zone). The area of CO-free region increases with applied current. Finally, for currents higher than 1.8 mA, the low coverage state spreads over the entire probed region. Detailed quantitative analysis reveals also, on some pictures, broad transition regions between the CO-covered and CO-free zones.

Careful reader of Sect. 1.2 will realize that electrooxidation of CO on Pt, as the process exhibiting S-NDR characteristics, could be a candidate for the formation of spatial structures which exactly meet the criteria of electrochemical Turing patterns. In fact, such patterns were obtained for this system, they are described further in Sect. 2.10.

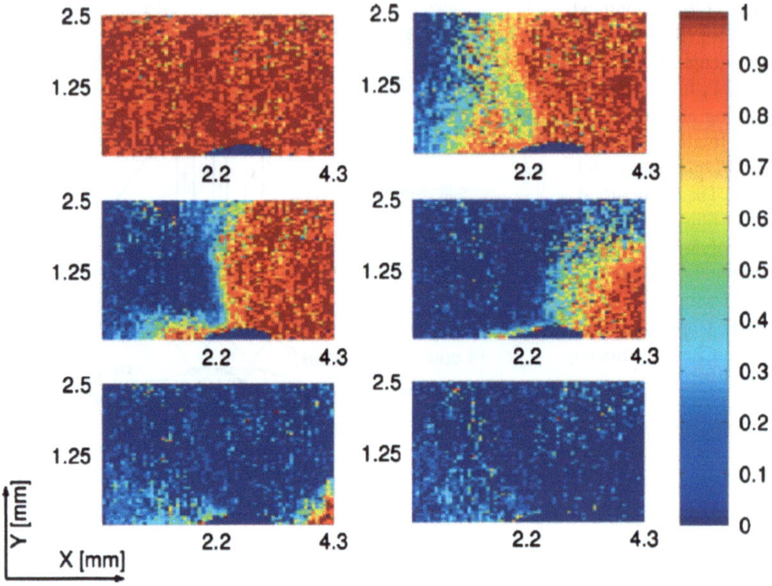

Fig. 2.38 Integrated CO peak plotted versus space for the following currents: 0.30, 1.40, 1.50, 1.70, 1.75, and 1.80 mA (*left* to *right* and *top* to *bottom*). *Red* indicates a high coverage, and *blue* a CO free surface. Note that the *blue region* in the *middle* of the *lower edge* indicates an area where the film hardly showed any CO peak even for complete coverage. Therefore, this region was set to 0 in all images. Reprinted with permission from [50]. Copyright 2008 American Chemical Society

2.8 Patterns in HCOOH Oxidation

The oscillatory oxidation of formic acid on Pt electrodes, being a HN-NDR type oscillator was described in detail in Chap. 5 of volume I. Under appropriate conditions, this process can also give rise to spatial and spatiotemporal patterns. In 1999, Christoph et al. [51] have reported an interesting phenomenon, observed during the potentiostatic oxidation of formic acid on a Pt ring (see Fig. 2.39), under conditions when the electrode process exhibited bistability, i.e., the coexistence of the passive (OH-poisoned, low-current) and active (high-current) states. An appropriate perturbation at one location of the ring caused the emergence of a wave on the opposite side, thus the *remote triggering* of the electrochemical wave was observed.

The Pt working electrode was initially prepared in the passive state by scanning the voltage to a high value of 750 mV (vs. Hg/Hg_2SO_4 reference electrode placed in the center of the ring) and then the voltage was lowered to a value close to (just before) the value of 220 mV at which the return to the active state would occur; in this way the rather metastable state was created. When the perturbation, in the form of a positive pulse, was applied to trigger Pt electrode (+3 V vs. the working electrodes potential), the system switched locally to the active (low-potential) state and, as expected, two fronts of activation started to propagate along opposite

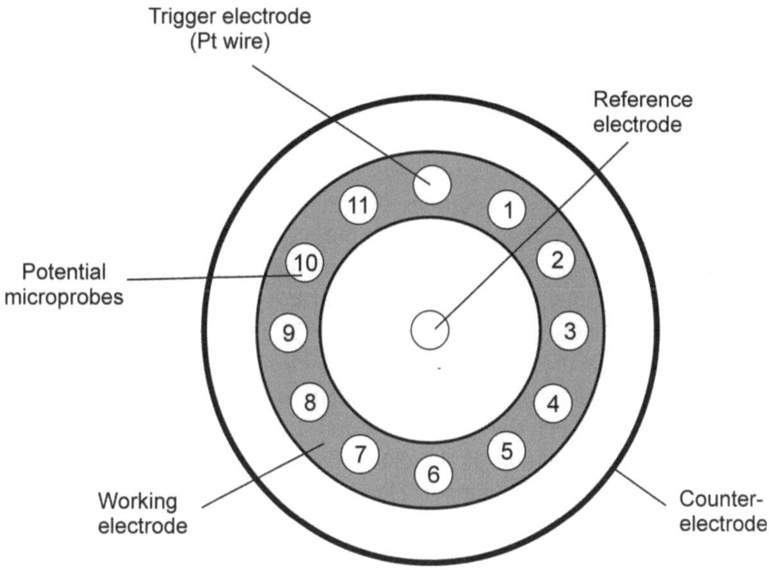

Fig. 2.39 Schematic view of the electrode setup. The potential microprobes (1 through 11) are separated by 30° angles, and the perturbations were applied with a Pt wire. After [51]. Reprinted from [51] with permission of AAAS

directions of the ring until they met after ca. 200 ms. However, if the negative pulse was applied, instead of expected local stabilization of the double layer potential, its value increased only slightly at positions 1 and 11, but at the opposite site (electrode No. 6) activation occurred, followed by decelerated front propagation. This is the above-mentioned *remote triggering* of the wave.

Of course, such remote response is an experimental proof for a long-range (nonlocal) coupling between different parts of the working electrode, realized through the ionic migration in the solution, under the influence of the electric field. The almost instantaneous manifestation of this coupling is a consequence of a significant difference in time scales of two processes which is the condition for occurrence of such effects: while switching the electrode surface to the active state, requiring the removal of passivating OH species, requires ca. 100 ms, the electric field effects spread with the velocity of light, i.e., require less than 10^{-9} s across the system. Numerical calculations based on theory developed earlier by Christoph et al. [52] and involving the coupling function H [see description of Eqs. (2.7)–(2.10) for more details] fully confirmed the role of long-range migration coupling in the reported experimental phenomena. Such calculations clearly show how the local perturbation causes the onset of the front propagation from this place, but at a remote distance the perturbation of the reversed direction is induced.

In the course of further studies, Strasser et al. [53] have described the formation of *standing wave oscillations* of the interfacial potential across the ring polycrystalline Pt electrode, monitored by means of equispaced potential microprobes along the electrode, as for the construction shown in Fig. 2.39, with the exception for the absence of the triggering electrode. The Hg/Hg_2SO_4 reference electrode was placed in the Luggin capillary, the tip of which was located in the center of the working electrode ring, thus in a configuration ensuring the same potential distribution between the RE and the working electrode, independently of the azimuthal angle. A concentric Pt wire ring served as a counter electrode. Each of 11 potential microprobes consisted of a glass tube with a microcapillary (100–200 μm diameter), hosted a Hg/Hg_2SO_4 reference electrode and was filled with 0.5 M Na_2SO_4 solution.

Under such potentiostatic conditions that the oxidation current of formic acid exhibited spontaneous oscillations, the distribution of the electrode potential along the ring electrode exhibited inhomogeneities, both in time and space, with one half of the electrodes in high-potential (passive) state, and the other half—in the low-potential (active) state, periodically interchanging in time. The whole pattern had a form of a standing wave, with the spatial node located close to the position of electrode number 6, for which the amplitude of the oscillations was very low (Fig. 2.40).

The correlation between the shape of period-2 cycle observed in the course of total current and the variation of the electrode potential along the entire ring is shown in Fig. 2.41. This comparison clearly shows that large total currents are associated with pronounced spatial inhomogeneities along the working electrodes, whereas low catalytic activity (low current) correspond to nearly homogeneous potential profiles.

By varying the concentration of a supporting electrolyte (i.e., the solution conductivity) and the external potential one can obtain various patterns, including aperiodic temporal variations of total current associated with corresponding turbulent spatiotemporal patterns along the electrode (but even in this complex regime, the antiphase behavior between opposite points along the working electrode was maintained).

Finally, upon insertion of sufficiently high serial resistance ($R_{ex} > 4\ \Omega$) the synchronization of the potential profile was observed, i.e., any spatial patterns practically ceased and the total current exhibited oscillations correlating with the temporal variations of the electrode potential.

The reader familiar with the content of Sect. 1.2 will realize that conditions of both the pattern formation (potentiostatic control with the reference electrode approaching the WE with the Luggin capillary) and the pattern destruction (increasing external resistance) corresponds exactly to the situation when negative global coupling operates and turns into positive one. Thus, the construction of the theoretical model [53] of the observed experimental phenomena refers to the essential mathematical description of this kind of coupling, in a formalism developed earlier by Christoph et al. [52, 54]. This approach introduces the spatial coupling function H_B which represents the coupling between two points x and x' along the working electrode, and due to the symmetry of the ring electrode geometry it is here the function of $|x–x'|$. In terms of this approach, the dynamics of the interfacial potential drop at the working electrode is described in terms of the following dependence:

$$C_{\mathrm{d}}\frac{\partial \phi_{\mathrm{dl}}}{\partial t} = -i_{\mathrm{f}}(\phi_{\mathrm{dl}}) + \frac{U-\phi_{\mathrm{dl}}}{R_{\mathrm{tot}}} + \sigma \int_0^1 H_B(|x-x'|)[\phi_{\mathrm{dl}}(x',t) - \phi_{\mathrm{dl}}(x,t)]\mathrm{d}x' \quad (2.7)$$

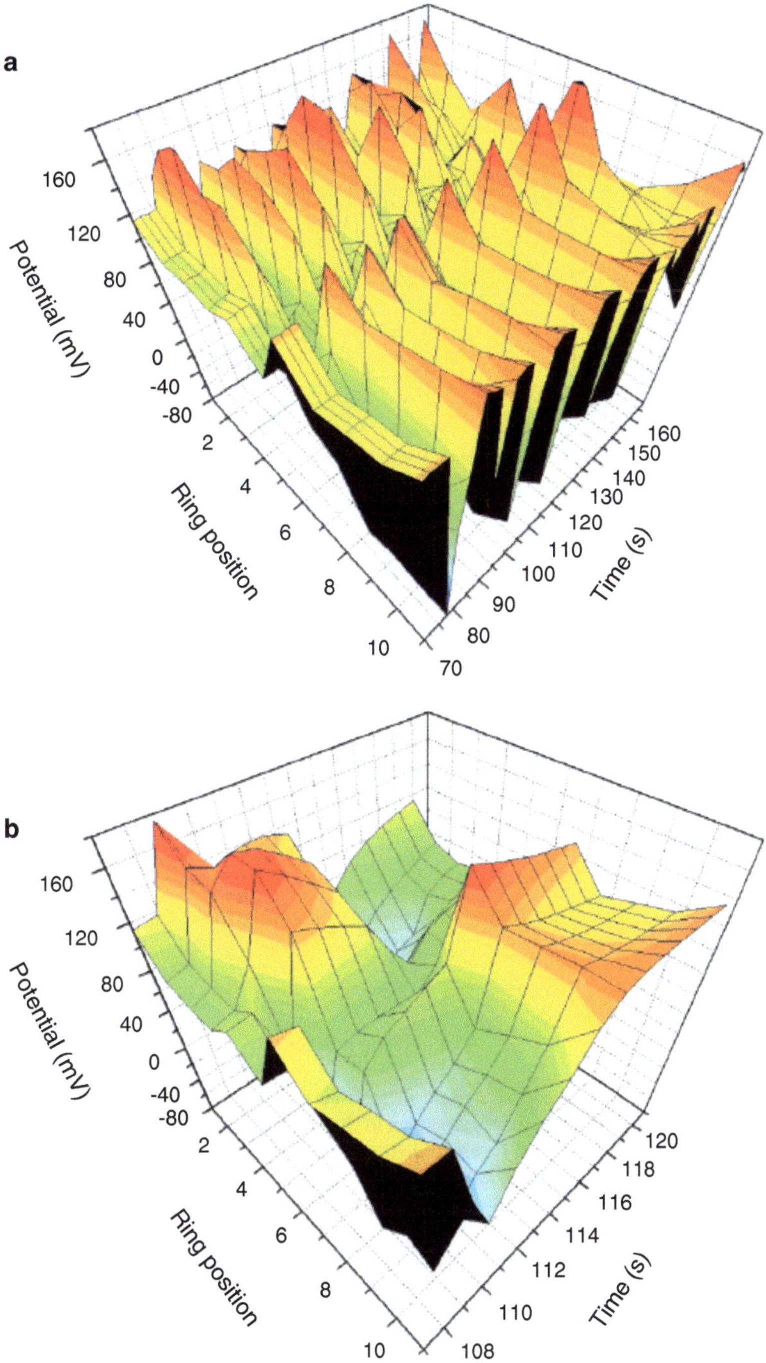

Fig. 2.40 (**a**) Spatiotemporal evolution of the local potential in front of the Pt WE during formic acid oxidation under oscillatory total currents. The potential distribution displays an oscillatory standing wave pattern. $U_{ex} = 325$ mV, 0.05 M HCOONa/0.025 M H_2SO_4, no stirring, the

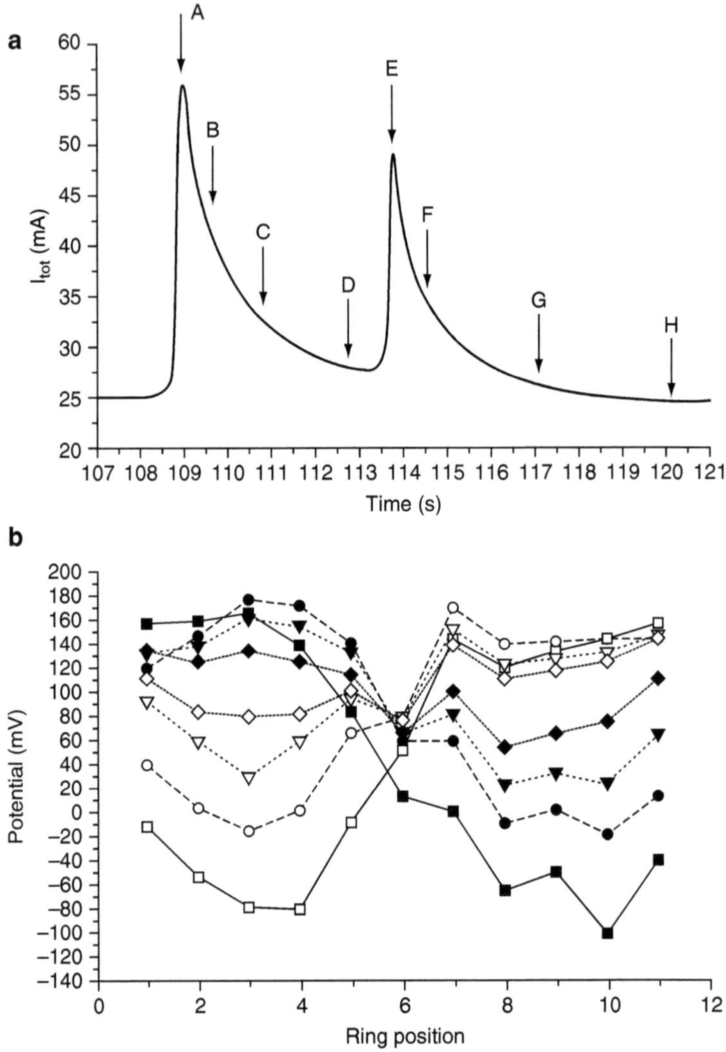

Fig. 2.41 (**a**) Blowup of a period-2 cycle in total current I_{tot}. The letters indicate eight time instances for which the spatial profiles are given in (**b**). (**b**) Spatial potential profiles at subsequent time distances corresponding to a. A (*filled squares*), B (*filled circles*), C (*filled triangles*), D (*filled diamonds*), E (*open diamonds*), F (*open triangles*), G (*open circles*), and H (*open squares*). Reprinted with permission from [53]. Copyright 2000 American Chemical Society

---

Fig. 2.40 (continued) oscillations have been induced by a homogeneous 200 ms/+200 mV potential pulse applied to the WE. (**b**) Blowup of one spatiotemporal period given in (**a**). Reprinted with permission from [53]. Copyright 2000 American Chemical Society where i_f is the faradaic current, U is the external voltage (vs. the reference electrode), and R_{tot} is the total resistance. The integral part constitutes the mathematical formulation of a spatial coupling which vanishes for the homogeneous electrode state (since then $\phi_{dl}(x', t) = \phi_{dl}(x, t)$ for any x and x'). On the other hand, for the inhomogeneous state, the value of coupling function quantitatively describes the effect that any location x' affects the dynamics at the position x. The value of H is determined by parameter B which depends on the external serial resistance R_{ex}, on the aspect ratio of WE and RE (β), being the distance between RE and WE normalized to diameter of the WE, on the inner diameter of the annulus (A) and the solution conductivity (σ):

$$H_B(|x - x'|) = H_0(|x - x'|) + \frac{h^2 B}{1 + hB} \tag{2.8}$$

with

$$h = -\int_0^1 H_0(|x - x'|dx') \tag{2.9}$$

and

$$B = R_{ex}\pi\sigma(1 - A^2) - \sqrt{1 + \beta^2} + \sqrt{A^2 + B^2} \tag{2.10}$$

Figure 2.42a shows the course of the $H_B(|x - x'|)$ function for three different values of parameter B: $B = 0$ (or slightly negative), approximating the condition of the absence of external resistance, with large distance between WE and RE; $B < 0$ for sufficiently small distance between WE and RE, creating conditions for significant negative global coupling and $B > 0$, corresponding to the presence of a sufficiently large external resistance. In fact, the negative sign of H indicates the regions of the working electrode subject to negative global coupling which is thus manifesting itself for $B < 0$ at the remote locations of the WE. Accordingly, Fig. 2.42b shows the theoretical oscillating *standing wave* of the interfacial potential, calculated from the model involving two dynamical variables: the interfacial potential (fast variable) and the surface concentration of an electroactive species (slow variable), for $B < 0$ (see caption for detailed values of particular parameters).

From the point of view of nonlinear dynamics, the birth of the oscillatory standing wave occurs via the *wave bifurcation* (Turing's second bifurcation) of the stationary state that precedes the Hopf bifurcation. Most probably this is the first report on the formation of oscillatory standing waves in electrochemical systems. These and other dynamical regimes are collected on a one-parameter bifurcation diagram constructed from the dependence of the maximum/minimum total current density i_m of the respective patterns on the external voltage U_{ext}, at constant total dimensionless resistance R_{tot} (Fig. 2.43); the region of occurrence of standing wave region is there abbreviated by sw. The wave bifurcation occurs at $U_{ex} \approx 200$, the first Hopf

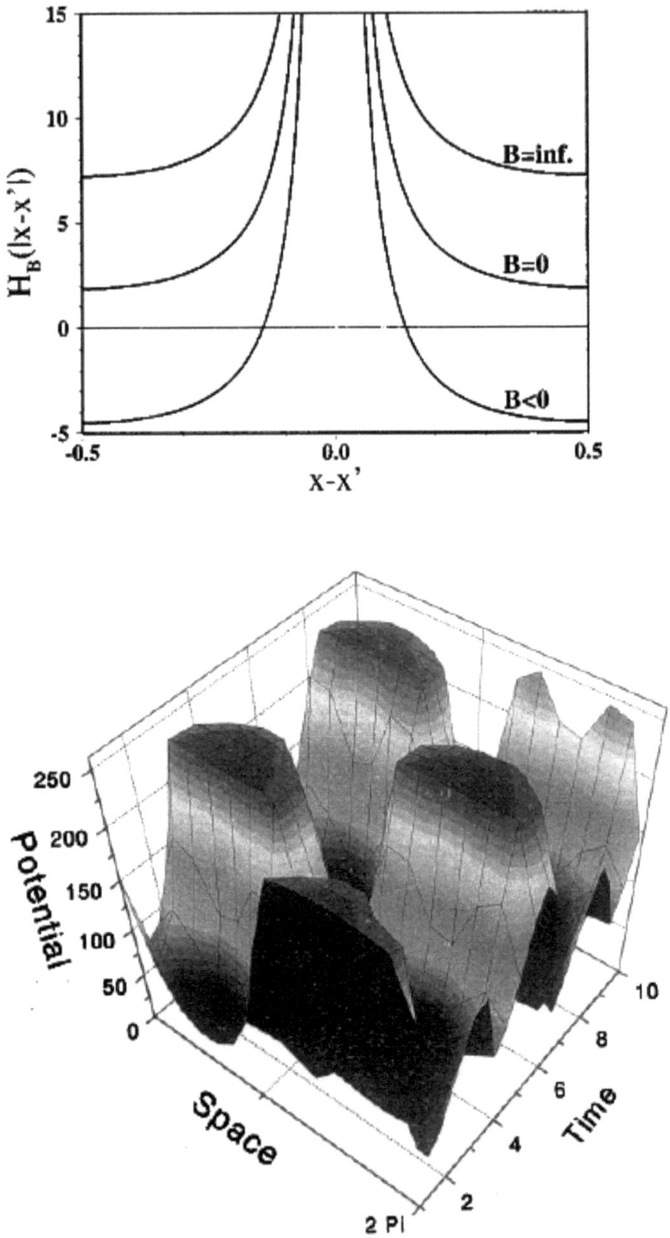

Fig. 2.42 (**a**) Coupling function H_B for different values of the parameter B; H_B represents the coupling between two points x and x' along the interface with distance $x–x'$. (**b**) Calculated oscillating standing wave of the interfacial potential in the two-variable reaction–migration model past a wave bifurcation. For the mathematical form of the function and the parameters, see original reference. Reprinted with permission from [53]. Copyright 2000 American Chemical Society

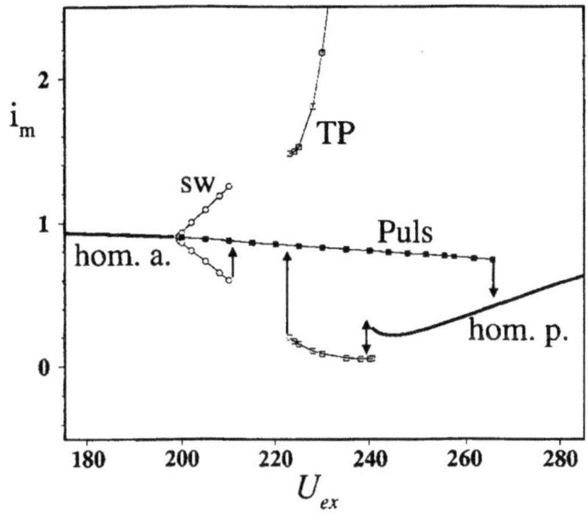

Fig. 2.43 One-parameter bifurcation diagram of the model discussed, at $R_{tot} = 40$. The maxima and minima of the total current density i_m of the occurring patterns are plotted against the dimensionless outer applied potential; sw—oscillatory standing waves, hom. a. = homogeneous active stationary state. TP—target patterns, Puls = pulse solution branch, hom.p. = homogeneous passive stationary state. All other parameters as for Fig. 2.42b. See [53] for further details. Reprinted with permission from [53]. Copyright 2000 American Chemical Society

bifurcation—at $U_{ex} \approx 217$, and a second Hopf bifurcation—at $U_{ex} \approx 241$ (both indicated by the arrows). In certain regions of external voltage, one observed also bistability: target patterns/pulse and pulse/homogeneous passive stationary state.

It is useful to comment in more detail the existence of the wave bifurcation in the above system. In principle, this is a second bifurcation predicted by Alan Turing in his seminal paper [55], after the first one predicting the stationary patterns, described in Section 1.1.4. In this second bifurcation, oscillatory standing waves are born, like in the experimental system and its model described earlier. One should note, however, that in chemical reaction-diffusion systems (RDS), for the wave bifurcation to occur two dynamical variables (species) are not sufficient—there should be present at least a third species, with a diffusion coefficient determining the intrinsic wavelength of the time-dependent solution of differential equations [56–58]. The present electrochemical system is analyzed in terms of two-variable model, so apparently the wave bifurcation should not occur. However, one can prove [53] that for electrochemical systems with negative global coupling, both Turing bifurcations require *one variable less* than in RDS, i.e., 1 and 2 instead of 2 and 3, respectively. The instabilities occur then with the largest possible wavelength because of the very fast coupling. As a further consequence of that, those complex behaviors that were predicted for three-variable RDS (i.e. target patterns, traveling pulses) should be obtained for the two-variable electrochemical reaction–migration system. The bifurcation diagram, shown in Fig. 2.43, confirms that conclusion. Another example is the electrodissolution of Co in acidic media [27, 52].

In further studies, Lee et al. [59] have analyzed the oscillatory oxidation of formic acid on a Pt electrode modified by Bi deposition, using the ring Pt electrode. In the presence of Bi adatoms, the current density increased about five times, since less adsorbed CO was then formed, being presumably removed by Bi–oxygen species. In spite of this modification of one of the reaction pathways, the galvanostatic oscillations, born via the Hopf bifurcation, were observed. Accordingly, impedance measurements indicated the hidden negative differential resistance, confirming that the oscillator falls into Class IV category (HN-NDR systems). The electrochemical source of this hidden NDR is presumably caused by the fast oxidation of Bi adatoms: $Bi_{ad}OH^- \rightarrow BiO_{ad} + 2e + H^+$. Cyclic voltammetric experiments revealed also the explicit NDR region which is presumably caused by, increasing with positive potential, adsorption of oxygen (this NDR gives rise to oscillations in the CV experiment in the presence of sufficient ohmic potential drops).

An interesting extension of these experiments was achieved again by placing 11 microprobes of the potential along the ring Pt electrode (cf. Fig. 2.39), in order to monitor the spatiotemporal evolution of the interfacial potential [59]. Figure 2.44 shows the temporal current response and a two-dimensional space–time plot of the local potential along the Pt ring electrode, for constant outer potential (U) applied. In particular, black and white zones mean that the narrow pulse domain of high activity (relatively low interfacial potential) is continuously traveling on the ring surface.

Comparison of Fig. 2.44a, c shows that the oscillations of the total current are associated with the periodical change of the size and shape of the traveling potential pulse (caused by imperfection of the electrode). This is an impressive result which shows the role of spatial inhomogeneities in the evolution of temporal dynamics.

Very recently, Christoph et al. [60] have developed theoretical and experimental studies of the oscillatory oxidation of formic acid on a thin Pt ribbon electrode which also showed the spatiotemporal self-organization. The source of the system's inhomogeneity is the enhancement of the current density near the edges of ribbon (as for all disk electrodes), so the interfacial potential drop is inhomogeneously distributed. The theory underlining the spatiotemporal phenomena involves the spatial distribution of both the faradaic and migration current in the three-dimensional space, assumed as a geometry of a ribbon electrode. The dual-path mechanism of the formic acid oxidation (cf. Chap. 5, volume I) was further incorporated in the model, i.e., the CO is treated as a surface poison which is removed by interaction with OH_{ads} formed at sufficiently positive potentials. For the detailed mathematical construction, the reader is advised to inspect the original reference. Here only exemplary results of the calculations of the spatiotemporal evolution of the interfacial potential drop are shown (Fig. 2.45).

For comparison of these theoretical predictions with the experiment, a specially designed electrode arrangement was assembled. A smooth polycrystalline Pt ribbon with a width of 4.0 mm and a length of 58.0 mm was used as the working electrode (WE). The counter electrodes (two Pt coils) were placed 80 mm above the WE. The tip of the Luggin capillary hosting a Hg/Hg_2SO_4, K_2SO_4 (saturated) reference electrode was placed at the symmetric axis above the center of the ribbon in various

Fig. 2.44 (**a**) Current profile at constant outer potential (*U*) of + 150 mV; (**b**) contour plot (two-dimensional) of the local interfacial potential during the current oscillations; and (**c**) time sequence of traveling pulse along the ring Pt electrode. *Black* and *white* represent the activating and passivating state of the Pt electrode, respectively. Electrolyte is 0.1 M HCOONa/0.033 M H_2SO_4 [59]. Reprinted from [59], Copyright 2001, with permission from Elsevier

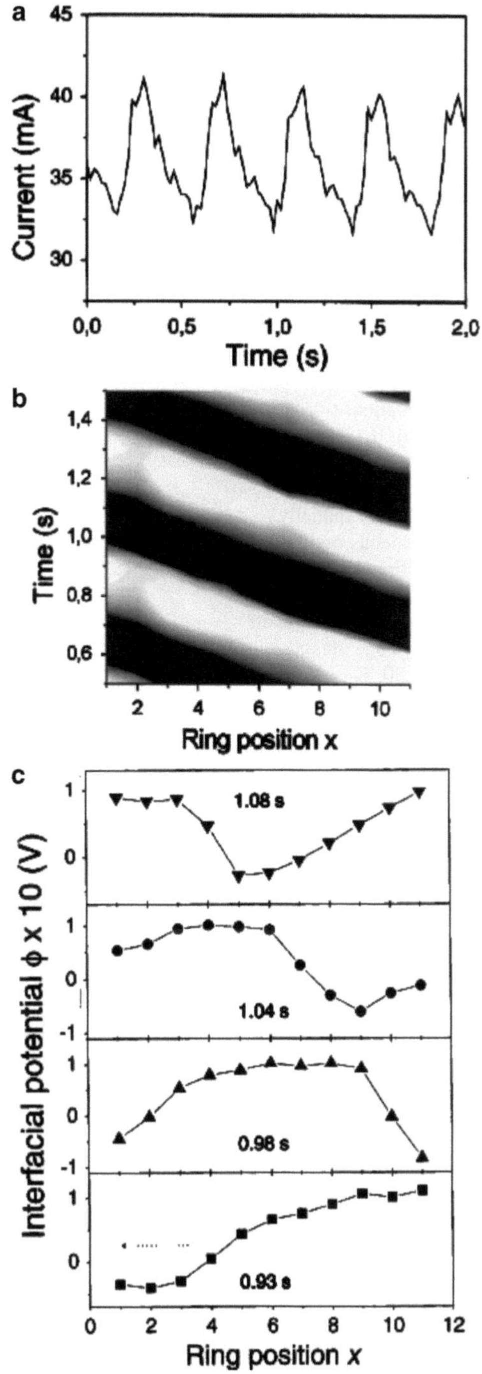

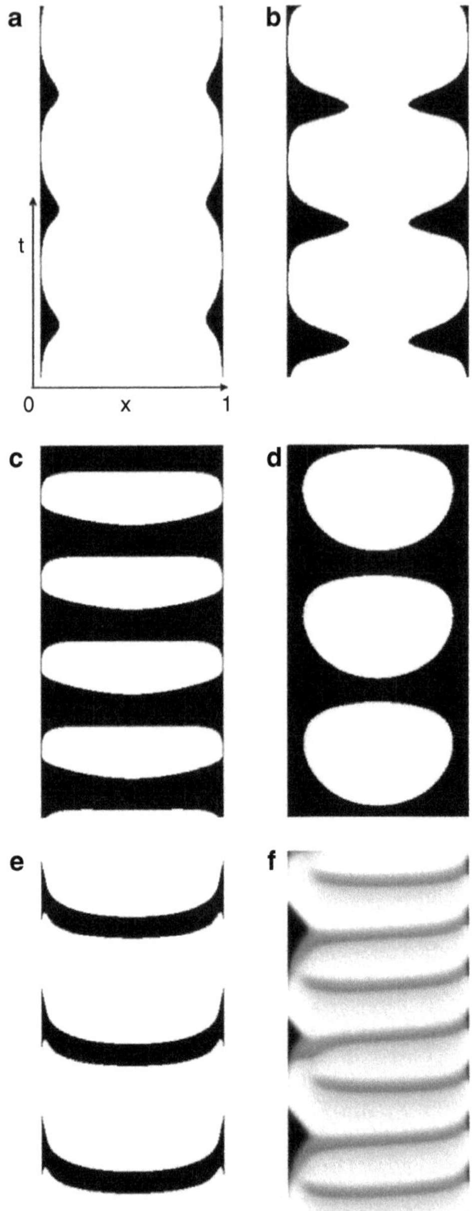

Fig. 2.45 Spatiotemporal evolution of the double layer potential on a thin ribbon electrode for a distant position of the RE. Simulations were carried out with 200 equidistant grid points using the Euler algorithm. The passive phase corresponds to the *black color regions* unless stated otherwise, external voltage E_0 is given in the units of mV, the average resistance $<\rho>$ in units of Ω cm^2. (**a**) In-phase edge oscillations ($<\rho> = 50$, $E_0 = 440$); (**b**) larger edge oscillations at increased outer potential ($<\rho> = 50$, $E_0 = 442$); (**c**) spatially inhomogeneous oscillations on the entire electrode ($<\rho> = 50$, $E_0 = 454$); (**d**) center oscillations with stationary edge areas ($<\rho> = 30$, $E_0 = 428$); (**e**) oscillations with active and passive fronts emerging from the center (target pattern with source point $x_{SP} = 0.5$ at $<\rho> = 300$, $E_0 = 640$); (**f**) coexisting asymmetric mixed oscillations. The passive state ($E \approx E_0 \equiv U$) near the (*left*) edge at $x = 0$ is depicted with black color, whereas the passive/active phase of the limit cycle is described with *gray/white color* ($<\rho> = 300$, $E_0 = 640$). See [60] for more details. Reprinted from [60], Copyright 2009, with permission from Elsevier

distances. The pulse electrode (PE), a Pt wire of 2 mm length (with remaining part sealed in glass), was placed above the WE up to ca. 1 mm. For the measurements of the time-dependent local interfacial potentials on WE, eight microprobes were placed along the ribbon electrode (at the distance of 0.2 mm). The perturbation

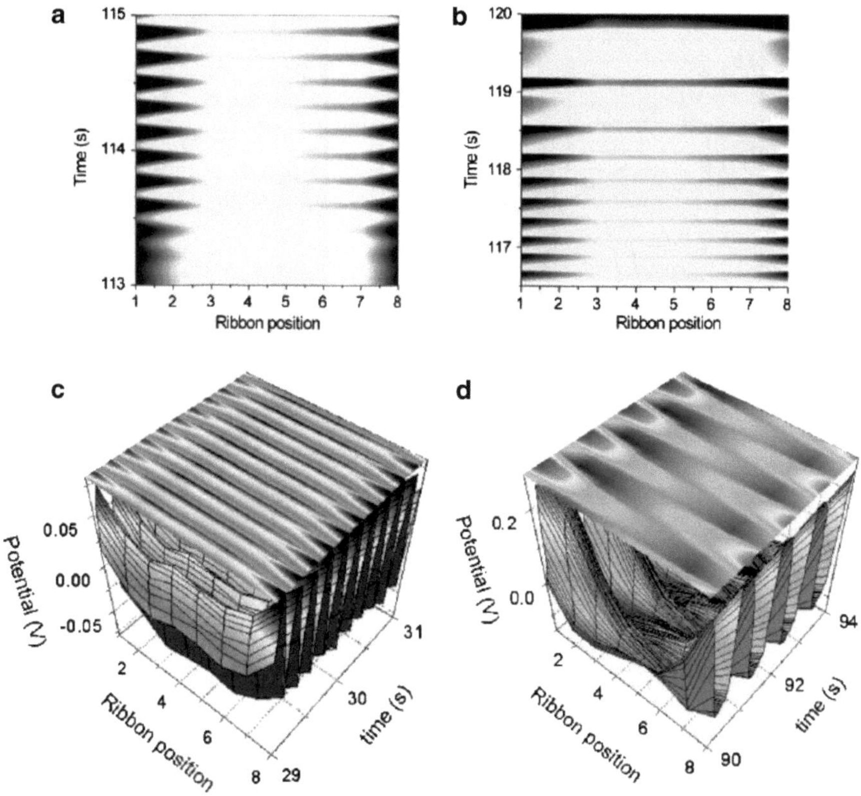

Fig. 2.46 Experimentally observed spatiotemporal distribution of the double layer potential. (**a**) Distant RE (measured by the vertical distance of RE to the WE, normalized to the WE ribbon length, $\beta = 1.5$) and low outer potential ($E_0 = 540$ mV) leading to edge oscillations; cf. Fig. 2.45a. (**b**) Distant RE ($\beta = 1.5$) and higher outer potential ($E_0 = 580$ mV) with a larger oscillatory region; cf. Fig. 2.45b; (**c**) in-phase oscillations at a decreased distance of the RE ($\beta = 0.5$); cf. Fig. 2.45c; (**d**) sufficiently small distance of the RE ($\beta = 0.3$) inducing negative coupling and therefore creating antiphase oscillations, $E_0 = 215$ mV (see [60] for more details). Reprinted from [60], Copyright 2009, with permission from Elsevier

of the steady state was achieved by applying a short pulse at appropriate position. Experimental results, shown in Fig. 2.46, were interpreted as revealing good concordance with essential theoretical predictions. Again, these modelings and experiments emphasize the necessity of consideration of the spatiotemporal inhomogeneities in the system dynamics, which effect would remain hidden, if only the variations of the average, total current or the effective electrode potential were considered. The reader interested in more detailed characteristics of the spatiotemporal self-organization on ribbon electrodes is advised to consult the original paper by Christoph et al. [60].

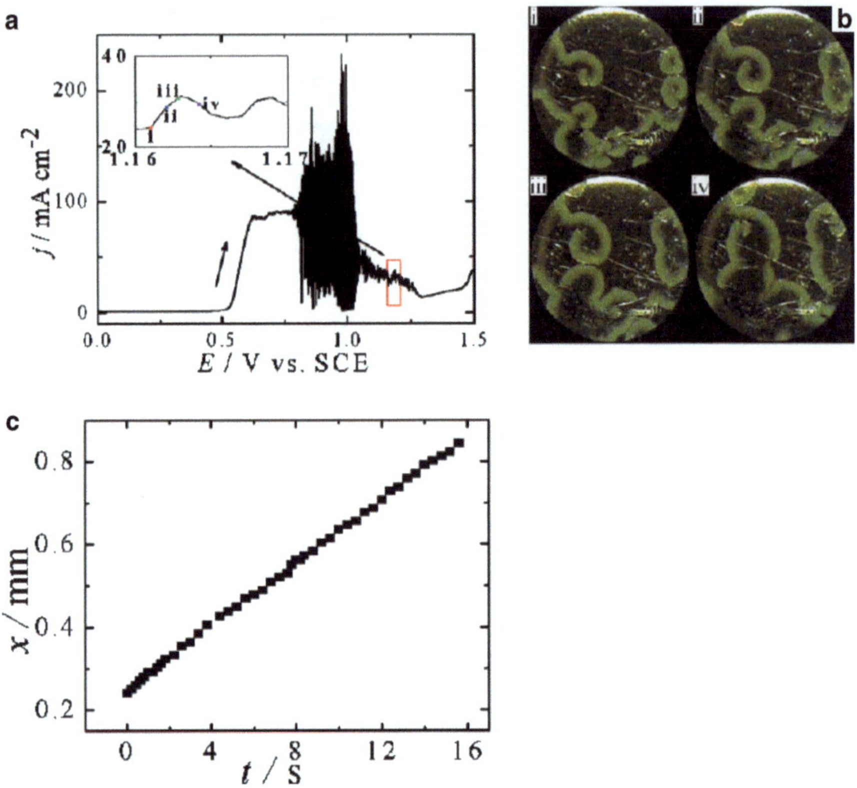

Fig. 2.47 Linear potential sweep, snapshots of the evolution of pulses, and position–time plots of pulses. The scan rate of applied voltage is 0.1 mV s^{-1}. The time interval between i and ii is 9 s, ii and iii is 6 s, iii and iv is 16 s. Reprinted with permission from [61]. Copyright 2011 American Chemical Society

2.9 Spatiotemporal Patterns in Sulfide Electrooxidation

The complex oscillatory electrooxidation of S^{2-} ions, described in Sect. 5.7 of volume I, can become a source of various spatiotemporal patterns at the Pt electrode (2 or 5 mm diameter), as Varela, Epstein et al. have recently reported [61]. The Pt wire and SCE were employed as the counter and reference electrodes, respectively, and the solution studied was 1 M Na_2S at 20°C. During the electrochemical experiment, the working electrode surface was monitored with a CCD camera. Figure 2.47 shows representative experimental results: when the Pt potential entered the N-NDR oscillatory region, pulses and spirals emerged, which corresponded, respectively, to points i–iv in Fig. 2.47a. The propagation velocity of the pulses was estimated as 0.039 mm s^{-1}. Depending on experimental conditions also fronts, labyrinths, twinkling eye-like patterns (presumably observed for the first time as the experimental, not only theoretically predicted phenomena), and homogeneous oscillations were reported. High scan rates and high series resistances favored pulses, whereas low scan rates and external resistance led to homogeneous oscillations of deposition and dissolution, with twinkling eyes and labyrinthine patterns occurring in parameter regions between synchronized oscillations and pulse waves [61].

2.10 Turing Patterns in Electrochemical Systems

One of the most recent impressive discoveries in the area of electrochemical systems is finding the conditions of the formation of stationary patterns of the Turing type. The principles of formation of such patterns due to the coupling of chemical kinetics with diffusion of reagents exhibiting various diffusion coefficients were described in Sect. 1.1.4. In terms of the concept of the activator–inhibitor dynamical system, the Turing patterns emerge when the inhibitor, the production of which is accelerated by the activator, diffuses faster than the activator and in this way the spatial progress of the activator becomes hindered. In electrochemical systems, as described in Sect. 1.2.3.4, such patterns can emerge as a consequence of the interaction between the electrochemical process with S-NDR characteristics and the long-range, migration coupling, as Mazouz and Krischer have predicted only in the year 2000 [62]. The principle of the concept was that the condition of different rate transport of the activator and the inhibitor is particularly easy to realize in electrochemical systems, when besides diffusion, the migration (usually much faster transport) is present. In S-NDR systems the electrode potential is typically a fast inhibitor, while the chemical species involved in the autocatalytic loop is an activator which is transported by diffusion.

At this point it is useful to realize that stationary character of the patterns observed in the experiment does not mean that the Turing patterns were formed, so not every stationary pattern is a Turing pattern. For example, stationary patterns observed during the reduction of $S_2O_8^{2-}$, which is not the S-NDR, but the N-NDR system (cf. Sect. 2.2) [24] cannot be classified as Turing ones. As described earlier, in the case of peroxodisulfate reduction, the formation of two spatial domains on the working electrode was due to the negative global coupling, caused by the external circuit, and not only by the migrations of ions in the electrolyte which is one of the major factors deciding the emergence of Turing patterns.

Coming back to Turing-like patterns, theoretical predictions of the conditions of their occurrence in electrochemical systems were soon followed by their experimental verification, reported in 2001 by the same group of researchers [63]. For the experimental detection of such patterns, the surface plasmon microscopy was applied using methodology described in [1] and in the setup shown in Fig. 2.48.

The periodate reduction on Au(111) electrodes in the presence of camphor as an adsorbing inhibitor was chosen as an electrochemical system. Camphor adsorbed on Au(111) surface exhibits two first-order phase transitions, one around +0.1 V and the other one around −0.75 V (vs. Hg–Hg_2SO_4 reference electrode). In the potential region from −0.3 to −0.7 V, camphor forms a condensed film on the electrode surface, causing the strong inhibition of the reduction of periodate ions, while at more negative potentials the electrode coverage is very low. Upon

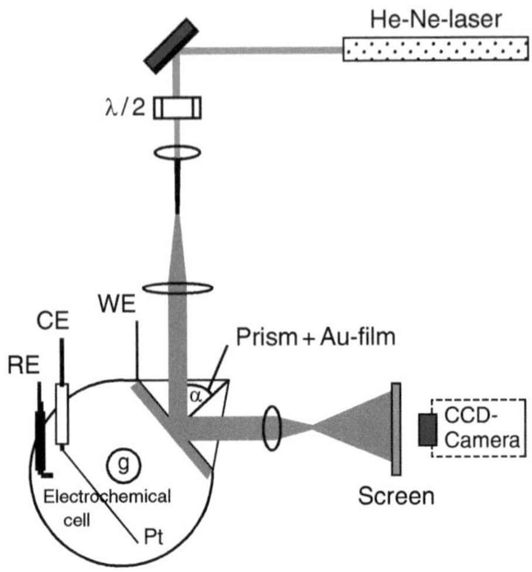

Fig. 2.48 Experimental setup for imaging patterns at the electrode/electrolyte interface by means of surface plasmon microscopy (*top view*). WE: working electrode [a ~50-nm-thick Au film evaporated onto a glass plate (B270) that was brought into optical contact with a glass prism; the area of the WE exposed to the electrolyte was 8 mm by 30 mm]. CE: counter electrode (a Pt wire, wound to a spiral of the same size as the working electrode and placed parallel to it at a distance of ~4 cm). RE: Hg/Hg_2SO_4 reference electrode. (All voltages are given with respect to this reference electrode). G: gas inlet for Ar bubbling. (The mass transport was achieved by means of gas bubbling). The electrode was irradiated from behind by a broadened, p-polarized He–Ne laser beam and imaged with a charge-coupled device (CCD) camera. Before each experiment, the film was flame annealed several times in a gentle butane flame, resulting in a film with a high degree of (111) facets, as revealed by electrochemical characterization. From [63]. Reprinted with permission of AAAS

cyclic variation of the electrode potential in the regions coverings either the first or the second phase transition, one observes the hysteresis in cyclic voltammetry of periodate ions. According to the general characteristics of the S-NDR systems, the electrode coverage with camphor is here an activator, while the electrode potential is an inhibitor in this activator–inhibitor dynamical system. Experimentally it is manifested in the following way: upon the shift of the electrode potential from negative towards positive values, an increase of camphor coverage occurs until the critical value, at which sharp first-order transition to a dense camphor film takes place, due to attracting lateral interactions between the adsorbate particles. This causes the inhibition of periodate reduction, i.e., the decrease in its faradaic current which in turn, under potentiostatic conditions and *in the presence of uncompensated serial resistance*, makes the electrode potential more negative. At such potential camphor is desorbed, so the negative feedback loop is created: the autocatalytic growth of camphor film causes, through the changes of the electrode potential, the decrease in surface camphor concentration.

For this experimental system, *stationary patterns* were detected on the electrode surface, using surface plasmon microscopy for such diluted $NaClO_4$ solution as a supporting electrolyte that the ohmic potential drop exceeded the width of the bistable region on the *I*–*E* characteristics of periodates. Then, this dependence took a quasi-linear shape of a positive slope (cf. concordance with the theoretical predictions of Turing patterns in Sect. 1.2.3.4 and Fig. 2.10, vol. I). In other words, patterns were observed indeed under conditions when the S-shaped *I–E* region has deformed into the single-valued dependence. Representative examples of such "spot-like" patterns are shown in Fig. 2.49. It is important to note that these patterns remained unchanged for the fixed external potential, hence they were stationary and stable.

By varying the voltage and specific conductivity of the solution, the diagram of occurrence of those patterns was constructed, reasonably well concordant with theoretical calculations, made according to the model, outlined in Sect. 1.2.3.4.

It is here useful to remember that one of the essential features of the Turing patterns is the independence of their wavelength of the size of the system[1], but the dependence only on the kinetic and diffusion parameters of the system. In the opinion of the authors of [63], the formation of such Turing patterns can find potential applications in manufacturing structured electrodes (like those developed for fabrication of (bio)sensors). In such procedure, one could adjust the wavelength of the designed patterns by choosing different faradaic reactions, adsorbates, or modifying the temperature (the faster the faradaic reaction and the dynamics of the adsorbate system, the smaller are the patterns). If successful, such an approach would be an example of application of nonlinear dynamics in the materials science. Also, of course, recognizing the mechanisms of formation of Turing patterns in electrochemical systems can deepen our understanding of pattern formation in biological systems, in which the gradient of electric potential exist. This problem was not existing in the description of Turing patterns for purely chemical systems without migration transport explicitly considered.

The next example of Turing-like patterns in electrochemical systems was found in 2003, also by the group of Krischer [64, 65], for the CO electrooxidation on Pt which also exhibits the S-shaped stationary *I–E* dependence under potential control. As described in Sect. 5.2 of volume I, in this case the S-NDR characteristics were explained in terms of an autocatalytic reaction cycle, caused by the competitive Langmuir–Hinshelwood mechanism between surface-bonded CO and OH species. The experimental setup [65] involved a rotating ring polycrystalline Pt electrode ($\omega = 1{,}200$ rpm), of a 1 mm width and 85 mm mean circumference, embedded in cylindrical Teflon pieces. The local electric potential in the electrolyte close to WE was recorded along the angular direction by means of a stationary microprobe. The WE rotated over the microprobe, which allowed the temporal evolution of the angular potential distribution in front of the Pt ring to be obtained in situ. Experiments were performed for low-conductivity electrolytes, e.g., 2 mM perchloric acid [64, 65] which ensured relatively high ohmic drops and appropriate deformation of the *I–U* dependence, being in certain range even quasi-linear, as controlled predominantly by the solution resistance. In this way, the conditions for the formation of spatial Turing patterns were achieved. In fact, at $U = 0.2$ V (vs. $Hg–Hg_2SO_4$ electrode) when the rate of electrooxidation of CO attained significant

[1] An extension of the Turing model for the case when the system's size affects the pattern formation, was described by Murray [9].

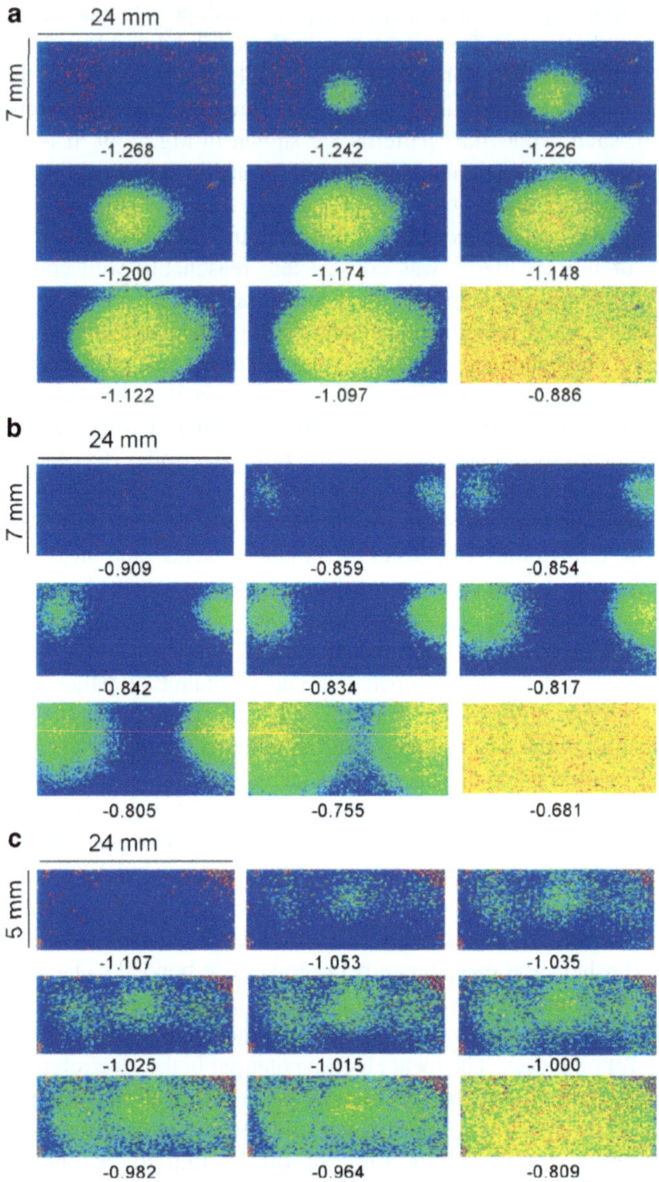

Fig. 2.49 Patterns emerging in different electrolyte compositions: (**a**) 5 mM camphor, 0.5 mM $NaClO_4$, and 0.5 mM $NaIO_4$; (**b**) 5 mM camphor, 1 mM $NaClO_4$, and 1 mM $NaIO_4$; and (**c**) 5 mM camphor, 0 mM $NaClO_4$, and 1 mM $NaIO_4$. In all three cases, the corresponding *I*–*U* curve was single valued. The numbers given below the images are the potential values at which the images were obtained during a potential scan. When holding the potential at those values, the patterns did not change with time, i.e., under these conditions, stationary patterns existed on the electrode. *Blue colors* indicate a nearly camphor-free electrode; orange and *yellow colors* indicate a high camphor coverage (The measured intensities are normalized at each point). From [63]. Reprinted with permission of AAAS

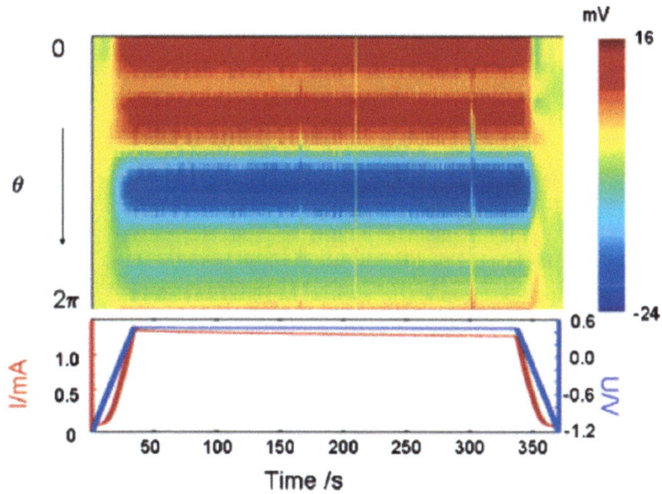

Fig. 2.50 *Top*: Position–time plot of the inhomogeneous part of the potential probe signal during bulk CO electrooxidation in 2 mM KOH. *Bottom*: Corresponding applied voltage (*blue line*) and total current (*red line*) vs. time. Rotation speed of the electrode $\omega = 1{,}200$ rpm. Reprinted with permission from [65]. Copyright 2005 American Chemical Society

values, stationary potential patterns formed spontaneously. Both small (ca. 15 mV peak to peak) and large amplitude (up to 200 mV) potential patterns were found depending on the applied voltage. It was also possible to model these patterns in terms of spatially extended mechanism for CO electrooxidation proposed by Koper et al. [49] and outlined in Sect. 5.2 of volume I. These phenomena were studied more thoroughly for both diluted acidic and basic supporting electrolytes [65]. Figure 2.50 shows representative experimental result, whereas Fig. 2.51 shows successful modeling of relevant phenomena. In the model, based again on achievements by Koper et al. [49] and Strasser et al. [66], the diffusion transport of CO was implemented in a way involving two diffusion coefficients: D_1 for the transport from the bulk solution and D_2 for its radial diffusion at the electrode, with $D_2/D_1 = 100$:

$$\frac{\partial c_s}{\partial t} = -\frac{2S_{tot}}{\delta} v_{CO,ads} + \frac{2D_1}{\delta^2}(c_b - c_s) + D_2 \frac{\partial^2 c_s}{\partial x^2} \tag{2.11}$$

where c_s—surface concentration of CO on the solution side, c_b—bulk concentration of CO, $v_{CO,\,ads}$—the rate of CO adsorption, S_{tot}—maximum surface concentration of adsorption sites (2.2×10^{-9} mol cm^{-2}), δ—thickness of the diffusion layer, $D_1 = 5 \times 10^{-5}$ cm^2 s^{-1} and $D_2 = 5 \times 10^{-3}$ cm^2 s^{-1}; coordinate x is parallel to the surface of the working electrode, while the concentration gradient in the solution developed along the perpendicular z coordinate. The other three partial differential equations describe the surface dynamics of the interfacial potential drop $\phi_{dl}(x, t)$, the electrode coverage $\theta_{CO}(x, t)$, and the electrode coverage $\theta_{OH}(x, t)$, respectively. Further details of the model the reader can find in original reference [65].

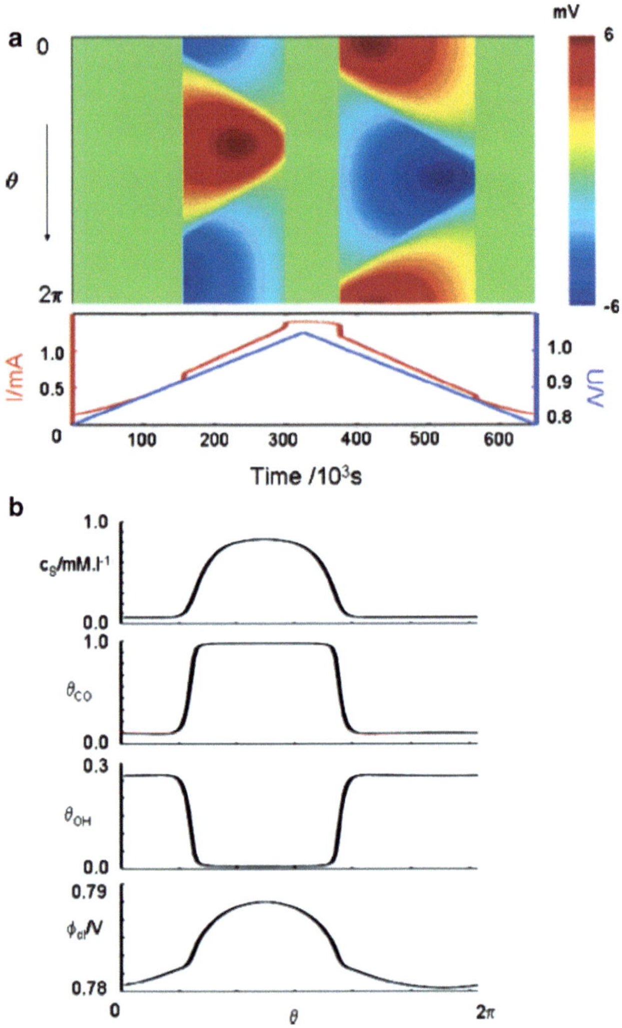

Fig. 2.51 (**a**) *Top*: Position–time plot of the inhomogeneous part of the calculated stationary electrode potential, ϕ_{dl}, during a cyclic change in U. $\delta = 0.0067$ cm (diffusion layer thickness). See [65] for remaining parameters. *Bottom*: corresponding total current in *red* and of U in *blue*. (**b**) Calculated spatial distribution of the different variables: $c_s(\theta)$, $\phi_{\mathrm{dl}}(\theta)$, $\theta_{\mathrm{OH}}(\theta)$ and $\theta_{\mathrm{CO}}(\theta)$ and plotted for $U = 0.975$ V. Reprinted with permission from [65]. Copyright 2005 American Chemical Society

2.11 Dendritic Patterns in Metal Electrodeposition

2.11.1 Dendritic Deposition on Solid Surfaces

Studies of electrodeposition of metals on different surfaces are obviously an important problem in electrochemical industry, and finding the factors determining the morphology of the deposits is of primary importance. Here we shall focus on the self-organized patterns of deposits, which can take, among others, the fractal shape. Such patterns can be important also for understanding of biological morphogenesis, as they resemble some structures observed in a living nature. Since they develop under nonequilibrium conditions, i.e., in a mass diffusion field, a thermal diffusion field, or an electrostatic potential field, etc., generally described in terms of Laplace equations, the phenomena described below can be understood as electrochemical model of morphogenesis in the Laplacian field [67, 68]. With respect to biological systems, the growth of bacterial colonies in Petri dish is a representative example of such patterns [69, 70].

The formation of fractal- or dendrite-like patterns of electrodeposited metals is a well-known phenomenon. Systematic studies of them, reported for the electrochemical cells of radial geometry, have been published, among others, by Sawada et al. [71] and Grier et al. [72]. More recently, Kuhn and Argoul [73] have described, both experimentally and theoretically (in terms of the diffusion-limited aggregation approach, see also below), the ramified silver aggregates produced in thin-gap cells using a redox reaction between metallic copper and Ag^+ ions in solution. In turn, Trigueros et al. [74] have presented a study of dendritic patterns formed during the zinc electrodeposition for parallel arrangement of flat electrodes. Depending on experimental conditions, e.g., voltage applied, electrolyte concentration, solution thickness, electrode geometry, etc., different morphologies of electrodeposits were obtained, ranging from largely disordered fractals to well-ordered anisotropic dendritic structures. These results will be briefly summarized here.

The thin-layer electrolytic cell consisted of the electrolyte (Zn^{2+} salt) solution sandwiched between two 5 mm thick glass plates, and confined between the cathode and anode, being two parallel copper wires 3.5 cm long, separated by a distance of 3 cm. The growth of zinc electrodeposits occurred after applying an appropriate constant voltage between these electrodes. The deposited patterns were recorded through a stereo zoom microscope connected to a standard video camera and transferred to a PC-based image analysis system. The digitization was performed by means of real-time image digitizer. Figures 2.52–2.54 show exemplary patterns of deposited Zn: the so-called homogeneous ones, meaning the tree-like, densely branched deposits with a well-defined front and more complex, dendritic patterns.

Systematic studies of these patterns allowed to elaborate the morphology diagram shown in Fig. 2.55. Also mixtures of those patterns and transitions between them, during prolonged electrolysis, were reported. The review of all possible patterns of electrodeposited zinc includes its following types [74]: (a) compact deposits obtained at low potentials; (b) ramified structures with a well-defined outer front, called above as "homogeneous" ones, and formed at low Zn^{2+} concentrations over a wide range of applied potentials; (c) dendrites, formed at higher Zn^{2+} concentrations; and either (d) open ramified, fractal-like deposits, or (e) mixed patterns with characteristics of dendritic backbones and disorderly ramified side branches, both observed at the highest Zn^{2+} concentrations. Calculations of the

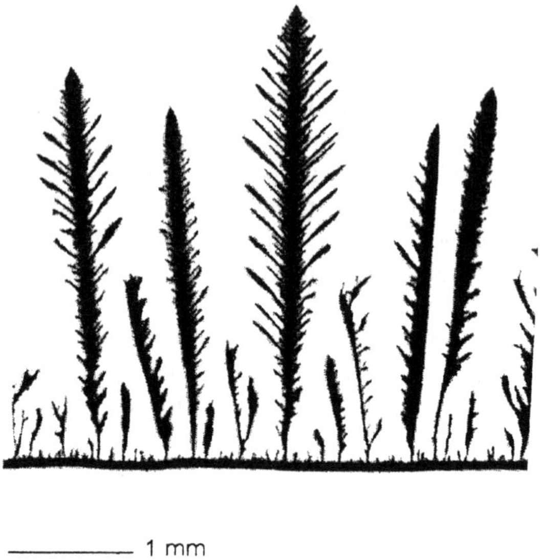

Fig. 2.52 Dendritic pattern observed at $[Zn^{2+}] = 0.07$ M and $\Delta V = 9.0$ V; growth time $t = 5$ min. Reprinted from [74], Copyright 1991, with permission from Elsevier

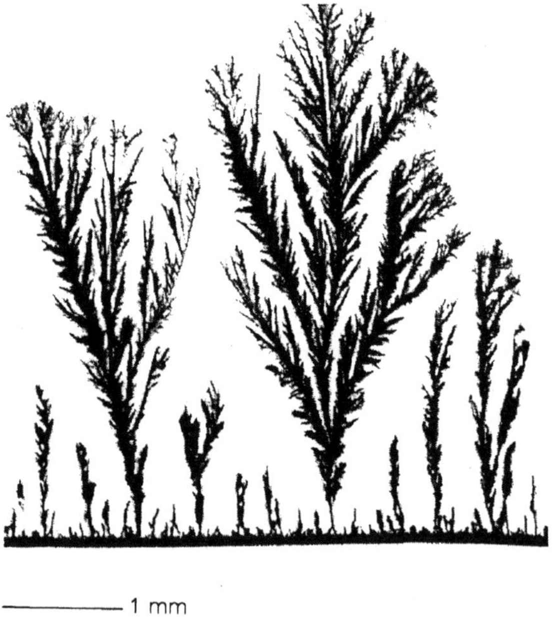

Fig. 2.53 Self-similar, open ramified pattern observed at $[Zn^{2+}] = 0.4$ M and $\Delta V = 6.0$ V; growth time $t = 20$ min. Reprinted from [74], Copyright 1991, with permission from Elsevier

fractal dimensions for various patterns yielded the (typical for this phenomena) average value $D_q = 1.61 \pm 0.02$.

Deposited patterns of such morphology are typically numerically modeled in terms of the DLA (*Diffusion Limited Aggregation*) formalism based on the

Fig. 2.54 Mixed pattern observed at $[Zn^{2+}] = 0.4$ M and $\Delta V = 10$ V; growth time $t = 10$ min. Reprinted from [74], Copyright 1991, with permission from Elsevier

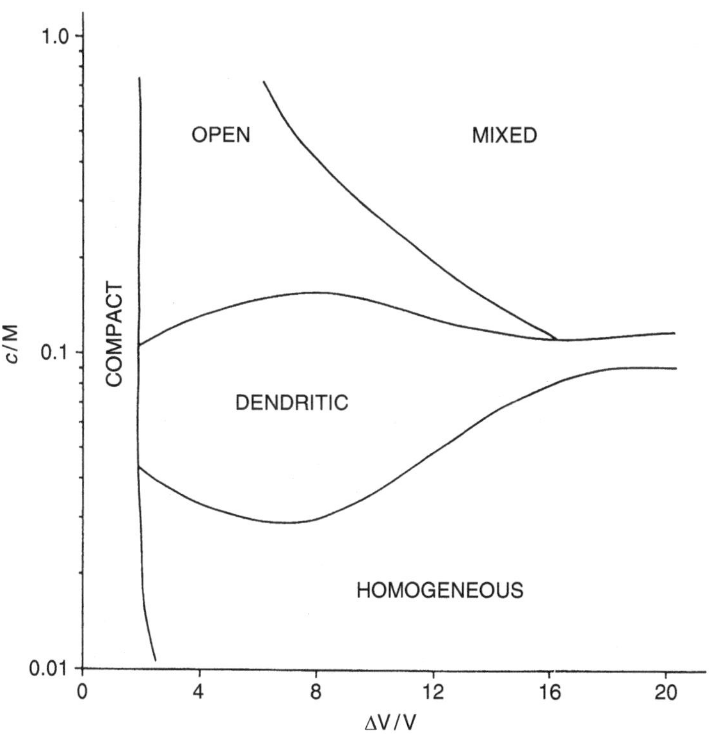

Fig. 2.55 Morphological diagram for the various patterns obtained in the zinc electrodeposition under appropriate experimental conditions. Reprinted from [74], Copyright 1991, with permission from Elsevier

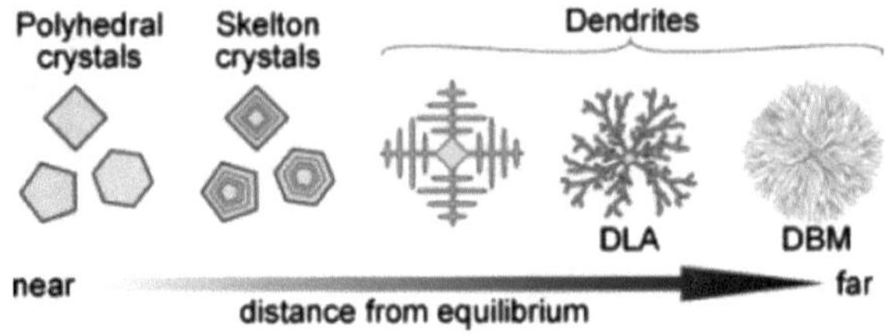

Fig. 2.56 Schematic illustration of a correlation between the distance of the formation conditions from the equilibrium and the morphologies of formed crystals. Reprinted with permission from [83] Copyright 2007 American Chemical Society

following two assumptions. The first one postulates the random walk of ions on the two-dimensional plane, while the second one claims that at every collision between the walking ion and the electrode, the ion is immediately reduced and stays there in a form of neutral atom, composing the metal phase (cf. e.g., [71, 75–77].

One can pose a question, whether it is possible to report synchronization of the dendrite electrodeposition with the oscillatory course of the electrode process. Recently, Nakanishi et al. have reported that the electrodeposition of some metals under diffusion-limited conditions gives rise to nonlinear electrochemical oscillations, accompanied by formation of dendrites with ordered microstructures, which occur in synchronization with cycles of the oscillations [78–81]. In addition, the spatial period (size) of the ordered microstructures could be tuned by changing the oscillation amplitude and/or period, through variation in the current density or the concentration of the electroactive species [82]. These results led later to the elaboration of a general mechanism for such synchronization, causing the formation of ordered two- and three-dimensional microstructures [83].

As an introduction to such considerations it is useful to realize a more general, typical assumption that the morphology of crystals strongly depends on the "thermodynamic distance" between the actual conditions of crystallization and the thermodynamic equilibrium [84, 85]. Schematically this dependence is shown in Fig. 2.56 which also indicates that the dendrites are formed at extremely nonequilibrium conditions, and the growth of such patterns occurs via diffusion-limited aggregation (DLA) or dense branching morphologies (DBM). This sequence is explained by the fact that with increasing driving force for crystallization, the growing fronts of crystals with flat surfaces become destabilized due to an increasing contribution of mass and heat diffusion; the increasing surface instability leads then to formation of dendrites [83].

Experimental conditions of formation of such patterns involved both potentiostatic and galvanostatic conditions. For the *potentiostatic* mode of operation, representative examples of experimentally obtained three different types of deposits are shown in Fig. 2.57 (where OM stands for optical digital microscopy and SEM—for a high-resolution scanning electron microscope as the methods of obtaining the

Fig. 2.57 Dendrites of (**a**) tin (Sn), (**b**) zinc (Zn), and (**c**) copper (Cu), formed in the electrodeposition under the diffusion-limited conditions, where (**a**) is an OM image and (**b**) and (**c**) are SEM images. The electrolyte: (**a**) 0.2 M Sn(II) + 4.0 M NaOH, (**b**) 0.2 M Zn(II) + 4.0 M NaOH, and (**c**) 0.6 M $CuSO_4$ + 3.0 M lactic acid + a small amount of NaOH to adjust the pH to 9.0. The applied potential: (**a**) −1.7 V, (**b**) −1.9 V, and (**c**) −1.5 V vs. Ag–AgCl. Reprinted with permission from [83] Copyright 2007 American Chemical Society

images). Type I (needle-like) pattern can be qualified as one-dimensional patterns, type II (wafer-like) as two-dimensional patterns, and type III (round-particle like) as three-dimensional patterns (note the different in the characteristic sizes).

In the explanation of the growth of such patterns under diffusion-limited conditions, one should realize its *autocatalytic nature*, caused by the faster spherical diffusion of ions near the peaked parts of the substrate, caused in turn by steeper concentration gradient, compared to flat regions of the electrode (Fig. 2.58). As a consequence, peak parts grow autocatalytically more and more, which effect means, in other words, the destabilization of the flat surface.

In turn, under *galvanostatic* conditions, formation of dendritic patterns can occur also if the imposed current exceeds the diffusion-limited value and is often accompanied with the oscillations of the electrode potential. Furthermore, patterns occurring in this oscillatory regime often exhibit periodic spatial structure, indicating some kind of synchronization with those oscillations. Then one stage of the lattice is produced by one cycle of the oscillation. Figure 2.59 shows representative examples of such correlation between the galvanostatic oscillations of the electrode potential and simultaneous formation of patterns on the electrode surface [83].

These results suggest the existence of essentially common mechanism of synchronization for various metals, which was further studied using also in situ phase-contrast optical microscopic (POM) measurements of the electrode–electrolyte interface and electron backscattering diffraction (EBSD) analysis of the deposit structure. The essential construction of the proposed mechanism is briefly described below, taking Sn deposition as the process considered in detail.

The autocatalytic needle growth was observed to occur at the diffusion-limited conditions. Due to this pattern growth, the effective electrode area (roughness)

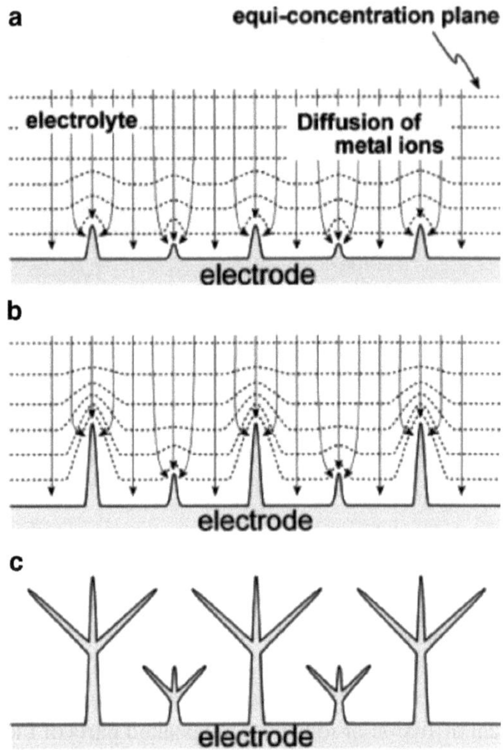

Fig. 2.58 Schematic illustration of autocatalytic crystal growth producing dendrites under diffusion-limited conditions. (**a**) Spherical diffusion layers for deposited metal ions are formed near the peaked parts on the substrate, where the concentration gradient is thus steeper than in the electrolyte near flat parts; (**b**) in consequence, much faster diffusion of the metal ions to the peaked parts results in much faster (autocatalytic) growth of those parts; (**c**) the sharp crystal growth, causing roughening of the initial substrate surface, is accompanied with bifurcations in crystallographically equivalent directions, which leads to the formation of dendrites. Reprinted with permission from [83] Copyright 2007 American Chemical Society

increases, and therefore the electrode potential shifts to more positive values, in order to maintain the imposed fixed (total) current. This positive potential shift continues until the conditions become no longer diffusion controlled. Then the autocatalytic growth of needles is stopped and, instead of needles, the Sn cuboid crystals are formed, surrounded with the thermodynamically stable (100) and (010) faces. The reader interested in further details of this mechanism is advised to consult [83].

The explanation of the mechanism underlying the return of the electrode potential back to negative values is the following. Based on inspection of video movies showing the growth of Sn latticework and numerical simulations, involving finite difference model, it was suggested that the negative potential shift is caused by the exhaustion of Sn(II) ions in the electrolyte, near (or a little inside) the tip of the latticework. This should cause the decrease of the current inside the latticework, the effect equivalent to the decrease in the effective surface area which causes the negative potential shift, in order to maintain the fixed current. The extension of the above discussion on other types of deposits the reader can find in an original reference [83]. One should emphasize that the understanding of the mechanism of

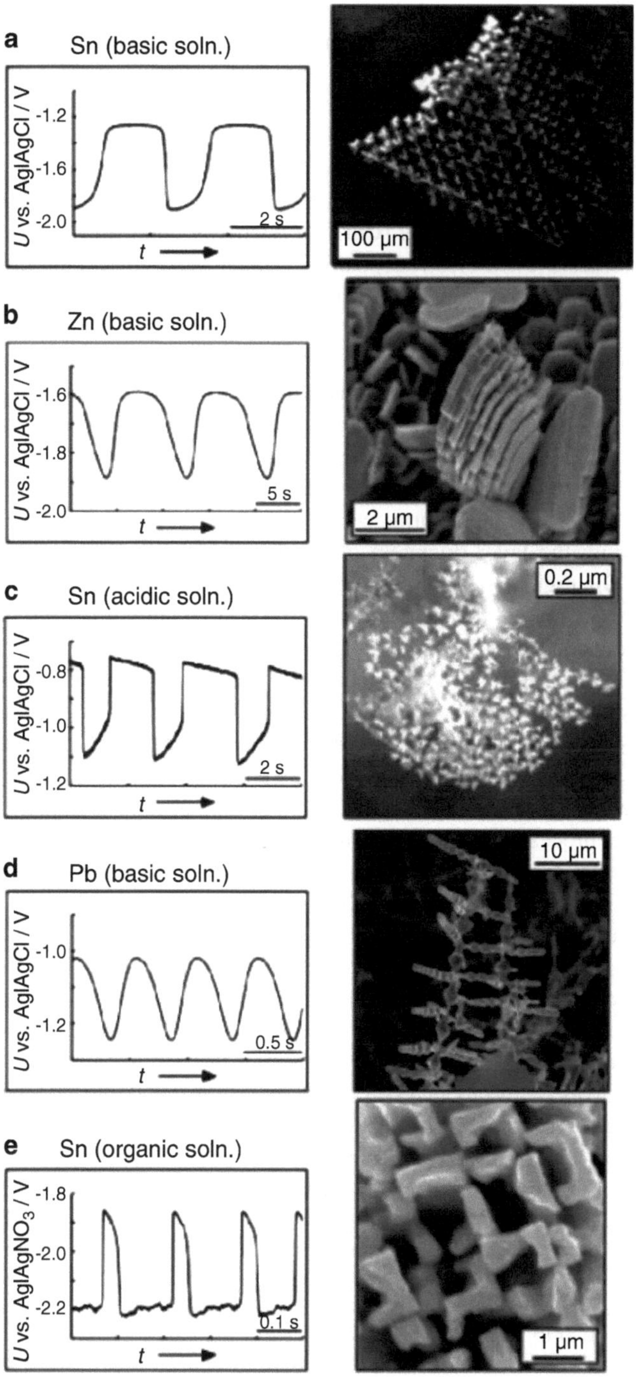

Fig. 2.59 Examples of (*left*) potential oscillations under the constant-current conditions and (*right*) dendrites formed during the oscillations. The electrolyte and the applied current density:

formation of such deposits and the control of their shape and characteristic size should have important potential applications in the materials science for the preparation of designed and controlled micro- and nanostructures at solid surfaces.

Similar studies of pattern formation were conducted also for other metals (including their alloys). Nakanishi et al. [86] have studied the layer-by-layer electrodeposition of copper at a polycrystalline Au electrode, in the presence of *o*-phenanthroline (*o*-phen). This system exhibits the NDR region caused by the adsorption of the reduced form of $[Cu(II)(o\text{-phen})_2]^{2+}$ complex, $[Cu(I)(o\text{-phen})_2]^{+}$, which suppresses the Cu electrodeposition. Under such conditions the electrodeposition exhibits both potential and current oscillations, suggesting that the system can be classified as the hidden NDR (HN-NDR) oscillator. Simultaneously with these oscillations, the color and surface morphology of Cu deposits varied synchronically. Microscopic investigations allowed to correlate the morphology of deposits with the stage of the oscillation cycle: in the positive side of the potential oscillations (high-current state of the current oscillation) the dense round Cu leaflets, which look gray, grow, while in the opposite-side stages of the oscillations, thin Cu leaflets form, which sequence of events produces a layered Cu deposit with the layer thickness of ca. 5 μm. One should note that a way in which the NDR region is partly hidden is very specific: due to growth of thin Cu leaflets there increases the effective electrode area and this causes a current increase that masks the NDR. Since this mechanism of hiding the NDR is different from the previously proposed ones (see Sect. 3.4, volume I), the authors consider the system studied as the *new type* of the HN-NDR oscillator. In others words, the former classification of oscillators, which assumed implicitly that the electrode surface is geometrically uniform, is now extended for the systems in which the effective electrode surface area, considered a key variable, increases. A detailed description of processes composing this mechanism is given in [86].

As another example of the dendritic growth one can invoke deposition of metallic lithium in symmetrical lithium/polymer/lithium cells, as Chazalviel et al. have reported [87]. In this case, however, the formation of dendrites is a highly undesirable phenomenon, since it affects the charge efficiency of the lithium anode. Thus, the understanding of this process, attributed to the existence and/or formation of local inhomogeneities on the electrode surface, served here as a way to determine the conditions, under which the dendrites are *not* formed. In more recent work, Chazalviel et al. [88] have presented the experimental and theoretical study of the onset of the growth of copper deposit from $CuSO_4$ solution. It was found that, in order to avoid (as above) irregularity in the deposit morphology (e.g., formation of the dendrites), the vertical, thin-layer cell, with the

---

Fig. 2.59 (continued) (**a**) 0.2 M Sn(II) + 4.0 M NaOH, -36 mA cm^{-2}; (**b**) 0.2 M Zn(II) + 4.0 M NaOH, -21 mA cm^{-2}; (**c**) 0.3 M $SnSO_4$ + 0.3 M H_2SO_4 + 1.0 g L^{-1} gelatin, scan rate 1.67 mA s^{-1}; (**d**) 0.15 M $PbSO_4$ + 4.0 NaOH, -14 mA cm^{-2}; (**e**) 0.1 M $SnCl_2$ + 0.1 M $NaClO_4$, -7.2 mA cm^{-2}. The solvent for the electrolytes was water, except that dimethylformamide is used in (**e**). Reprinted with permission from [83] Copyright 2007 American Chemical Society cathode on top, should be used, since in this configuration the stability of the electrochemical system is enhanced. However, when the imposed current was increasing, close to the limiting current density, the cell voltage exhibited oscillatory variations of rather unclear origin. In his theoretical work, Chazalviel [89] has shown earlier that the growth of ramified metallic deposits by electrodeposition from dilute salt solutions and in a high electric field is a direct consequence of the creation, under such conditions, of a space charge upon anion depletion in the vicinity of the cathode. The front of the ramified deposit was predicted to advance at a speed just equal to the velocity of the anions in the applied electric field: $v_a = -\mu_a E_0$, meaning the product of the ionic mobility and the electric field intensity, respectively. Of course, such dilute electrolyte solutions are not often encountered in electrochemical practice [for examples related to nonlinear dynamics, cf. Sects. 4.1 (volume I), 2.10, and 5.9.4].

Concerning further the alloys, Nakanishi et al. [90] have studied the electrodeposition from an acidic solution containing Cu^{2+}, Sn^{2+}, and a cationic surfactant. The *I–E* characteristic of such system exhibited negative differential resistance, caused by the potential-dependent adsorption of the surfactant (acting as an inhibitor for diffusion of metal ions) on the alloy surface. Based on scanning Auger microscopic measurements it was found that alloy films deposited during the oscillations had an alternate multilayer structure composed of two alloy layers of different compositions. The multilayer had the period of thickness of 40–90 nm and was uniform over an area of ca. 1 × 1 mm. In another work, the formation of nano-scale layered structures in induced co-deposition of unique, widely used in industries, Ni–P, Ni–W, and Co–W alloys was investigated using an in situ electrochemical quartz crystal microbalance technique [91]. Also in this case the negative differential resistance and its crucial role in the onset of oscillations, and in consequence in the layer-structure formation, were found. Strictly speaking, the NDR region could overlap with the hydrogen evolution current and then the oscillator could be qualified as the hidden (HN-NDR) one. In this case, the important role is played by such electrolyte components as $H_2PO_2^-$ and WO_4^{2-} which adsorb on the electrode and act as a promoter for the co-deposition reaction; also the NDR arises from desorption of adsorbed promoter, when the potential moves into negative direction. In the opinion of the authors, the detailed scheme of processes elaborated by them constitutes the general mechanism for the induced co-deposition of some iron-group alloys. Furthermore, understanding of such processes should allow to control the chemical composition, the size of the layered structures, and various properties of the alloys for high performances in industrial applications [91].

Recently, Ihara et al. [92] have described a large interfacial energy gradient produced at a front of electrochemical wave appearing during the CuSn alloy oscillatory electrodeposition from an aqueous solution. This was observed as the directional lateral motion of an oil (nitrobenzene) droplet put on an electrode surface. During the oscillation, the surface composition of the electrodeposit changed periodically between Cu- and Sn-rich alloys, and the transition between these phases initiated the propagation of the electrochemical wave over the entire

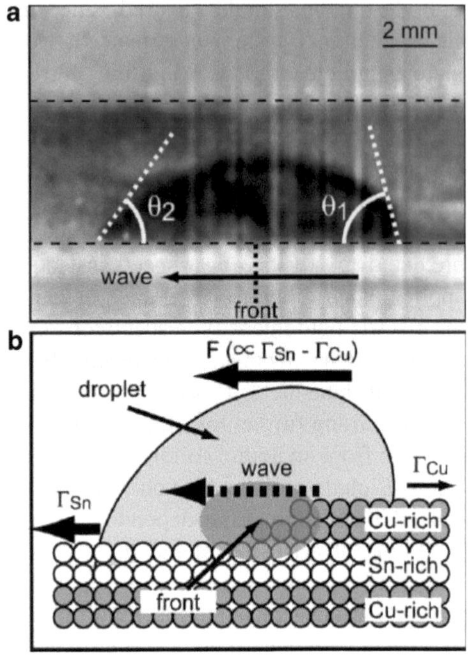

Fig. 2.60 (**a**) Picture taken at the moment when the wave just reaches to the right edge of the droplet. (**b**) Schematic illustration of the mechanism of the droplet motion. Reprinted from [92], Copyright 2009, with permission from Elsevier

electrode surface. The fact that the force produced is able to transport such macroscopic object, like the oil droplet, proves the existence of extremely large imbalance of the interfacial energy (interfacial tension)—see Fig. 2.60. In this way, the reaction–diffusion (migration) system exhibits coupling with convection.

2.11.2 *Dendritic Deposition on Liquid/Liquid Interface*

In recent studies, the traditional solid/liquid interface was replaced by liquid/liquid interface. For example, Efrima et al. [93–96] have studied the patterns of silver electrodeposited at the water–air and water–organic liquid interfaces, under both dc and ac conditions, showing e.g., the formation of fractal "metal leaves" (Fig. 2.61) [93]. In turn, Fig. 2.62 shows reported shapes of Ag deposit at the water/air interface [96].

The morphology of patterns was discussed in terms of the Wagner number which in its original form is defined as $W = R_f/R_s$ [97], where R_f is the faradaic (charge-transfer) resistance of the electrode process and R_s is the ohmic resistance of the solution across a characteristic distance in the electrochemical cell. The small Wagner number means that the morphology of the deposit is controlled by the ohmic resistance of the solution, and then the instabilities set in, causing that the deposits tend to be ramified. In contrast, when the Wagner number is large, the

Fig. 2.61 Typical deposits at 5 V for different solvents with a water phase containing ammoniacal 0.05 M $AgNO_3$, 0.1% anisic acid, and 0.03% FC143 surfactant, for (**a**, *top left*) air, (**b**, *bottom left*) carbon tetrachloride, (**c**, *top right*) dichloromethane, and (**d**, *bottom right*) bromobenzene. Reprinted with permission from [93]. Copyright 1996 American Chemical Society

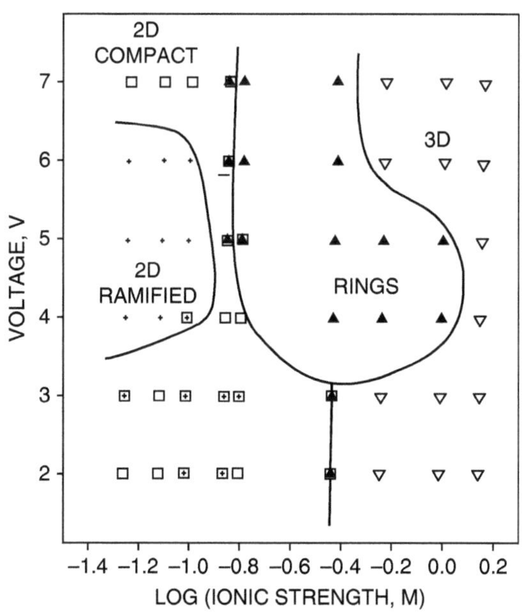

Fig. 2.62 Shape diagram for silver electrodeposition at the water/air surface. The silver ion concentration is 0.05 M. Reprinted with permission from [96]. Copyright 1997 American Chemical Society

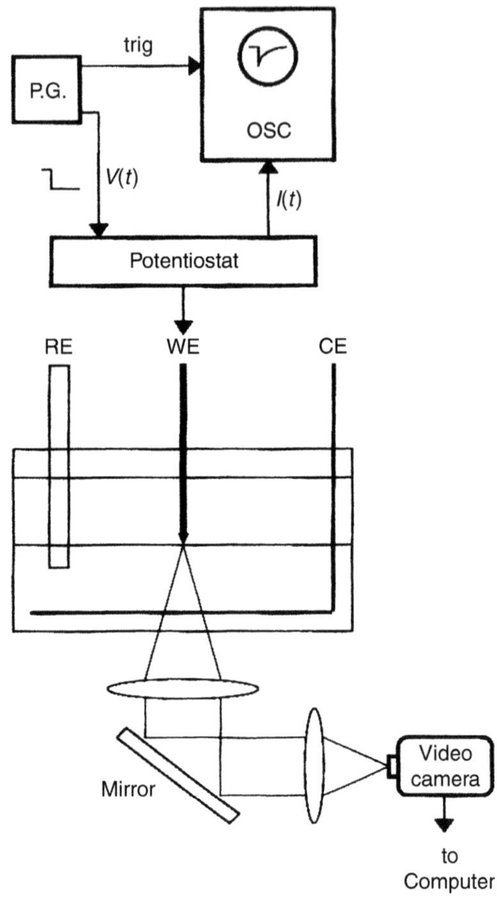

Fig. 2.63 Schematic representation of the experimental setup for the electrochemical and the magneto-electrochemical measurements. PG and OSC represent pulse generator and oscilloscope, respectively. Reprinted from [99], Copyright 1999, with permission from Elsevier

electron-transfer step at the interface controls the current distribution and then the deposit turns out to be compact [96]. For the description of the morphology of such patterns, Efrima has derived a generalized Wagner number, which indicates whether the pattern will be a compact deposit or whether it will be irregular, ramified, and dendritic (for details, see the original reference [98]).

As a continuation of these works, Nakabayashi et al. [99] have studied the patterns of zinc electrodeposited at the interface between two practically immiscible liquids: 4-methyl-2-pentanone and aqueous electrolyte solution (for the oscillators based on the characteristics of such interfaces, cf. Chap. 6). The aim of these studies was to explain the details of the anisotropic growth of the metal phase, by investigating the effect of the electrode potential and of the magnetic field, causing magnetohydrodynamic effects. The experimental setup that allowed the studies for both the air–liquid and the liquid–liquid interfaces is shown in Fig. 2.63. In particular, the time course of the shape change of the metal deposit was observed through the magnification optics focused on the top of the working

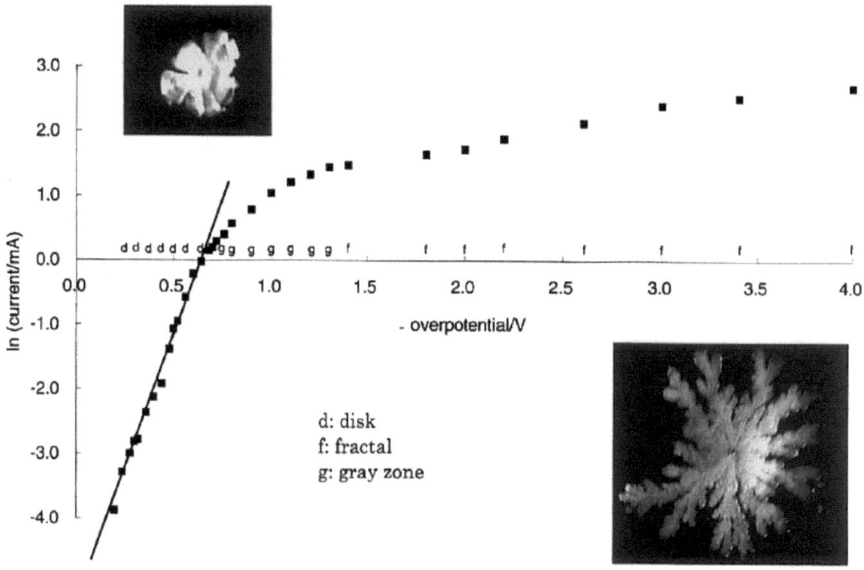

Fig. 2.64 Logarithm of the sampling current as a function of the final potential. "d," "f," and "g" above the potential axis represent the pattern of the deposit and the pictures at the upper and lower right are typical shapes of the deposit, disk (d) and fractal (f), respectively. Reprinted from [99], Copyright 1999, with permission from Elsevier

electrode from the bottom of the cell, and recorded with the video camera. As the working electrode (cathode) the carbon rod with a diameter of 0.5 mm was used, sharpened to ca. 0.1 mm in diameter at the top. The counter electrode was a ring of zinc wire dipped in the aqueous phase and as a reference electrode, the conventional SCE was used. The aqueous solution contained 2 M zinc sulfate and this was the lower phase in the experiments with the liquid/liquid interface (the upper phase being then 4-methyl-2-pentanone).

In the series of experiments performed for the liquid–liquid interface, the potential was stepped from −1.2 V (where the deposition of zinc began) to more negative values. In Fig. 2.64, the logarithm of current was plotted as a function of the final potential of the step, expressed versus the initial −1.2 V (the current flowing through the first 5 ms was neglected, in order to eliminate the capacitive current). This plot indicates simultaneously the shape of the zinc deposits, varying from the disk, through gray zone, up to fractal dendrites at sufficiently negative potentials. Since the disk shape occurred for the logarithmic dependence of the current on the potential (Butler–Volmer kinetics), one can conclude that this shape corresponds to the conditions, under which the rate of zinc deposition is controlled by the rate of the electron-transfer step. In turn, since fractal shapes were formed under more negative potentials, they were associated with the process of zinc deposition controlled by the rate of mass (Zn^{2+} ions) transport, in concordance with the assumption underlying the applicability of the DLA formalism.

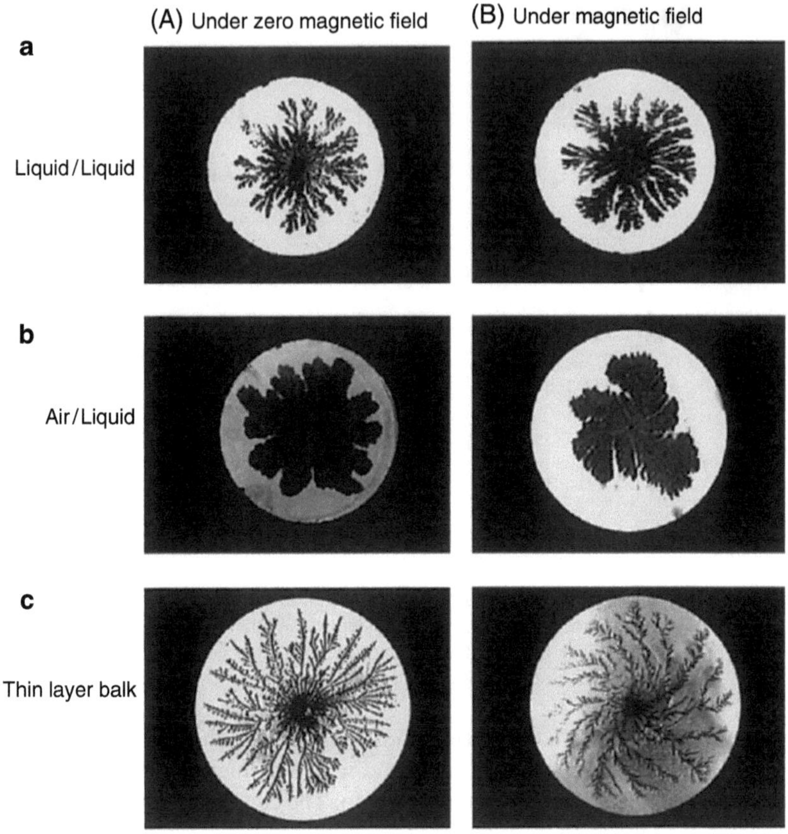

Fig. 2.65 Typical two-dimensional fractal deposit at (**a**) the 4-methyl-2-pentanone–aqueous electrolyte interface, (**b**) the air–aqueous electrolyte interface and in (**c**) the thin layer electrolyte. The magnetic field strengths are 1,200, 80, and 320 G for (**a**)–(**c**), respectively. Reprinted from [99], Copyright 1999, with permission from Elsevier

Placing the electrochemical cell in the magnetic field affects the motion of the electrolyte. Such magnetohydrodynamic effects were studied after the placing of the electrochemical cell just above the bore of the helium-free superconducting magnet; in this configuration, the direction of magnetic field was vertical and upward. In some experiments, oriented on the formation of quasi-two-dimensional deposits, the reaction proceeded in a thin (0.5 mm) 0.05 M zinc sulfate solution layer, placed between two plastic plates. In the presence of magnetic field, the Lorenz force acts on the ions towards the tangential direction, causing the spiral distortion of the deposited zinc patterns. Figure 2.65 shows such effects for different experimental conditions, but in each case for the electrode potential so negative that the deposition is controlled by the rate of mass transfer.

A bit surprising is the fact that of all interfaces studied, the liquid–liquid one was most resistant to the effect of the magnetic field (the morphology of the zinc pattern remained practically intact up to the field of 1,200 G, but above this value the whole pattern was destroyed by rotational flow).

The liquid–liquid interface as the localization of electrodeposition of metal films was a subject of also more recent studies by Nakanishi et al. [100]. In this work there was studied the electrodeposition of zinc at an aqueous $ZnSO_4$–*n*-butylacetate (BuAc) interface, under conditions when potential oscillations were observed in the region of the current density exceeding the diffusion-limited value (first reported by Tada et al. [101]). In situ optical microscopic observations showed that the formation of two-dimensional Zn film with a concentric pattern at the liquid–liquid interface was accompanied with the synchronic meniscus oscillation of this interface. Furthermore, the vigorous growth of the deposits occurred only when the shape of the meniscus became hollow on the negative potential side of the potential oscillation, while on the positive side, the meniscus became almost flat and the deposits formed in the preceding stage were thickened. The mechanism of those phenomena invoked the fact that the interfacial tension at the growing metal/aqueous solution interface was extremely large. In brief, the electrochemical oscillations were found to occur due to the cooperation of various processes: autocatalytic electrodeposition under diffusion-controlled conditions, increase and decrease of the electrode/aqueous phase interfacial tension and meniscus oscillation. For description of analogous interface dynamics, see also Sect. 6.1.

In turn, the oscillatory gold electrodeposition, occurring in the form of thin film formed at a liquid/air interface was described first by Saliba et al. [102]. It was further studied by Fukami et al. [103], using the Raman scattering spectroscopy (SERS) under in situ conditions. This technique allowed to monitor the dynamic nanostructural changes of the deposit. This system exhibited only potential oscillations under galvanostatic conditions. It was found that the oscillations occurred when the applied current exceeded the diffusion-limited current for Au electrodeposition, with this process being charge-transfer limited when the potential corresponded to the positive side of the oscillation (then the dendritic growth of crystals occurs) and switching to the control by diffusion, when the potential moved to negative values in the oscillatory peak (then the thickening of dendritic crystals occurs). The mechanism of oscillations involves, starting from the diffusion-limited conditions, the autocatalytic character of electrodeposition, caused by enhanced diffusion of $AuCl_4^-$ ions to the circular or spherical diffusion layer formed at the interface and resulting in the growth of dendritic needle-like crystals. This leads to a large increase in the effective surface area of the Au film and thus to a large decrease in the effective current density, causing a positive shift in the electrode potential, until the diffusion-controlled conditions switch into the control by the charge-transfer rate at the interface. Then the needle growth stops and the electrodeposition proceeds steadily. The needles are thickened and coalesced with each other, resulting in a rather continuous film. The corresponding decrease in the effective electrode surface area causes a back increase of the effective current density, until the electrode potential shifts to so negative values that the maintaining of the externally imposed current requires the contribution from the evolution of

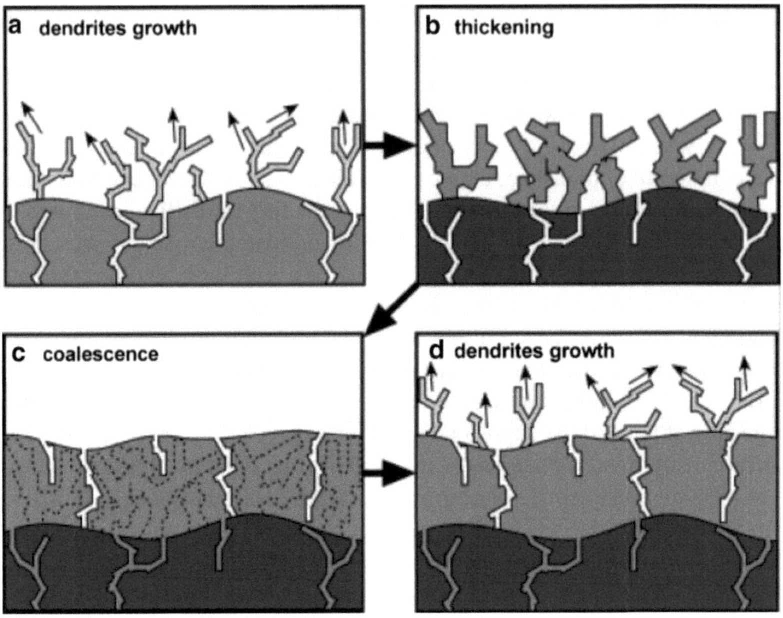

Fig. 2.66 Schematic drawings for explaining the mechanism for the oscillatory growth of the Au film and associated oscillations of the electrode potential under galvanostatic conditions. (**a**) Very fast, diffusion-limited electrodeposition of Au at the negative end of the potential oscillation leads to autocatalytic growth of the dendrites. The associated growth of effective surface area, causing the decrease in effective current density $|j_{eff}|$, results in continuous positive shift in the electrode potential; (**b**,**c**) At the attained high-potential state, when the electron transfer-limited condition is attained, the dendritic needles stop to grow and, instead, they are thickened (**b**) and coalesced (**c**) with each other, resulting in a rather continuous film; (**d**) Due to decrease in effective surface area, the effective current density $|j_{eff}|$ increases and when exceeds the diffusion-limited current density of $AuCl_4^-$ electroreduction, the electrode potential suddenly shifts to negative direction in order to maintain the externally controlled current by addition of hydrogen evolution current; in this way, the low-potential state, characterized with the growth of dendrites, is reached again, Reprinted with permission from [103] Copyright 2007 American Chemical Society

hydrogen. The sequence of these steps of the oscillations is schematically shown in Fig. 2.66.

Other examples of dendritic morphology of deposited metals, associated with the oscillations driven by convective motion, are described in Sect. 5.8.3.

2.12 Dendritic Patterns in Silicon Electrodissolution

Fractal patterns are not specific for metallic deposits only. Recently, Lublow and Lewerenz [104] have described the fractal etch structures formed on *n*-type Si photoelectrodes under anodic polarization in concentrated ammonium fluoride

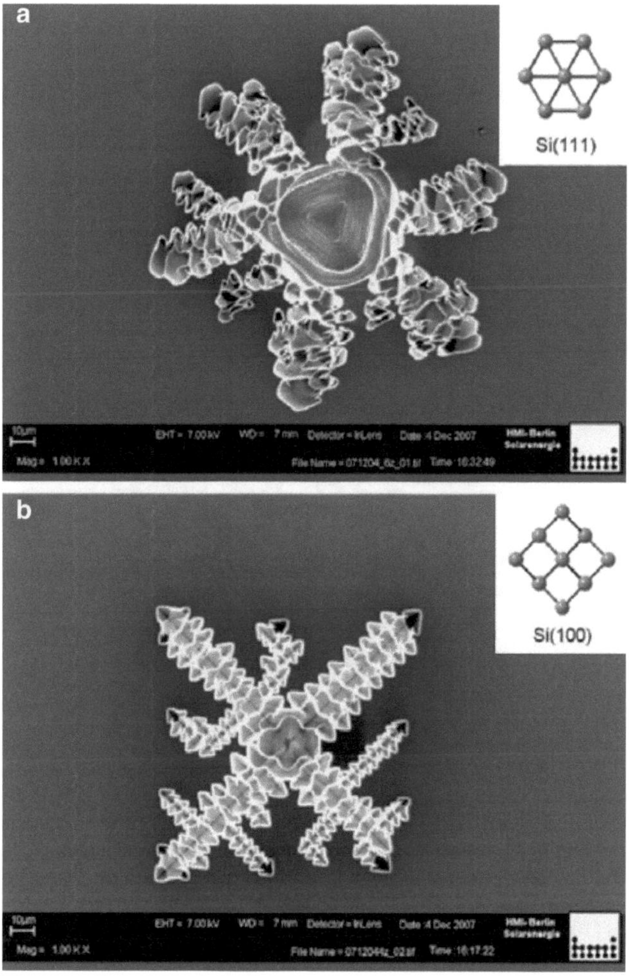

Fig. 2.67 HR-SEM images of structures obtained on (**a**) Si(111) and (**b**) Si(100) after dissolution in 40% NH_4F for 10 min using a light intensity of about 7 mW cm^{-2}. The insets indicate the surface lattice and the relative orientation of the structures to the lattice. Reprinted from [104], Copyright 2009, with permission from Elsevier

solutions. However, in this case the propagating branches of these structures generally reflect the surface lattice geometry of the substrates on a micrometer scale while inner topographies are characterized by ensembles of slow-etching planes. Figure 2.67 shows the images of exemplary patterns formed on Si(111) and Si(100) surfaces.

In the mechanism of formation of these patterns, not only the dissolution of illuminated n-Si:

$$\mathrm{Si} + 6\mathrm{HF} + 4\mathrm{h}^{+}_{\mathrm{VB}}(h\nu) \rightarrow \mathrm{SiF}_6^{2-} + 6\mathrm{H}^{+} \quad (2.12)$$

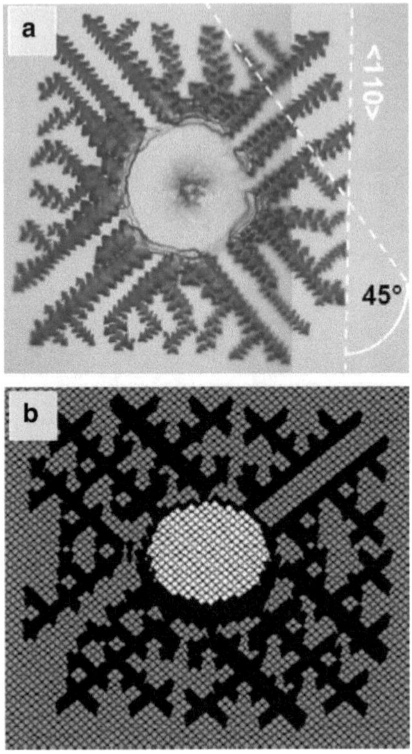

Fig. 2.68 Comparison of an experimental structure obtained in 40% NH_4F after 10 min at $U = 6$ V (light intensity 7 mW cm^{-2}) with a simulated structure. (**a**) The orientation of the fractal structure with respect to the surface lattice is indicated by an angle which refers to the primary flat of the wafer. (**b**) Simulated structure assuming uniform propagation of the branches which extend from initially cracked sites located at the boundary of an adsorbed oxygen bubble (see [104]). Reprinted from [104], Copyright 2009, with permission from Elsevier

comprising first electrochemical formation of SiO_2, followed by its pH-dependent dissolution by HF and HF_2^- to SiF_6^{2-}, was taken into account. Also, the parallel oxygen evolution reaction (OER):

$$2H_2O \rightarrow O_2 + 4H^+ + 4e \qquad (2.13)$$

was assumed to play an important role for the topographical structure formation. Oxygen gas bubbles are rapidly forming in the initial phase of the photoelectrochemical decomposition process and they are adsorbed to surface sites which coincide with the centers of fractal structures. In order to model such phenomena, simplified simulation procedures were derived from quantitative in-plane stress analysis. Comparison of experimental and simulated patterns is shown in Figs. 2.68 and 2.69. Other types of spatial patterns emerging during etching of semiconductors are described in Chap. 4.

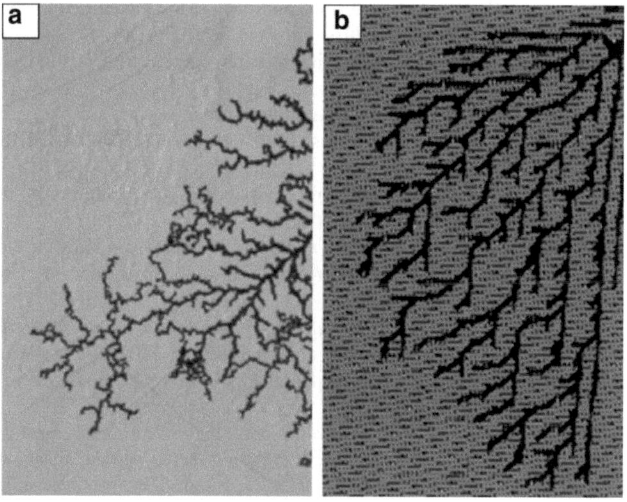

Fig. 2.69 Comparison of an experimental and simulated structure for the case of lower light intensity (below 1 mW cm^{-2}). According to model considerations, the lateral propagation velocities of the branches are varying with time and resulting thus in less regular structures. Reprinted from [104], Copyright 2009, with permission from Elsevier

References

1. Flätgen G, Krischer K, Pettinger P, Doblhofer K, Junkes H, Ertl G (1995) Two-dimensional imaging of potential waves in electrochemical systems by surface plasmon microscopy. Science 269:668–671
2. Bonhoeffer KF (1948) Activation of passive iron as a model for the excitation of nerve. J Gen Physiol 32:69–91
3. Heatcote HL (1902) Z physik Chem 37:368
4. Heatcote HL (1907) J Chem Soc Ind 26:899
5. Lillie RS (1918) Science 48:51
6. Lillie RS (1930) J Gen Physiol 14:344
7. Lillie RS (1936) Biol Rev Camb Philos Soc 16:216
8. Murray JD (2002) Mathematical biology. I. An introduction, 3rd edn. Springer, New York, NY
9. Murray JD (2003) Mathematical biology. II. Spatial models and biomedical applications, 3rd edn. Springer, New York, NY
10. Franck UF (1989) Periodische Strukturen und Vorgänge in gleichgewichtsfernen physikalisch-chemischen Systemen. Nova Acta Leopoldina NF 60:109–131
11. Nakabayashi S, Baba R (1998) Spatiotemporal propagation of a non-linear electrochemical reaction over an iron electrode. Chem Phys Lett 287:632–638
12. Baba R, Shiomi Y, Nakabayashi S (2000) Spatiotemporal reaction propagation of electrochemically controlled non-linear iron current oscillator. Chem Eng Sci 55:217–222
13. Agladze K, Steinbock O (2000) Waves and vortices of rust on the surface of corroding steel. J Phys Chem A 104:9816–9819
14. Agladze K, Thouvenel-Romans S, Steinbock O (2001) Electrochemical waves on patterned surfaces: propagation through narrow gaps and channels. J Phys Chem A 105:7356–7363
15. Hudson JL, Tabora J, Krischer K, Kevrekidis IG (1993) Spatiotemporal period doubling during the electrodissolution of iron. Phys Lett A 179:355–363

16. Koper MTM, Sluyters JH (1993) A simplified approach to the modeling of wave propagation at electrode/electrolyte interfaces. Electrochim Acta 38:1535–1544
17. Kristev I, Nikolova M, Nakada I (1989) Spiral structures in electrodeposited silver-antimony alloys. Electrochim Acta 34:1219–1223
18. Gladyshev VP, Kovaleva SV (1997) Autooscillatory processes and surface periodic patterns, emerging at anodic oxidation of amalgams of alkaline metals. Zh Obshch Khim 67:1743, in Rusian
19. Krischer K (2001) Spontaneous formation of spatiotemporal patterns at the electrode–electrolyte interface. J Electroanal Chem 501:1–21
20. Krischer K (1999) Principles of temporal and spatial pattern formation in electrochemical systems. In: Conway BE, Bockris JO'M, White R (eds) Modern aspects of electrochemistry. Plenum, New York, NY
21. Krischer K (2003) Nonlinear dynamics in electrochemical systems. In: Alkire RC, Kolb DM (eds) Advances in electrochemical science and engineering. Wiley-VCH, Weinheim
22. Flätgen G, Krischer K (1995) Accelerating fronts in an electrochemical system due to global coupling. Phys Rev E 51:3997–4004
23. Flätgen G, Krischer K, Ertl G (1996) Spatio-temporal pattern formation during the reduction of peroxodisulfate in the bistable and oscillatory regime: a surface plasmon microscopy study. J Electroanal Chem 409:183–194
24. Grauel P, Christoph J, Flätgen G, Krischer K (1998) Stationary potential patterns during the reduction of peroxodisulfate at Ag ring electrodes. J Phys Chem B 102:10264–10271
25. Otterstedt R, Plath PJ, Jaeger NI, Sayer JC, Hudson JL (1996) Accelerating fronts during the electrodissolution of cobalt. Chem Eng Sci 51:1747–1756
26. Otterstedt R, Plath PJ, Jaeger NI, Hudson JL (1996) Modulated electrochemical waves. Phys Rev E 54:3744–3751
27. Otterstedt RD, Plath PJ, Jaeger NI, Hudson JL (1996) Rotating waves on disk and ring electrodes. J Chem Soc Faraday Trans 92:2933–2939
28. Lev O, Sheintuch M, Pismen LM, Yarnitzky Ch (1988) Standing and propagating wave oscillations in the anodic dissolution of nickel. Nature 336:458–459
29. Lev O, Sheintuch M, Yarnitsky H, Pismen LM (1990) Spatial current distribution during nickel anodic dissolution in sulfuric acid. Chem Eng Sci 45:839–847
30. Haim D, Lev O, Pismen LM, Sheintuch M (1992) Modelling spatiotemporal patterns in anodic nickel dissolution. Chem Eng Sci 47:3907–3913
31. Bîrzu A, Green BJ, Otterstedt RD, Jaeger NI, Hudson JL (2000) Modeling of spatiotemporal patterns during metal electrodissolution in a cell with a point reference electrode. Phys Chem Chem Phys 2:2715–2724
32. Bîrzu A, Green BJ, Jaeger NI, Hudson JL (2001) Spatiotemporal patterns during electrodissolution of a metal ring: three-dimensional simulations. J Electroanal Chem 504:126–136
33. Jaeger NI, Otterstedt RD, Bîrzu A, Green BJ, Hudson JL (2002) Evolution of spatiotemporal patterns during the electrodissolution of metals: experiments and simulations. Chaos 12:231–239
34. Bîrzu A, Plenge F, Jaeger NI, Hudson JL, Krischer K (2003) Excitable dynamics during electrodissolution of a metal disk electrode: model calculations. Phys Chem Chem Phys 5:3724–3731
35. Bîrzu A, Krischer K (2006) Two-dimensional electrochemical turbulence during the electrodissolution of metal disk electrodes: model calculations. Phys Chem Chem Phys 8:3659–3668
36. Holmes P, Lumley J, Berkooz G (1996) Turbulence, coherent structures, dynamical systems and symmetry. Cambridge University Press, Cambridge, MA
37. Grauel P, Krischer K (2001) Fronts and stationary domains during electrochemical H_2 oxidation on Pt: the impact of the position of the reference electrode on the spatiotemporal behavior. Phys Chem Chem Phys 3:2497–2502
38. Grauel P, Varela H, Krischer K (2001) Spatial bifurcations of fixed points and limit cycles during the electrochemical oxidation of H_2 on Pt ring-electrodes. Faraday Discuss 120:165–178

39. Varela H, Beta C, Bonnefont A, Krischer K (2005) A hierarchy of global coupling induced cluster patterns during the oscillatory H_2-electrooxidation reaction on a Pt ring-electrode. Phys Chem Chem Phys 7:2429–2439
40. Plenge F, Varela H, Krischer K (2005) Asymmetric target patterns in one-dimensional oscillatory media with genuine nonlocal coupling. Phys Rev Lett 94:198301-1–198301-4
41. Varela H, Beta C, Bonnefont A, Krischer K (2005) Transitions to electrochemical turbulence. Phys Rev Lett 94:174104-1–174104-4
42. Mikhailov AS (1994) Foundation of synergetics I. Springer, Berlin
43. Benjamin TB, Feir J (1967) The disintegration of wave trains on deep water. Part 1. Theory. J Fluid Mech 27:417–430
44. Shraiman BI, Pumir A, van Saarloos W, Hohenberg PC, Chaté H, Holen M (1992) Spatiotemporal chaos in the one-dimensional complex Ginzburg-Landau equation. Physica D 57:241–248
45. Krischer K, Varela H, Bîrzu A, Plenge F, Bonnefont A (2003) Stability of uniform electrode states in the presence of ohmic drop compensation. Electrochim Acta 49:103–115
46. Plenge F, Li YJ, Krischer K (2004) Spatial bifurcations in the generic N-NDR electrochemical oscillator with negative global coupling: theory and surface plasmon experiment. J Phys Chem B 108:14255–14264
47. Baba N, Krischer K (2008) Mixed-mode oscillations and cluster patterns in an electrochemical relaxation oscillator under galvanostatic control. Chaos 18:015103-1–015103-9
48. Bonnefont A, Morschl R, Bauer P, Krischer K (2009) Electrochemical impedance spectroscopy of patterned steady-states on electrode surfaces. Electrochim Acta 55:410–415
49. Koper MTM, Schmidt TJ, Marković RPN (2001) Potential oscillations and S-shaped polarization curve in the continuous electro-oxidation of CO on platinum single-crystal electrodes. J Phys Chem B 105:8381–8386
50. Morschl R, Bolten J, Bonnefont A, Krischer K (2008) Pattern formation during CO electroxidation on thin Pt films studied with spatially resolved infrared absorption spectroscopy. J Phys Chem C 112:9548–9551
51. Christoph J, Strasser P, Eiswirth M, Ertl G (1999) Remote triggering of waves in an electrochemical system. Science 284:291–293 doi:10.1126/science.284.5412.291
52. Christoph J, Otterstedt R, Eiswirth M, Jaeger NI, Hudson JL (1999) Negative coupling during oscillatory pattern formation on a ring electrode. J Chem Phys 110:8614–8621
53. Strasser P, Christoph J, Lin WF, Eiswirth M, Hudson JL (2000) Standing wave oscillations in an electrocatalytic reaction. J Phys Chem A 104:1854–1860
54. Christoph J, Eiswirth M (2002) Theory of electrochemical pattern formation. Chaos 12:215–230
55. Turing AM (1952) The chemical basis for morphogenesis. Phil Trans R Soc London B 327:37–72
56. Krömker S (1998) Wave bifurcation in models for heterogeneous catalysis. Acta Math Univ Comenianae 67:83–100
57. Zhabotinsky AM, Dolnik M, Epstein IR (1995) Pattern formation arising from wave instability in a simple reaction–diffusion system. J Chem Phys 103:10306–10314
58. Dolnik M, Zhabotinsky AM, Epstein IR (1996) Modulated standing waves in a short reaction–diffusion system. J Phys Chem 100:6604–6607
59. Lee J, Strasser P, Eiswirth M, Ertl G (2001) On the origin of oscillations in the electrocatalytic oxidation of HCOOH on a Pt electrode modified by Bi deposition. Electrochim Acta 47:501–508
60. Christoph J, Noh T-G, Lee J, Strasser P, Eiswirth M (2009) Spatiotemporal self-organization in the oscillatory HCOOH oxidation on a Pt ribbon electrode—theory and experiments. Surf Sci 603:1652–1661
61. Zhao Y, Wang S, Varela H, Gao Q, Hu X, Yang J, Epstein IR (2011) Spatiotemporal pattern formation in the oscillatory electro-oxidation of sulfide on a platinum disk. J Phys Chem C 115:12965–12971

62. Mazouz N, Krischer K (2000) A theoretical study on Turing patterns in electrochemical systems. J Phys Chem B 104:6081–6090
63. Li YJ, Oslonovitch J, Mazouz N, Plenge F, Krischer K, Ertl G (2001) Turing-type patterns on electrode surfaces. Science 291:2395–2398 doi:10.1126/science.1057830
64. Bonnefont A, Varela H, Krischer K (2003) Stationary small and large amplitude patterns during bulk CO electrooxidation on Pt. Chem Phys Chem 4:1260–1263
65. Bonnefont A, Varela H, Krischer K (2005) Stationary spatial patterns during bulk CO electrooxidation on Pt. J Phys Chem B 109:3408–3415
66. Strasser P, Eiswirth M, Ertl G (1997) Oscillatory instabilities during formic acid oxidation on Pt(100), Pt(110) and Pt(111) under potentiostatic control. II. Model calculations. J Chem Phys 107:991–1003
67. Ihle T, Müller-Krumbhaar H (1994) Fractal and compact growth morphologies in phase transitions with diffusion transport. Phys Rev E 49:2972–2991
68. Arneodo A, Argoul F, Couder Y, Rabaud M (1991) Anisotropic Laplacian growths: from diffusion-limited aggregates to dendritic fractals. Phys Rev Lett 66:2332–2335
69. Matsuyama T, Matsushita M (1993) Fractal morphogenesis by a bacterial cell population. Crit Rev Microbiol 19:117–135
70. Ben-Jacob E, Cohen I, Gutnick DL (1998) Cooperative organization of bacterial colonies: from genotype to morphotype. Annu Rev Microbiol 52:779–806
71. Sawada Y, Dougherty A, Gollub JP (1986) Dendritic and fractal patterns in electrolytic metal deposits. Phys Rev Lett 56:1260–1263
72. Grier DG, Ben-Jacob E, Clarke R, Sander LM (1986) Morphology and microstructure in electrochemical deposition of zinc. Phys Rev Lett 56:1264–1267
73. Kuhn A, Argoul F (1995) Diffusion-limited kinetics in thin-gap electroless deposition. J Electroanal Chem 397:93–104
74. Trigueros PP, Claret J, Mas F, Sagués F (1991) Pattern morphologies in zinc electrodeposition. J Electroanal Chem 312:219–235
75. Garik P, Barkey D, Ben-Jacob E, Bochner E, Broxholm N, Miller B, Orn B, Zamir R (1989) Laplace- and diffusion-field-controlled growth in electrochemical deposition. Phys Rev Lett 62:2703–2706
76. Matsushita M, Sano M, Hayakawa Y, Honjo H, Sawada Y (1984) Fractal structures of zinc metal leaves grown by electrodeposition. Phys Rev Lett 53:286–289
77. Hurd AJ, Schaefer DW (1985) Diffusion-limited aggregation in two dimensions. Phys Rev Lett 54:1043–1046
78. Nakanishi S, Fukami K, Sakai SI, Nakato Y (2002) New autocatalytic mechanism for metal electrodeposition leading to oscillations and fern-leaf-shaped deposits. Chem Lett (Japan) 31:636–637
79. Fukami K, Nakanishi S, Sakai SI, Nakato Y (2003) Mechanism of oscillatory electrodeposition of zinc, revealed by microscopic inspection of dendritic deposits during the oscillation. Chem Lett 32:532–533
80. Fukami K, Nakanishi S, Tada T, Yamasaki H, Fukushima S, Sakai SI, Nakato Y (2005) Self-organized periodic growth of stacked hexagonal wafers in synchronization with a potential oscillation in zinc electrodeposition. J Electrochem Soc 152:C493–C497
81. Nakanishi S, Fukami K, Tada T, Nakato Y (2004) Metal latticeworks formed by self-organization in oscillatory electrodeposition. J Am Chem Soc 126:9556–9557
82. Tada T, Fukami K, Nakanishi S, Yamasaki H, Fukushima S, Nagai T, Sakai SI, Nakato Y (2005) Tuning of the spacing and thickness of metal latticeworks by modulation of self-organized potential oscillations in tin (Sn) electrodeposition. Electrochim Acta 50:5050–5055
83. Fukami K, Nakanishi S, Yamasaki H, Tada T, Sonoda K, Kamikawa N, Tsuji N, Sakaguchi H, Nakato Y (2007) General mechanism for the synchronization of electrochemical oscillations and self-organized dendrite electrodeposition of metals with ordered 2D and 3D microstructures. J Phys Chem C 111:1150–1160

84. Kuroda T, Irisawa T, Ookawa A (1977) Growth of a polyhedral crystal from solution and its morphological stability. J Cryst Growth 42:41–46
85. Oaki Y, Imai H (2003) Experimental demonstration for the morphological evolution of crystals grown in gel media. Cryst Growth Des 3:711–716
86. Nakanishi S, Sakai S, Nishimura K, Nakato Y (2005) Layer-by-layer electrodeposition of copper in the presence of *o*-phenanthroline, caused by a new type of hidden NDR oscillation with the effective electrode surface area as the key variable. J Phys Chem B 109:18846–18851
87. Rosso M, Gobron T, Brissot C, Chazalviel JN, Lascaud S (2001) Onset of dendritic growth in lithium/polymer cells. J Power Sources 97–98:804–806
88. González G, Rosso M, Chassaing E, Chazalviel JN (2007) Experimental and theoretical study of the onset of the growth of an irregular metal electrodeposit. Electrochim Acta 53:141–144
89. Chazalviel JN (1990) Electrochemical aspects of the generation of ramified metallic electrodeposits. Phys Rev A 42:7355–7367
90. Nakanishi S, Sakai S, Nagai T, Nakato Y (2005) Macroscopically uniform nanoperiod alloy multilayers formed by coupling of electrodeposition with current oscillations. J Phys Chem B 109:1750–1755
91. Sakai S, Nakanishi S, Nakato Y (2006) Mechanisms of oscillations and formation of nano-scale layered structures in induced co-deposition of some iron-group alloys (Ni–P, Ni–W, and Co–W), studied by an in situ electrochemical quartz crystal microbalance technique. J Phys Chem B 110:11944–11949
92. Ihara D, Nagai T, Yamada R, Nakanishi S (2009) Interfacial energy gradient at a front of an electrochemical wave appearing in CuSn-alloy oscillatory electrodeposition. Electrochim Acta 55:358–362
93. Zeiri L, Efrima S, Deutsch M (1996) Interfacial electrodeposition of silver: the role of wetting. Langmuir 12:5180–5187
94. Zeiri L, Efrima S (1997) Ac driven interfacial electrodeposition of silver. Langmuir 13:4722–4728
95. Zeiri L, Efrima S, Deutsch M (1997) Electroaggregation of silver interfacial colloids. J Phys Chem B 101:9757–9766
96. Zeiri L, Younes O, Efrima S, Deutsch M (1997) Interfacial electrodeposition of silver. J Phys Chem B 101:9299–9308
97. Wagner C (1954) Contribution to the theory of electropolishing. J Electrochem Soc 101:225–228
98. Efrima S (1997) Morphology of quasi-two dimensional electrodeposits—a generalized Wagner number. Langmuir 13:3550–3556
99. Nakabayashi S, Aogaki R, Karantonis A, Iguchi U, Ushida K, Nawa M (1999) Two-dimensional metal deposition at the liquid–liquid interface; potential and magnetohydrodynamic pattern transition. J Electroanal Chem 473:54–58
100. Nakanishi S, Nagai T, Fukami K, Sonoda K, Oka N, Ihara D, Nakato Y (2008) Oscillatory electrodeposition of metal films at liquid/liquid interfaces induced by the large surface energy of growing deposits. Langmuir 24:2564–2568
101. Tada E, Oishi Y, Kaneko H (2007) Electrochemical oscillation during electrodeposition of zinc at the interface between two immiscible liquids. Electrochemistry (Japan) 75:731–733
102. Saliba R, Mingotaud C, Argoul F, Ravaine S (2002) Spontaneous oscillations in gold electrodeposition. Electrochem Commun 4:629–632
103. Fukami K, Nakanishi S, Sawai Y, Sonoda K, Murakoshi K, Nakato Y (2007) In situ probing of dynamic nanostructural change of electrodeposits in the course of oscillatory growth using SERS. J Phys Chem C 111:3216–3219
104. Lublow M, Lewerenz HJ (2009) Scaling effects upon fractal etch pattern formation on silicon photoelectrodes. Electrochim Acta 55:340–349

Chapter 3
Cooperative Dynamics of Coupled and Forced Oscillators

3.1 Coupled Oscillators

3.1.1 Outline Theoretical Aspects of Coupling the Electrochemical Oscillators

The formation of dissipative patterns on electrode surfaces is related to various kinds of spatial couplings that interact with the electrochemical process in a way dependent, among others, on the geometrical arrangement of the electrodes and the operational mode of the electrochemical experiments. A natural development of the concept of individual reaction sites engaged into such couplings is the case of coupled oscillators. It is thus not surprising that the treatment of coupling of the individual oscillators and spatiotemporal phenomena in a single oscillator may have common points. The analysis of such complex systems allows to understand better the general mechanisms underlying the spatiotemporal phenomena, including the coupling that exist between oscillators in living systems, e.g., in the biological cell membranes.

The coupling of chemical or electrochemical oscillators can lead to either more complex or simpler oscillatory regimes, in the latter case including even ceasing of the oscillations. A special variant of such coupling can be realized through the external perturbation of a single oscillator or of a set of them.

Karantonis and Nakabayashi [1] have analyzed the case of coupling of two limit cycle electrochemical oscillators, described by only two dynamical variables. In this work the diffusive coupling in one dynamical variable is considered which, under appropriate conditions, can lead to dephasing of the oscillators. This particular problem is related to the suggestion by Kuramoto [2] that chemical spatiotemporal chaos might arise where diffusive coupling leads to dephasing. The problem is important since it was shown, in terms of the model approach, that the diffusive coupling of the transmembrane voltage, in the presence of a strong deformation of the phase flow in the vicinity of the limit cycle, can cause the dephasing of oscillators and bursting oscillations [3]. The model of electrochemical

M. Orlik, *Self-Organization in Electrochemical Systems II*,
Monographs in Electrochemistry, DOI 10.1007/978-3-642-27627-9_3,

oscillator, described by Karantonis and Nakabayashi, is simplified: only one electroactive species is present which is produced on the electrode surface and also carries all current in the solution via diffusion and migration. The linear approximation of the concentration distribution in the diffusion layer (Nernst concept) is also applied. The electrochemical process is characterized with the N-NDR region, and the total current consists of the faradaic and the capacitive currents. Accordingly, as described in Chap. 2 of volume I, the dynamics of such a system is ruled by the two ordinary differential equations of a general form, in terms of dimensionless variables (note that for consistence with the figures, the original notation of variables, different from typically used in this monograph, was kept: u is now the electrode potential, denoted usually as E, while v is now an external voltage, denoted usually by U; c and R denote surface concentration and ohmic resistance):

$$\varepsilon \frac{\mathrm{d}u}{\mathrm{d}t} = \frac{v-u}{R} - f(u,c) \tag{3.1}$$

$$\frac{\mathrm{d}c}{\mathrm{d}t} = (1-c) - \frac{v-u}{R} + \alpha f(u,c) \tag{3.2}$$

In these equations, parameter $\varepsilon \ll 1$ (related to the double layer capacitance) determines the difference in time scales of the temporal variations of the electrode potential u and surface concentration of the reactant c (cf. Sect. 2.2.3, volume I). In Eq. (3.2), being the mass balance equation, the first term on the right-hand side corresponds to the diffusion transport of electroactive species, the second term—to the contribution from the migration current through the solution, and $f(u,\ c) = c(a_1u + a_1u^2 + a_3u^3)$ is a model polynomial description of the electrochemical process, exhibiting the N-NDR characteristics. In the calculations, the following parameters were kept constant: $r = 10$, $\alpha = 0.1$, $a_1 = 1.125$, $a_2 = -0.075$, $a_3 = 0.00125$, while the external voltage v was varied as the bifurcation parameter. In terms of this model construction, the diffusive coupling between the ith and all other N oscillators was expressed by the introduction of the linear potential-coupling term:

$$\varepsilon \frac{\mathrm{d}u_i}{\mathrm{d}t} = \frac{v-u_i}{R} - f(u_i,c) + \frac{1}{N}\sum_{j=1}^{N} k\left(u_j - u_i\right) \tag{3.3}$$

$$\frac{\mathrm{d}c_i}{\mathrm{d}t} = (1-c_i) - \frac{v-u_i}{r} + \alpha f(u_i,c_i) \tag{3.4}$$

where k represents the coupling strength.

Figure 3.1a collects bifurcations of the uncoupled dynamical system [Eqs. (3.1) and (3.2)] for $\varepsilon = 0.055$. Figure 3.1b, c show the nullclines of this system (together with the limit cycle), for $v = 29.27$, whereas Fig. 3.1c, d show the results of weak coupling of such oscillators, for $k = 0.05$ [Eqs. (3.3) and (3.4)].

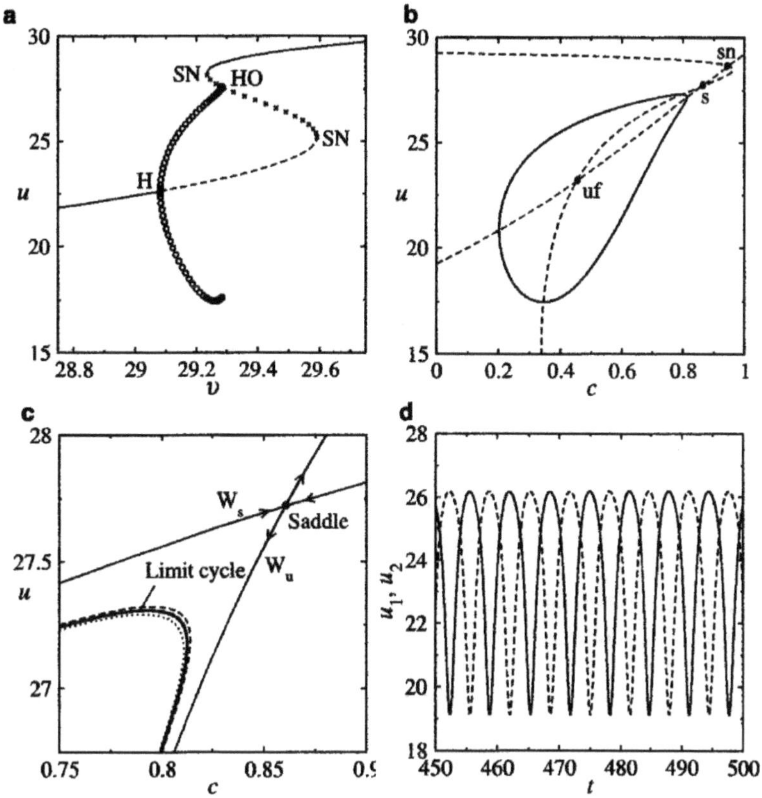

Fig. 3.1 (**a**) Bifurcation diagram for $\varepsilon = 0.055$. Stable steady state (*full line*), unstable steady state (*dashed line*), saddle (x), stable limit cycle (o), Hopf bifurcation (H), saddle-node bifurcation (SN), homoclinic bifurcation (HO). (**b**) Limit cycle and the nullclines for $\varepsilon = 0.055$, $v = 29.27$. Unstable focus (uf), saddle (s), and stable node (sn). (**c**) Enlargement of (**b**) with the saddle as well as the stable and unstable manifolds. Uncoupled limit cycle (*thick line*), interacting trajectories inside (*dotted line*) and outside (*dashed line*) the limit cycle. $K = 0.05$, $N = 2$ and $v = 29.27$. (**d**) Time series of the electrode potential of two interacting oscillators for $k = 0.05$ and $v = 29.27$. Reprinted from [1], Copyright 2001, with permission from Elsevier

One should note that due to the presence of saddle point near a part of the limit cycle in the phase space, the trajectories (phase flows) of the system undergo there strong deformation (cf. the characteristics of the homoclinic bifurcation in Chap. 1 of volume I). Now, when two such oscillators (or limit cycles) are *weakly* coupled and they initially exhibit a small phase difference ($\delta\phi \approx 0.01\pi$ for Fig. 3.1d), the dephasing occurs, i.e., the phase difference increases or decreases, with the effect of increase dominating. Finally, the dephasing leads to stabilization of the phase difference at a value of π, similarly to the model system of coupled neural oscillators [3]. An inspection of Fig. 3.1c shows how the coupling causes one oscillator to be pushed outside the limit cycle, and the other one—inside the limit cycle. The system closer at these times to the saddle point is characterized with slower velocities, so it is delayed with respect to the other one.

Intuitively, it is understandable that if the number of coupled oscillators increases, their coupling should lead also to more complex dynamic regimes. For example, for $N = 100$ coupled oscillators, and the coupling strength greater than in the above example, $k = 0.1$ (one can call it the *moderate coupling*) the current response, defined as:

$$I = \frac{1}{N} \sum_{i=1}^{N} \frac{v - u_i}{R} \tag{3.5}$$

exhibited the bursts of oscillations. This means that the moderate coupling allowed the system to behave in a way which for a single oscillator would be possible only under condition of its extension for at least one more dynamical variable, yielding the three-dimensional system.

Finally, for other differences in time scale, also other bifurcation scenarios occur. For example, for $\varepsilon = 0.03$, the heteroclinic orbits emerge in the phase space and two coupled oscillators can exhibit complete synchronization (heteroclinic orbits mean the collection of trajectories that connect sequences of, e.g., fixed points via saddle-sink). The relevant bifurcation diagram, nullclines, limit cycles, and the manifestation of such synchronization are shown in Fig. 3.2.

Of course, the model is limited to diffusive coupling, and in the electrochemical systems also the migration coupling, i.e., through the electric field, should be taken into account, similarly as for neural systems. For that reason the above model considerations constitute rather an introduction to the role of coupling of the electrochemical oscillators.

3.1.2 Single and Coupled H_2O_2 Oscillators

Fukushima et al. [4] have developed studies of spatiotemporal patterns in electroreduction of H_2O_2, involving two or more Pt electrodes. In Sect. 4.7 of volume I, recent studies of temporal oscillations for that process exhibiting NDR characteristics were described, and the reported oscillations were divided into five types: A–E. The present work is a continuation of those studies, applied for the oscillations of type A. The electrode arrangement, used in the presently described experiments, is shown in Fig. 3.3. As the particular electrodes were employed: polycrystalline Pt ring(s) as the working electrode(s) (WE), the Au ring as the counterelectrode (CE), and Au wires covered with glass tube except for the tips served as probe electrodes (PE). The spatiotemporal reaction waves were obtained by simultaneous measuring of the potential of the PE vs. the WE potential, which potential could be assumed to be nearly proportional to the local current density near the PE.

Figure 3.4a, b collects the essential features of the H_2O_2 reduction which exhibits NDR region within −0.30 to −0.36 V, due to underpotential deposition of hydrogen, the layer of which inhibits the dissociative adsorption of H_2O_2: 2Pt +

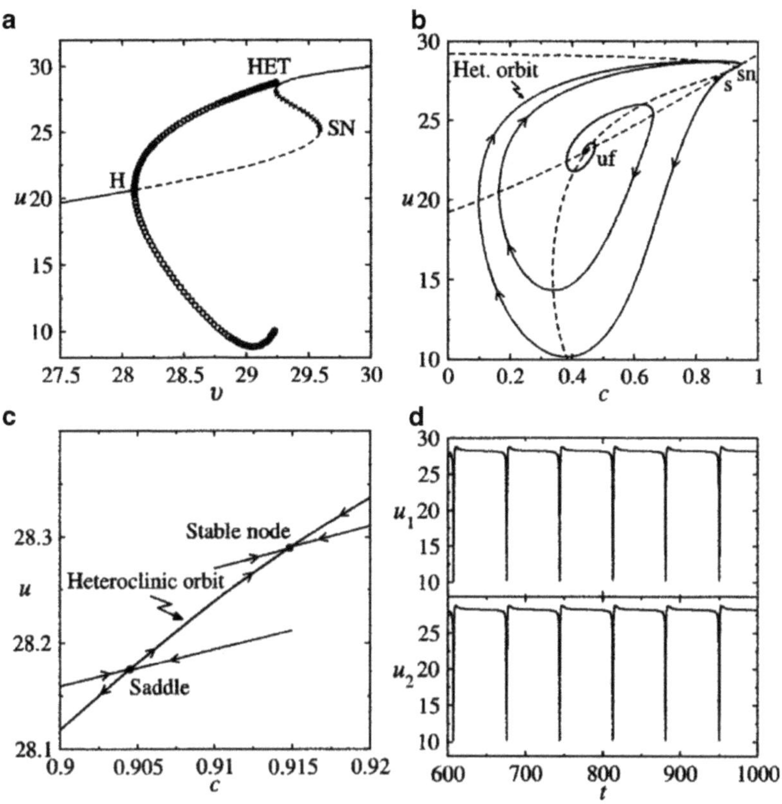

Fig. 3.2 (**a**) Bifurcation diagram for $\varepsilon = 0.03$. Stable steady state (*full line*), unstable steady state (*dashed line*), saddle (x), stable limit cycle (o), Hopf bifurcation (H), saddle-node bifurcation (SN), heteroclinic bifurcation (HET). (**b**) Heteroclinic orbit and the nullclines for $\varepsilon = 0.03$, $v = 29.236$. Unstable focus (uf), saddle (s), and stable node (sn). (**c**) Enlargement of (**b**) with the saddle as well as the stable and unstable manifolds and the heteroclinic orbit connecting the saddle and the node. (**d**) Time series of the electrode potential of two interacting oscillators for $k = 0.05$ and $v = 29.235$. Reprinted from [1], Copyright 2001, with permission from Elsevier

$H_2O_2 \rightarrow 2Pt\text{–}OH$, followed by electrochemical reduction of adsorbed OH to H_2O: $Pt\text{–}OH + H^+ + e \rightarrow Pt + H_2O$. In turn, Fig. 3.4c–e shows spatiotemporal reaction waves for a *single* Pt-ring working electrode, for various positions of RE indicated by the numbers (to correlate with positions indicated in Fig. 3.3) and various external voltages. The indicated current densities were calculated based on the above assumed proportionality with respect to the observed potential of the PE. The position on the WE was indicated by the angle θ, cf. Fig. 3.3.

Figure 3.4c–e shows the following modes of evolution of the spatiotemporal patterns. For the reference electrode placed far from the WE (position 1), a *spatially homogeneous wave* appears, in which the current at every θ oscillates in an in-phase mode (Fig. 3.4c). If RE is placed close to WE (position 2), depending on U, either a *standing wave* appears, in which the current oscillates in antiphase mode in two

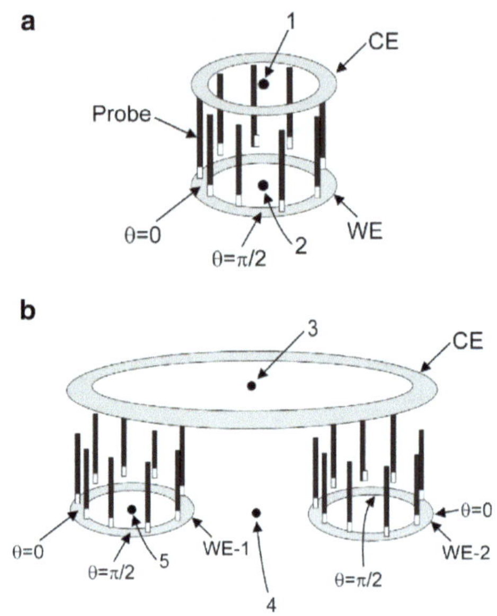

Fig. 3.3 Schematic illustrations of the electrode arrangement in a cylindrical cell, WE, WE-1, and WE-2: the working electrodes, CE: the counterelectrode, and Probe: the probe electrodes (PE's). *Black dots* with numbers 1–5 indicate the position of the reference electrode. Based on original drawing, reprinted from [4], Copyright 2005, with permission from Elsevier

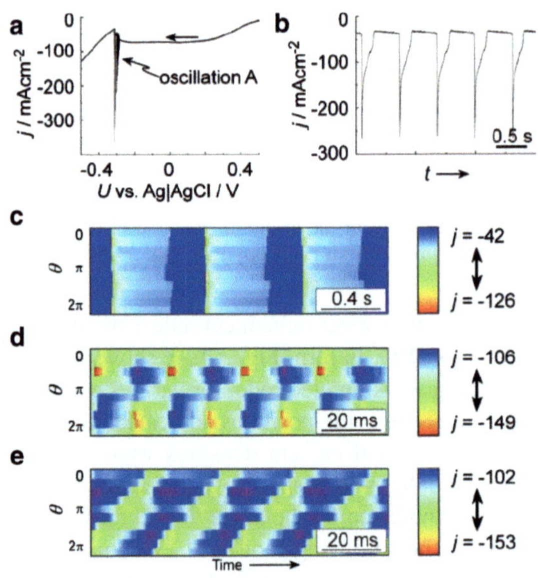

Fig. 3.4 (**a**) Potential-regulated j vs. U obtained at a scan rate of 10 mV/s. (**b**) j vs. t at $U = -0.30$ V. (**c**)–(**e**) Observed spatiotemporal waves (**c**) at $U = -0.33$ V with the RE placed at position-1, (**d**) at $U = -0.25$ V with RE at position-2, and (**e**) $U = -0.30$ V with the RE at position-2. Electrolyte: (**a**)–(**c**) 0.3 M H_2O_2 + 0.5M H_2SO_4 and (**d**) and (**e**) 0.4 M H_2O_2 + 0.3 M H_2SO_4. The unit of $j = \text{mA cm}^{-2}$. Reprinted from [4], Copyright 2005, with permission from Elsevier

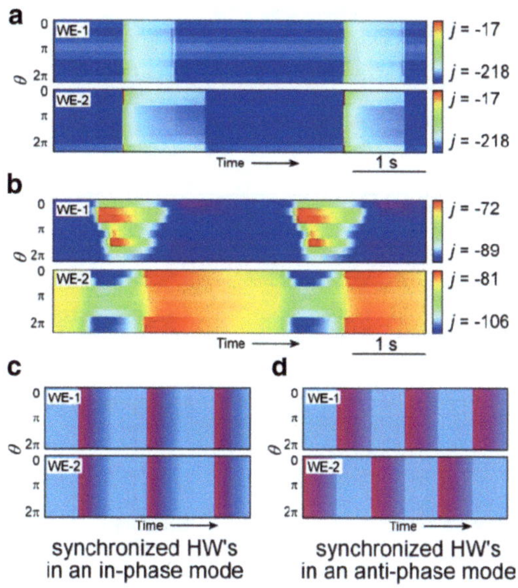

Fig. 3.5 (**a**) Observed spatiotemporal waves at U(WE-1) = U(WE-2) = −0.30 V with the RE placed at position-3, and (**b**) those at U(WE-1) = −0.20 V and U(WE-2) = −0.15 V with the RE at position-4. Electrolyte: 0.3 M H_2O_2 + 0.5 M H_2SO_4. (**c**) and (**d**) Schematic illustrations of (**a**) and (**b**), respectively, for easy understanding. The unit j is mA cm^{-2}. Reprinted from [4], Copyright 2005, with permission from Elsevier

domains of the Pt-ring electrode (Fig. 3.4d), or the traveling wave develops, in which the local high-current (active) region travels (rotates) on the Pt-ring electrode (Fig. 3.4e).

Figure 3.5 shows the patterns that develop in the system of *coupled* electrodes, as sketched in Fig. 3.3b. Figure 3.5a (and corresponding Fig. 3.5c) indicate that for the large distance between RE and WE (position 3), and equal applied potentials, wholly synchronized homogeneous waves in an in-phase mode appear on each of WE. In turn, Fig. 3.5b (and corresponding Fig. 3.5d) show that for close position of RE close to WE-1 and WE-2 (position 4), and different applied potentials, synchronized, but in antiphase mode homogeneous waves are observed.

In turn, for the position 5, i.e., when the RE was placed at the center of WE-1 and various individual potentials of the working electrodes, the spatiotemporal patterns, shown in Fig. 3.6, were observed.

Then a homogeneous stationary wave appears on the WE-2 which exhibits correlation with the phenomena occurring at WE-1: when the WE-2 is in high-current state (red color), a stationary wave appears on the WE-1, and when the WE-2 is in low-current state, either a standing wave (Fig. 3.6a, b) or traveling wave (Fig. 3.6c, d) develops.

The explanation of the formation of such synchronized waves in coupled separate oscillators was explained in terms of two kinds of coupling that emerge

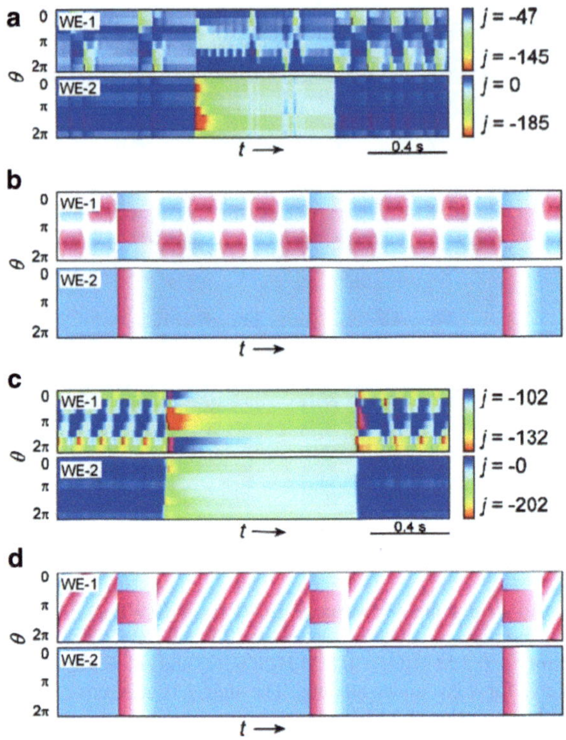

Fig. 3.6 (**a**) Observed spatiotemporal waves at U(WE-1) = −0.20 V and U(WE-2) = −0.25 V, and (**c**) those at U(WE-1) = −0.21 V and U(WE-2) = −0.27 V, both with the RE places at position-5. Electrolyte: 0.3 M H_2O_2 + 0.5 M H_2SO_4. (**b**) and (**d**) Schematic illustrations of (**a**) and (**c**), respectively, for easy understanding. The unit j is mA cm^{-2}. Reprinted from [4], Copyright 2005, with permission from Elsevier

in such system: the *positive local* coupling (PLC) and the *negative global* coupling (NGC) effects. The PLC is induced by electrical interaction between two oscillators through the ohmic drops in the electrolyte solution—a current oscillation in one oscillator causes an oscillatory ohmic drop in the solution which induces an in-phase modulation of the true electrode potential (double layer potential) of the other oscillator, thus leading to synchronized oscillations. This interaction is most effective when the RE is located close to the CE (or far from WE). In turn, the NGC effect, significant when RE is located close to WE, enhances the inhomogeneity: an appearance of a local high negative current near the RE causes a large negative shift of the potential of the solution at the RE with respect to CE owing to the ohmic drop. Under potentiostatic conditions this causes a large negative shift in the true electrode potential in the low current (NDR) region. This means further decrease of current in this passive region. In the present system both types of couplings cooperate in a way shown schematically in Fig. 3.7.

Thus, under conditions when the RE is close to WE (position 2), if one region of the working electrode surface becomes high current ("active," denoted by red

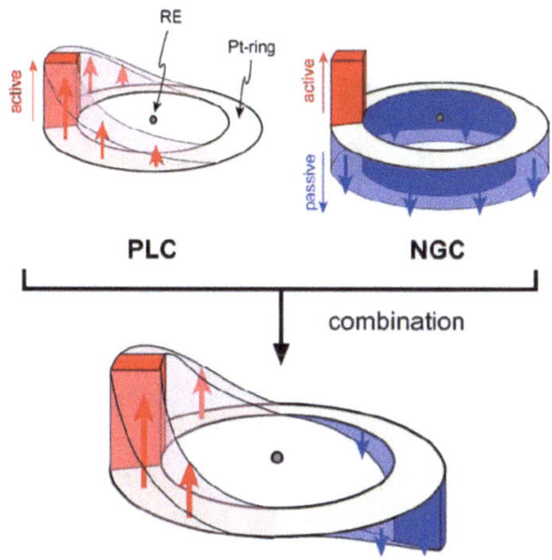

Fig. 3.7 Schematic illustration for explaining the PLC and the NGC effects and their combination, for position 2 between WE and RE. Reprinted from [4], Copyright 2005, with permission from Elsevier

color), the PLC tends to spread this region over the wider area of the electrode surface, while the NGC makes the lower current ("passive") regions even more passive. In consequence, when the strength of both couplings is comparable, either the stationary (SW) or traveling (TW) waves are born. This is why such patterns are formed under conditions corresponding to Fig. 3.4d, e, while the homogeneous wave in Fig. 3.4c, for RE placed far from WE (position 1), occurs due to predominating PLC effect. Analogous considerations on the interplay of both couplings can be applied to the explanation of spatiotemporal patterns in the system of two coupled oscillators [4].

As a continuation of this work, Nakanishi et al. [5] have employed the electrochemical reduction of H_2O_2 on a Pt-ring electrode as a model system for elucidation the pattern selection rule among oscillatory patterns with spatial nonuniformity. The content of these works refers also to Sect. 1.2. The present analysis required the formulation of the mathematical model of the above-mentioned electrochemical process and its detailed nonlinear bifurcation analysis. In principle, it can be considered a new model of spatiotemporal patterns in electrochemical systems, which allowed to produce uniform oscillations (UO), standing waves (SW), and rotating waves (RW), in concordance with the modes reported in the experiment. At present stage of understanding of spatiotemporal patterns in electrochemical systems it is thus clear that for the N-NDR oscillator it is easy to switch between the spatially uniform and nonuniform spatial modes just by changing the position of the RE between the WE and CE, i.e., by tuning in this way the extent of the NGC. This conclusion was recently enriched by interesting finding by Nakanishi et al. [6] that the spatially uniform and nonuniform modes, observed for H_2O_2 reduction on a Pt-ring electrode, are separated by the region of bistability. The experimental results were concordant with theoretical calculations.

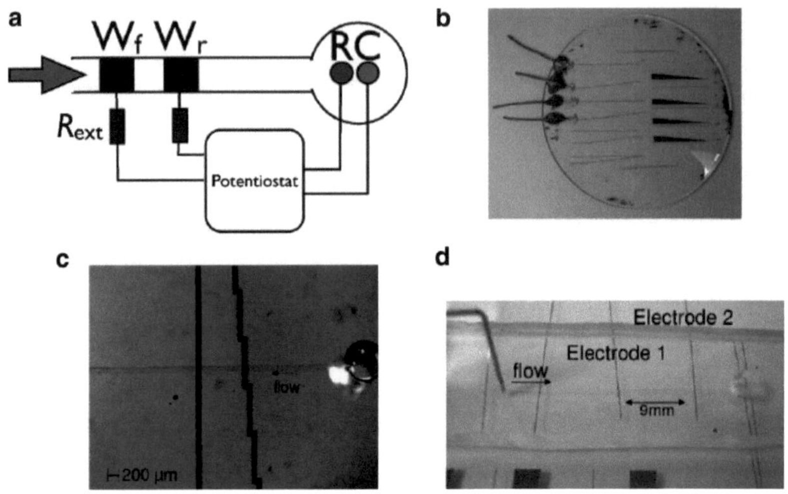

Fig. 3.8 Experimental setup. (**a**) Schematic of dual-electrode flow cell. W_f, W_r, R, and C are rear working, reference, and counterelectrodes, respectively. *Arrow* indicates flow of electrolyte. R_{ext}: external resistors. (**b**) Glass round containing Pt electrodes of varying sizes (100–12 mm) and distances. (**c**) Dual-Pt electrodes with PDMS-based flow channel. (**d**) Assembled microfluidic dual-electrode flow cell (without reference and counter electrode). Reprinted from [7], Copyright 2009, with permission from Elsevier

3.1.3 Coupling in the Oscillatory Oxidation of Formic Acid

In Sect. 5.4 of volume I, there was described the oscillatory oxidation of formic acid/formate ions and its electrochemical mechanism. Recently, a novel approach to the studies of the role of coupling in the electrooxidation of formic acid on Pt electrodes was described by Kiss et al. [7] who have applied a microchip-based dual-electrode flow cell, with microfluidic flow control (Fig. 3.8). The flow channel had dimensions 120 μm (width) × 100 μm (depth) and the flow rate 0.5–1.5 μL/min, as in typical analytical measurements (cf., e.g., [8]).

By varying the relative positions of the electrodes it was possible to observe various dynamical behaviors. When the reference/counter electrodes were placed far away from the microelectrodes, nearly in-phase synchronized current oscillations were observed (Fig. 3.9). In turn, this synchronization practically disappeared with close reference/counter electrode placements.

As the cause for the synchronization there was suggested the weak, albeit important, bidirectional electrical coupling between the electrodes (thus, the unidirectional mass transfer interactions could be considered negligible). A natural consequence of the present studies would be the development of analogous studies with a multielectrode array setup.

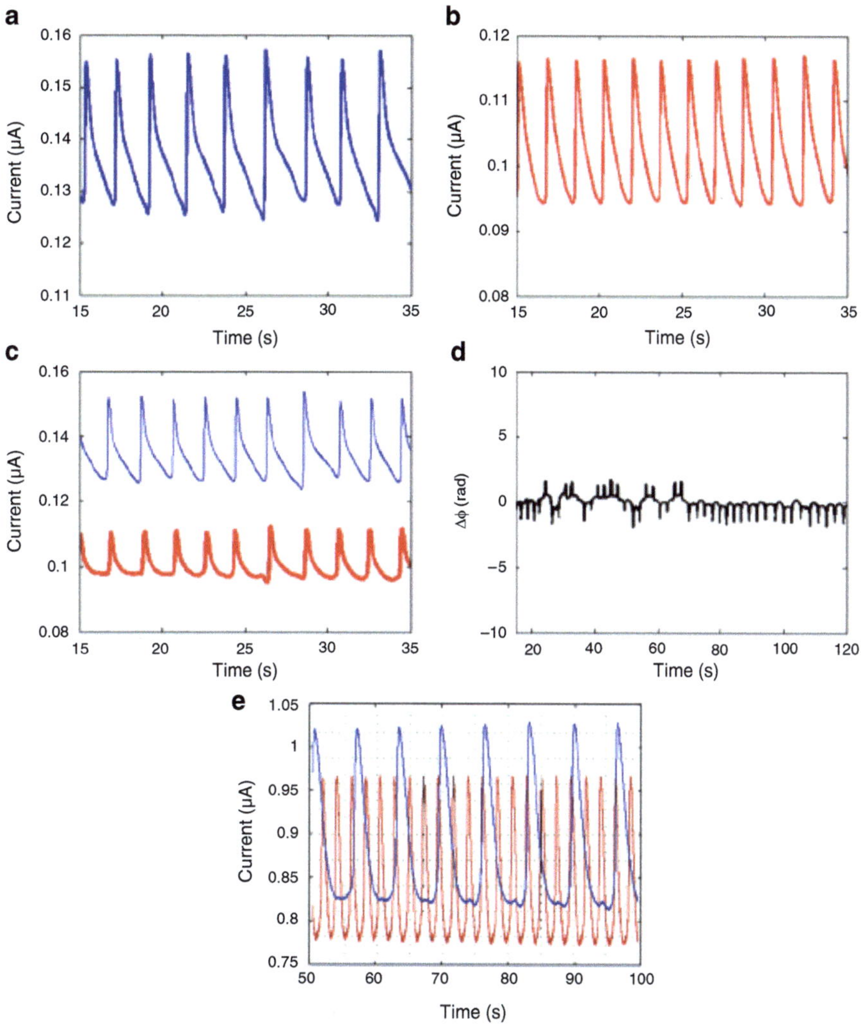

Fig. 3.9 Synchronization of current oscillations in dual-electrode cell with far reference electrode placement ($d = 11.0$ mm). (**a**) Time series of current oscillations of the (single) front working electrode. ω (front) = 0.439 Hz. (**b**) Time series of current oscillations of the (single) rear working electrode. ω (rear) = 0.581 Hz. (**c**) Synchronized current oscillation in the dual-electrode setup, ω (front) = ω (rear) = 0.485 Hz. *Solid line*: front electrode, *bold line*: rear electrode. (**d**) Phase difference between the oscillations of the rear and front electrodes vs. time in the dual-electrode setup. $Q = 1.5$ μL/min. Front electrode: $U = 1.2$ V, $R_{ext} = 10$ MΩ. Rear electrode: $U = 1.3$ V, $R_{ext} = 12$ MΩ, $c(Bi^{3+}) = 2\times10^{-6}$ M. (**e**) 1:3 synchronized current oscillations. $Q = 2$ μL/min. $U = 0.8$ V, $R_{ext} = 10$ MΩ, $c(Bi^{3+}) = 2\times10^{-7}$ M, $c(H_2SO_4) = 0.05$ M, $c(HCOOH) = 0.1$ M. Reprinted from [7], Copyright 2009, with permission from Elsevier

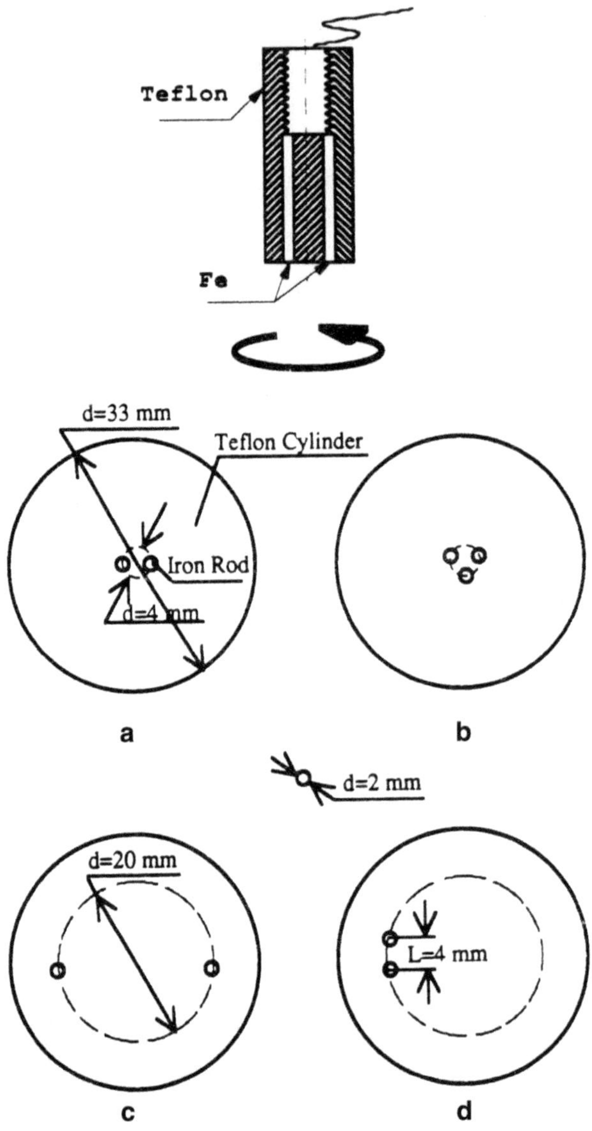

Fig. 3.10 Schematic of Fe electrode configuration. Reprinted with permission from [9]. Copyright 1992 American Chemical Society

3.1.4 Coupled Fe/H_2SO_4 Oscillators

Wang and Hudson [9] have described the coupling in the Fe electrode system consisting of one, two, or three Fe rods (diameter 2 mm), embedded in the Teflon cylinder in a way shown by schematic in Fig. 3.10.

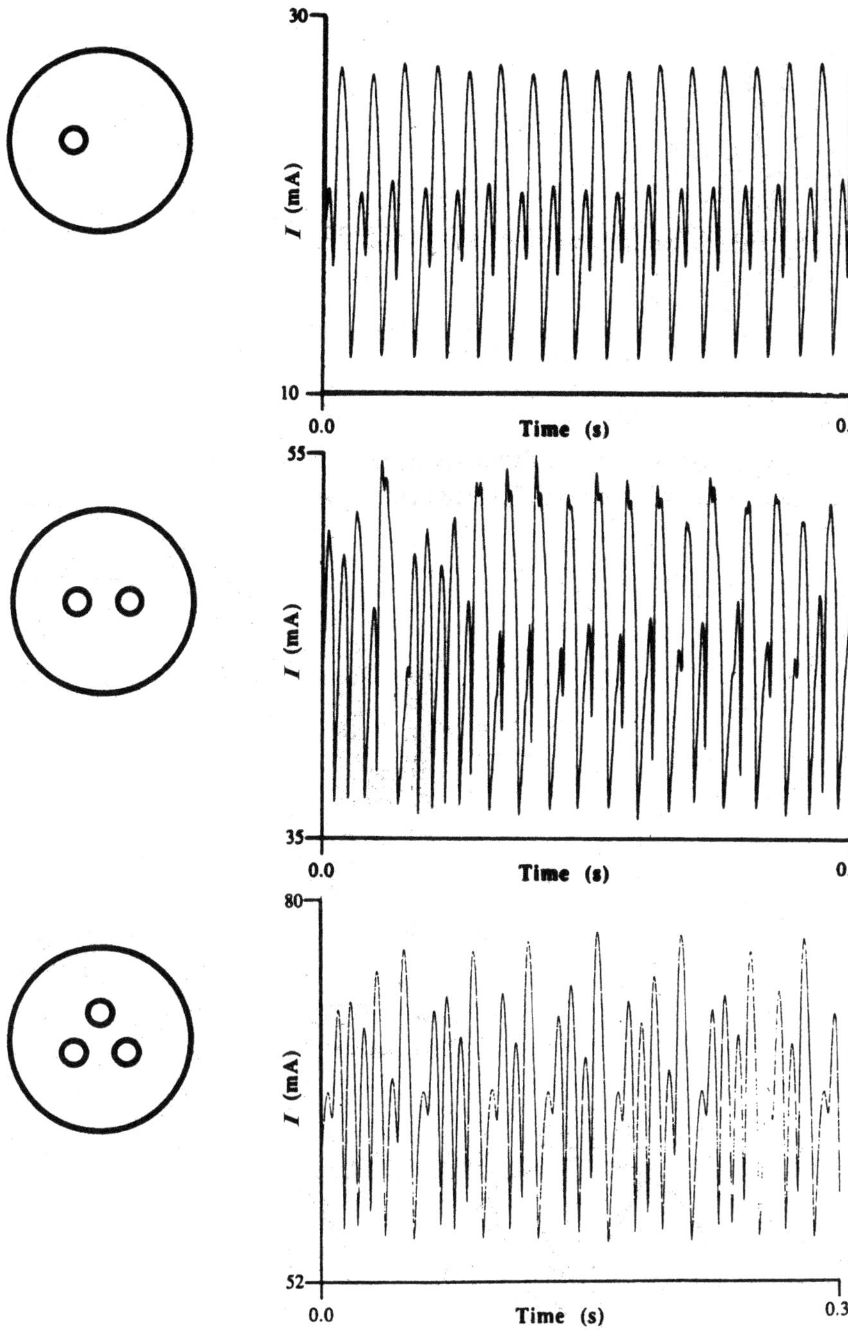

Fig. 3.11 Time series for single, double, and triple system of coupled Fe/H_2SO_4 oscillators. $R = 2$ mm, $E = -0.37$ V (vs. Hg/Hg_2SO_4), 400 rpm. Reprinted with permission from [9]. Copyright 1992 American Chemical Society

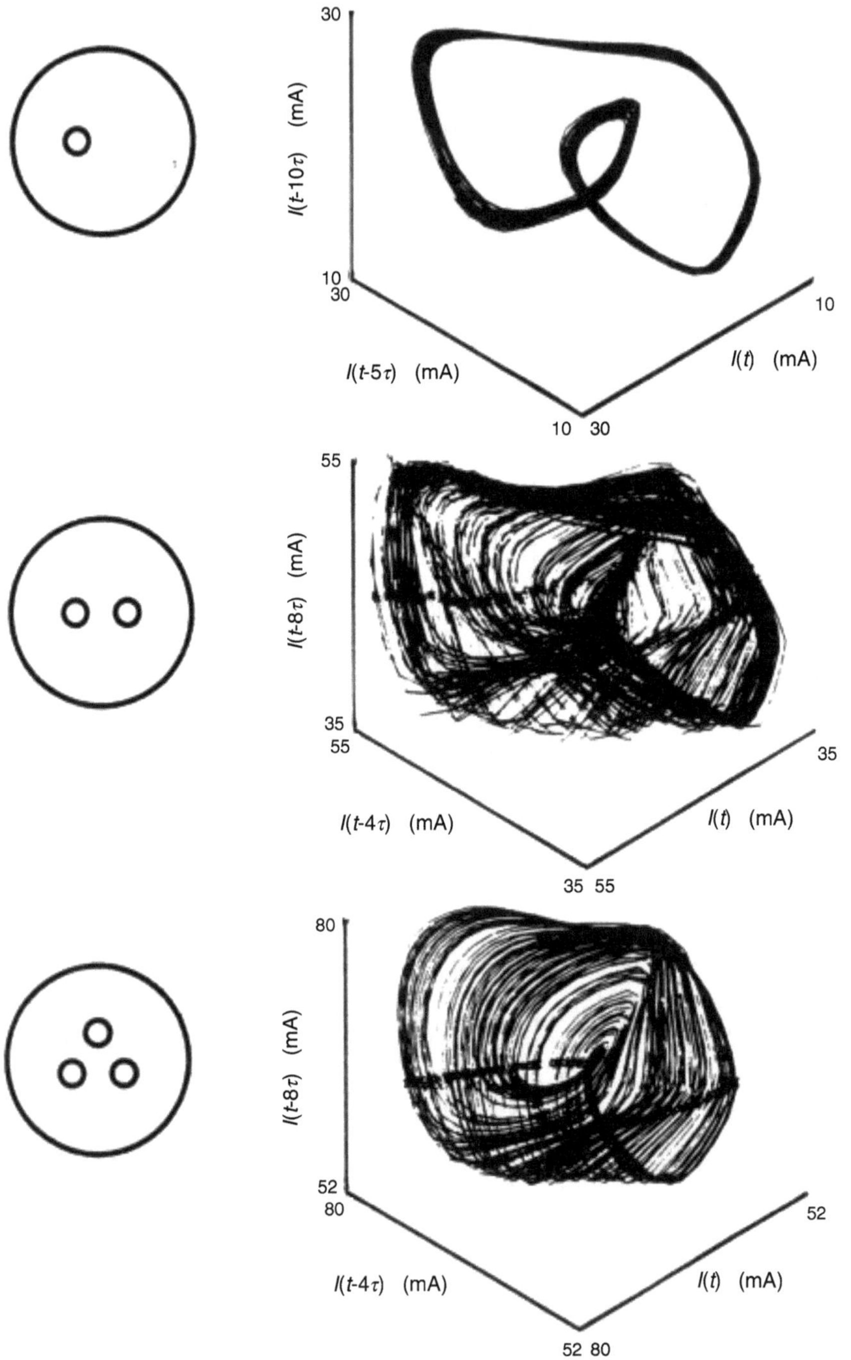

Fig. 3.12 Attractors corresponding to Fig. 3.11: $\tau = 0.4$ ms. The plane used to make the sections is defined by the equation $I(t - 8\tau) = 45$ mA and $I(t - 8\tau) = 64$ mA for the two and three

All the electrodes were kept at the same potential and only the total current was measured. Figure 3.11 shows that when the single electrode produces periodic oscillations, the coupling of two or three electrodes is a source of chaos, analyzed in terms of the phase trajectories (forming then strange attractors) and Poincaré sections (Fig. 3.12).

Furthermore, coupling of single chaotic oscillators leads to even more complicated, higher order chaos (as proved by the calculated fractal dimension of chaotic attractors, increasing from 2.2 for the single oscillator, through 2.6 for the sets of two, up to 3.3 for the three oscillators).

The synchronization of two current Fe/H_2SO_4 oscillators was later studied also by Nakabayashi et al. [10, 11]. The extent of synchronization, increasing upon reducing the interelectrode distance, was illustrated by the maps of the electric potential distribution in the solution, indicating the role of ohmic potential drops in these couplings. Also, it was reported the change of the oscillation frequency upon irradiation of the iron electrode with visible light.

In later work, illustrative for both single and coupled Fe/H_2SO_4 oscillators, Karantonis et al. [12] have proved that during the oscillatory electrodissolution of Fe in H_2SO_4 solution, the formation of traveling activation/passivation fronts occurs spontaneously (cf. Sect. 2.1). The experimental system involved either a single Fe ribbon (3 cm long and 1 mm wide) or this electrode covered in the middle with the insulating film of various thickness, or even separated with the acrylic partition (Fig. 3.13). Introducing the insulator changed the single Fe/H_2SO_4 oscillator into the system of two identical oscillators, coupled through the electrolyte solution. The propagation of the reaction front along the electrode surface was measured with the 16-channel logic analyzed, with the platinum potential probes arranged 1 mm above the electrode surface.

Even for the continuous electrode surface the oscillatory response of current was accompanied with the development of the *accelerated* potential front, i.e., the state of the electrode surface did not change uniformly during one oscillatory peak (see Sects. 1.2 and 2.2). In other words, the individual areas of the electrode surface were not in the same state at the same time. Quantitatively, the progress of the potential along the electrode surface was very fast, taking ca. 30 ms, which was three orders of magnitude lower time than the period of oscillations.

For relatively small separation of two electrode areas with the insulating film (e.g., 3 mm length) the response of the systems of two oscillators, named A and B, is periodic and the potential front travels along the electrode surface in a similar fashion than for the continuous electrode. The drastic change in this response occurs for *sufficiently large* separation of both electrode segments, when, e.g., the sequence of small current peaks (corresponding to passive-to-active transitions) and large peaks (corresponding to active-to-passive transitions of both region A and B of the

Fig. 3.12 (continued) electrode configurations, respectively. Two Poincaré sections are shown, one for each direction crossing the plane. Reprinted with permission from [9]. Copyright 1992 American Chemical Society

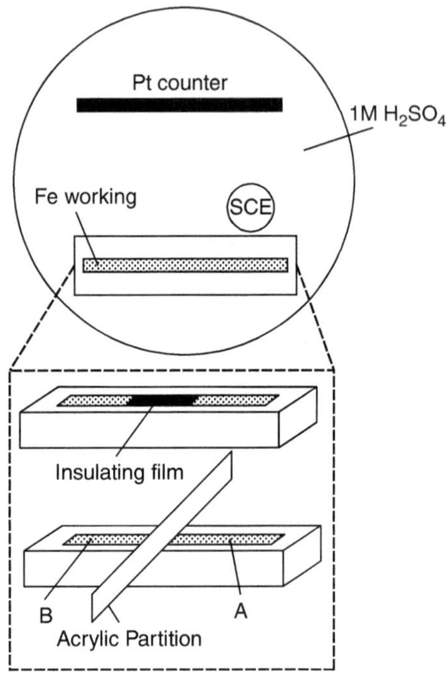

Fig. 3.13 Schematic representation of the experimental setup for the construction of two Fe/H_2SO_4 oscillators coupled through the electrolyte solution. Reprinted from [12], Copyright 2000, with permission from Elsevier

electrode) takes place. They attain either periodic or aperiodic course in time. The observed behaviors included simple in-phase synchronization, k/n phase locking (where k is the number of peaks of region A of the electrode and n is the number of peaks in region B), out-of-phase locking or more complex ones (see Fig. 3.14 as an example). Furthermore, the detailed sequence of events appeared to be *strongly dependent* on the position of the reference electrode, the fact being a strong premise for the understanding of coupling emerging in the process (see Sect. 1.2).

If the electrodes (and the adjacent solutions) are strongly separated by the acrylic partition, the two oscillatory subsystems A and B behave like independent oscillators, which exhibit different local electrolyte conditions. The total current response is chaotic and the two electrode regions oscillate independently with different phases. Karantonis et al. [12] have supported their experimental observations with the appropriate three-dimensional model, being extension of the approach by Koper and Sluyters for active/passive transitions of solid electrodes (cf. Sect. 6.1.2.2, volume I), extended now for inclusion of the presence of a salt precipitant during the oscillatory response and the presence of the set of the coupled oscillators. The dynamic variables were the interfacial potential drop of the working electrode, the surface concentration of H^+ ions at the surface concentration of Fe^{2+} ions, for the ith oscillator ($i = 1$ or 2). The coupling term was introduced to the equation describing the dynamics of electrode potential, according to the assumption that this coupling is realized mainly through the migration current flowing along the electrode surface. Results of numerical simulations were found to be well

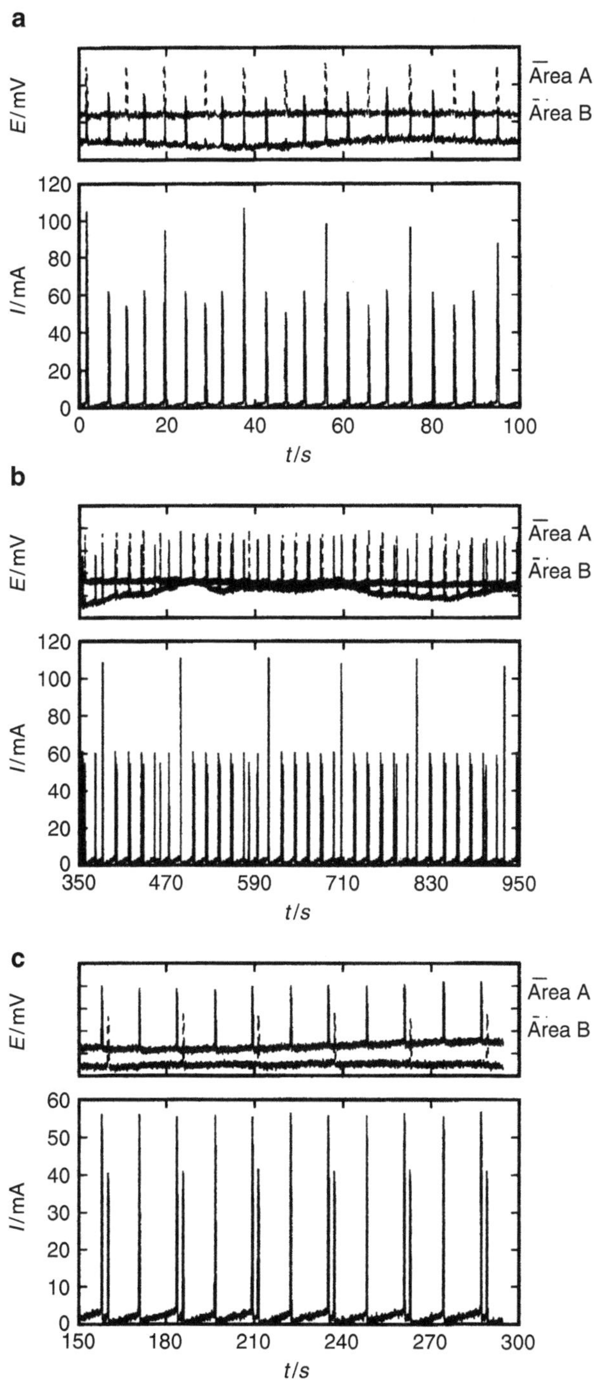

Fig. 3.14 Coupling patterns for film length 22 mm. (**a**) In-phase 3/2 response for $U = 272$ mV, (**b**) In-phase 6/5 pattern for $U = 280$ mV. (**c**) Out-of-phase 2/1 pattern for $U = 282$ mV. Potential scale 50 mV div^{-1}. Reprinted from [12], Copyright 2000, with permission from Elsevier

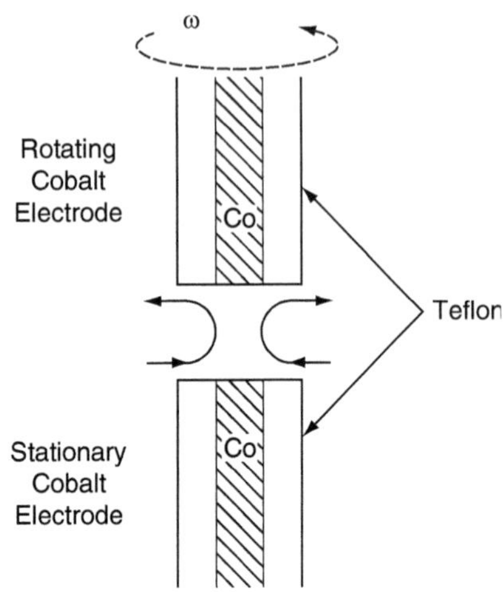

Fig. 3.15 Schematic of coupled Co electrode system. Reprinted with permission from [15]. Copyright 1992 American Chemical Society

concordant with experimental results. One should emphasize that the phenomena described above constitute an introduction to extremely important and modern problem of coupling of electrochemical oscillators, which is related to the modeling of neural networks, outlined later in this chapter.

In complement to the above considerations it is useful to note that systems with a passive barrier were studied also for the case of only diffusional coupling. For example, Sielewiesiuk and Górecki [13] have analyzed theoretically the dynamics of two excitable systems, separated by a narrow passive barrier, in terms of the FitzHugh–Nagumo model for excitable characteristics of nerve tissue, with diffusion transport included. It was shown that at the barrier a periodic train of pulses may be transformed into a complex output signal, with the ratio of frequencies of the output and input signals, plotted as a function of barrier's width or of the input signal frequency, having a fractal, devil's staircase-like shape (for the explanation of the latter term, cf., e.g., http://mathworld.wolfram.com/DevilsStaircase.html).

3.1.5 Coupled Co/HCl + CrO_3 Oscillators

The simplest manifestation of coupling between the two cobalt electrodes, immersed in HCl + CrO_3 solution and exhibiting oscillations of their potentials, is described by the following experiment. If those electrodes are not connected, they usually differ somewhat in the frequencies of the oscillations, but after connecting (e.g., through a low resistance or the capacitance) these frequencies become almost identical and close to the average of the two original, independent values. Also, after a few cycles the oscillations acquire the same phase. In a particular case when the electrodes are not connected through the external circuit, but just put close to each other in the same solution, the coupling results in common frequency higher than either of the original frequencies. In other words, the change of the system's state occurs now more quickly than in the absence of coupling and this happens due to the distribution of the local currents which can partially enter the neighboring electrode which is better conductor than the electrolyte solution [14].

Systematic studies of dynamics of coupled oscillating cobalt electrodes were described by Hudson et al. [15, 16]. For that purpose, one rotating (upper) and one stationary (lower) Co disk, immersed in the same HCl + CrO_3 electrolyte (Fig. 3.15), were employed. The coupling between the electrodes is realized through the transport of reactants among the electrodes and through the ohmic potential drop in the electrolyte.

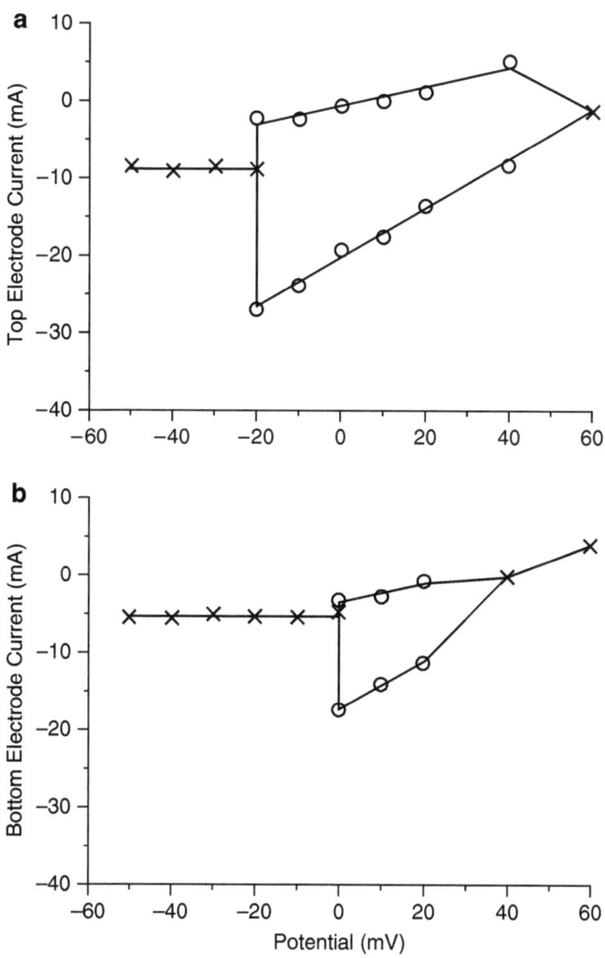

Fig. 3.16 Current vs. potential for each electrode with opposite electrode blank: (×) steady state, (o) minimum and maximum of current oscillations; 0.6 M CrO_3/1.0 N HCl, 75 rpm, 4-mm gap. (**a**) Top electrode; (**b**) bottom electrode. Reprinted with permission from [15]. Copyright 1992 American Chemical Society

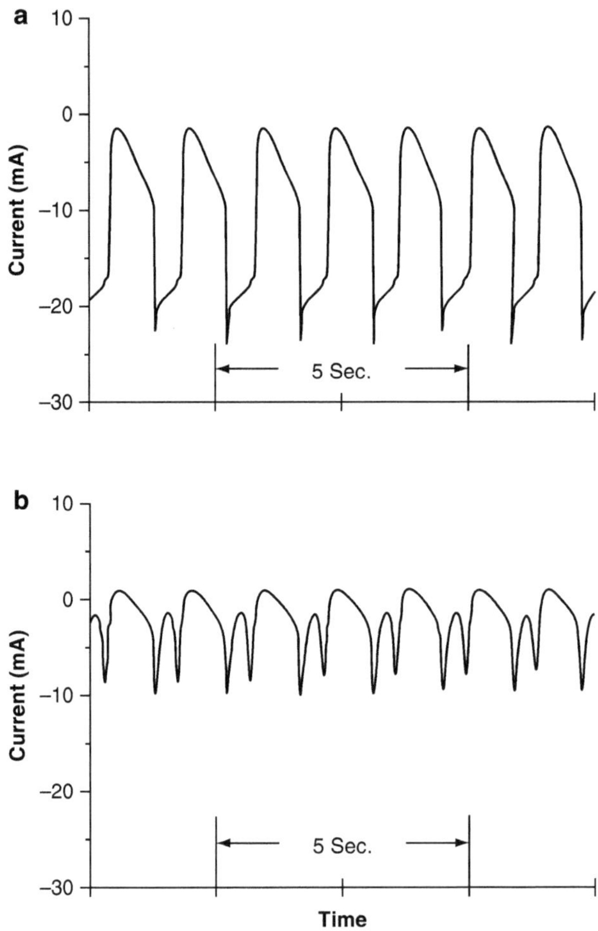

Fig. 3.17 Oscillations: (**a**) top electrode, $E = -50$ mV, below single disk oscillatory range; (**b**) bottom electrode, $E = +40$ mV, above single disk oscillatory range. Reprinted with permission from [15]. Copyright 1992 American Chemical Society

The current which flows is a net current produced by anodic dissolution of cobalt and cathodic reduction of chromic acid. The currents of the upper and the lower disks for given potential were recorded separately. Also the potential of each disk could be recorded separately. In the first series of experiments, the blank disk was placed opposite to each actually studied disk and the current of the latter one was recorded as a function of the electrode potential. The interelectrode distance of only 4 mm affected the hydrodynamic flows in this region and this effect, now isolated, will contribute to the electrochemical characteristics of the systems, when both disks become working electrodes. Figure 3.16 shows that for both disks the oscillatory region was found, being narrower for the stationary disk. More

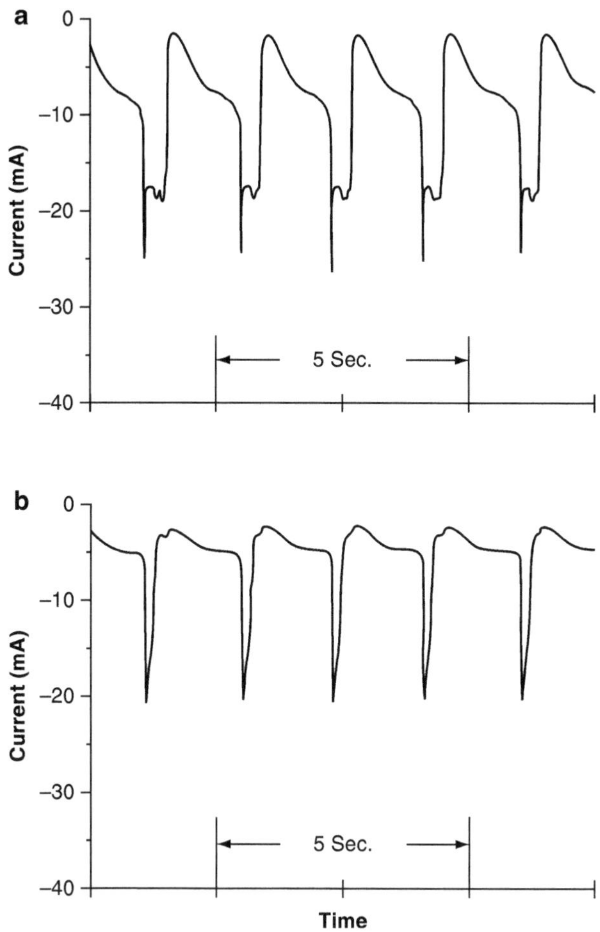

Fig. 3.18 Oscillations: (**a**) top electrode, $E = -30$ mV, below single disk oscillatory range; (**b**) bottom electrode, $E = -30$ mV, below single disk oscillatory range. Reprinted with permission from [15]. Copyright 1992 American Chemical Society

precisely, the region of oscillations corresponded to such behavior which established after the transient, nonoscillatory regime, for given potential.

Next, both disks were active and kept at fixed potential, and the current was measured separately for each of them. Then a variety of states—steady states, periodic and chaotic oscillations were observed. Furthermore, if the amplitude of current oscillations at one disk is much higher than at the other one, one can consider the first disk driving the other one which, as a separate electrode, would not exhibit oscillations. Under different conditions, both electrodes can exhibit oscillations of substantial amplitude, although separately none of them would exhibit oscillations, at the potentials below and above single disk oscillatory range, respectively (Fig. 3.17).

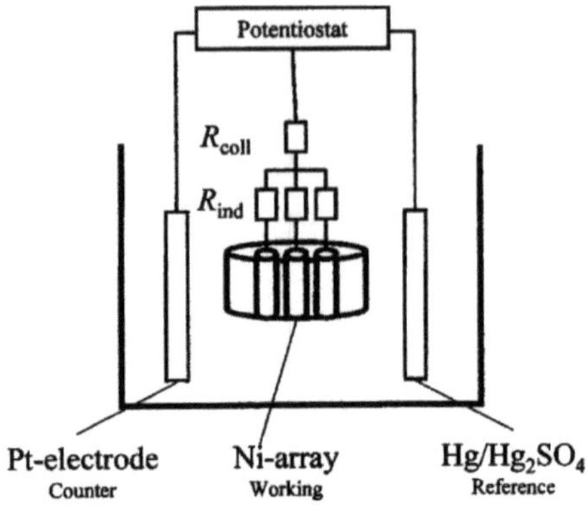

Fig. 3.19 Experimental setup for studying coupled Ni/H_2SO_4 oscillators. Reprinted with permission from [17]. Copyright 1999 American Chemical Society

Furthermore, it can also happen that both electrodes exhibit oscillations, if both kept separately at the potential below the oscillatory regime (Fig. 3.18).

Finally, when both electrodes exhibit independent oscillations, their coupling leads to chaotic courses.

3.1.6 Coupled Ni/H_2SO_4 Oscillators

In view of spatiotemporal patterns reported for the single Ni/H_2SO_4 oscillator (Sect. 2.4) it was justified to study the effects of systematic variation of the extent of global coupling on the course of Ni dissolution. In recent work, Kiss et al. [17] have analyzed the dynamics of the periodically oscillating electrode arrays, consisting of from 1 to 64 nickel electrodes immersed in sulfuric acid solution. The great advantage of using the electrode arrays was the possibility to measure the individual current at each electrode, allowing to determine conveniently the distribution of the electrochemical reaction rate as a function of both time and space. Also, due to long-range (migration) coupling through the electrolyte solution, the total set of individual electrodes can exhibit, under appropriate conditions, an overall behavior similar to that of a single electrode of the same area (experiments proving that for the iron-sulfuric acid system were performed by Hudson et al. [18]). The experimental methodology used in the present work aimed to introduce the global coupling to the process of Ni electrooxidation, with the fundamental assumption that only the strength of this coupling will be changed, with all other parameters (like the electrolyte concentration or cell geometry) held constant. This was achieved through the use of external resistors connected to every single electrodes, which resistances were varied, but the total external resistance was kept constant. Figure 3.19 shows schematically the relevant experimental setup in which there are visualized two ways of connecting the external resistors: the individual (R_{ind}) to every electrode, and the collective (R_{coll}) to all the electrodes simultaneously. Figure 3.20 represents the exemplary current–potential recordings for a *single* electrode. Particular regions

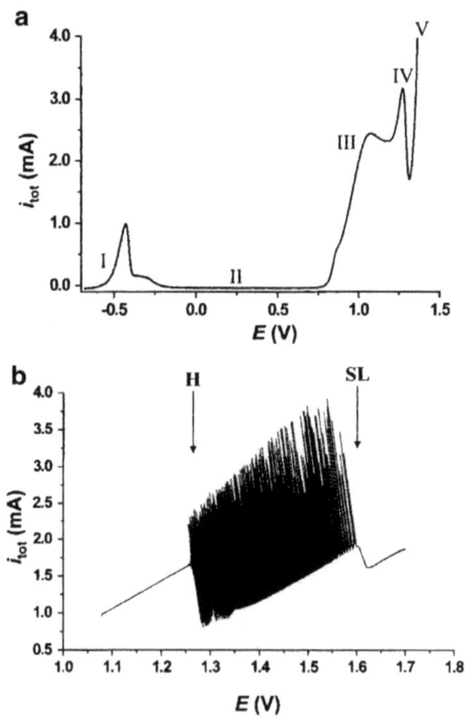

Fig. 3.20 Anodic polarization curves of the Ni-sulfuric acid system. (**a**) *IR*-compensated scan (compensated resistance: 2.5 Ω) with a scan rate of 10 mV/s. (**b**) Scan for single electrode using $R_{eq} = 200\ \Omega$ external resistance. Scan rate of the potential was 1 mV/s. *E* is the true potential of the working electrode obtained by subtracting the *IR* drop through external (200 Ω) and solution resistance (3 Ω) from the applied potential *U*. Reprinted with permission from [17]. Copyright 1999 American Chemical Society

of the *I–E* curve in Fig. 3.20a with *IR* compensation correspond to: (I) active dissolution; (II) primary passivity with the formation of an oxide film; (III) secondary dissolution in the transpassive region, developing into (IV) regime of secondary passivity and (V) O_2 evolution. The instabilities, corresponding to the characteristics of the HN-NDR oscillator, set in only in the presence of external resistance, and Fig. 3.20b shows that with increasing electrode potential harmonic oscillations are born through the Hopf bifurcation, develop further into relaxation oscillations, and eventually cease due to saddle-loop bifurcation.

According to the scheme of connections shown in Fig. 3.19, there are two ways of connecting the external resistors to the cell: the individually (R_{ind}) and the collectively (R_{coll}). The total resistance R_{tot} is defined as:

$$R_{tot} = R_{coll} + \frac{R_{ind}}{n} \tag{3.6}$$

where n is the number of electrodes.

Also, it can be convenient to define the equivalent resistance (R_{eq}):

$$R_{eq} = nR_{tot} \tag{3.7}$$

which gives the equivalent individual resistor that should be inserted into the circuit to observe the same potential drop on the electrode. R_{eq} is useful for intercomparison of experimental results with different numbers of electrodes in the array.

The collective resistor creates the global coupling, since the change of current flowing through one electrode affects the dynamics of the other electrodes, due to equal change of their potentials. The extent of global coupling can be quantitatively described by the ratio:

$$\varepsilon = \frac{R_{coll}}{R_{tot}} \tag{3.8}$$

which attains zero for no additional global coupling through external resistance and unity for the maximal external global coupling. Studies of increasing ε covered the array of 2 up to 64 electrodes, for the oscillations occurring in the potential region close to either the Hopf, or the saddle-loop bifurcation. In the former case, when the oscillations are small amplitude and harmonic, the increasing global coupling leads to synchronization of all individual oscillators without qualitative change in their dynamics. However, in the case of relaxation oscillations, occurring at potentials close to the saddle-loop bifurcation, the sequence of events is more complex even for only 2 electrodes. With increasing extent of global coupling the synchronization is first observed, but later it breaks up and irregular behavior sets in, including transient clusters; at stronger coupling the clusters stabilize and in this region even multistability, i.e., coexistence of many cluster configurations, was observed. With the cluster we understand here the pattern constructed by the distribution of individual currents between the electrodes. As even stronger coupling antiphase oscillations occur, which correspond to clusters made of (approximately) equal number of oscillators. Finally, strong coupling synchronizes all the oscillators into a simple behavior, resembling the dynamics in the absence of coupling. Essential experimental observations were reproduced by the relevant mathematical model [17].

Figure 3.21 illustrates the role of increasing global coupling for the set of two electrodes by showing the individual currents at every electrode (left column) and total current (right column).

In turn, Fig. 3.22 shows the patterns of current (clusters) reported for the set of 64 electrodes. One should note that the reported formation of clusters is an electrochemical example of the phenomena studied also in simulation of coupled chaotic maps [19] and differential equations [20].

Later Kiss et al. [21] have performed similar studies for the smaller sets of Ni electrodes: an array consisting of two to eight elements. The emphasis was on the interactions among chaotic oscillators and the effect of the coupling on the complexity of individual and total dynamics. It was observed (with some exceptions) that upon increasing the global coupling strength, the overall (total) current dynamics

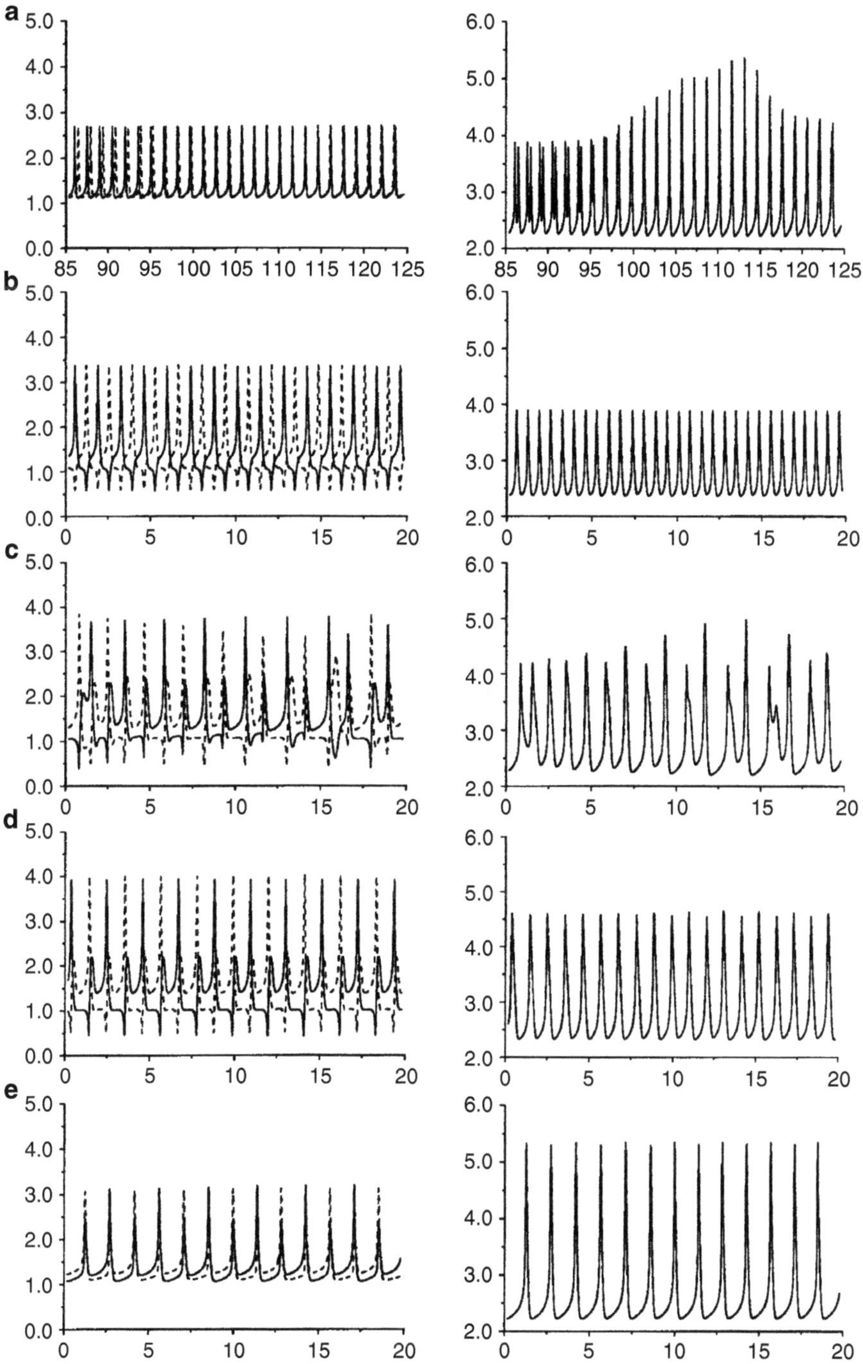

Fig. 3.21 Two electrodes near saddle-node bifurcation (U = 1.480 V, R_{eq} = 260 Ω). Individual (*left column*) and total (*right column*) currents (mA) vs. time (s). (**a**) $\varepsilon = 0.0$; (**b**) $\varepsilon = 0.6$; (**c**) $\varepsilon = 0.8$; (**d**) $\varepsilon = 0.9$; (**e**) $\varepsilon = 1.0$. Reprinted with permission from [17]. Copyright 1999 American Chemical Society

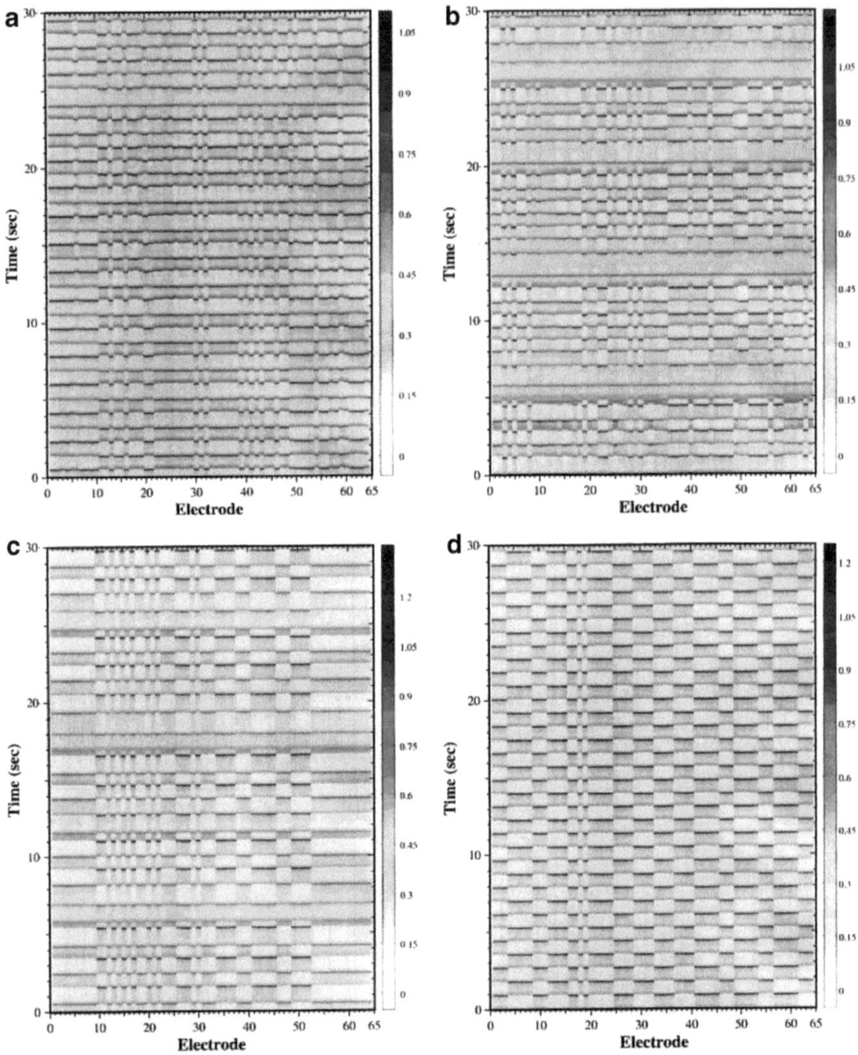

Fig. 3.22 64-electrode array, relaxation oscillations ($U = 1.53$ V, $R_{eq} = 906\ \Omega$). Current as function of space/time; dark corresponds to high current. (**a**) $\varepsilon = 0.67$; (**b**) $\varepsilon = 0.78$; (**c**) $\varepsilon = 0.89$; (**d**) $\varepsilon = 0.95$. Reprinted with permission from [17]. Copyright 1999 American Chemical Society

became simpler, while the complexity of the individual currents passed through a maximum at intermediate values of the coupling strength (i.e., at conditions in which interactions occur but the coupling is not yet strong enough to produce synchronization). Those general trends with increasing coupling strength were interrupted by clustering that occurs with the four and eight electrode arrays.

In these considerations, calculations of the *information dimension* [22] were used for quantitative description of the complexity of the signals.

Further studies in this area, aiming to review experimental results on the collective dynamics of sets of globally coupled chaotic electrochemical oscillators, were published also by Kiss, Hudson et al. [23, 24]. In these experiments, involving arrays of *chaotic* Ni/H_2SO_4 electrochemical oscillators, besides global coupling, also periodic forcing[1] and a feedback, in which a signal bases on the total current was fed back to the applied potential, were applied, so the effects of all these actions could be compared. For the case of very weakly coupled set of chaotic oscillators, transition to a synchronized state with sufficiently large values of coupling strength was observed. At intermediate values, both intermittent and stable chaotic cluster states were found. Cluster formation and synchronization were also obtained by applying feedback and forcing to a moderately coupled base state. The feedback and forcing also produced periodic cluster states and more than two clusters, with the configurations of two (chaotic) clusters and two, three, or four (periodic) clusters were observed. Using the same array of 64 Ni/H_2SO_4 oscillators under conditions of increasing coupling strength it was possible to verify the theory of Kuramoto [25], predicting that global coupling in a set of smooth limit-cycle oscillator with different frequencies should produce a phase transition in which some of the elements synchronize; the studies were extended also for relaxation and chaotic oscillators [26].

In line with increasing interest in the investigation of systems consisting of interacting chemical oscillators, which can resemble the principle of coherent dynamics occurring in biological systems, Mikhailov et al. [27] have presented both numerical simulations and laboratory experiments of a system of phase oscillators. A phase oscillator is a simple cyclic element (a clock) specified only by its natural frequency. In the model, a population of 1,000 of such oscillators with a strongly nonuniform distribution of natural frequencies displaying several maxima was chosen. When interactions among the elements were included, formation of synchronous oscillator groups was observed. For sufficiently high interaction strengths, most of the population was entrained into several large clusters. Eventually, the dynamics of the entire system became fully synchronized. The results of simulations are shown in Fig. 3.23. The simulation results were compared with the experimental studies, involving an array of 64 nickel electrodes, dissolved anodically in sulfuric acid medium. The reader interested in further aspects of those phenomena is advised to consult the original reference.

[1] See Sect. 3.2 for examples of forced oscillators.

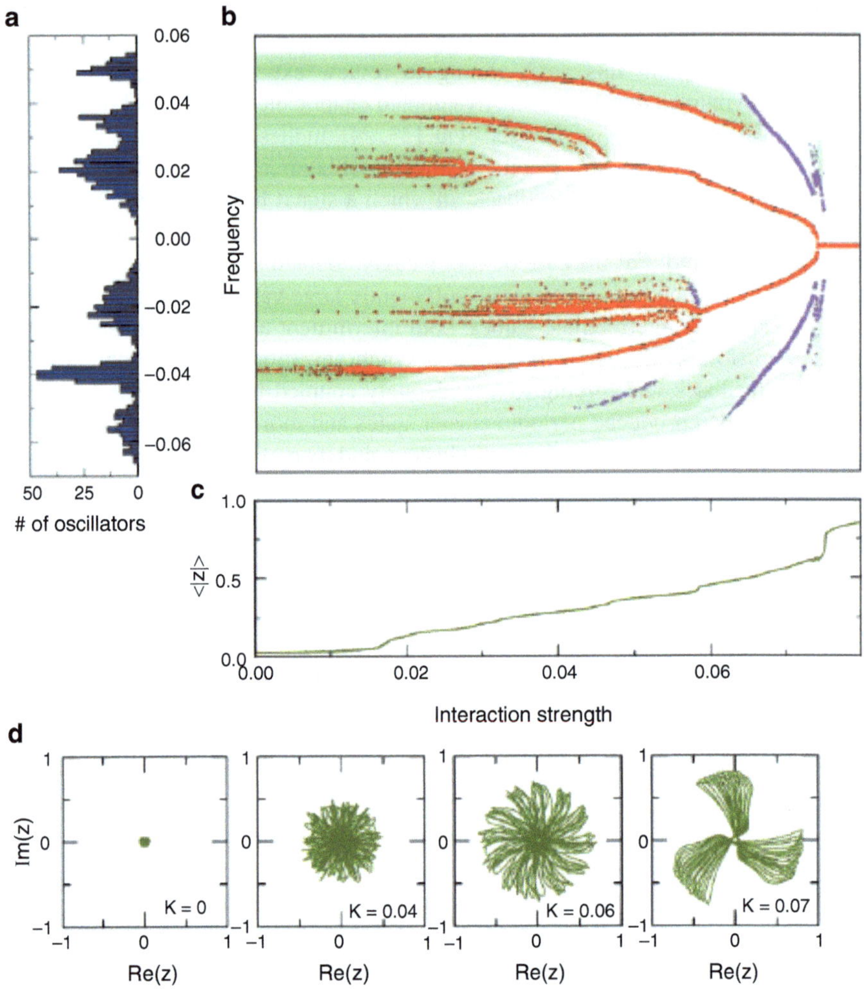

Fig. 3.23 Simulation for a population of $n = 1{,}000$ interacting phase oscillators. Evolution of phases $\phi_j(t)$ is governed by equations $d\phi_j/dt = \omega_j + (K/N)\sum_{k=1}^{N}\sin(\phi_k - \phi_j)$ with $j = 1, 2, \ldots, N$, where K is the interaction strength. These equations were numerically integrated by using an Euler scheme with time step $\Delta t = 1$. (**a**) Distribution of natural frequencies, ω_j, in the studied population. (**b**) Distribution of effective frequencies as a function of the interaction strength K. In the color code, *darker shades of green* indicate the higher local oscillator density. Additionally, clusters are displayed by using *red* and *magenta*. A cluster is defined as a group of oscillators whose effective frequencies differ by $<10^{-5}$. *Small dots* indicate clusters of a size between 5 and 10; *large dots* correspond to clusters of a size >10. Small clusters of a size <5 are not displayed. *Magenta dots* indicate resonant clusters. Effective frequencies for each oscillator were computed by averaging $d\phi_j/dt$ over a time interval of 6×10^4 units. (**c**) The average amplitude $<|z(t)|>$ of the complex collective signal as a function of the interaction strength. (**d**) Trajectories of $z(t)$ in the complex plane [Re(Z), Im(Z)] for several values of the interaction strength K, each for a time interval of 3,000. From left to right, $K = 0$, 0.04, 0.06, and 0.07. Reprinted with permission from [27]

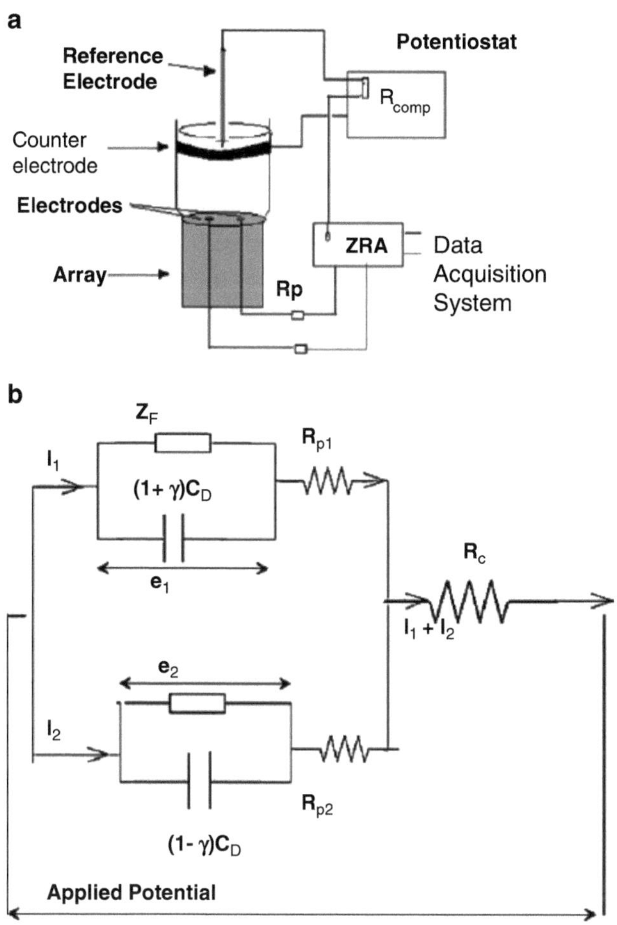

Fig. 3.24 (**a**) Experimental setup. (**b**) Equivalent circuit for the two electrodes. Reprinted from [29], Copyright 2009, with permission from Elsevier

In course of further development of complex dynamical systems, Kiss et al. [28] have described a methodology for the *design of complex dynamical structures* which could be tuned to desired states. For this purpose, weak, nondestructive signals were used to alter interactions among nonlinear rhythmic elements. Such an approach is analogous to mild control in living systems, in which the desired behavior (e.g., altering pathological behavior) should be attained without destroying its fundamental nature. In particular, the phenomena described in [28] can reveal correspondence with the neural systems. In the present work, the population of N oscillators was set into desired behavior through the imposition of a nonlinear, time-delayed feedback. As an efficient tool for the analysis of the collective behavior of such coupled oscillator, the *phase models*, which describe each rhythmic unit by a single variable, the phase, were employed. As an experimental system again the electrodissolution of Ni electrode in sulfuric acid was used. The following results were obtained: (1) the tuning of desired arbitrary phase differences between two dissimilar oscillators, (2) the generation of complex patterns in the system of four dissimilar oscillators that include self-organized switching between unstable dynamical states (the system sequentially visits unstable two-cluster states with two oscillators in each cluster), and (3) desynchronization of system of 64 coupled relaxation oscillators. The latter case can have important biological significance, as the way of destruction of pathological synchronization of interacting oscillators, leading to disease.

3.1.7 Coupled Oscillators and IR Compensation

The role of ohmic drops in creation of negative coupling that can lead to pattern formation was explained in Sect. 1.2, for the case of a single working electrode. Recent experimental and theoretical work [29] has shown that the role of ohmic drop *compensation* is quite sophisticated if, instead of a single electrode, a dual-electrode metal dissolution system is used (taking Ni or Fe electrode in H_2SO_4 medium as an example). Figure 3.24 shows the experimental, two-electrode setup and the equivalent circuit.

In Figure 3.24b, R_{p1} and R_{p2} mean two parallel resistors intentionally added in most experiments, while R_c is the collective resistance in series with the electrodes. If $R_c > 0$, then it models the ohmic potential drop in the electrolyte, while $R_c < 0$ corresponds to negative coupling induced by *IR* compensation. If $R_{p1} = R_{p2} = R_p$, then the equivalent resistance of the circuit is given by $R_{eq} = R_p + 2R_c$.

For studies of the effect of *IR* compensation on the bistable region, the nickel electrodissolution was used, while for analogous studies concerning the oscillatory regime, the electrodissolution of iron was employed. In both cases, the electrical coupling was expressed in terms of the respective model. Here we show, as an example, only the equations for the simpler case of electrical coupling in bistable system:

$$\frac{\mathrm{d}e_1}{\mathrm{d}t} = \frac{v - e_1}{R_{eq}} + i_f(E_1) + \frac{K}{2}(e_2 - e_1) + A\xi_1 \tag{3.9}$$

$$\frac{\mathrm{d}e_2}{\mathrm{d}t} = \frac{v - e_2}{R_{eq}} + i_f(e_2) + \frac{K}{2}(e_1 - e_2) + A\xi_2 \tag{3.10}$$

where e_1 and e_2 mean here the effective (double layer) potentials of the respective electrodes, v—the external voltage. The quantity K is the coupling strength given by:

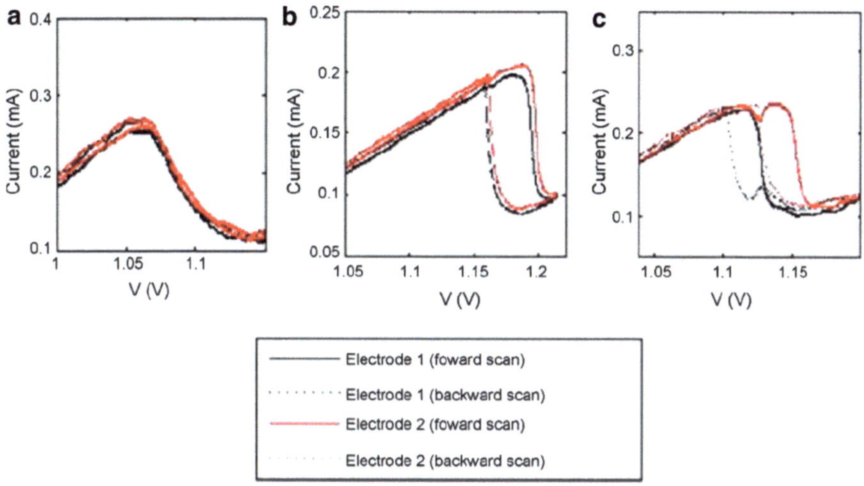

Fig. 3.25 Polarization scan curves of nickel electrodes at different values of added external resistance, R_p and *IR* compensation R_c; forward scan rate $= 1$ mV/s and backward scan rate $= -1$ mV/s. (**a**) $R_p = 0, R_c = 0$. (**b**) $R_p = 602\ \Omega, R_c = 0$; (**c**) $R_p = 602\ \Omega, R_c = -301\ \Omega$. Reprinted from [29], Copyright 2009, with permission from Elsevier

$$K = \frac{\varepsilon}{1 - \varepsilon} \frac{1}{R_{eq}} \tag{3.11}$$

where $\varepsilon = 2R_c/R_{eq}$ is the collective resistance fraction. If *IR* compensation is applied to both electrodes, $K < 0$. Finally, the last terms on the right-hand side of Eqs. (3.9) and (3.10) were introduced to account for unavoidable small noise existing in the experimental system.

In turn, for the modeling of coupled oscillators, the model of Koper and Sluyters [31], which involves dynamics of the electrode potential, of the surface hydrogen ion concentration and of the iron(II) ions concentration, was used. Respective equations are given in the original reference [29].

The most important results of these studies can be summarized as follows. Unlike the case of a single electrode, for the two-electrode system the *IR* compensation is not able to remove bistability and for large *IR* compensation the nonuniform (patterned) surface develops. This is illustrated by Fig. 3.25c, which shows that upon *IR* compensation, the *I–v* dependence corresponding to zero resistance (Fig. 3.25a) is NOT recovered: one electrode exhibits larger, the other electrode the lower current than it would be expected.

In turn, for the oscillatory system, addition of *IR* compensation produces aperiodic time series that are characterized by switching between oscillations with 1:1, 1:2, and 2:1 entrainment ratios; for large enough *IR* compensation the amplitude death occurs and either coexistence of oscillations with steady state or multiple antisymmetric steady states are observed. Figure 3.26 shows oscillatory courses for moderate added negative coupling strength, whereas Fig. 3.27 shows the relative contribution from different entrainment ratios, as a function of $R_{comp.}$

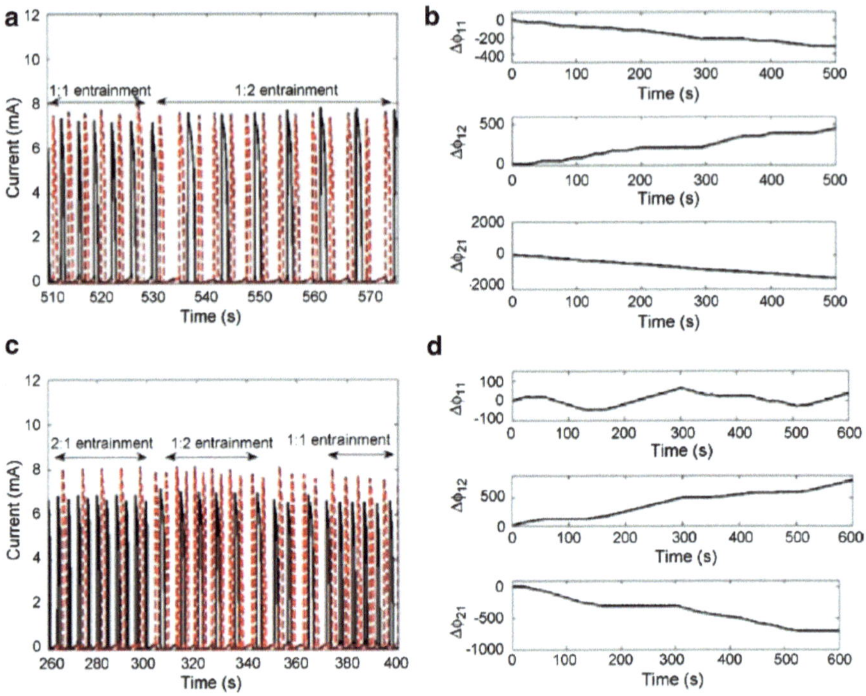

Fig. 3.26 Current time series (electrode 1: *solid line* and electrode 2: *broken line*) and phase difference for two iron electrodes at intermediate added negative coupling strengths at $V_{app} = -202$ mV. (**a**) Current for $R_{comp} = -30\ \Omega$ and $R_p = -2R_{comp} = 60\ \Omega$; regions of different entrainment are also marked. (**b**) Phase difference $\Delta\phi_{11}$ (*top panel*), $\Delta\phi_{12}$ (*middle panel*), and $\Delta\phi_{21}$ (*bottom panel*) as a function of time for current in (**a**). (**c**) Current for $R_{comp} = -40\ \Omega$ and $R_p = -2R_{comp} = 80\ \Omega$, (**d**) Phase difference $\Delta\phi_{11}$ (*top panel*), $\Delta\phi_{12}$ (*middle panel*), and $\Delta\phi_{21}$ (*bottom panel*) as a function of time for current in (**c**). Reprinted from [29], Copyright 2009, with permission from Elsevier

As in other examples of coupled oscillators, also in this case one can find analogies with the dynamics of biological systems. The behavior similar to switching among entrainment states was reported for the cardiorespiratory system, where the signals from the cardiac system and the respiratory system stay synchronized with a certain entrainment ratio for some time and then change the entrainment ratio [32].

3.1.8 Coupling the S-NDR Oscillators

S-NDR oscillators are definitely less common (and thus not so intensively studied) as the N-NDR oscillators. Therefore, it is noteworthy that Bîrzu and Gáspár [30] recently have undertaken theoretically the problem of S-NDR oscillators, synchronized via the global coupling. As the model system, the Lee–Jorné

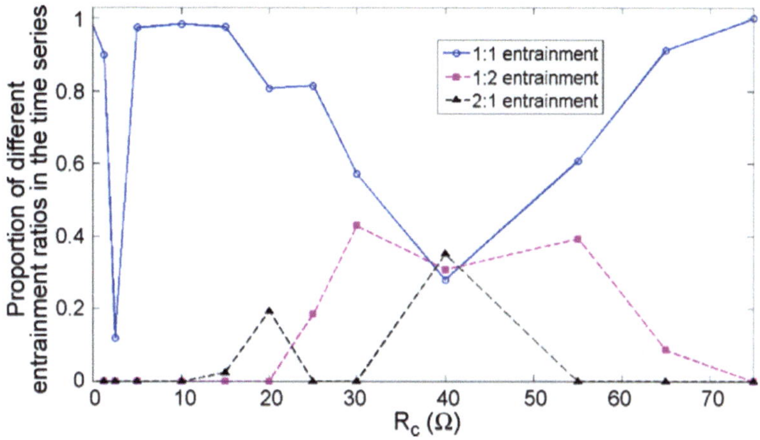

Fig. 3.27 Percentage of different entrainment ratios in the time series [(*open circle*) 1:1 entrainment, (*open triangle*) 2:1, (*filled square*) 1:2] as a function of R_{comp}. Reprinted from [29], Copyright 2009, with permission from Elsevier

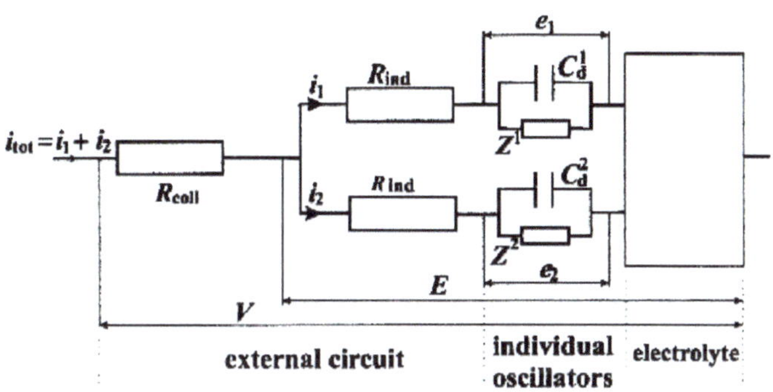

Fig. 3.28 Equivalent electric circuit of two globally coupled electrochemical oscillators. Reprinted from [30], Copyright 2009, with permission from Elsevier

mechanism [33] for the zinc electrodeposition was applied for the system of either 2 or an array of 128 globally coupled nonidentical electrodes. The strength of global coupling between the electrodes was systematically varied by changing collective and individual resistors in the circuit, shown in Fig. 3.28.

Using symbols from this scheme, the strength of coupling was expressed in terms of quantity κ, defined as:

$$\kappa = \frac{1}{A}\left(\frac{R_{\text{coll}}}{R_{\text{ind}}}\frac{1}{R_{\text{ekv}}}\right) \tag{3.12}$$

with $R_{\text{ekv}} = 2R_{\text{coll}} + R_{\text{ind}}$.

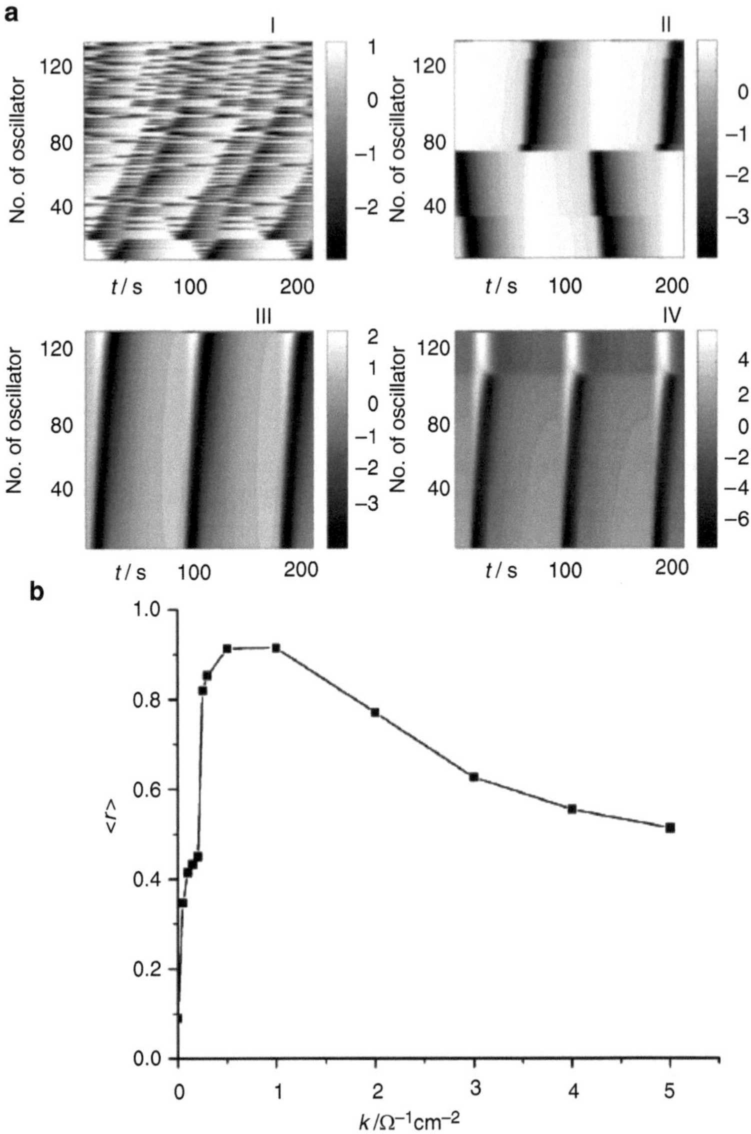

Fig. 3.29 "Wave-like" dynamics of current oscillations of 128 coupled electrodes. (**a**) Time evolution of the current of the individual electrodes at different values of coupling strength κ: 0 Ω^{-1} cm^{-2} (I), 0.15 Ω^{-1} cm^{-2} (II), 0.5 Ω^{-1} cm^{-2} (III), and 3.0 Ω^{-1} cm^{-2} (IV) and other parameters (see [30]). The electrodes have been made nonidentical [30]. The *gray scale* on the side of the frames scales the current values in mA. (**b**) The average order parameter $<r>$ as a function of the control parameter κ. Reprinted from [30], Copyright 2009, with permission from Elsevier

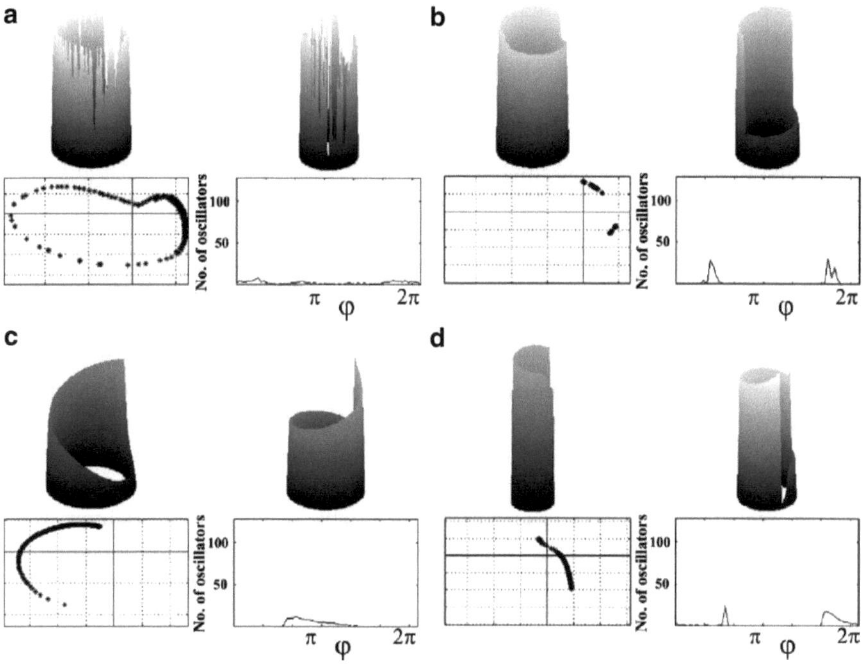

Fig. 3.30 Dynamics of 128 coupled oscillators showing "wave-like" dynamics of current oscillations (Fig. 3.29). In each frame (**a**)–(**d**): *upper left*—current of the individual oscillators: *upper right*—instantaneous phase of the individual oscillators; *lower left*—a characteristic phase portrait; lower right—distribution of instantaneous phase of the individual oscillators. Frames (**a**), (**b**), (**c**), and (**d**) correspond, respectively, to the state of the system at $t = 12.45$ s in (1), at $t = 31.42$ s in (II), at 12.45 s in (III), and at $t = 109.1$ s in (IV) of Fig. 3.29a. Reprinted from [30], Copyright 2009, with permission from Elsevier

By systematically varying the strength of global coupling, several different scenarios and partial synchronization have been generated. For the two coupled nonidentical Zn electrodes, the synchronized in-phase and antiphase oscillations, partial synchronization, and period-2 synchronization were observed. For 128 coupled electrodes, there were reported complete synchronization, dynamical cluster formation, and "wave-like" dynamics in the change of the amplitude and the instantaneous phases of the analytical signal of the individual electrodes. The latter type of synchronization is (as an example) illustrated in Figs. 3.29 and 3.30.

3.1.9 Coupled Electrochemical Oscillators and Neural Cells

Before the presentation of some electrochemical models of the interneural connections, it is useful to realize [1] that synaptic transmission between neural cells occurs either by electrical or chemical synapses. These two synapses differ with the particular way the interneural communication is realized. For electrical synapses, the electric current is produced in the terminals of the presynaptic neuron and flows through bridging channels, causing the change of the membrane potentials of the postsynaptic cells. Noteworthy, the current produced in the postsynaptic cell can flow back into presynaptic neuron, so the electrical synapses are bidirectional [35, 36]. For chemical synapses, the electric spike (action potential) in the presynaptic neuron leads to the secretion of a chemical transmitter that diffuses across the synaptic cleft and binds to ion channel at the postsynaptic cell; there the current is induced and the membrane potential is changed. Contrary to electrical synapses, the chemical ones act only unidirectionally, since chemical transmitters can be released only by the presynaptic neuron.

The way in which the postsynaptic membrane responses to a signal depends on the dynamic characteristics of that membrane, i.e., whether it is excitable or bistable. In the case of excitability, when the membrane potential is close to a saddle-node bifurcation, the sufficiently strong signal arriving from the presynaptic neuron causes a single large amplitude spike in the membrane potential. In turn, if the membrane is bistable, meaning in this case the coexistence of a stable steady state with a limit cycle, the stimulation can cause either the onset or the decay of the oscillatory membrane activity. It is thus clear that the understanding of the neural networks requires the application of the concepts of nonlinear dynamics (see Sect. 6.1.2, vol. I).

One can consider the case of a single electrochemical oscillator, mimicking the neuron, including its response to the external stimulus and the network of coupled oscillators which can also be perturbed. In a series of works, Karantonis et al. [12, 34, 37–39] have shown that the excitable or oscillatory electrochemical interfaces, placed in a close vicinity, interact through the electrolytic medium in a manner similar to neurons connected through electrical synapses. In particular, the bridging channels between the neurons are imitated by the electrolytic medium which plays a role of coupling ohmic resistance, and the communication between the electrochemical interfaces is bidirectional (compared to either bidirectional or unidirectional in case of neurons, see below), since the electrochemical interfaces in the same electrolyte interact.

Let us consider the spatiotemporal patterns which emerged through coupling of identical relaxation oscillatory electrode pairs, with each pair consisting of an iron anode (in Fe/H_2SO_4 system) and copper cathode (in $Cu/CuSO_4$ system), under conditions of the fixed external voltage [40] (Fig. 3.31). Thus, a single cell consists of only an iron ring anode and a copper coil cathode.

Various geometrical configurations of the cells were studied, including linear arrangement of two cells, a chain of three cells, regular hexagon set, and star sets. Only few oscillatory cycles appeared sufficient to achieve the synchronized state of such systems, similarly to collective phenomena occurring in biological systems. Of course, in the particular case studied, the coupling was realized physically through the electrolyte solution. Figure 3.32 shows as an example, synchronization occurring for two configurations of the set of three cells. In general, the characteristics of such phenomena closely resemble the response of coupled relaxation neural oscillators [40]. In particular, these assemblies exhibited excitatory or

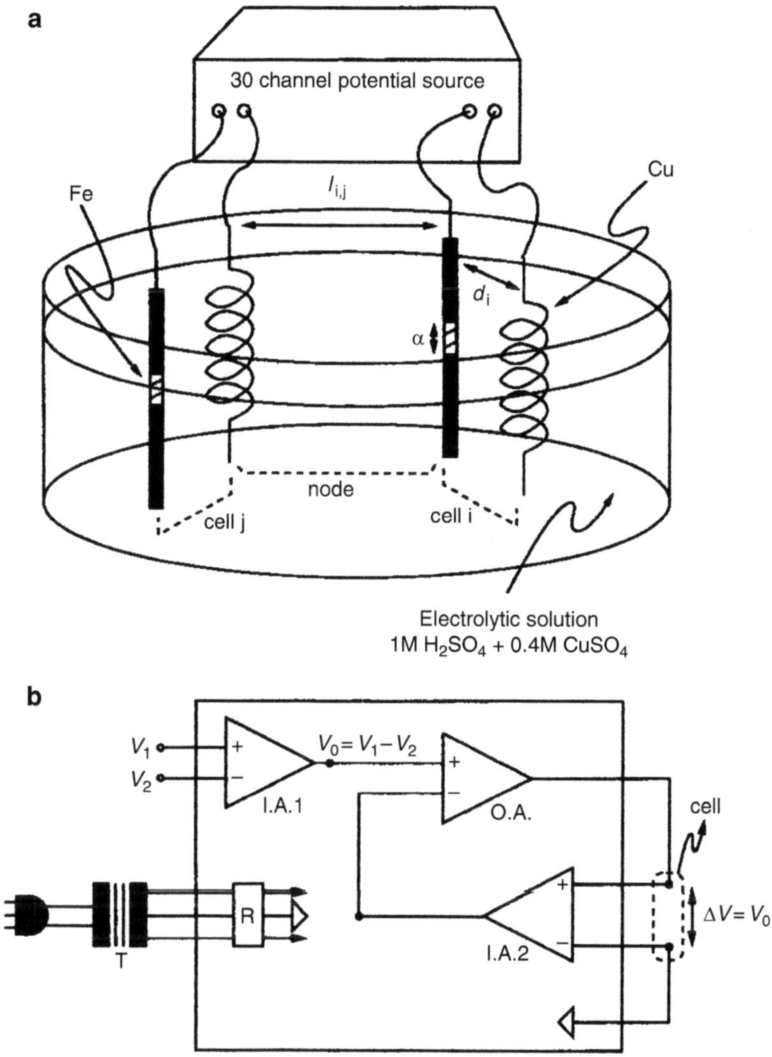

Fig. 3.31 (**a**) Schematic experimental setup for a set consisting of two cells ($N = 2$): The Fe and Cu electrodes are anodes and cathodes, respectively. The reacting surface of the Fe electrodes has height α. The distance between an anode and a cathode for cell i is d_i. The distance between adjacent electrodes is $l_{i,j}$. Electrodes within a distance $l_{i,j}$ define a node. (**b**) Simplified schematic representation of one channel of the potential source. T: transformer, R: voltage regulator. IA: instrument amplifier, OA: operational amplifier. Reprinted with permission from [40] http://link.aps.org/abstract/PRE/v65/p046213. Copyright 2002 by American Physical Society

inhibitory connections (i.e., mimicked the excitatory or inhibitory neurons), depending on whether electrode pairs were coupled through electrodes of the same kind, i.e., anode–anode or cathode–cathode (excitatory connection) or through electrodes of different kind, i.e., anode–cathode (inhibitory connection).

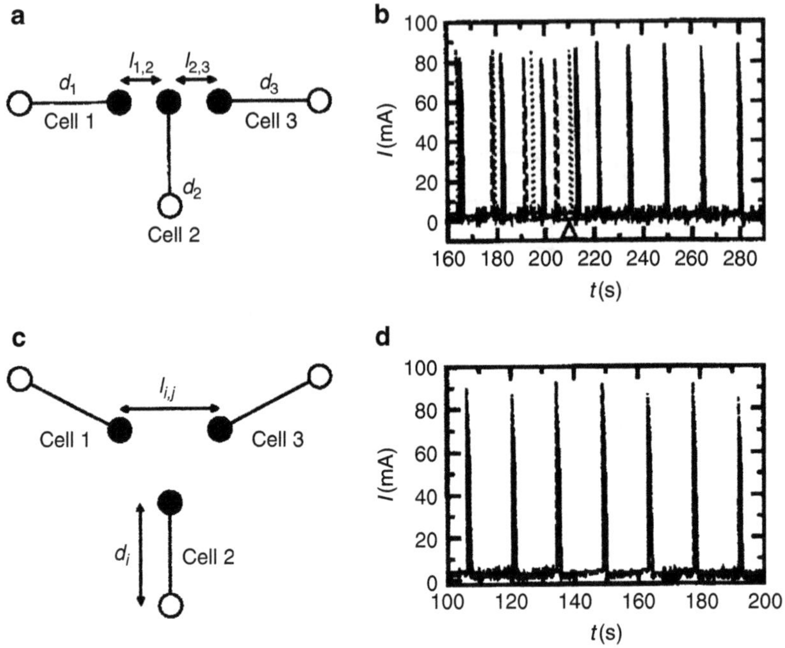

Fig. 3.32 (**a**) A chain of three cells, $l_{1,2} = l_{2,3} = 1.5$ cm, $d_i = 5$ cm and (**b**) transition to synchrony. The natural periods of the cells are $T_1{*} = 16.6$ s, $T_2{*} = 13.2$ s, and $T_3{*} = 15.8$ s. At the arrow, the physical obstacles are removed and coupling is effective. (**c**) A ring of three cells, $l_{ij} = 1.5$ cm, $d_i = 5$ cm, and (**d**) synchronized response (traces overlapped). Reprinted with permission from [40] http://link.aps.org/abstract/PRE/v65/p046213. Copyright 2002 by American Physical Society

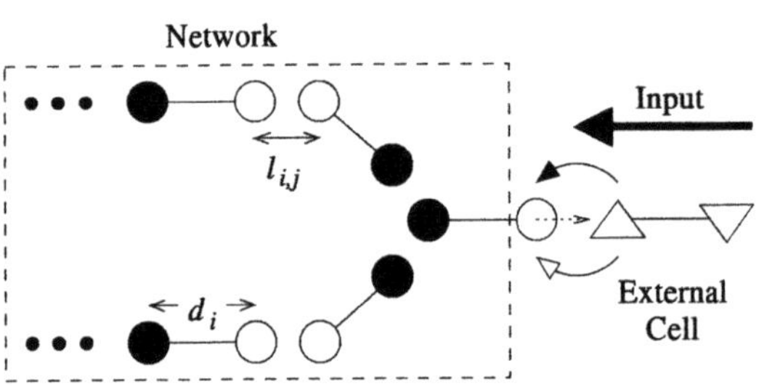

Fig. 3.33 Schematic representation of an electrochemical network accepting an input (*thick arrow*) through an external cell via excitatory (*black arrow*) or inhibitory (*white arrow*) connections. *White* and *black circles* represent anodes and cathodes, respectively. The effect of the network to the external cell (*dashed arrow*) is assumed to be small. Reprinted from [34], Copyright 2002, with permission from Elsevier

The connection is called excitatory if the averaged action of one oscillator facilitates the activity of another one, in the opposite case the connection is considered inhibitory, analogously to the characteristics of neurons, depending on the response of their postsynaptic membrane potential.

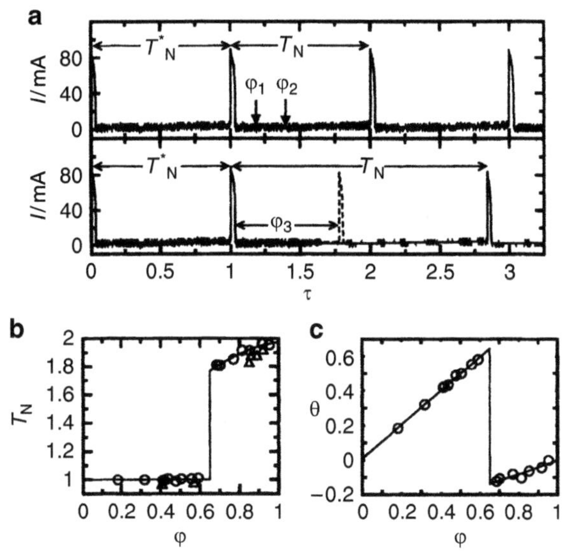

Fig. 3.34 (**a**) Response of a network cell to a pulse input through an excitatory connection. *Vertical arrows* represent pulse inputs at the corresponding phases, $\varphi_1 = 0.18$, $\varphi_2 = 0.41$, and $\varphi_3 = 0.76$. Pulse duration $\Delta t \approx 0.7$ s, $V_{\mathrm{ext}} = 50$ mV and $V = 230$ mV. *Dashed lines* represent induced peaks. (**b**) Relation of the new period, T_{N} vs. the phase φ for one network cell (*circles*) and a linear network (*triangles*). (**c**) Relation of the new phase θ vs. phase φ. Reprinted from [34], Copyright 2002, with permission from Elsevier

In view of these biological analogies, it becomes quite obvious to study the response of such assemblies to an external input. This problem was analyzed later by Karantonis et al. [34]. Compared to earlier experimental configuration, the present one includes also an external circuit which is a source of the input signal (Fig. 3.33).

Experiments involved constant or pulse input, through an excitatory (through the anode) or inhibitory (through the cathode) connection with an electrochemical cell, behaving as a relaxation periodic oscillator exhibiting its own period T^*. For the constant input the response means the decrease or increase in the oscillation period, depending on the excitatory or inhibitory connection. If the input has a form of the pulse, the response, for an excitatory connection, depends on the phase of relaxation oscillations—within the refractory region the network remains unaffected, whereas within the excitable region one observes an increase in the oscillation period, together with the induction of new peaks (Fig. 3.34). Analogous observations were made for a linear array of four cells.

The differences in responses mean that the simple electrochemical network is able to recognize the type of the input. Again, one can suggest that this behavior of an electrochemical system may be helpful in understanding of the behavior of simple networks of physiological origin.

One should add that Karantonis et al. [39] also have analyzed, both theoretically and experimentally, the case of a network of two *identical* discrete electrochemical

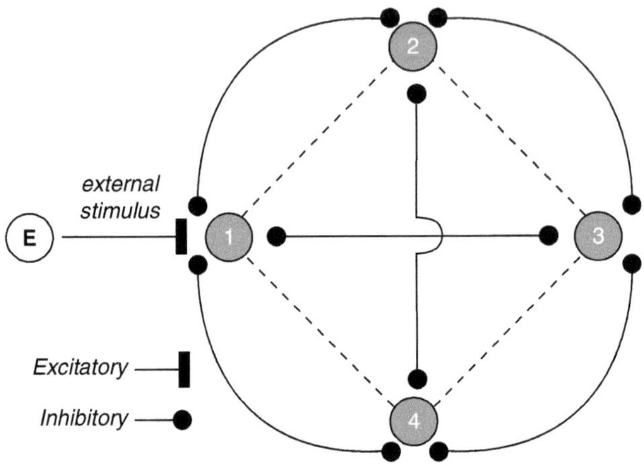

Fig. 3.35 Schematic representation of a network consisting of $N = 4$ oscillators (*gray circles*) connected through inhibitory connections, together with an external master (*white circle*) unidirectionally coupled with one of the oscillators through an excitatory connection. Reprinted with permission from [37] http://link.aps.org/abstract/PRE/v71/p056207. Copyright 2005 by American Physical Society

oscillators, for which the connections were altered from excitatory to inhibitory ones. The model included two point working electrodes, controlled potentiostatically, the point reference electrode and the counter electrode, the relative position of which was changed. It was shown that the type of connection—from the excitatory to the inhibitory one—depended on both the uncompensated and the solution resistance of the electrochemical cell. This is concordant with the common consideration that for coupled electrochemical oscillators, the total current flowing through the cell consists of mainly two components, one flowing from the working electrode to the reference electrode via the corresponding uncompensated resistance and the other, flowing from one working electrode to another one via a solution resistance. From the practical point of view it means that the type of connection can be altered by changing the relative values of the resistances involved, i.e. by changing the relative position of the working, counter and point reference electrodes. As a consequence, different in-phase and out-of-phase stable synchronized states were obtained, both in the experiments and respective numerical calculations.

In the following works [37, 41, 42], the network consisting of four ($N = 4$) bidirectionally all-to-all inhibitory coupled electrochemical oscillators of relaxation type was considered, with one or more of those N oscillators capable of receiving an external excitatory signal (i.e., being unidirectionally coupled with the external stimulus in an excitatory manner)—see Fig. 3.35.

In the experimental setup, N electrochemical oscillators can thus be driven and monitored, and perturbed selectively. The Fe/H_2SO_4 oscillator served as an experimental system, under such conditions that the oscillations exhibited relaxation character. The experimental setup [41] involved further the laser beam as a source of perturbation: the electrode was illuminated by frequency-doubled light ($\lambda = 532$ nm), with the pulse duration of 7-ns in full width half maximum. Results of these experiments consisted of a kind of spatiotemporal coding in the network considered. The network of N electrochemical oscillators, mutually coupled by all-to-all inhibitory connections, can have $(N - 1)!$ coexisting out-of-phase states, each being a permutation of a periodic spiking sequence and undergoing modification upon application of shots of laser pulse perturbation. The resulting phase relation is stored as a coded pattern. In conclusion, the network of such construction can function as a rewritable memory of $(N - 1)!$ different spatiotemporal patterns [42].

The same group of researchers analyzed also theoretically the dynamics of the ring networks (i.e., for which the boundaries are not present) of discrete, coupled relaxation electrochemical oscillators, indicating the existence of such synchronization modes, as in-phase, antiphase, and fractured synchrony [38].

3.1.10 The Pitting Corrosion of Steel as a Cooperative Process

Steel (as well as other passive-film-forming metals) is in common use, so better understanding of the mechanisms of its corrosion should help to develop more efficient methods of preventing the corrosion onset. Numerous studies were devoted to the studies of these phenomena, but here we shall focus on the selection of those works which are related to understanding of corrosion in terms of nonlinear dynamics.

Generally, the corrosion of a stainless steel has a nature of pitting corrosion which begins in the sites of MnS inclusions and is catalyzed by aggressive ions, like Cl^- (cf. Sect. 6.1.1, volume I). There are different literature suggestions on the detailed mechanisms of this process. Eklund [43, 44] has suggested the following dissolution mechanism, based on the concluded thermodynamic instability of MnS above the potential of ca. −100 mV (SHE):

$$MnS + 4H_2O \rightarrow Mn^{2+} + SO_4{}^{2-} + 8H^+ + 8e \tag{3.13}$$

$$MnS + 2H^+ \rightarrow Mn^{2+} + H_2S \tag{3.14}$$

$$H_2S \rightarrow S + 2H^+ + 2e \tag{3.15}$$

In turn, Wranglen [45] has suggested the following reaction sequence:

$$MnS \rightarrow Mn^{2+} + S + 2e \tag{3.16}$$

$$S + 3H_2O \rightarrow HSO_3{}^- + 5H^+ + 4e \tag{3.17}$$

$$HSO_3{}^- + H_2O \rightarrow SO_4{}^{2-} + 3H^+ + 2e \tag{3.18}$$

$$MnS + 2H^+ \rightarrow Mn^{2+} + H_2S \tag{3.19}$$

Furthermore, Castle and Ke have suggested the reaction scheme [46]:

$$MnS + 3H_2O \rightarrow Mn^{2+} + HSO_3^{-} + 5H^{+} + 6e \tag{3.20}$$

$$HSO_3^{-} + 5H^{+} + 4e \rightarrow S + 3H_2O \tag{3.21}$$

while Lott and Alkire [47] have proposed the following mechanism, consisting of:
electrochemical dissolution:

$$2MnS + 3H_2O \rightarrow 2Mn^{2+} + S_2O_3^{2-} + 6H^{+} + 8e \tag{3.22}$$

and
chemical dissolution:

$$MnS + H^{+} \rightarrow Mn^{2+} + HS^{-} \tag{3.23}$$

Later, Webb and Alkire [48–50] have performed experimental and model studies of the role of single sulfide inclusions in the pitting corrosion of stainless steel in contact with NaCl solutions, including the effect of thiosulfate ions.

The pitting corrosion of steel in NaCl solutions, monitored by the measurements of the current and electrode potential dynamics as a function of time, reveals irregularities that are usually termed the “electrochemical noise” (EN). EN is considered a useful and powerful tool for the understanding of the corrosion mechanisms and for the corrosion monitoring. García et al. [51] have analyzed EN in a system consisting of carbon steel rod as a working electrode, the SCE as the reference electrode, and the Pt wire as the counterelectrode. The irregular fluctuations of the steel electrode potential (vs. SCE) were measured at the open circuit potential (free corrosion potential), while for the measurements of current the working electrode was maintained at the corrosion potential in the three-electrode arrangement. From the measured *E*-*t* and *I*-*t* courses, the power spectra and chaotic phase portraits (strange attractors) were constructed. Three types of such attractors (defining three types of the EN in the studied system) were identified. In order to verify whether these phase trajectories indeed represent chaos, their static (dimensions) and dynamic (Lyapunov exponents) characteristics were determined. When though one of the Lyapunov exponents in the N-dimensional phase space is positive, the phase trajectories diverge to form chaos. The most informative is thus the largest (positive) of Lyapunov exponents which quantify the greatest rate at which the distance between two nearby trajectories increases exponentially in time (see Sect. 1.10.2, volume I). The 30-day long experiments, indicating the transition of the system between various, mentioned above, types of EN, allowed to determine the largest Lyapunov coefficients also varying as a function of time, reaching positive values largely between 0.1 and 0.3, thus confirming the chaotic nature of the reported fluctuations.

Let us now consider in more detail the microscopic and dynamic characteristics of the pitting corrosion of steel in view of recent works, with the reference to earlier studies. As Hudson, Mikhailov et al. [52, 53] have shown, both in terms of the model and experimental data, there is a certain "memory" in the pitting process, i.e., the formation of metastable pits can affect the subsequent pitting events. In other words, there is certain cooperation between the pits which causes an abrupt transition from metastable to stable pits. We shall now show that the pitting corrosion can be in fact considered a *highly nonlinear* phenomenon. Such a modern interpretation of the dynamics of the evolution of the pitting corrosion of stainless steel was recently suggested by Hudson, Mikhailov et al. [54]. One of the important premises for this mechanism was the observation of a sharp rise in the pitting corrosion rate, as the potential, solution concentration or temperature changed only slightly. Furthermore, real-time microscopic in situ visualizations allowed one to observe the nucleation and evolution of individual pits during the transition. These results led to the concept that the sudden onset of corrosion can be explained by an explosive autocatalytic growth in the number of metastable pits and that stabilization of individual pits takes place only later. In other words, the pits develop not independently of each other, but the onset of pitting corrosion is a *cooperative critical* phenomenon resulting from *interactions* among metastable pits. Such a concept was corroborated by the appropriate theoretical model.

In the electrochemical experiment, monitoring the state of the steel anode, the formation of an individual pit manifests itself as a transient small peak of anodic current, which persists for a few seconds and then dies out. It is usually assumed that such pit forms in the place of the surface defect or inclusion. Upon small change of conditions (applied anodic potential, corrodant concentration, temperature, etc.) the pitting corrosion shows a sharp acceleration, meaning the sudden transition from a low-activity regime (with a few metastable pits) to a state with high pitting activity [55, 56]. These phenomena are well known, but their mechanistic explanation can be different, in spite of numerous studies. Various mechanisms, known up to 80s of XX century have been published in several reviews, cf., e.g., [57–59]. Later in 1985, Williams et al. have presented the stochastic models of pitting corrosion of stainless steels based on the idea of stabilization of individual pits [60]. Apart from mathematical details which can be found in original references, it is here most important to invoke the basic assumptions of this model: (1) pitting events are nucleated with certain frequency and have certain probability of dying, (2) events which survive beyond a critical age do not die, and (3) each event has an induction time, during which the local current, caused by this event, does not increase, but during which the event may die. According to these models, pits are unstable when they are first nucleated and a pit becomes stable only after it has survived past the critical age. The significant rise of current is caused by the appearance of a stable pit with high current that does not repassivate. In the following work, Williams et al. [61] have collected experimental results of the studies of pitting corrosion of stainless steel in contact with NaCl solution, and have developed further the microscopic model, emphasizing the role of local acidification in the initiation of pitting corrosion. According to the original explanation, the production, on the scale of the surface roughness of the specimen, of a gradient of acidity, as a consequence of the small passive dissolution current of

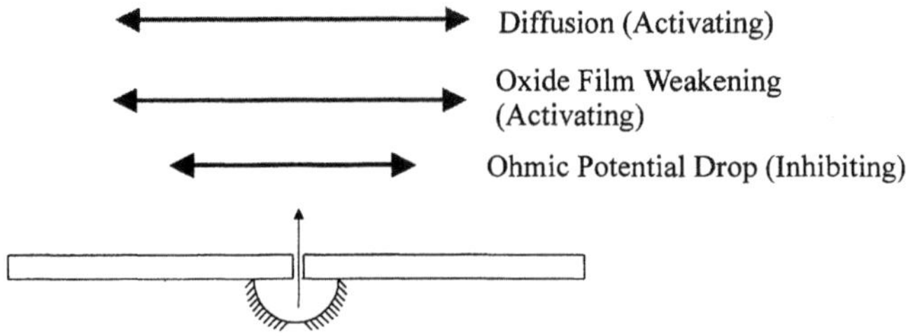

Fig. 3.36 Schematic of the local changes that occur near a pit site. Included are changes in the local solution chemistry by diffusion, oxide film damage, and ohmic potential drop in solution. Reproduced from [72] by permission of The Electrochemical Society

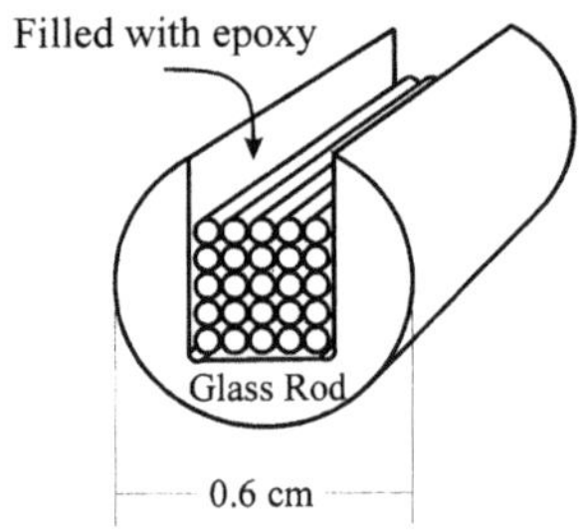

Fig. 3.37 Schematic of the 5 × 5 electrode array configuration with 0.09-mm spacing. The electrode wires were positioned as close together as possible within a notch of a glass rod, then covered with epoxy. Reproduced from [72] by permission of The Electrochemical Society

the metal, triggers off a local instability, which is maintained by the local potential gradient. In other words, the initiation of the pitting corrosion is associated with the production and persistence of gradients of acidity and electrode potential on the scale of surface roughness of the metal. The birth and the death of the pitting events is caused by the fluctuations in these gradients, caused by fluctuations in the thickness of the hydrodynamic (solution) boundary layer at the metal surface. There was also introduced a concept of a critical penetration or critical defect size for the onset of stable pitting forms. In the opinion of the authors, the stochastic approach allowed to deepen the understanding of the phenomenon of a pitting corrosion.

Nevertheless, the formation of spatiotemporal patterns in the concentration and potential fields during pitting remained largely unexplained. That is why more recent works, which invoke the concepts of nonlinear dynamics, are worth considering here. In such recent approach, the emphasis was put on an earlier neglected *cooperation* of single pits. A first premise for that was finding of certain statistical correlation between the current spikes which led to the indication of some *memory* in the pitting process [52]. Thus, the pitting corrosion was found to be not a completely random phenomenon.

What can be a source of communication between the pits? When a pit begins to form (and current flows), its local environment is altered, i.e., when the metal

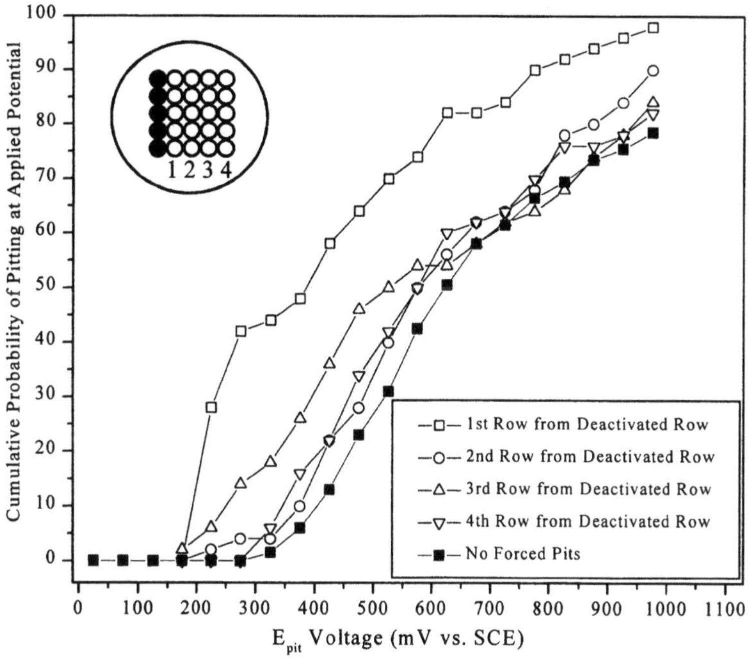

Fig. 3.38 Cumulative pitting probabilities when all electrodes in row no. 0 were held at 1 V vs. SCE for 10 min and then returned to OCP to create deactivated pit sizes. The effect of the just-deactivated pit sites can be seen to most strongly alter the pitting probabilities on row no. 1. Reproduced from [72] by permission of The Electrochemical Society

dissolves and its cation enters the solution, the interaction with water molecules (hydrolysis) enhances the local acidity, and the electroneutrality condition causes the local increase in anion (chloride) ion concentration. Already in 1991, Harb and Alkire [62] have shown that such concentration gradients develop in the solution for a distance of up to three pit diameters. Furthermore, Sugimoto et al. [63], Xu et al. [64], and Hoar et al. [65] have proved that the presence of a pit site can alter the nearby oxide film. Moreover, it was found [66] that the formation of a pit can dissolve or alter nearby sulfide inclusions which factor can also affect the formation of nearby pits. Finally, near a pit site the ohmic potential drop, associated with the flow of a local current, can reach even 75–100 mV which means significant change in the local spatial distribution of the electric field (cf. e.g., [67–69]). Newman et al. [70] and Harb and Alkire [71] have carried out numerical calculations of the potential distribution in the solution. It is important to emphasize that formation of a pit, i.e., the flow of local current, which causes in the solution an ohmic potential drop and decreases the interfacial potential (i.e., makes it less anodic) in the vicinity of the first pit, prevents the formation of other pits around it. Thus, the whole process of nucleation and autocatalytic increase of further pits is a result of competition of several factors that should be taken into account in a realistic model of these phenomena (see Fig. 3.36).

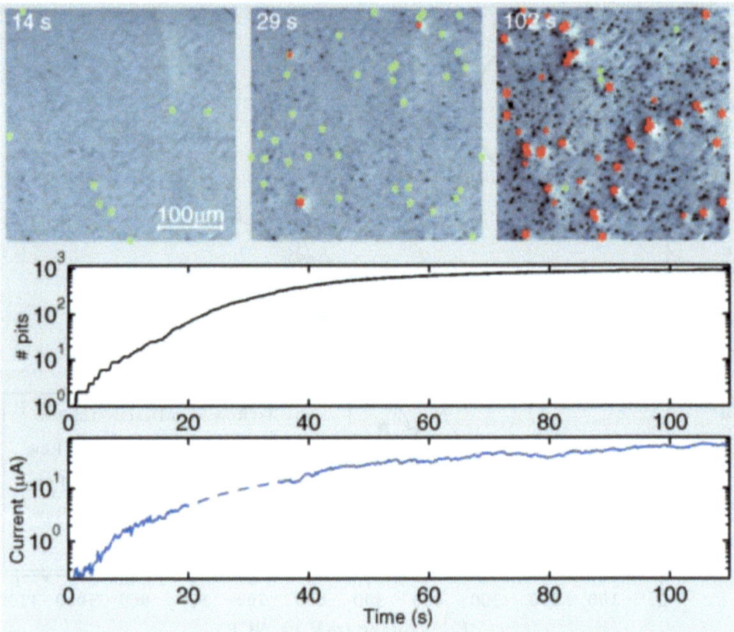

Fig. 3.39 Sudden onset of pitting corrosion observed with contrast-enhanced microscopy. (*Top*) Three snapshots of computer-processed optical microscopy images. *Green stars* show the nucleation of new pits. *Red stars* point out highly active pits (with halo). A full video sequence is available (movie S2 [54]). (*Bottom*) Total number of pits and total current on logarithmic scales as functions of time. The reaction occurred in 0.05 M NaCl at 22 °C; the potential scanned from 693 mV_{NHE} at 1 mV/s. From [54]. Reprinted with permission of AAAS

These facts, constituting a set of sound premises for the concept of cooperative interactions between the pits, led further to the experimental and model elaboration of the spatial interactions among localized corrosion sites [72]. The experimental setup—an array of 25 electrodes arranged in a 5×5 square configuration of 0.25 mm diameter 316 stainless steel wires—is shown in Fig. 3.37.

A series of smart potentiodynamic experiments were performed, with the aim to check whether the formation of stable pit at a given electrode(s) affected the induction times for pitting at the remaining electrodes. Figure 3.38 shows a representative example—a dependence of the cumulative probability of pitting at applied potential in the potentiodynamic experiment, for the particular rows of the electrodes, for the case when row No. 0 (most-left) was first held at 1V to cause stable pits and then returned to open-circuit potential (OCP), in order to create deactivated pit sites. Evidently, the pitting probability is not the same for all rows, and is the highest for row No. 1, i.e., the closest neighbor to row No. 0. Also, the probability of pitting was the lowest, if no prepitting was performed prior to the potentiodynamic experiments.

The principles of the model, whose first steps were described in [72] are the following: appearance of a pit in a defect place is associated with chemical reactions that change local concentrations of species (ions), causing weakening of the passive film in the surroundings. Thus, the formation of every pit enhances the probability of appearance of further pits in the neighborhood. This is the cause for the autocatalytic, explosive growth in the number of active pits. Their stabilization occurs only after this sudden transition. In order to verify this concept, it was necessary to apply two techniques which allow to monitor formation and evolution of the pits in a relatively short time scale. One of them was *ellipso-microscopy for surface imaging* (EMSI) based on an ellipsometric effect, but faster than classical ellipsometry and therefore permitting the real-time observation of ultra-thin layers on the entire surface area [73]. However, since the spatial resolution of EMSI is restricted to 12 μm which does not allow to resolve individual pits, for that purpose it was necessary to apply a complimentary method—specially adapted high-resolution contrast-enhanced optical microscopy.

For the presently discussed mechanism it is useful to notice that such measurements have revealed, among others, a bright zone around the sufficiently strong pit, which presumably reflected the zone of a hemispherical concentration gradient of ions. Such pits with "halos" will be termed "highly active pits."

Figure 3.39 shows the contrast-enhanced microscopic observations of the sudden onset of pitting corrosion (note the logarithmic scale of the ordinates) occurring under potential scan, starting from 693 mV, at scan rate 1 mV/s. The exponential growth of the number of pits, correlating with the rise of the anodic current, was observed during first ca. 20 s and then both dependences tend to the plateau.

Further analysis has shown that primary (green) pits appear with significant rate before the plateau of the current (or total number of pits) is reached. Later, when plateau is formed, the number of pits remains roughly fixed since all the places on the passivated metal surface (defects and inclusion sites) are already involved in the pit formation. The only dynamics occurring then is the reactivation/repassivation of previously formed pits. The nearest-neighbor distance between these pits was found to be of the order of a few micrometers.

These results suggest validity of the essential construction of the model of the onset of corrosion that was already outlined earlier in this section and will be now summarized in a bit more quantitative way. It was assumed [72] that the reactions in a metastable pit produce aggressive species which move laterally away and into the bulk. These species, acting chemically on the surrounding zone of the pits, weaken the protective layer, until the pits repassivates and thus stops to emit the aggressive species; then the damaged portion of the passivated film slowly repassivates. Aggressive agents can, however, enhance further pitting expansion over distances of tens of micrometers from an active pit, which phenomenon can persist for relatively long time, i.e., even tens of seconds after the disappearance of the pit being a source of this phenomenon. The pitting corrosion is thus an autocatalytic process analogous to a chain reaction: each metastable pit, through the local production of the aggressive species, weakens the protective film within time T_0 (being the recovery time of the passive layer), in the surrounding area of radius l_0 [being the characteristic lateral diffusion length of the aggressive species in the electrolyte, $l_0 \sim (DT_0)^{1/2}$]. If inside this diffusion layer the rate of pit generation is equal to w_1, being *much higher* than the rate of spontaneous pitting, the probability of the nucleation of a new pit is $p \approx w_1 l_0^2 T_0$. If this probability p is sufficiently large, of the order of unity as an approximate critical value, each pit gives rise to many pits around it, so the autocatalytic (chain) acceleration of the pits formation sets in [54]. This description postulates the existence of only metastable pits and not explicitly invokes their (quite possible in real systems) final transformation into stable pits.

The chemical processes associated with the autocatalytic growth of pits can be explained in detail in the following way [74, 75], assuming that stainless steel remains in contact with an aqueous NaCl solution. A first pit forms in the preferred site for that which is the Mn(Fe, Cr)S inclusion, i.e., the sulfide-rich inclusion, like MnS, in the Fe–Cr–Ni and Fe–Cr–Ni–Mo alloys. Such Mn(Fe, Cr)S inclusions in stainless steel dissolve chemically and/or electrochemically through a series of reactions resulting in microcrevices between the inclusions and the metal matrix. Chemically or electrochemically produced metal cations appear thus in the surrounding solution and, upon interaction with water molecules, undergo acidic hydrolysis, as a result of which in the pit and its surrounding the concentration of H^+ ions increases. Due to electroneutrality condition, the formation of metal and hydrogen cations causes the migration of Cl^- ions to this region. Chloride ions act as a catalyst to speed up the dissolution. The local solution, becoming significantly enriched in all these ions, is thus more aggressive to the steel than the solution in the bulk. The passive film around the pit undergoes thinning (partial dissolution) by strongly acidic solution and the second pit nucleates at the boundary between the film-thinned region and the original oxide film. Besides these chemical processes one should note simultaneous significant changes in the distribution of the electric potential in the solution.

Calculations based on the proposed model led to results similar to experimental observation from Fig. 3.39. Not only the shapes of the respective plots, but also the characteristic time and length scales, as well as the pit densities are close to those in the experiment.

In further model calculations it was clearly shown how the random formation of initial pits turns into formation of clusters of subsequent pits. Figure 3.40 shows the temporal evolution of metastable pits, the cluster of which is formed in the lower right corner of the model square surface.

One should note that the above model assumes homogeneous electrode surface which does not correspond to the real characteristics of the passivated surface of solid electrodes. Therefore, Organ et al. [75] have extended their approach to heterogeneous surfaces with randomly distributed inclusion sites in an otherwise passive electrode. Of course, the pitting corrosion is favored in these sites, so the metastable pits were assumed to form only at initially distributed inclusions. Using this model, it was possible to study the effect of the spatial density of inclusions for the explosive growth of metastable pits, leading to pitting corrosion. It was also found that the critical density of inclusions can be altered by the convective mass transport, since the cooperative interactions among metastable pits could be decreased by stirring the solution, causing the enhanced removal of aggressive reaction products from the surface into the bulk solution. Of these effects, we shall

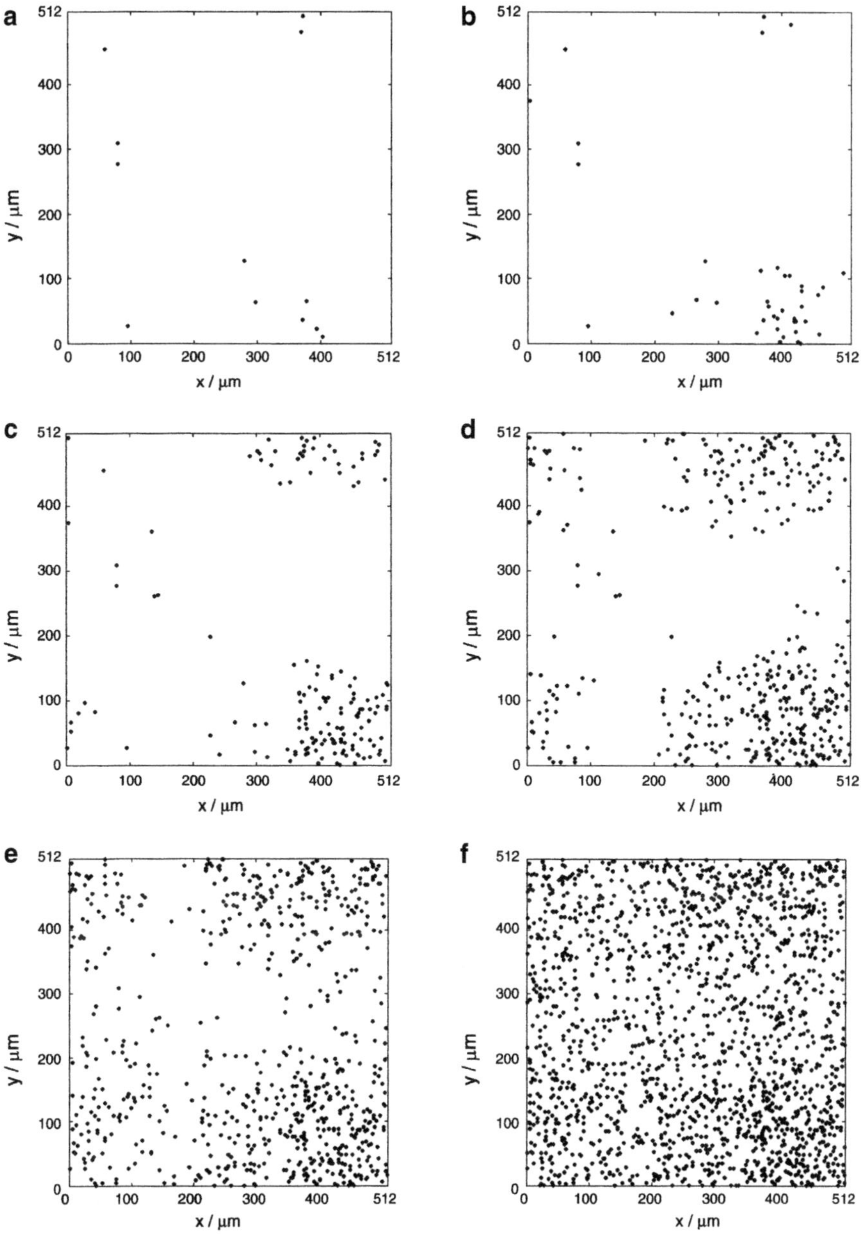

Fig. 3.40 Snapshots of cumulative metastable pits at (**a**) 150 s (12 pits); (**b**) 200 s (40 pits); (**c**) 250 s (160 pits); (**d**) 300 s (432 pits); (**e**) 325 s (697 pits); (**f**) 400 s (1,704 pits). Reprinted from [74], Copyright 2005, with permission from Elsevier

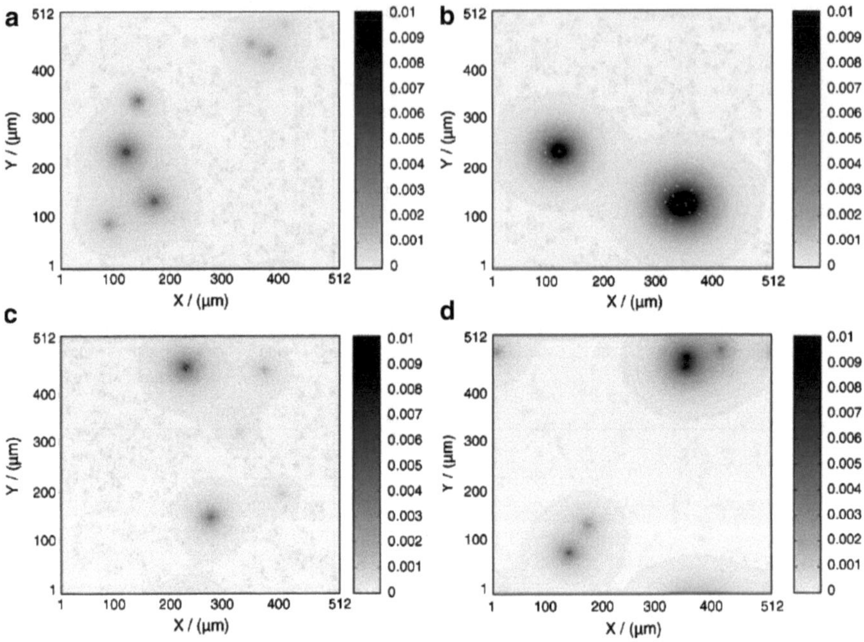

Fig. 3.41 Snapshots of film damage at different times: (**a**) $t = 400$ s; (**b**) $t = 800$ s; (**c**) $t = 1{,}200$ s; (**d**) $t = 1{,}501$ s. Inclusion density just *below* the critical value. See [75] for details. Reprinted from [75], Copyright 2007, with permission from Elsevier

present here as an example the snapshots of the simulated damage of the oxide passive film, when the inclusion density is either below the critical value (the pit initiation events die off and the damaged film regions heal with time, Fig. 3.41) or just above the critical value (rapid evolution of corrosion is observed, Fig. 3.42).

The physical sense of existence of this critical density for the autocatalytic progress of pitting corrosion is that it corresponds to the condition when the mean spacing between the inclusions is equal to the boundary layer thickness which also defines the radius of the region around an active pit that is significantly affected by its activity. Further simulations showed also that the pitting potential, i.e., the value of the potential, at which the anodic current begins to rise sharply in the potentiodynamic experiment (due to explosive growth of metastable pits at the microscopic level) decreases with increasing inclusion density, i.e., applying the higher anodic potential plays the role analogous to the increase of the inclusion density. In other words, if the inclusion density is low, the sufficiently high rate of interaction between them can be achieved by applying a sufficiently positive potential.

Finally, increasing stirring of the solution, which enhances the removal of aggressive reactions products from the interface to the bulk of the solution, can be expressed as the decrease in the boundary layer thickness at the interface. Simulations proved that there exists a critical value also of this layer thickness (or, equivalently, the stirring rate), at which the transition to high corrosion rate

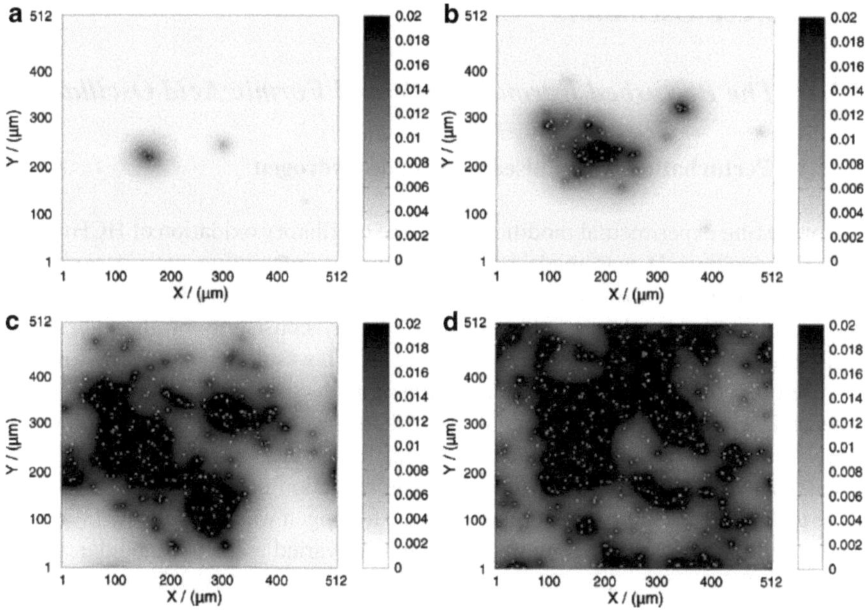

Fig. 3.42 Snapshots of film damage at different times: (**a**) $t = 400$ s; (**b**) $t = 600$ s; (**c**) $t = 800$ s; (**d**) $t = 1{,}200$ s. Inclusion density just *above* the critical value. Reprinted from [75], Copyright 2007, with permission from Elsevier

occurs. Of course, for increasing inclusion density a higher stirring rate is necessary to suppress the onset of high pitting interactions and thus high corrosion rate.

These model results correspond well with the characteristics of real stainless steel materials, for which it was found that alloys with low inclusion densities exhibited higher critical potentials associated with explosive growth in pit sites. The idea of cooperation of the pits was discussed also by Sasaki and Isaacs [76] who have analyzed the electrochemical noise associated with the pitting corrosion of the passivated Al surface.

The practical conclusion that can be drawn from the above analysis is that the destructive autocatalytic, explosive corrosion of stainless steel and similar materials can be suppressed by (1) enhanced transport of the aggressive agents that are formed in the pits, toward the solution bulk; (2) addition of corrosion inhibitors, accelerating the repassivation of the pits, and (3) improvements of the structure of the metal phase by reducing the number of defects. The reader interested in further details of this concept is advised to consult the original references [74, 75, 77].

At the end of this subsection one should mention, that in Sect. 6.3 there is described a liquid membrane oscillator (cf. Fig. 6.14) which is interpreted as composed of two coupled oscillators, in order to explain its complex dynamics.

3.2 Forced Oscillators

3.2.1 The Perturbed Formaldehyde and Formic Acid Oscillators

3.2.1.1 Perturbation with Pulsed Stream of Hydrogen

An interesting experimental modification of the oscillatory oxidation of HCHO (see Sect. 5.5, volume I) was made by introducing its interaction with periodically produced gaseous hydrogen, as described by Nakabayashi and Kira [78]. In this way a kind of forced electrochemical oscillator, consisting of the inherently oscillating system driven by external periodic signal, was designed. Its construction involved the application of the rotating ring-disk electrode (RRDE), allowing the independent control of the potential applied to the disk ($\phi = 5.5$ mm) and the current imposed on the ring (inner $\phi = 6.5$ mm, outer $\phi = 9$ mm). The self-sustained oscillatory process of formaldehyde oxidation occurred at the ring electrode under galvanostatic conditions, while the hydrogen was periodically produced at the central disk, the potential of which was varied with a triangular wave, between -0.4 V (corresponding to hydrogen evolution) and +0.1 V (no electrode reactions), with the scan rate 0.5 V s^{-1}. Due to electrode rotation, gaseous hydrogen was transported from the disk to the area of ring and in this way it could interfere with the formaldehyde oscillator, to a degree dependent on the rate and frequency of its formation. Just like for typical forced nonlinear oscillators, after the onset of H_2 evolution the transient behavior of the HCHO/Pt oscillator is observed, before the asymptotic regime is eventually established. Exemplary results of experiments showing the effect of duration of the pulse H_2 evolution on the asymptotic dynamics of the HCHO/Pt are depicted in Fig. 3.43.

Evidently, periodic perturbation with H_2 modifies the course of oscillations of the formaldehyde system. The details of this interaction are better visualized on schematic in Fig. 3.44 which suggests three seemingly different patterns of interference: α, β, and γ. Noteworthy, the maxima of H_2 production (peaks of the disk current) correlate with local maxima of the ring potential.

One should note that the shape of the resulting oscillations is never a simple superposition of original courses, so the coupling is by no means trivial and requires description through the appropriate kinetic mechanism. A relevant model is presented in Fig. 3.45.

Two possible ways of the H_2 effect: A and B, engaged in indirect and direct pathways of HCOOH oxidation, respectively, and both leading to increase in electrode coverage with passivating CO, are considered. Path A involves chemical reduction of the adsorbed OH species, which causes the decrease in the rate of removal of CO from Pt surface [cf. Eq. (5.16), vol. I]. In turn, path B postulates chemical reduction of adsorbed COOH intermediate to adsorbed CO. The resulting increase in the CO surface concentration, causing the passivation of the electrode, shifts its potential to positive values, in order to keep the imposed total current. This explains why the maxima of H_2 production, causing temporal increase of electrode

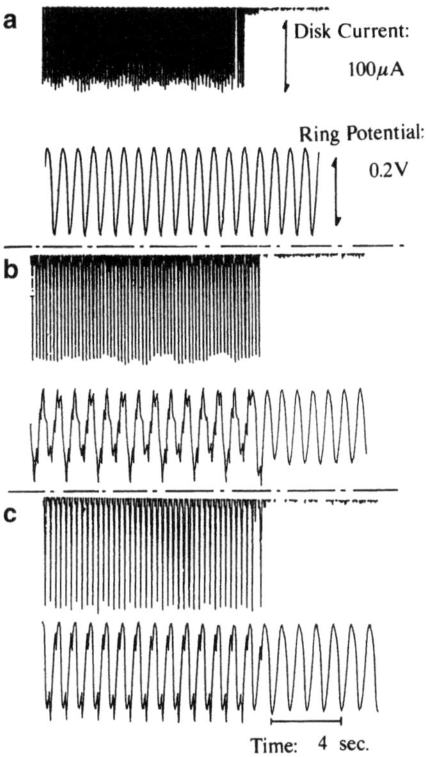

Fig. 3.43 Forced oscillations for hydrogen-pulse intervals of (**a**) 0.1, (**b**) 0.2, and (**c**) 0.3 at a ring current of 100 μA. Solution: 0.4 M formaldehyde + 2 M H_2SO_4. Reprinted from [78]. Copyright 1992 American Chemical Society

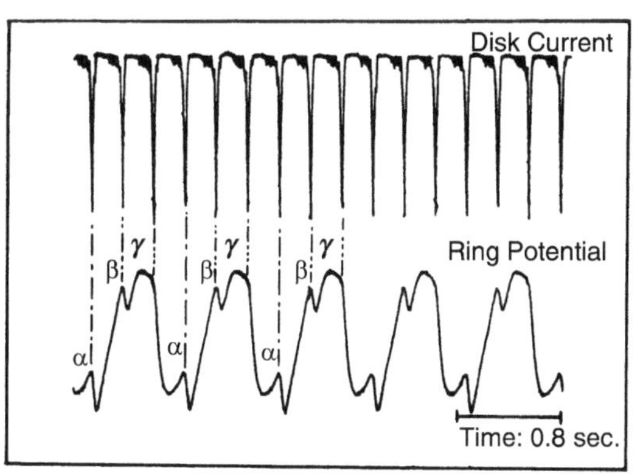

Fig. 3.44 A part of Fig. 3.43c expanded demonstrating the modulation of the ring-potential oscillation by the disk current pulse. Reprinted from [78]. Copyright 1992 American Chemical Society

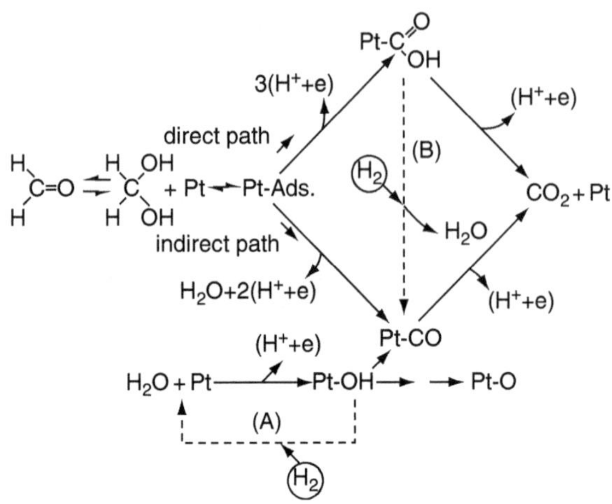

Fig. 3.45 A scheme for the self-sustained potential oscillation (*solid arrows*) and for processes of the interference of hydrogen (*dashed arrow*). Reprinted from [78]. Copyright 1992 American Chemical Society

coverage with poisoning CO, correlate with local positive bumps in the oscillatory peaks of HCOOH oxidation.

3.2.1.2 Perturbation-Induced Optimization of Formic Acid Oxidation

As a concise complement to the above system, one should know that, according to Schell et al. [79], if the galvanostatic oxidation of *formic acid* at Pt electrodes is modified so that the sinusoidal component is added to the current, then, for a range of frequencies of perturbation, the average electrode potential is less positive than for constant imposed current. It means that under perturbed current condition the inhibiting effect of intermediate carbon monoxide can be decreased substantially, since the imposed current can then be attained for smaller polarization of the anode. In terms of mechanism of formate ion oxidation (cf. Sect. 5.4, volume I), this effect was explained in terms of more formic acid following the direct route than in the unperturbed system, even though the average value of the applied current was the same for both systems. The mechanism underlying these phenomena is related to the mechanism of autonomous oscillations in formate ions oxidation. Thus, under perturbed current conditions, the average current–potential characteristics of formic acid oxidation appeared to be more favorable for practical applications of this process, including, e.g., the fuel cells [79].

3.2.2 *The Forced Fe/H_2SO_4 Oscillator*

The effect of sinusoidal forcing (modulation of the electrode potential) for the theoretical Franck–FitzHugh model of the anodic dissolution of Fe in H_2SO_4

medium was studied by Sazou et al. [80, 81]. Later in the same group, the oscillatory dissolution of Fe in H_2SO_4 was studied as an experimental example of the forced oscillator, when the electrode potential of the Fe/2M H_2SO_4 system (of inherent oscillation frequency ω_0) was perturbed by external sinusoidal signal of amplitude δE and frequency ω [82, 83]. Noteworthy, the response of this particular system studied as a function of varying δE and the frequency ratio (ω_0/ω) appeared to be significantly different from the characteristics of other forced oscillators. In particular, there was observed a formation of current spikes, resembling the bursting phenomena. Furthermore, by appropriate tuning of the parameters, one observed replacement of the spike generation pattern by a quasiperiodic behavior. All the dynamical responses were presented by using time-delay reconstructions of the attractors, Poincaré maps and Fourier power spectra. Such observation may serve as a test of validity of earlier electrochemical mechanism of oscillations in the Fe/H_2SO_4 system (see Sect. 6.1.2, volume I), as they should be able to explain also the effects of external perturbations.

In a recent paper, Karantonis et al. [84] have described the bursts of electrochemical oscillations in this system, composed of either of single or two Fe electrodes. In this approach, the behavior of the Fe/H_2SO_4 oscillator was compared with the neural cell membranes which, besides excitability, bistability, and simple oscillations, can also exhibit bursts of oscillations which are considered to be very important in neural information processing. Therefore, building and studying of the networks of coupled bursting cells have recently attracted attention of researchers in biological sciences, including theoretical biology. In consequence of these studies, there were found three different regimes of synchronization occurring in such systems: synchronization of individual spikes, synchronization of bursts, and complete synchrony where both spike and burst synchronization occur simultaneously [84].

The presently described electrochemical experiments were performed with both single Fe electrode and the couple of such electrodes, engaged in a three-electrode potentiostatic system. For a single Fe electrode, at external voltage $V_{SH} \approx 225$ mV (vs. SCE) the system exhibited the subcritical Hopf bifurcation, associated with transition from active dissolution to oscillations, while the reverse transition occurred at $V_{FC} \approx 175$ mV via the fold-limit cycle bifurcation (called also the saddle-node bifurcation of cycles, cf. Fig. 1.22, volume I). For the single Fe electrode, if the external voltage was varied in a cyclic manner, with the amplitude chosen so that the system was periodically passing through these bifurcation points, the periodic Fe/H_2SO_4 oscillator under such dynamical external conditions turned into the periodic burster (Fig. 3.46).

Analogous bursts, although with roughly constant amplitude, were observed when the cyclically changing voltage was reaching a more positive value, at which the oscillations decay via the homoclinic bifurcation. In both cases, the bursts were caused by a slow modulation of the voltage (bifurcation parameter), crossing the respective bifurcation points. In fact, bursts of oscillations are generally understood as a result of coupling between the fast periodic, oscillatory

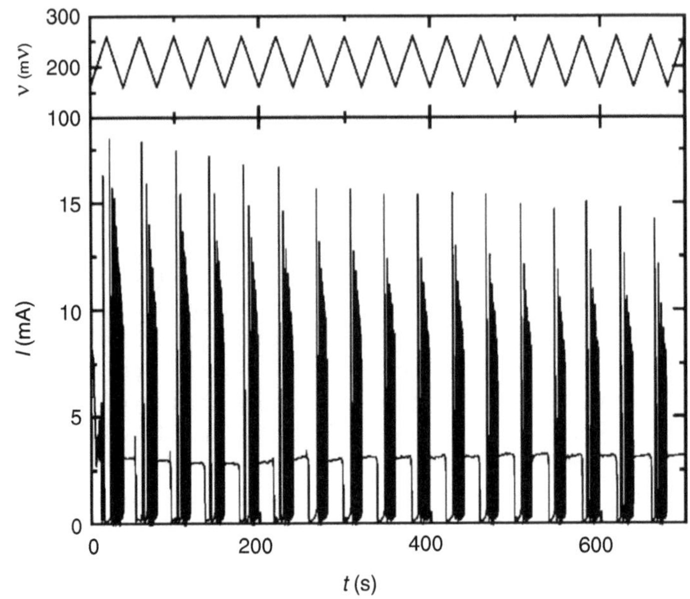

Fig. 3.46 Emergence of bursting due to the slow variation of the potential V close to the subcritical Hopf bifurcation. $V_{min} = 160$ mV and $V_{max} = 260$ mV. Period of triangular wave $\varepsilon T = 40$ s (where T is the period of the autonomous oscillations and ε is a constant: $V(t) = f(t) = f(t + \varepsilon T)$. Potential values are versus the SCE. Reprinted from [84], Copyright 2009, with permission from Elsevier

subsystem and a slow subsystem causing the repetitive transition of the bifurcation parameter of the fast system through the bifurcation point [85].

At this point one should mention that the periodically forced Fe/H_2SO_4 oscillator could have been transformed to the *spontaneous* bursting oscillator, under potentiostatic conditions, if halide ions were added to the sample [86] (note that in the presence of such ions the pitting corrosion of passive layer is induced, cf. Sect. 6.1.1, volume I). When the couple of Fe electrodes was applied (see Fig. 3.47), this meant the system of two electrochemical bursters.

The type of coupling was determined by the distance (l) between the reference and working electrodes, while its strength depended on the distance, d, between the two working electrodes. Then, depending on particular experimental conditions, including the distance between the reference and the working electrodes, either out-of-phase (inhibitory) or in-phase (excitatory) synchronizations in the bursting regime were observed.

In view of the effect of halide ions on the transformation of the periodic oscillator in the periodic burster, it is now not clear if the role of, e.g., Cl^- can be explained only in terms of the pitting corrosion of the passive layer they cause purely chemically. In order to understand the variation of the effective electrode potential, underlying the repetitive trip of the system through the bifurcation points, it may be necessary to verify the possibility of participation of such ions in the

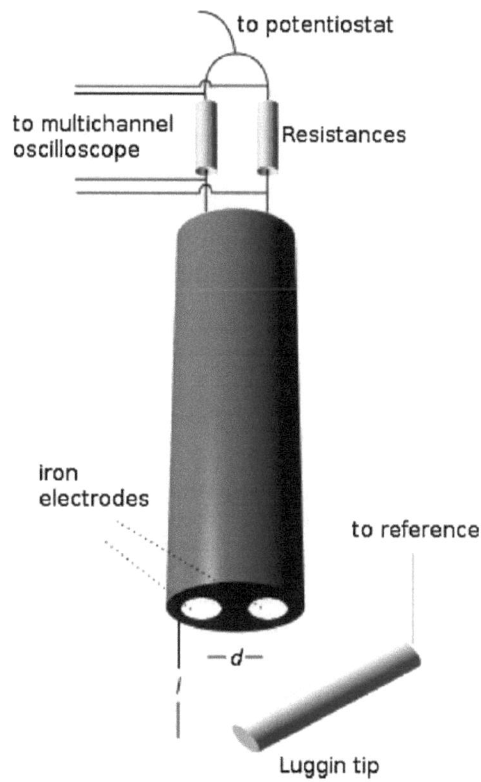

Fig. 3.47 Schematic representation of the experimental setup. Distance between iron electrodes *d* and distance between the Luggin capillary tip and the iron electrodes, *l*. Reprinted from [84], Copyright 2009, with permission from Elsevier

faradaic process, as Chin and Nobe have suggested [87]. Anyway, these simple experiments with the Fe/H_2SO_4 oscillator seem to exhibit substantial analogies with the dynamics of bursting neural systems, including the common bifurcation characteristics.

3.2.3 The Laser-Perturbed Fe/H_2SO_4 Oscillator

The laser perturbation of the oscillatory system can be considered the specific variant of creating the coupling between the nonlinear dynamical systems. In this case the "slave" oscillator can be, e.g., the excitable or oscillatory electrochemical Fe/H_2SO_4 system, while the "master" oscillator means the periodically produced laser beam, irradiating the electrode surface, either locally or globally. Such way of perturbing the Fe/H_2SO_4 oscillator was described, both experimentally and theoretically by Karantonis et al. [88–90]. Here we shall summarize the principles of such measurements and exemplary results. The experimental setup is shown in Fig. 3.48 [90].

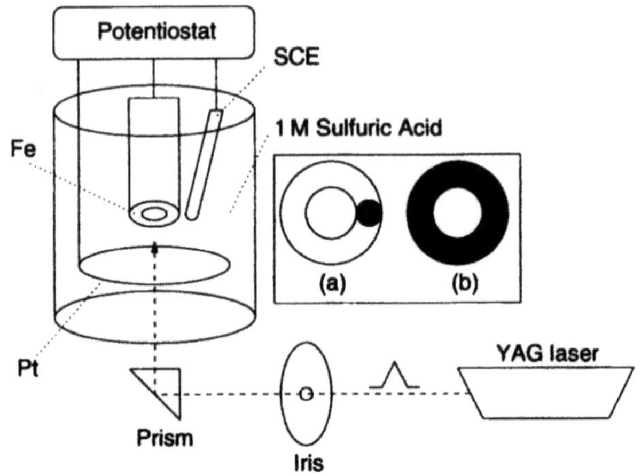

Fig. 3.48 Experimental setup for the laser beam perturbed electrochemical oscillator. An iron working electrode, a reference saturated calomel electrode (SCE), and a counter platinum electrode are immersed in 1 M H_2SO_4. The working electrode surface is perturbed by a laser beam produced by a YAG laser [$\lambda = 532$ nm (2nd) plus 1,064 nm, ~0.6 J cm^{-2} per pulse]. (**a**) Ring working electrode, od 8 mm, id 4 mm, diameter of irradiated area 1.5 mm and (**b**) ring working electrode, od 8 mm, id 4 mm, irradiated area od 8 mm, id 4 mm. Reproduced from [90] by permission of the PCCP Owner Societies

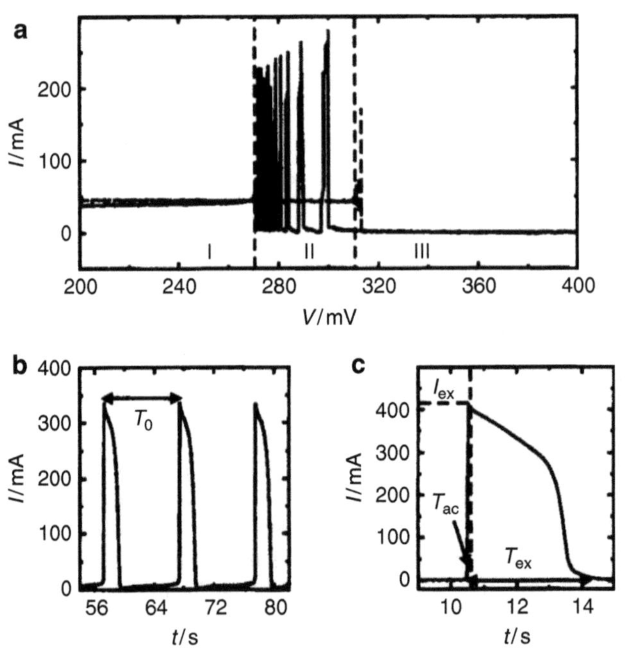

Fig. 3.49 (**a**) Polarization curve of the Fe/1 M H_2SO_4 system with scan rate 1 mV s^{-1}. Region I is the active region, region II—the oscillatory region, and region III—the passive (excitable) region. *Dashed line*: forward scan, *full line*: backward scan. (**b**) Oscillatory response for $V = 305$ mV. Period, $T_0 \approx 10$ s. (**c**) Excitable behavior for $V = 340$ mV. Excitability time, $\tau_{ex} = 3.75$ s, excitability amplitude, $I_{ex} = 411.4$ mA and activation time, $\tau_{ac} \approx 0.03$ s. Reproduced from [90] by permission of the PCCP Owner Societies

The description of the effect of a laser beam should be preceded with the electrochemical characteristics of the Fe/H_2SO_4 system under given conditions which is shown in Fig. 3.49.

In Fig. 3.49a, region of passivity (III) is excitable. Experiments showed that the local perturbation of this excitable system with laser beam causes a front of activation that spreads over the entire electrode which after that gradually returns to the passive state. The effect of laser beam, dissolving locally the passive layer, is explained in terms of the n-type semiconducting properties of the oxide film [91, 92]. It is thus clear that if the Fe/H_2SO_4 system oscillates spontaneously between the passive and the active state of the iron electrode, the laser beam can affect its dynamics only if it is applied while the system is in a passive state. If the laser beam is applied locally at certain part of the electrode, a new current peak is generated and also the oscillatory period is temporarily changed. When the perturbation with the laser beam is periodic, the response of the system is periodic. It can be expressed in terms of the p/q ratio, where p is the number of peaks of the system (autonomous plus induced) and q is the number of perturbations. For example, the $p/q = 2/1$ phase locking was observed. These phenomena were modeled in terms of the model analogous to the approach of Koper and Sluyters, being a development of the Franck–FitzHugh approach. Three dynamical variables were taken into account: the surface concentration of H^+ ions, the surface concentration of metal ions (M^+), and the electrode coverage of passive layer (θ). The mathematical construction of the spatially one-dimensional model, due to linear approximation of the Nernst diffusion layer, involved ordinary differential equations and in spite of this simplification, neglecting the temporal evolution of the diffusion layer, quite a good concordance with experimental findings was obtained [90]. In the following work, Karantonis et al. [93] have developed a more sophisticated, spatially also one-dimensional model (idealized thin ring electrode), involving partial differential equations for the double layer potential drop and hydrogen ions concentration, coupled with an ordinary differential equation for the electrode coverage by the passivating iron hydroxide. This approach refers also to the model of spatiotemporal patterns, developed by Flätgen and Krischer (see Sect. 1.2) [94, 95]. These equations were solved by the economical numerical method developed by Bieniasz [96, 97]: the finite difference patch-adaptive strategy (abbreviated further by PAS), used also earlier for the simulation of wave propagation along ring electrodes [98]. As for finite difference methods of integration of partial differential equations, both the temporal and spatial coordinates are divided into intervals, forming the respective grids. The above-mentioned algorithm automatically concentrates the spatial and temporal grids in the critical regions which adaptation is particularly useful for modeling, e.g., accelerating fronts. Calculations made according to this approach not only reproduced essential experimental observations, but also indicated that not only the double layer potential drop, but also the ionic concentrations may affect the development of the spatiotemporal patterns. Figure 3.50 shows exemplary calculated effect of periodic forcing applied to dynamical variables within a single oscillatory period.

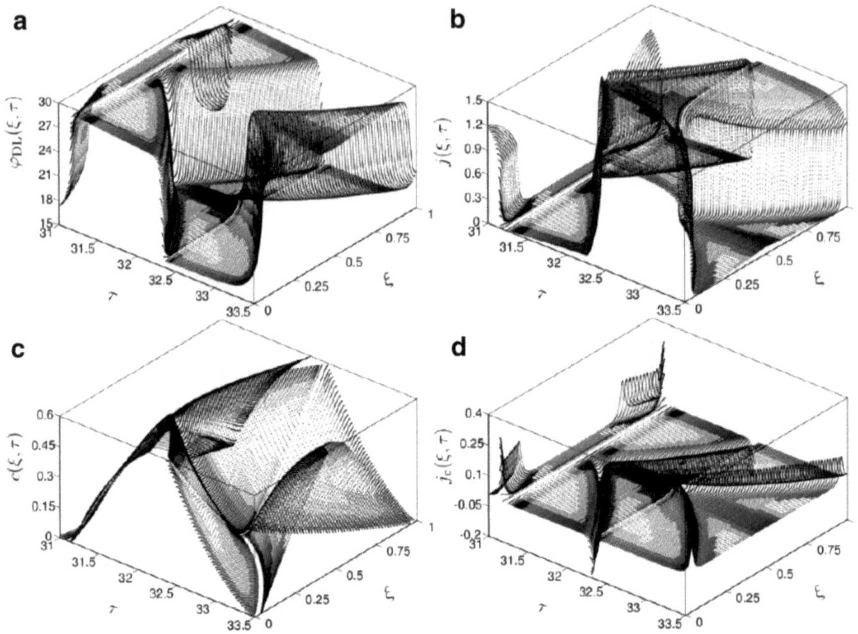

Fig. 3.50 Simulated effect of the periodic forcing on the dynamical variables within a single oscillation period. (**a**) Potential drop $\phi_{DL}(\xi, \tau)$ across the double layer; (**b**) current density $j(\xi, \tau)$, (**c**) concentration $c(\xi, \tau)$; (**d**) capacitive current density $j_C(\xi, \tau)$; ξ and τ are dimensionless distance and time, respectively. Reproduced from [93] by permission of the PCCP Owner Societies

Finally, Fig. 3.51 shows the corresponding effect of the periodic forcing, simulated on a longer time scale [the holes mean the places in which laser beam locally destroyed the $M(OH)_n$ layer, according to the kinetic equation: rate of $M(OH)_n$ decay ~ $\theta(x, t) \times C_{holes}(x, t)$, where θ is the electrode coverage with $M(OH)_n$ and C_{holes} is the surface concentration of laser-generated holes].

One should add that the application of lasers in perturbing the electrochemical oscillatory systems, at least exhibiting active/passive transitions, i.e., interacting with laser in a passive state, has also a more general application: by combining the steady-state electrochemical response and laser perturbation experiments with the bifurcation characteristics of the system it is possible to reveal the stable and unstable states of the system [88].

3.2.4 The Forced Ni/H_2SO_4 Oscillator

The anodic dissolution of nickel in H_2SO_4 under galvanostatic conditions was modified to become the forced oscillator, in a way described by Berthier et al. [99]. This system was chosen as the one showing the typical features of classical

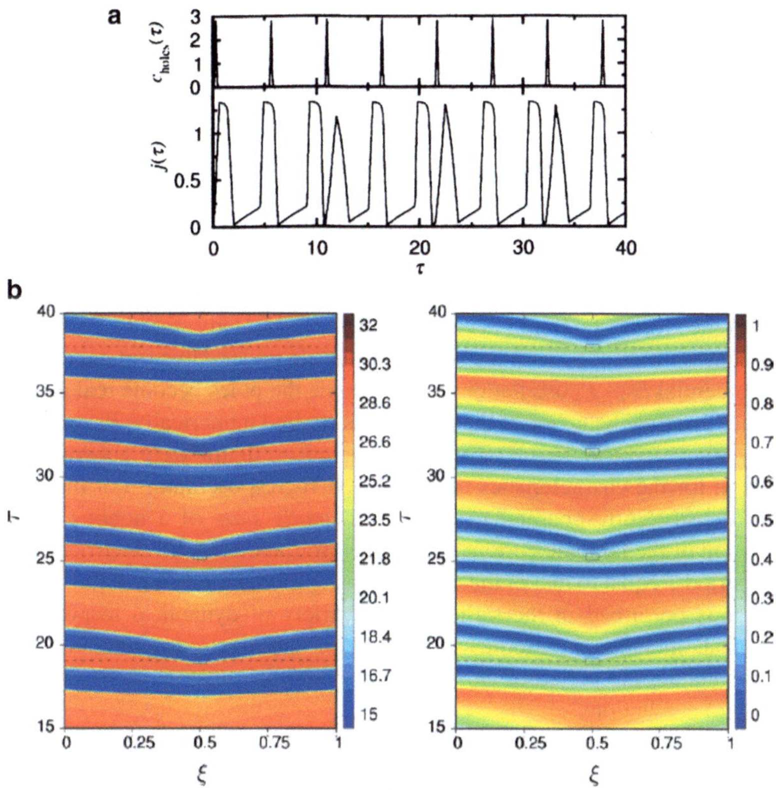

Fig. 3.51 Simulated effect of the periodic forcing on the dynamical variables. (**a**) Time series of the total current $j(\tau)$, together with the corresponding temporal variations of the average concentration $c_{\mathrm{holes}}(\tau)$ of (laser-generated) holes. (**b**) Spatiotemporal evolution of the potential drop $\phi_{\mathrm{DL}}(\xi, \tau)$ across the double layer. (**c**) Spatiotemporal evolution of the concentration $c(\xi, \tau)$. Reproduced from [93] by permission of the PCCP Owner Societies

forced self-oscillating system. The analysis consists of the theoretical and experimental part and refers to the Ni/1 M H_2SO_4 oscillator which, under galvanostatic conditions, is perturbed by the small amplitude, sinusoidal variations of current. The system's response is theoretically and experimentally studied with electrochemical impedance spectroscopy. For theoretical analysis, the two-variable kinetic *Schuhmann model* of anodic dissolution-passivation (called further Sdp model) [100] was employed. This model (one of the simplest that can explain galvanostatic oscillations) involves an intermediate adsorbed species X produced by the first step and converted into a passivating species Q in the second step or used as a catalyst in the third and final step:

$$\mathrm{M, s} \rightleftarrows \mathrm{X, s} + 2\mathrm{e} \tag{3.24}$$

$$X, s \rightleftarrows Q, s + 2e \tag{3.25}$$

$$X, s + A \rightarrow X, s + B + 2e \tag{3.26}$$

where M, s, A, and B denote the metal site at the electrode surface and two soluble species, respectively. For this mechanism, the expression for the faradaic impedance was derived and the linear stability was performed which in this galvanostatic case means searching for the poles of this impedance (see Sect. 3.7, volume I). At the potential $E = E_H$ (when the real part of the complex conjugate poles becomes positive) the Hopf bifurcation occurs, i.e., the stable-steady state loses its stability and becomes surrounded with the stable limit cycle under galvanostatic conditions. At the potential $E > E_r$ oscillations vanish, since the poles become real again. The nonlinear dynamic behavior of the model was studied numerically. For such model system, dynamic behaviors typical of the forced oscillator were generated as the prediction for the characteristics of the real Ni/H_2SO_4 oscillator. For the imposed current density corresponding to the *stable steady state* of a nonlinear system, modulation of the faradaic current caused the oscillations of the electrode potential with a period T equal to that of the current density modulation. When the oscillatory regime was perturbed in the same way, a variety of dynamical responses were determined, including: periodic oscillations of the electrode potential with the period matching the period of current density modulation, and quasi-periodic (i.e. biperiodic) oscillations.

Corresponding experimental dc and ac measurements were carried out under galvanostatic conditions in the transpassive branch of the polarization curve. Oscillations of the electrode potential were born in a supercritical Hopf bifurcation above certain critical value of the current density (Fig. 3.52) as described in Sect. 6.1.4 of volume I [101].

The effect of periodic perturbation of the externally imposed current was studied around the value close to Hopf bifurcation point, which corresponded to the critical current $I_{fb} \approx 0.10$ mA (corresponding current density was equal to ca. 3.2 mA cm^{-2}). Accordingly, the imposed current was set at an overcritical value $I_f = 0.12$ mA, where the frequency of the potential oscillations of unperturbed Ni/1 M H_2SO_4 system was equal to $f_0 = 0.228$ Hz. In the following, f will mean, as above, the frequency of the external current source (input signal). In series of the experiments, the input signal (current) was imposed with constant amplitude $\delta I_f = 20$ μA and various frequencies f. The following dynamics were observed: biperiodic behavior ($f = 0.62$ Hz), period-2 behavior ($f = 0.55$ Hz and 0.54 Hz), period-3 behavior ($f = 0.68$ Hz). Additionally, if the amplitude of perturbation was increased to $\delta I_f = 30$ μA, with the frequency kept at the value $f = 0.68$ Hz, the period-1 behavior was recovered. The latter results are quite understandable and of general applicability: if one wants to force the oscillatory systems to follow the period of external periodic signal, its amplitude has to be sufficiently high. Figure 3.53 shows, as an example, the period-3 behavior.

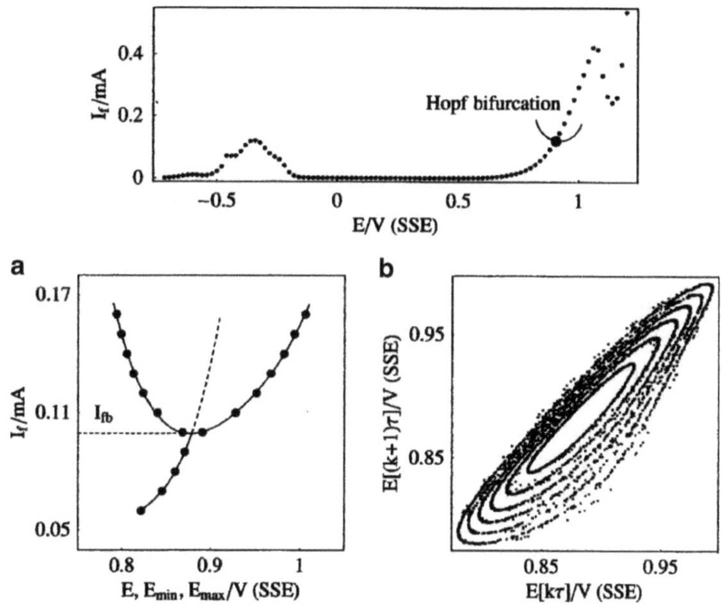

Fig. 3.52 *Top*: steady-state polarization curve of the Ni/1 M H_2SO_4 electrode ($\phi = 2$ mm) plotted at ambient temperature. *Bottom*: (**a**) experimental bifurcation diagram showing the minima and maxima of electrode potential oscillations. *Dots*: experimental points, *solid lines*: stable steady states and minima and maxima of potential oscillations (fit curves), *dashed curve*: unstable steady states (extrapolation of the fit curve). (**b**) Phase portrait for I_f(mA) $= 0.11, 0.12, 0.13, 0.14, 0.15, 0.16$; $\tau = 0.2$ s. Reprinted from [99], Copyright 2004, with permission from Elsevier

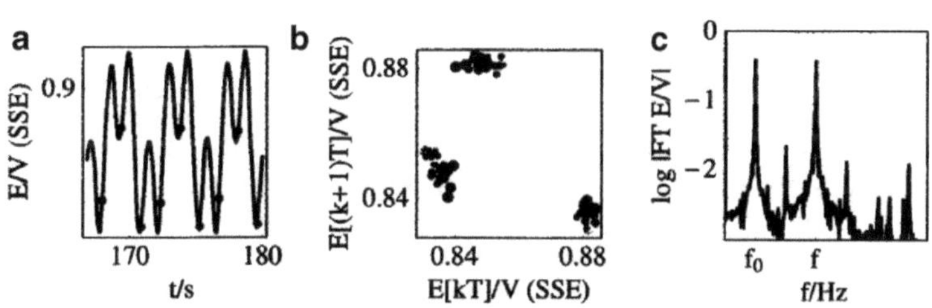

Fig. 3.53 Electrode potential response to a sinusoidal current modulation for the Ni/1 M H_2SO_4 system, showing the period-3 behavior. (**a**) Electrode potential oscillations, (**b**) Poincaré section in the $E[(k+1)T]$ vs. $E(kT)$ plane, (**c**) Fourier power spectrum of potential oscillations after the initial transients die away; $I_f = 0.12$ mA, $\delta I_f = 30$ μA, $f = 0.68$ Hz. Reprinted from [99], Copyright 2004, with permission from Elsevier

In the opinion of the authors, the forced Ni/1M H_2SO_4 oscillator is also the first experimental manifestation of the galvanostatic electrochemical oscillator which exhibits the Neimark-Sacker torus bifurcation (or Hopf bifurcation of a limit cycle, predicted for time-periodic forcing near a Hopf bifurcation), meaning the contraction of the two torus to a closed curve. The chaotic (aperiodic) behavior was however not found, either in the forced theoretical Sdp model, or in the experimental forced Ni/1 M H_2SO_4 system, although generation of chaos in this way was theoretically described for the driven van der Pol oscillator [102]. Chaotic dynamics was found experimentally only for (unperturbed) Ni/H_2SO_4 oscillator with high H_2SO_4 concentration [103].

Further details of the theoretical and experimental characteristics of the Ni/1 M H_2SO_4 oscillator, including, among others, the description of mode locking in terms of the so-called Arnold tounges (here, the areas plotted in the δj_f vs. ω/ω_0 representation), and the unexpected impedance spectra recorded for controlled current conditions in the biperiodic region, the reader can find in the original reference [99].

Concerning further aspects of dynamics of coupled electrochemical oscillators, one should expect the discovery of *chimera states*, reported recently for the (mainly) model and some physical systems. According to Abrams and Strogatz [104], who refer to the discovery of Shima and Kuramoto [105], the *chimera state* means the dynamical regime in which an array of *identical* oscillators splits into two domains: one coherent and phase-locked, and the other incoherent and desynchronized. The name of that unusual phenomenon, not reported before for *identical* oscillators, refers to Greek mythology, in which the chimera was fire-breathing monster having a lion's head, a goat's body, and a serpent tail. Today, the chimera state should generally be understood as anything composed of incongruous parts [104].

References

1. Karantonis A, Nakabayashi S (2001) Phase flow deformations and coupled electrochemical oscillators. Chem Phys Lett 347:133–137
2. Kuramoto Y (1984) Chemical oscillators, waves and turbulence. Springer, Berlin
3. Han SK, Kurrer C, Kuramoto Y (1995) Dephasing and bursting in coupled neural oscillators. Phys Rev Lett 75:3190–3193
4. Fukushima S, Nakanishi S, Fukami K, Sakai S, Nagai T, Tada T, Nakato Y (2005) Observation of synchronized spatiotemporal reaction waves in coupled electrochemical oscillations of an NDR type. Electrochem Commun 7:411–415
5. Fukushima S, Nakanishi S, Nakato Y, Ogawa T (2008) Selection principle for various modes of spatially nonuniform electrochemical oscillations. J Chem Phys 128:014714-1–014714-10
6. Fukushima S, Ogawa T, Nakato Y, Nakanishi S (2008) Bifurcation analysis of bistability between spatially uniform and non-uniform electrochemical oscillations. Chem Phys Lett 453:35–39
7. Kiss IZ, Munjal N, Martin RS (2009) Synchronized current oscillations of formic acid electrooxidation in a microchip-based dual-electrode flow cell. Electrochim Acta 55:395–403
8. Moehlenbrock MJ, Martin RS (2007) Development of an on-chip injector for microchip-based flow analyses using laminar flow. Lab Chip 7:1589–1596
9. Wang Y, Hudson JL (1992) Experiments on interacting electrochemical oscillators. J Phys Chem 96:8667–8671
10. Nakabayashi S, Zama K, Uosaki K (1996) Nonlinear iron electrochemical oscillator: coupling and photo effects. J Electrochem Soc 143:2258–2262
11. Nakabayashi S, Baba R, Shiomi Y (1998) Spatiotemporal propagation of a non-linear electrochemical reaction over an iron electrode. Chem Phys Lett 287:632–638

12. Karantonis A, Shiomi Y, Nakabayashi S (2000) Coherence and coupling during oscillatory metal electrodissolution. J Electroanal Chem 493:57–67
13. Sielewiesiuk J, Górecki J (2002) Complex transformations of chemical signals passing through a passive barrier. Phys Rev E 66:016212-1–016212-9
14. Wojtowicz J (1973) Oscillatory behavior in electrochemical systems. In: Bockris JO'M, Conway BE, White RE (eds) Modern aspects of electrochemistry, vol 8. Plenum, New York, pp 47–120
15. Bell JC, Jaeger NI, Hudson JL (1992) Coupled oscillating cobalt electrodes. J Phys Chem 96:8671–8676
16. Hudson JL, Tsotsis TT (1994) Electrochemical reaction dynamics: a review. Chem Eng Sci 49:1493–1572
17. Kiss IZ, Wang W, Hudson JL (1999) Experiments on arrays of globally coupled periodic electrochemical oscillators. J Phys Chem B 103:11433–11444
18. Fei Z, Kelly R, Hudson JL (1996) Spatiotemporal patterns on electrode arrays. J Phys Chem 100:18986–18991
19. Kaneko K (1989) Pattern dynamics in spatiotemporal chaos: pattern selection, diffusion of defect and pattern competition intermittency. Physica D 34:1–41
20. Zanette DH, Mikhailov AS (1998) Condensation in globally coupled population of chaotic dynamics. Phys Rev E 57:276–281
21. Kiss IZ, Wang W, Hudson JL (2000) Complexity of globally coupled chaotic electrochemical oscillators. Phys Chem Chem Phys 2:3847–3854
22. Badii R, Politi A (1985) Statistical description of chaotic attractors: the dimension function. J Stat Phys 40:725–750
23. Kiss IZ, Wang W, Hudson JL (2002) Populations of coupled electrochemical oscillators. Chaos 12:252–263
24. Wang W, Green BJ, Hudson JL (2001) Periodic forcing of chaotic electrochemical oscillators. J Phys Chem B 105:7336–7373
25. Kuramoto Y (1984) Cooperative dynamics of oscillator community. Prog Theor Phys Suppl 79:223–240
26. Kiss IZ, Zhai Y, Hudson JL (2002) Emerging coherence in a population of chemical oscillators. Science 296:1676–1678
27. Mikhailov AS, Zanette DH, Zhai YM, Kiss IZ, Hudson JL (2004) Cooperative action of coherent groups in broadly heterogeneous populations of interacting chemical oscillators. Proc Natl Acad Sci USA 101:10890–10894
28. Kiss IZ, Rusin CG, Kori H, Hudson JL (2007) Engineering complex dynamical structures: sequential patterns and desynchronization. Science 316:1886–1889
29. Jain S, Kiss IZ, Breidenich J, Hudson JL (2009) The effect of IR compensation on stationary and oscillatory patterns in dual-electrode metal dissolution. Electrochim Acta 55:363–373
30. Bîrzu A, Gáspár V (2009) Synchronization of electrochemical oscillators of S-NDR type. Electrochim Acta 55:383–394
31. Koper MTM, Sluyters JH (1993) A mathematical model for current oscillations at the active-passive transition in metal electrodissolution. J Electroanal Chem 347:31–48
32. Schäfer C, Rosenblum MG, Kurths J, Abel HH (1998) Heartbeat synchronized with ventilation. Nature 392:239–240
33. Lee MG, Jorné J (1992) On the kinetic mechanism of zinc electrodeposition in the region of negative polarization resistance. J Electrochem Soc 139:2841–2844
34. Miyakita Y, Karantonis A, Nakabayashi S (2002) Response of relaxation oscillatory electrochemical networks to external input. Chem Phys Lett 362:461–466
35. Kandel ER (1976) Cellular basis of behavior. W.H. Freeman and Co., San Francisco, CA
36. Hall Z (1992) An introduction to molecular neurobiology. Sinauer, Sunderland, MA
37. Miyakita Y, Nakabayashi S, Karantonis A (2005) Spatiotemporal coding in an electrochemical oscillatory network. Phys Rev E 71:056207-1–056207-9

38. Karantonis A, Pagitsas M, Miyakita Y, Nakabayashi S (2004) In-phase, anti-phase and fractured synchrony in ring networks of coupled relaxation electrochemical oscillators. J Phys Chem B 108:5836–5846
39. Karantonis A, Pagitsas M, Miyakita Y, Nakabayashi S (2003) From excitatory to inhibitory connections in networks of discrete electrochemical oscillators. J Phys Chem B 107:14622–14630
40. Karantonis A, Miyakita Y, Nakabayashi S (2002) Synchronization of coupled assemblies of relaxation oscillatory electrode pairs. Phys Rev E 65:046213-1–046213-9
41. Karantonis A, Pagitsas M, Miyakita Y, Nakabayashi S (2005) Manipulation of spatio-temporal patterns in networks of relaxation electrochemical oscillators. Electrochim Acta 50:5056–5064
42. Karantonis A, Pagitsas M, Miyakita Y, Nakabayashi S (2006) Synchronization phenomena in networks of coupled relaxation electrochemical oscillations. Int J Bifurcat Chaos 16:1951–1960
43. Eklund G (1972) On the initiation of corrosion on carbon steels. Scand J Metall 1:331–336
44. Eklund G (1974) Initiation of pitting at sulfide inclusions in stainless steel. J Electrochem Soc 121:467–473
45. Wranglen G (1974) Pitting and sulphide inclusions in steel. Corros Sci 14:331–349
46. Castle JE, Ke R (1990) Studies by Auger spectroscopy of pit initiation at the site of inclusions in stainless steel. Corros Sci 30:409–428
47. Lott SE, Alkire RC (1989) The role of inclusions on initiation of crevice corrosion of stainless steel. J Electrochem Soc 136:973–979
48. Webb EG, Alkire RC (2002) Pit initiation at single sulfide inclusions in stainless steel. I. Electrochemical microcell measurements. J Electrochem Soc 149:B272–B279
49. Webb EG, Alkire RC (2002) Pit initiation at single sulfide inclusions in stainless steel. II Detection of local pH, sulfide, and thiosulfate. J Electrochem Soc 149:B280–B285
50. Webb EG, Alkire RC (2002) Pit initiation at single sulfide inclusions in stainless steel. III. Mathematical model. J Electrochem Soc 149:B286–B295
51. García E, Hernández MA, Rodriguez FJ, Genescá J, Boerio FJ (2003) Oscillation and chaos in pitting corrosion of steel. Corrosion 59:50–58
52. Wu B, Scully JR, Hudson JL, Mikhailov AS (1997) Cooperative stochastic behavior in localized corrosion. I. Model. J Electrochem Soc 144:1614–1620
53. Lunt TT, Pride ST, Scully JR, Hudson JL, Mikhailov AS (1997) Cooperative stochastic behavior in localized corrosion. II Experiments. J Electrochem Soc 144:1620–1629
54. Punckt C, Bölscher M, Rotermund HH, Mikhailov AS, Organ L, Budiansky N, Scully JR, Hudson JL (2004) Sudden onset of pitting corrosion on stainless steel as a critical phenomenon. Science 305:1133–1136 doi:10.1126/science.1101358
55. Isaacs HS, Ishikawa Y (1985) Current and potential transients during localized corrosion of stainless steel. J Electrochem Soc 132:1288–1293
56. Pistorius PC, Burstein GT (1992) Metastable pitting corrosion of stainless steel and the transition to stability. Philos Trans R Soc Ser A 341:531–559
57. Galvele JR (1983) Corrosion – aqueous processes and passive films. In: Scully JC (ed) Treatise on materials science and technology, vol 23. Academic, London, p 1
58. Janik-Czachor M (1981) An assessment of the processes leading to pit nucleation on iron. J Electrochem Soc 128:513C–519C
59. Janik-Czachor M, Wood GC, Thompson GE (1980) An assessment of the processes leading to the nucleation of pits. Br Corros J 15:154–160
60. Williams DE, Westcott C, Fleischmann M (1985) Stochastic models of pitting corrosion of stainless steel. I. Modeling of the initiation and growth of pits at constant potential. J Electrochem Soc 133:1796–1804
61. Williams DE, Westcott C, Fleischmann M (1985) Stochastic models of pitting corrosion of stainless steel. II. Measurement and interpretation of data at constant potential. J Electrochem Soc 133:1804–1811

62. Harb JN, Alkire RC (1991) Transport and reaction during pitting corrosion of Ni in 0.5 M NaCl. I. Stagnant fluid. J Electrochem Soc 138:2594–2600
63. Sugimoto K, Matsuda S, Ogiwara Y, Kitamura K (1985) Microscopic ellipsometric observation of the change in passive film on 18Cr-8Ni stainless steel with the initiation and growth of pit. J Electrochem Soc 132:1791–1795
64. Xu Y, Wang M, Pickering HW (1993) On electric field induced breakdown of passive films and the mechanism of pitting corrosion. J Electrochem Soc 140:3448–3457
65. Hoar TP, Mears DC, Rothwell GP (1965) The relationships between anodic passivity, brightening and pitting. Corros Sci 5:279–289
66. Brossia CS (1998) In: Kelly RG, Frankel GS, Natishan PM, Newman RC (eds) Critical factors in localized corrosion III, vol PV 98-17. The Electrochemical Society Proceedings Series, Pennington, NJ, p 326
67. Lunt TT, Brusamarello V, Scully JR, Hudson JL (2000) Interactions among localized corrosion sites investigated with electrode arrays. Electrochem Solid State Lett 3:271–274
68. Butler G, Kolson HC, Mercer AD (1971) Br Corros J (London) 16:31
69. Rosenfeld IL, Danilov IS (1967) Electrochemical aspects of pitting. Corros Sci 7:129–132
70. Newman J, Hanson DN, Vetter V (1977) Potential distribution in a corroding pit. Electrochim Acta 22:829–831
71. Harb JN, Alkire RC (1991) Transport and reaction during pitting corrosion of Ni in 0.5 M NaCl. II. Flowing fluid. J Electrochem Soc 138:3568–3575
72. Lunt TT, Scully JR, Brusamarello V, Mikhailov AS, Hudson JL (2002) Spatial interactions among localized corrosion sites. Experiments and modeling. J Electrochem Soc 149: B163–B173
73. Rotermund HH, Haas G, Franz RU, Tromp RM, Ertl G (1995) Imaging pattern formation in surface reactions from ultrahigh vacuum to atmospheric pressure. Science 270:608–610
74. Organ L, Scully JR, Mikhailov AS, Hudson JL (2005) A spatiotemporal model of interactions among metastable pits and the transition to pitting corrosion. Electrochim Acta 51:225–241, and references cited therein
75. Organ L, Tiwary Y, Scully JR, Mikhailov AS, Hudson JL (2007) Interactions among metastable pits on heterogeneous electrodes. Electrochim Acta 52:6784–6792
76. Sasaki K, Isaacs HS (2004) Origins of electrochemical noise during pitting corrosion of aluminum. J Electrochem Soc 151:B124–B133
77. Mikhailov AS, Jain S, Organ L, Hudson JL (2006) Cooperative stochastic behavior in the onset of localized corrosion. Chaos 16:037104-1–037104-9
78. Nakabayashi S, Kira A (1992) Forced electrochemical nonlinear oscillator assembled on rotating ring-disk electrodes. J Phys Chem 92:1021–1023
79. Schell M, Albahadily FN, Safar J (1993) Decreases in the inhibition of the electrocatalyzed oxidation of formic acid by carbon monoxide. J Electroanal Chem 353:303–313
80. Pagitsas M, Sazou D (1991) The improved Frank-FitzHugh model for the electrodissolution of iron in sulphuric acid solutions: linear stability and bifurcation analysis. Derivation of the kinetic equations for the forced Franck-FitzHugh model. Electrochim Acta 36:1301–1308
81. Pagitsas M, Karantonis A, Sazou D (1992) Application of periodic forcing on the simplified Franck-FitzHugh model for the electrochemical oscillations observed during the electrodissolution of iron in sulphuric acid solutions. Electrochim Acta 37:1047–1059
82. Sazou D, Karantonis A, Pagitsas M (1993) Generalized Hopf, saddle-node infinite period bifurcations and excitability during the electrodissolution and passivation of iron in a sulfuric acid solution. Int J Bifurcat Chaos 4:981–997
83. Karantonis A, Pagitsas M, Sazou D (1993) Dynamical response of the sinusoidally perturbed electrodissolution/passivation of iron in sulfuric acid solution: entrainment, spike generation, and quasiperiodicity. Chaos 3:243–255
84. Karantonis A, Koutsaftis D, Koulombi N (2009) Single and coupled electrochemical bursters during the electrodissolution/passivation of iron. Electrochim Acta 55:374–382
85. Izhikevich EM (2000) Neural excitability, spiking and bursting. Int J Bifurcat Chaos 10:1171–1266

86. Koutsaftis D, Karantonis A, Pagitsas M, Kouloumbi N (2007) Transient and persistent electrochemical bursting induced by halide ions. J Phys Chem C 111:13579–13585
87. Chin RJ, Nobe K (1972) Electrodissolution kinetics of iron in chloride solutions. J Electrochem Soc 119:1457–1461
88. Karantonis A, Shiomi Y, Nakabayashi S (2001) Laser experiments and theoretical modeling for the diagnosis of bifurcation sequences of an oscillating electrode reaction. Int J Bifurcat Chaos 11:1275–1294
89. Karantonis A, Shiomi Y, Nakabayashi S (2001) One-dimensional discrete maps for periodic laser forcing on electrochemical oscillators. Chem Phys Lett 335:221–226
90. Shiomi Y, Karantonis A, Nakabayashi S (2001) Spatio-temporal laser forcing on an oscillatory/excitable electrochemical system. Phys Chem Chem Phys 3:479–488
91. Hardee KL, Bard AJ (1976) Semiconductor electrodes. V. The application of chemically vapor deposited iron oxide films to photosensitized electrolysis. J Electrochem Soc 123:1024–1026
92. Yeh LSR, Hackerman N (1977) Iron oxide semiconductor electrodes in photoassisted electrolysis of water. J Electrochem Soc 124:833–836
93. Karantonis A, Bieniasz L, Nakabayashi S (2003) The combined unidirectional and local coupling in a spatially one-dimensional model of oscillatory metal electrodissolution. Patch-adaptive simulaton study. Phys Chem Chem Phys 5:1831–1841
94. Flätgen G, Krischer K (1995) A general model for pattern formation in electrode reactions. J Chem Phys 103:5428–5436
95. Flätgen G, Krischer K (1995) Accelerating fronts in an electrochemical system due to global coupling. Phys Rev E 51:3997–4004
96. Bieniasz LK (2000) Use of dynamically adaptive grid techniques for the solution of electrochemical kinetic equations. Part V. A finite-difference, adaptive space/time grid strategy based on a patch-type local uniform spatial grid refinement, for kinetic models in one-dimensional space geometry. J Electroanal Chem 481:115–133
97. Bieniasz LK (2002) Use of dynamically adaptive grid techniques for the solution of electrochemical kinetic equations. Part 10. Extension of the patch-adaptive strategy to kinetic models involving spatially localized unknowns at the boundaries, multiple space intervals, and non-local boundary conditions, in one-dimensional space geometry. J Electroanal Chem 527:1–10
98. Bieniasz LK (2002) Use of dynamically adaptive grid techniques for the solution of electrochemical kinetic equations. Part 13. Patch-adaptive simulation of wave propagation along ring electrodes: one-dimensional approximation. J Electroanal Chem 529:51–58
99. Berthier F, Diard JP, Le Gorrec B, Montella C (2004) Study of the forced Ni/1 M H_2SO_4 oscillator. J Electroanal Chem 572:267–281
100. Schuhmann D (1968) Etude phenomenologique a l'aide de schemas reactionnels des impedances faradiques contenant des resistances negatives et des inductances. J Electroanal Chem 17:45–59
101. Haim D, Lev O, Pismen LM, Sheintuch M (1992) Modeling periodic and chaotic dynamics in anodic nickel dissolution. J Phys Chem 96:2676–2681
102. Mettin R, Parlitz U, Lauterborn W (1993) Bifurcation structure of the driven van der Pol oscillator. Int J Bifurcat Chaos 3:1529–1555
103. Lev O, Wolffberg A, Pismen LM, Sheintuch M (1989) The structure of complex behavior in anodic nickel dissolution. J Phys Chem 93:1661–1666
104. Abrams DM, Strogatz SH (2004) Chimera states for coupled oscillators. Phys Rev Lett 93:174102–174105
105. Shima SI, Kuramoto Y (2002) Coexistence of coherence and incoherence in nonlocally coupled phase oscillators. Nonlin Phenom Complex Syst 5:380–385

Chapter 4
Spatial and Spatiotemporal Patterns in Anodized Semiconductors

4.1 Spatiotemporal Nature of Silicon Electrooxidation and the Origin of Oscillations

As indicated in Sect. 6.3 of volume I, where the works of, among others, Gerischer, Chazalviel, Ozanam et al. were invoked, understanding of the oscillatory dissolution of silicon in fluoride media requires taking into account the spatiotemporal aspect of that process. The recent relevant models of the oscillatory Si electrodissolution were developed by groups of H.-J. Lewerenz and H. Föll. The principles of these approaches are briefly described below and further confronted with most recent ideas.

4.1.1 The Model of Two Oxides

An advanced theoretical model of the oscillatory Si dissolution was elaborated by the group of *Lewerenz* in the year 2000 [1, 2]. Mathematical construction of their approach refers to the Markov processes [3]. The chemical basis of this model is the (assumed also in earlier mechanisms) cyclic increase and decrease in the amount of SiO_2 covering the Si surface. The detailed construction refers to the idea of self-oscillating domains, suggested by Chazalviel et al. [4], although in the present model the external periodic excitation was not necessary to obtain the oscillations. Both the domains and the complete SiO_2 layer were described by sets of *thickness oscillators*, meaning the oscillator with time-dependent thickness of SiO_2 layer. For each point of the Si surface there was defined one such thickness oscillator, and the infinite set of all thickness oscillators modeled the behavior of the overall SiO_2 layer. In the *synchronization* mechanism of single oscillators, the key role is ascribed to the pores in the oxide layer. The sequence of calculations corresponds to the following events [1]. First, the bare surface of Si is assumed, with all thickness oscillators set to zero. Next, at a specific starting time, when the anodic

M. Orlik, *Self-Organization in Electrochemical Systems II*,
Monographs in Electrochemistry, DOI 10.1007/978-3-642-27627-9_4,

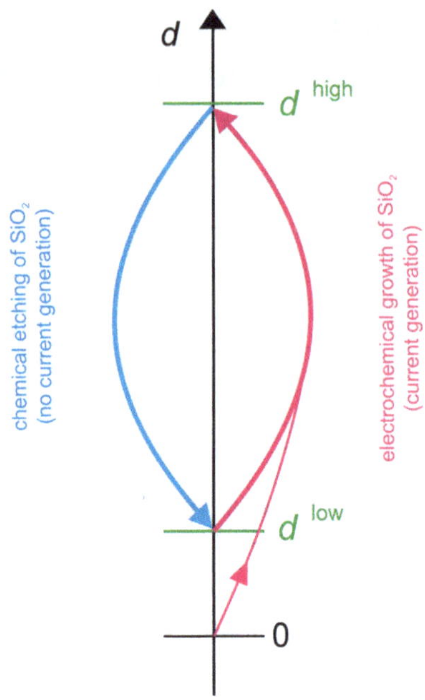

Fig. 4.1 The path of a thickness oscillator; d—thickness of SiO_2 layer at point (x, y) on the plane parallel to Si surface, starting from 0 and upon applying anodic potential increasing and oscillating between d^{low} (lower inflection point) and d^{high} (upper inflection point). Based on [1]

potential is applied to Si electrode, all these oscillators begin to increase in thickness (reflecting the oxide growth connected with current flow until maximal oxide thickness is reached). During this time, the oxidation process dominates over the chemical dissolution (etching) of SiO_2. When the electrochemical growth of oxide layer stops, the currentless chemical dissolution makes the SiO_2 layer thinner. Hence, each oscillator undergoes cycles which include the Si oxidation as well as the SiO_2 etching. Consequently, the period of a thickness oscillator consists of two parts: the time for oxide growth (first part) and the time for etching the oxide (second part). It was assumed that the first part was nearly constant, while the second part varied due to the inhomogeneity of SiO_2 layer [1]. Thus, the porosity of the SiO_2 layer was incorporated in the model in this way that the period of the SiO_2 dissolution time was spatially distributed. As a consequence, a probability distribution for the period of a thickness oscillator was introduced as an essential element of the model. The cycle of the thickness oscillator is illustrated in Fig. 4.1.

In this figure, d means the thickness of the SiO_2 layer corresponding to point (x, y) on the plane parallel to Si surface at the time t, i.e. $d = d(x, y, t)$. The thickness oscillator d oscillates between two states, the lower inflection point d^{low} and the upper inflection point d^{high}. The first state d^{low} corresponds to the start of the increase of the oxide coverage (thus also the increase of anodic current) [cf. Eq. (6.95), volume I]. The second state d^{high} is characterized by the end of the increase and the beginning of the (currentless) decrease of the SiO_2 layer [Eqs. (6.96) and (6.97), volume I). The time for the motion of d, beginning at d^{low}, passing d^{high} and ending at d^{low}, is called a period of a thickness oscillator, which varies along the (x, y) plane since the corresponding SiO_2 layer can change with respect to its stability due to the etching process.

It is defined a *synchronization state* to characterize all thickness oscillators belonging to the same cycle. The synchronization state defines, for each time, the (differential) number of thickness oscillators passing the lower inflection point simultaneously. A convolution of a synchronization state with the probability distribution for the oscillation periods creates the synchronization state of the next cycle. In order to calculate the current of the entire model electrochemical system from the synchronization states, the time-dependent elementary current peak was introduced that characterized the shape of the current flow resulting from one thickness oscillator during the first, oxide growth phase. Since all starting points of the thickness oscillators (all synchronization states) and the elementary current peak are known, a convolution of the sum of all synchronization states with the elementary current peak gives the current of the electrochemical system. The reader interested in mathematical details of the model is advised to consult the original references [1, 2]. Here, we shall briefly describe the (selected) specific physical and morphological assumptions for varying pore density and feedback mechanism [2]. Thus, if it is assumed that during one oscillatory cycle an increase in the pore density takes place, a corresponding increasing overlap of pores will occur. Concerning the nucleation and growth mechanism of oxide islands on Si, the following assumption was made: oxide clusters that are formed initially can grow into a nonperturbed silicon surface. With the growth of such oxide clusters (oxide I), the remaining silicon surface between the clusters will be exposed to an increasing strain due to the lattice mismatch between silicon oxide and silicon. The oxide formed in these regions (oxide II) will then exhibit more defects (or pores) than the initially grown oxide, i.e., this oxide will be etched faster than the initially grown one. The structural properties of oxide, which affect the etching rate, are dependent also on the electrode potential, according to earlier suggestion of Gerischer and Lübke [5], confirmed later by Lewerenz et al. in surface analysis experiments using photoelectron spectroscopy [6]. The model representation of the correlation between the variations of the oxide layer and the current oscillations, showing the crucial assumption of two oxides I and II, is sketched in Fig. 4.2. A detailed analysis of the relevant feedback mechanism is given in [2] (see Fig. 6 there).

At this point it is important to indicate that atomic force microscope (AFM) measurements of the anodic dissolution of p-Si(111) confirmed the reliability of the above reasoning (Fig. 4.3).

Leaving other mathematical and physical details of this mechanism [1, 2] to the interested reader, we shall note here that this model is able to produce both sustained and damped oscillations of the anodic current. As an example, Figs. 4.4 and 4.5 illustrate a very good concordance between the experimentally measured and theoretically calculated current oscillations.

The current state of the proposed mechanism of electrochemical oscillations during Si anodization is summarized in a very recent review by Lewerenz et al. [7] which also indicates the importance of such processes for preparation of photoactive devices. Figure 4.6 shows the scheme of a nanoemitter solar device.

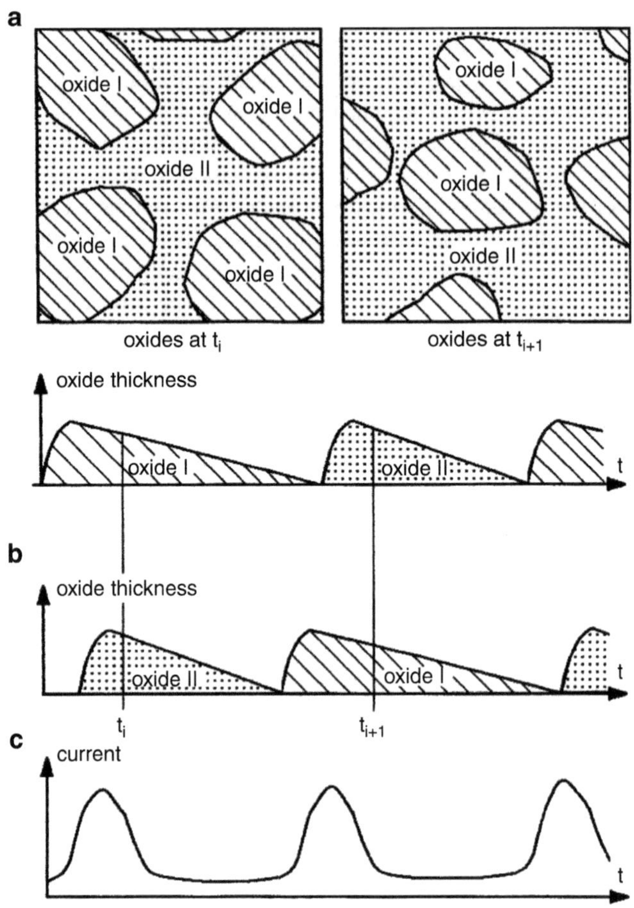

Fig. 4.2 Schematic representation of the synchronization mechanism and of the resulting current oscillations. *Shaded area*—earlier grown oxide I (less pores) which is more stable against the etching. *Dotted area*—later grown oxide II (more pores due to growing strain between silicon and its oxide) which is less stable against etching. (**a**) Lateral distribution of oxide I and II in two consecutive cycles at t_i and t_{i+1}. (**b**) Time-dependent oxide growth and etching of oxide I and II. (**c**) Resulting current oscillation. Reprinted from [2], Copyright 2000, with permission from Elsevier

In the oscillatory regime, defects (micro- and nanocracks) in the oxide layer can eventually develop into nanopores on the overall sample surface. They can be further deepened by alkaline etching and the voids are filled with metallic emitter materials for enhanced collection of minority carriers, generated by low-energy photons deeper in the silicon absorber.

Very recent developments of Lewerenz's, Grzanna's et al. model include extension to spatiotemporal resolution using cellular automata [8] and inclusion of nanopore morphology [9].

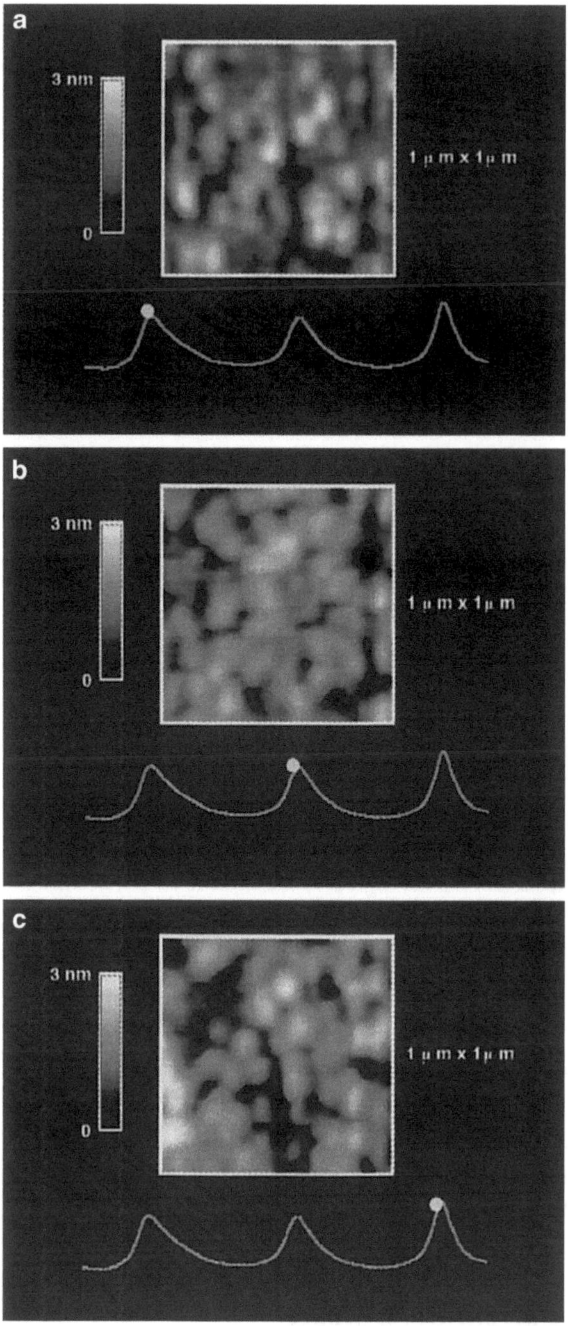

Fig. 4.3 AFM micrographs obtained at three different times (**a**–**c**) when the current nearly reaches its maximum (different positions of the *circles* in the oscillation graph). The oscillation period amounts to about 100 s. *Gray scale* indicates height. *p*-Si(111), 0.1 M NH_4F, pH = 4.2, applied potential 6 V vs. SCE. Reprinted from [2], Copyright 2000, with permission from Elsevier

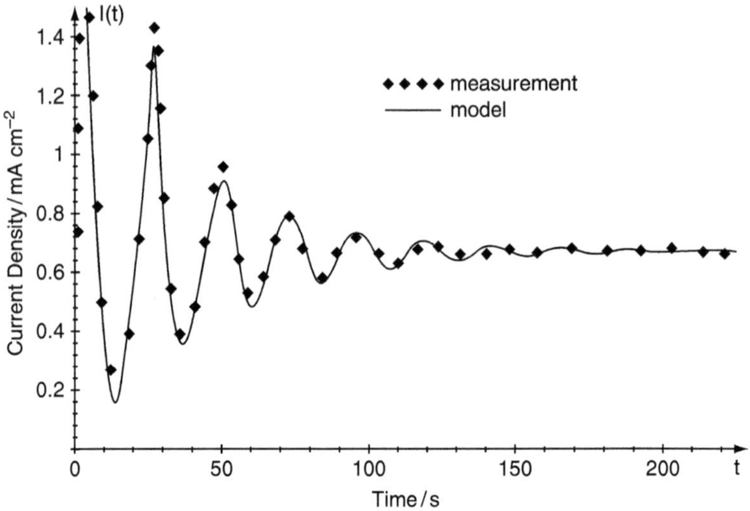

Fig. 4.4 Measured and calculated damped current oscillation for p-Si (111) dissolution. See [2] for the model construction and parameters. Reprinted from [2], Copyright 2000, with permission from Elsevier

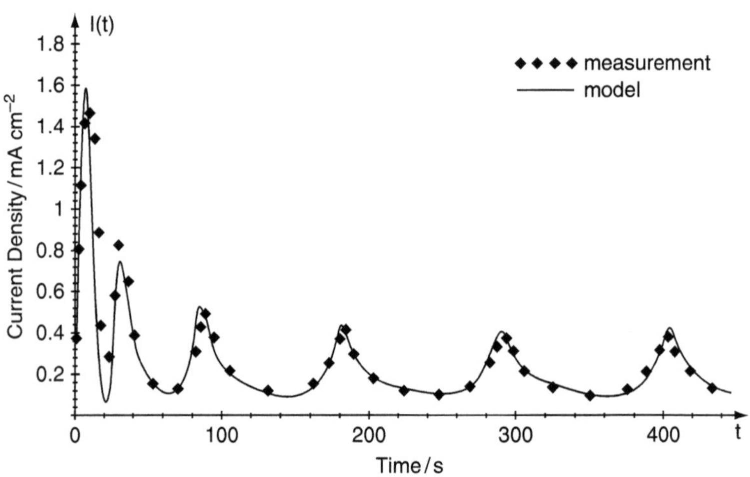

Fig. 4.5 Measured and calculated sustained current oscillation for p-Si (111) dissolution. See [2] for the model construction and parameters. Reprinted from [2], Copyright 2000, with permission from Elsevier

4.1.2 The Current-Burst Model

Alternative theoretical models of electrochemical oscillations occurring during Si anodization were elaborated in the *Föll's* group and are termed the "current-burst models" (CBM), according to Carstensen, Föll et al. [10, 11]. The current burst, CB,

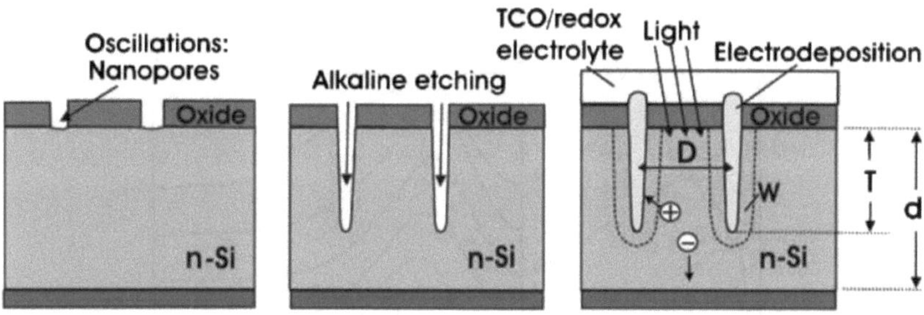

Fig. 4.6 Concept for a nanoemitter solar cell and its preparation. *Left*: nanopores are electrochemically formed in silicon oxide; *middle:* deepening of these pores by alkaline etching which selectively attacks silicon; *right*: filling of pores with an emitter material that forms a Schottky junction with Si, contacting the emitter with a transparent conductive oxide (TCO) or a redox electrolyte for solid-state or photoelectrochemical solar cells, respectively; see [7] for further details. Reprinted from [7] with kind permission from Springer Science + Business Media

means here an active, single localized charge transfer process. As the source of experimental data for constructing this model, there were analyzed the current transients for Si/HF system, measured after changing the anodic potential to open-circuit potential (OCP). Such an approach, found to be a valuable tool for obtaining data on the oxide present at different phases of the oscillation [12], has been by the present authors extended for a quantitative interpretation allowing to quantify parameters such as the dissolution rates, oxide thickness, distribution function of the oxide thickness, and the capacitance of the oxide. Figure 4.7 shows representative comparison of the oscillating current for constant $U = +4.5$ V (vs. SCE) with the transient $I - t$ courses recorded after stepping the potential to OCP, for different phases of the oscillation. Such measurements were compared with the in situ fast Fourier transform (FTT) impedance spectroscopy.

Under such *open-circuit* conditions the surface oxide is chemically dissolved, reducing the local thickness of the oxide layer with (assumed) constant velocity. When the suboxide (cf. Sect. 6.3, volume I) is reached, the current starts to flow because of the dissolution of the substoichiometric composition of the oxide layers near the bulk silicon. In the interpretation of the current transients, the following assumptions were made: (1) all of the transient current is generated by the charge of dissolving suboxide, (2) the released charge per unit area for the suboxide layer is constant and independent of the phase of the oscillation; (3) the thickness of the suboxide layer is small compared to the thickness of the oxide, so the time of dissolving the suboxide is appropriately small, compared with the time for dissolving the oxide.

The transient curves in Fig. 4.7b show the distribution of the oxide thickness at the phase of the oscillation where the transient current measurements were started and this relation can be treated quantitatively, in terms of appropriate mathematical formalism. Transient curves 6–9 in Fig. 4.7b, corresponding to the minimum of the oscillating current, form a relatively narrow peak, reflecting thus the appropriately narrow distribution $D(s)$ of the oxide, with $D(s)$ defined as the sum of all areas with

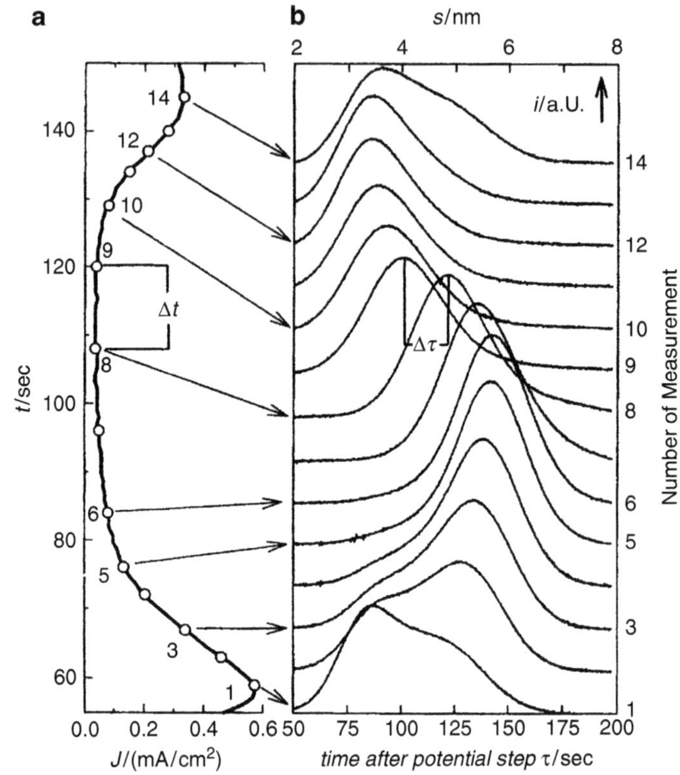

Fig. 4.7 (**a**) The oscillating current density J as a function of time t applying a voltage of 4.5 V vs. SCE. Electrolyte is 0.1 wt% HF + 1 M NH_4Cl (volume ratio: 1:1). (**b**) Transient currents after reducing the potential to OCP at several phases of the oscillation (indicated by the *arrows*) as a function of time τ. Reprinted from [10] with kind permission from Springer Science + Business Media

oxide thickness s at time $t = 0$. In other words, the lateral distribution of the oscillating oxide thickness is then relatively homogeneous, as well as its dissolution. In contrast, transient curves 1–3 show two peaks: a decreasing one for areas with thin oxide and a growing one for areas with thick oxide (correlating with the maximum in the oscillating current). This suggests that in some areas of the sample, the oxide surface is still dissolving, while in other areas the oxide thickness is growing. This indicates inhomogeneous lateral distribution of oscillating oxide thickness during the high-current phase of the oscillation [10]. Finally, since the current peak during the oscillation coincided with the growth of oscillating oxide, it allowed to formulate an additional simplifying assumption: that all of the currents build up silicon oxide [10].

The model of oscillations was based on the interpretation of the current transients representing the distribution function $D(s)$ of the oxide layer thickness s. Leaving the mathematical details to the interested reader, we shall briefly summarize its background [10]. The construction of the model required further (1) a mechanism for localized oscillations on a small area of the electrode, (2) a mechanism to synchronize these local oscillators, and also (3) a desynchronization mechanism. We now come back to Fig. 4.7 which shows (measurement 10) that the growth of oxide layer starts when its thickness s is small. The growing area of thick oxide corresponds to measurements 11–14 (or 1–6 after closing the cycle). There is an upper limit of thickness s, when the growth stops. On the other hand, the broadness of both the oscillatory and transient current peak is a measure of synchronization of the oxide dynamics, meaning that not all oxide areas start to grow at the same time. Furthermore, in the construction of the model, the idea of Gerischer [5] and Chazalviel [13] was invoked that the transport of ions through the oxide layer is the key to understanding of the oscillations in the Si/HF system. In fact, as for the very slow oscillations discussed in this work, all of the oscillating current is associated with the build-up of the oxide, thus current is a direct measure of the rate of ion transport through the oxide. In turn, since the driving force for this ion transport is the electric field across the oxide layer, it is further assumed that the oxide growth starts when the electric field in the oxide approaches a maximum value $E_{max} = U/s_{min}$, corresponding to a minimum value of the oxide layer thickness s_{min}. When the thickness gradually increases, electrical field gradually drops to a minimum value $E_{min} = U/s_{max}$, when the layer of thickness s_{max} stops to grow. This is a key assumption for the proposed model of a local oscillator, in which two limiting values of electric field serve as parameters of a model of "ionic breakthrough," in analogy to an electronic breakthrough [10]. It is assumed that after starting the ion transport through the oxide, the electric field can be reduced drastically, *without reducing* the ion current. At a more microscopic level, this assumption develops into the formation, in the oxide, of a narrow channel (pore) which conducts ions easily and which is opened (activated) at high field strength, and closed at sufficiently low field strength, according to the bumpy shape of the oxide layer, with $\Delta s = s_{max} - s_{min}$.

The idea of ion channel is further used in the intrinsic mechanism of *synchronization* of local oscillators. The lateral growth of the oxide spreading from an active channel must be of the same order of magnitude as the growing oxide thickness. The principle of synchronization between neighboring sites of hemispherical oxide growth is schematically illustrated in Fig. 4.8. If a new channel opens next to an already active one, it has to grow less oxide on one side before it turns off since the critically big layer thickness is achieved earlier, and so this channel is closed earlier (in this way it somewhat synchronized with the first channel).

To summarize, the model is generally defined by (1) an electric-field-induced local oscillator with highly nonlinear oxide growth caused by ionic breakthrough, (2) a coupling of next-neighbor channels with a strong synchronization effect, and (3) a roughness-dependent enhanced oxide dissolution.

In addition, the *desynchronization* mechanism was introduced, which is based on the fact that the current density around an active, single localized charge transfer process (current burst, CB) is very large. This leads to a significant decrease in the anodization potential across the oxide layer, i.e., to the substantial decrease in electric field *in the neighborhood*, through ohmic losses. This, in turn, decreases the lifetime

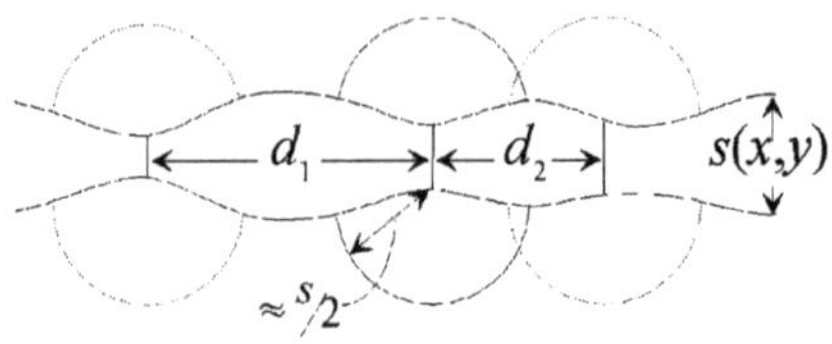

Fig. 4.8 The schematic cross-section through the oxide layer demonstrates some features of the oscillation model. Since the dynamics of oxide growth is defined by the electric field strength across the oxide, in a region of thin oxide (approximately s_{min}), an ion conducting channel opens. From the tip of the channel, a roughly hemispherical oxide inclusion is growing. The overlap between two channels is defined by the distance d between two channels and their growth radius (approximately $s_{\text{max}}/2$) and thus determines the coupling to neighboring areas. Reprinted from [10] with kind permission from Springer Science + Business Media

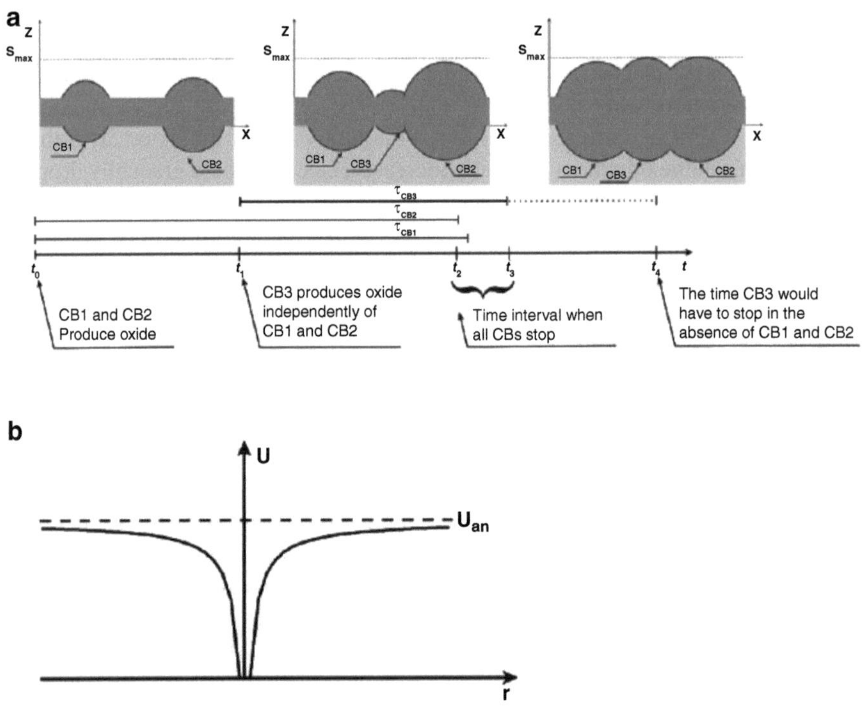

Fig. 4.9 Schematic representation of the synchronization (**a**) and desynchronization (**b**) mechanisms contained in the current-burst model (CBM). In (**a**) it can be seen that the "switching-off" time of current burst CB3 is correlated to those of CB1 and CB2. (**b**) shows schematically the voltage losses around an active CB, decreasing the nucleation probability of new CBs in the neighborhood. Reprinted from [14], Copyright 2007, with permission from Elsevier

of already existing CBs and decreases the probability for the nucleation of a new CB. This effect tends to destroy the correlation between CBs, since then they are more predestined to stay isolated from each other—and this means desynchronization. For nearly hemispherical symmetry, Fig. 4.9 collects the principles of both the synchronization (cf. Fig. 4.8) and the desynchronization mechanisms [14].

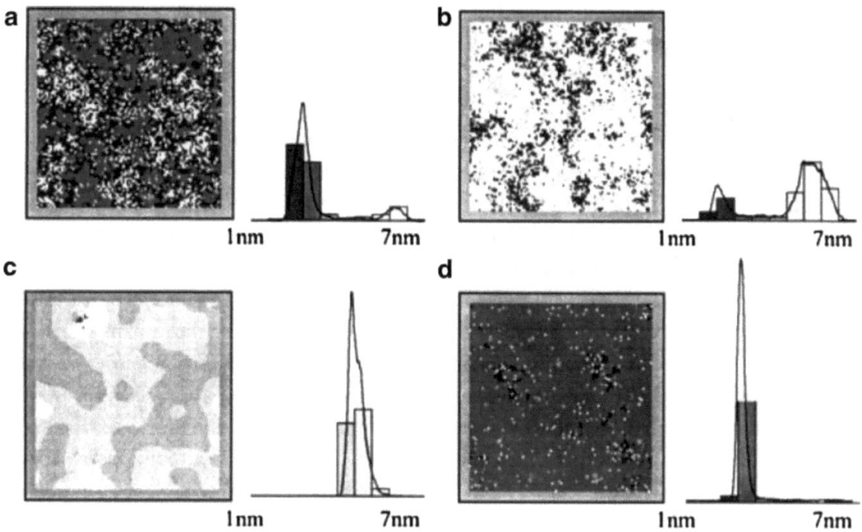

Fig. 4.10 (**a–c**) Oxide thickness distribution of a 200 nm × 200 nm area for different phases of the oscillation as a first result of the Monte Carlo simulation. The *gray colors* correspond to the thickness of the oxide as lined out in the adjacent histograms. (**a–c**) correspond to the phase of growing oxide showing a qualitative agreement with the distribution function of the oxide layer thickness in corresponding measurements. (**d**) demonstrates the homogeneous oxide thickness in the phase of purely chemical dissolution of oxide. (**a–c**) exhibit a percolative growth of the oxide layer, i.e., the growing regions are coupled (see [10] for further details). Reprinted from [10] with kind permission from Springer Science + Business Media

The mathematical dependences describing the principles of CBM model, at various stages of its development, were implemented as a Monte Carlo simulation algorithm. Numerical calculations made at an early stage [10] showed macroscopic current oscillations above a critical voltage U_{crit}, while below this voltage a nonoscillating macroscopic current was obtained, which was noisy because the local oscillators were still active, but no longer synchronized. These Monte Carlo calculations revealed also that the synchronization of many local oscillators proceeded by a percolation mechanism, i.e., a long-range connectivity in random systems which is a manifestation of a strongly nonlinear self-organization. This is illustrated by exemplary results of calculations visualized in Fig. 4.10. In other words, long-range synchronization, necessary for the macroscopic current oscillations to occur, requires the percolation.

Another conclusion of the calculations is that local oscillators are never perfectly synchronized—even when the oscillating current attains high values, oxide is grown in some (percolation) regions, whereas in other areas only chemical etching takes place. One should emphasize that simulation results are well concordant with experimental data, including both the information carried by the transient $I - t$ courses at OCP and in situ AFM investigations of the oscillating oxide, showing the magnitude of surface roughness in every phase of the oscillation.

In course of further development of the CBM approach, in 2000 Carstensen et al. [15] have proposed the mechanism accounting for also the formation of nano-, meso and macropores nucleation and growth, with their specific dependence on crystal orientation. Concerning the porous silicon, important for technical applications, one should note that *microporous* silicon was first observed by Uhlir [16, 17] (its strong luminescence at room temperature was described later by Canham [18]), while the *macroporous* structure was discovered by Lehmann and Föll [19]. In this model, the mean cycle between correlated current bursts was mostly given by the kinetics of oxide dissolution and hydrogen passivation (which introduces a strong surface orientation dependence). Structure generation at the Si electrode (current oscillations in the time domain or pore formation in the space domain) constituted a self-organizing process resulting from an interplay of synchronizing and desynchronizing mechanisms. As the synchronizing mechanisms, the enhanced nucleation probabilities on (100) surfaces, response to local oxides from another current burst, or coupling of current bursts by space charge region (SCR) effects were considered. In turn, as the source of desynchronizing effects, e.g., the quantum wire effects or local reduction of reactants or potential were taken into account. Qualitative agreement with at least most of modeled experimental phenomena was found.

On the other hand, the development of the CBM model for the description of Si surface covered always completely with silicon oxide (no pores), i.e., for the electropolishing regime, led to Monte Carlo calculations of the oscillations of not only the current, but also (for the first time) of the voltage [20] (see Fig. 4.11 for correlation of oscillating current with nucleation and synchronization of current bursts). Further representative simulation results of current and voltage oscillations, correlated with the variation of the oxide layer thickness, are shown in Figs. 4.12 and 4.13, respectively.

The current state of the CBM model, including details in the implementation of the Monte Carlo algorithm and critical analysis of the whole approach, together with exemplary simulated oscillations in the Si/HF system, was described in detail in a recently published extensive work by Foca, Carstensen and Föll [14] which is also a concise review of previous model approaches. The work shows the present capabilities of the software to calculate all local electrode features, e.g., current, potential, oxide thickness, interface roughness or capacitance as a function of time, in the form of color maps. Also, correlation lengths for certain domain features can be calculated; in more detail, the average free autocorrelation function (AFAF) measures the average magnitude of the correlation of two points separated in space by certain distance, i.e., it indicates the probability of finding the same oxide thickness at this distance. The reader interested in this particular subject is strongly advised to consult this reference. We shall here present two representative sets of results. Figure 4.14 shows the maps obtained for one time frame in a simulation run. Such maps were collected usually every 20 ms, with the time frame for an individual calculation being 5 ms.

In turn, Figure 4.15 shows calculation results for constant imposed potential $U = 6$ V and four different times of the oscillatory cycle. One can find the

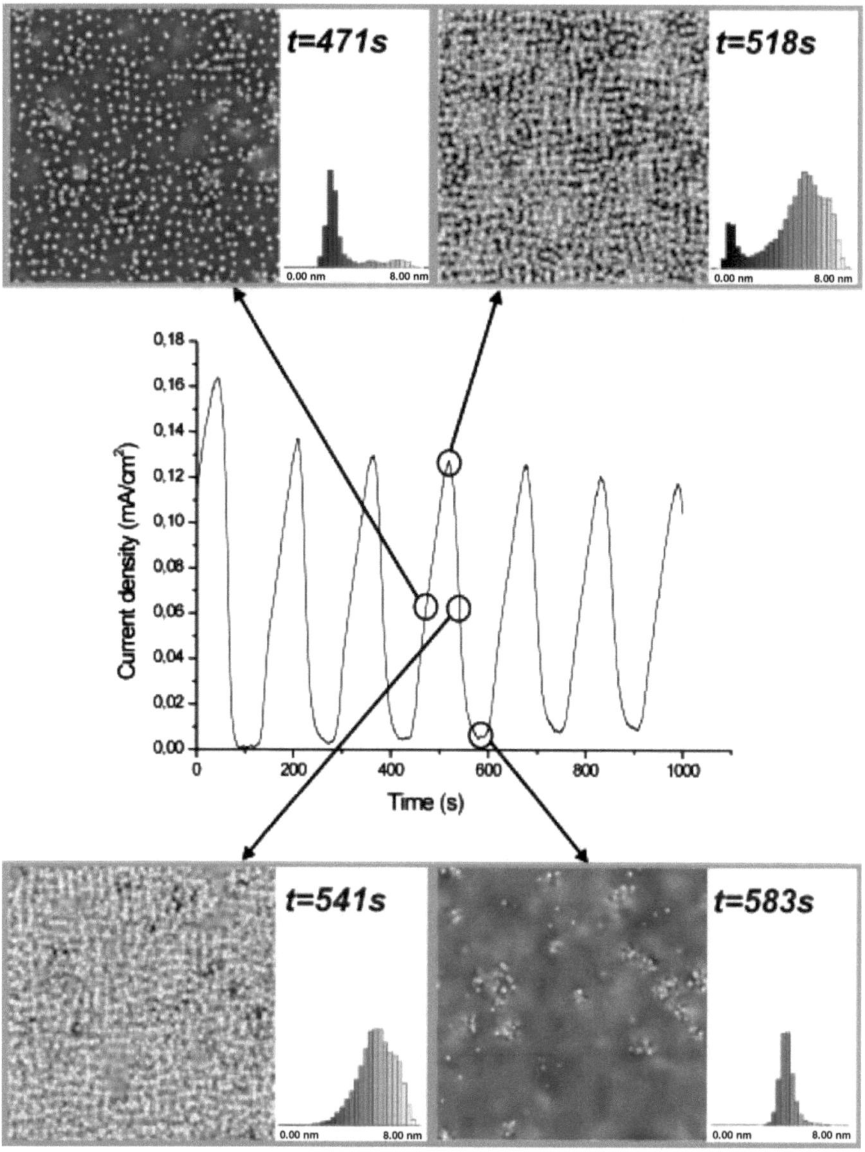

Fig. 4.11 Oscillation cycle will start with nucleation of single CBs (e.g., at $t = 471$ s), which will synchronize in time (e.g., at $t = 518$ s). After an upper limit of oxide thickness is achieved, the oxide formation will drastically decrease (e.g., at $t = 541$ s) and the dissolution process dominates, leading to formation of domains with the same oxide thickness (e.g., at $t = 583$ s). Reprinted from [20] with permission of John Wiley & Sons, Inc. Copyright 2005 Wiley-VCH Verlag GmbH & Co. KGaA

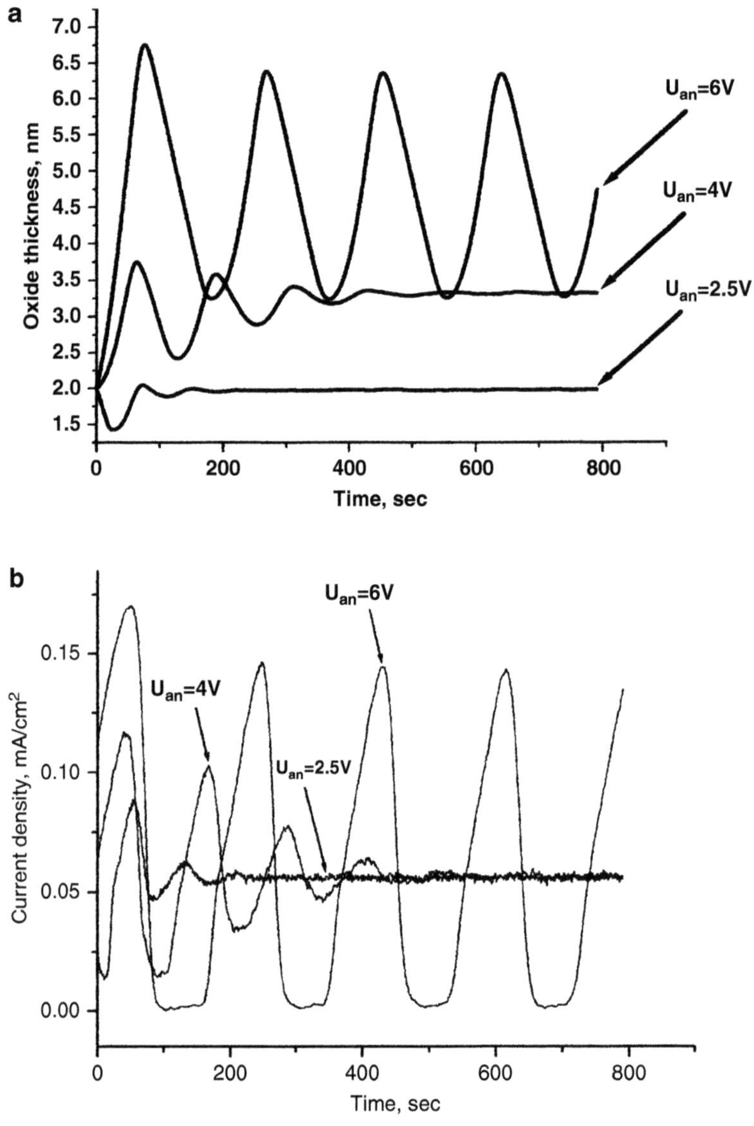

Fig. 4.12 (**a**) Oscillations of the mean oxide thickness for three different constant voltages. At potentials below a certain threshold only damped oscillations are obtained, in close agreement with the experiments. (**b**) The corresponding current density oscillations; again showing that stable oscillations only occur above a critical voltage. Reprinted from [20] with permission of John Wiley & Sons, Inc. Copyright 2005 Wiley-VCH Verlag GmbH & Co. KGaA

correlation between the shape (actual value) of the oscillating current and the dynamics of local current bursters, the oxide thickness and the voltage loss. A detailed discussion of these dependences is given in [14].

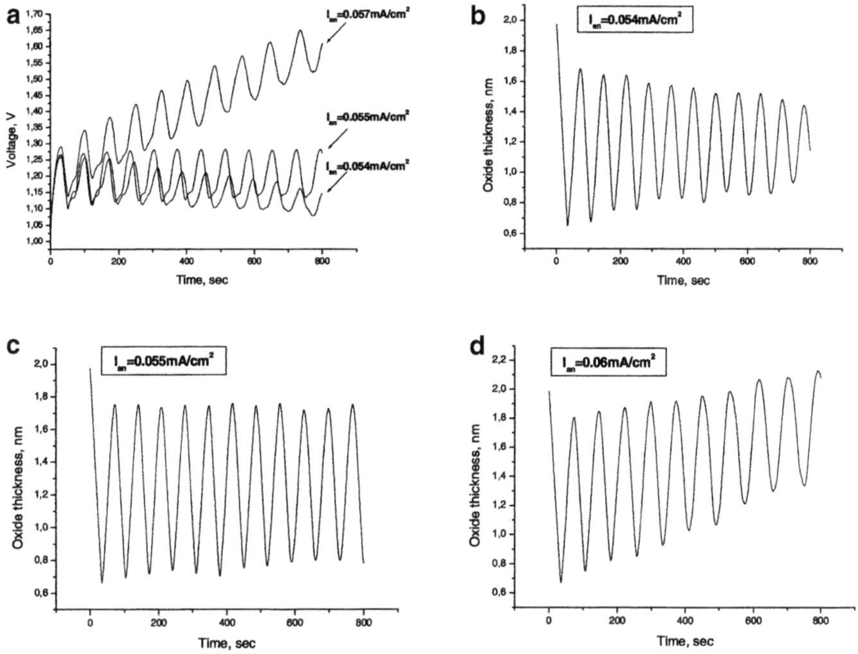

Fig. 4.13 (**a**) Voltage oscillations for several anodization currents. It is obvious that stable oscillations are only obtained in a narrow current interval, as known from experiments. (**b–c**) show the corresponding damped (**b**), stable (**c**) and unstable (**d**) oxide thickness oscillations. "Unstable" means that the average oxide thickness and voltage continuously increases with time, which in a real experiment leads to rapid termination of the effects because the limited voltage range of galvanostat quenches the oscillations (or the etching cell blows up). Reprinted from [20] with permission of John Wiley & Sons, Inc. Copyright 2005 Wiley-VCH Verlag GmbH & Co. KGaA

Calculations made for the case of damped current oscillations illustrate the expected loss of synchronization and the presence of several domains with random phases. Furthermore, the CBM approach is able to reproduce the birth of oscillations induced by extrinsic synchronization, i.e., by potential step following the decay of oscillations at previous potential value. Resetting the electrode to a new state produces the oscillations which Chazalviel et al. rated as belonging to "resonant" ones, for the case of external voltage perturbation, although the term "resonance" seems to be no longer necessary for the explanation of these phenomena. Next, the CBM model is able to reproduce fairly well experimental current oscillations, observed upon periodic (sinusoidal) perturbation of the external voltage (Fig. 4.16).

The possibility of the modeling of potential oscillations under galvanostatic conditions was mentioned already above. One should add here that due to variations of the electrode potential, and in view of very small values for the oxide thickness, the contribution from the capacitive current to the overall current is expected to be quite large. In the calculations of the capacitive current, due to variations of the

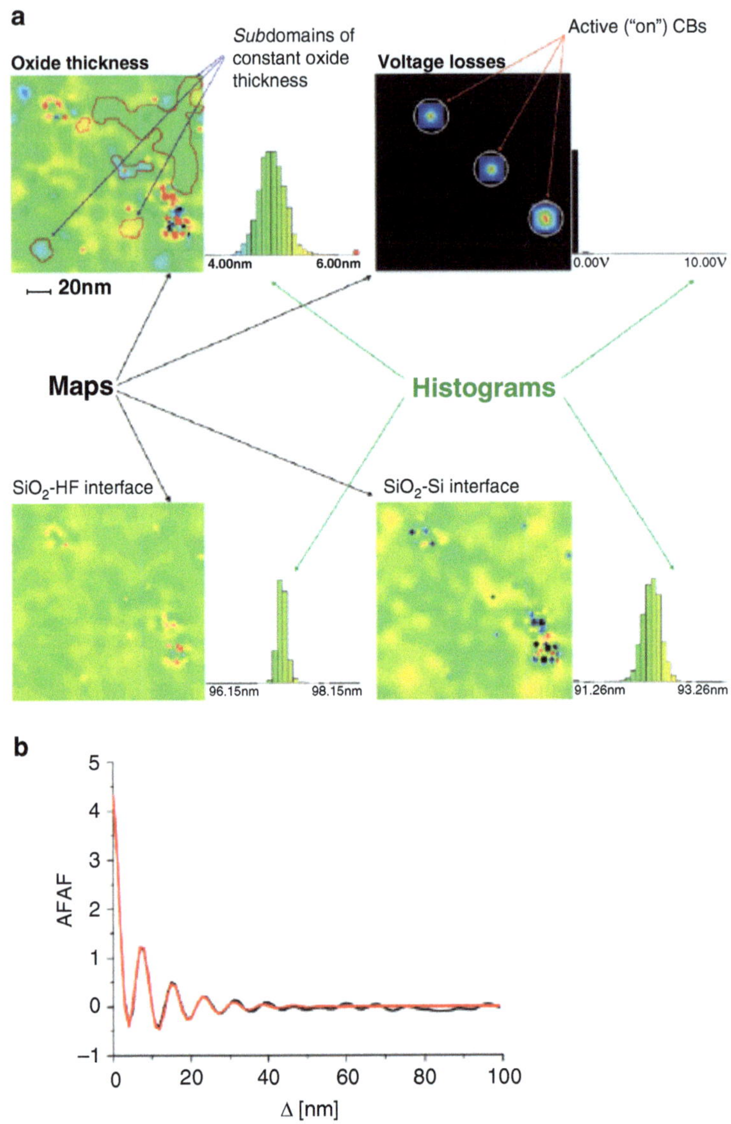

Fig. 4.14 Examples of maps generated for one time frame in a simulation run (**a**). On the oxide thickness map, some "smooth" subdomains are schematically indicated. The mottled areas can also be considered to be subdomains. The voltage loss map in this case shows directly the location of active CBs as indicated. For each calculated oxide thickness map, the average free autocorrelation function (AFAF) is calculated (*black curve*) and subsequently fitted (*red curve*) (**b**). Reprinted from [14], Copyright 2007, with permission from Elsevier

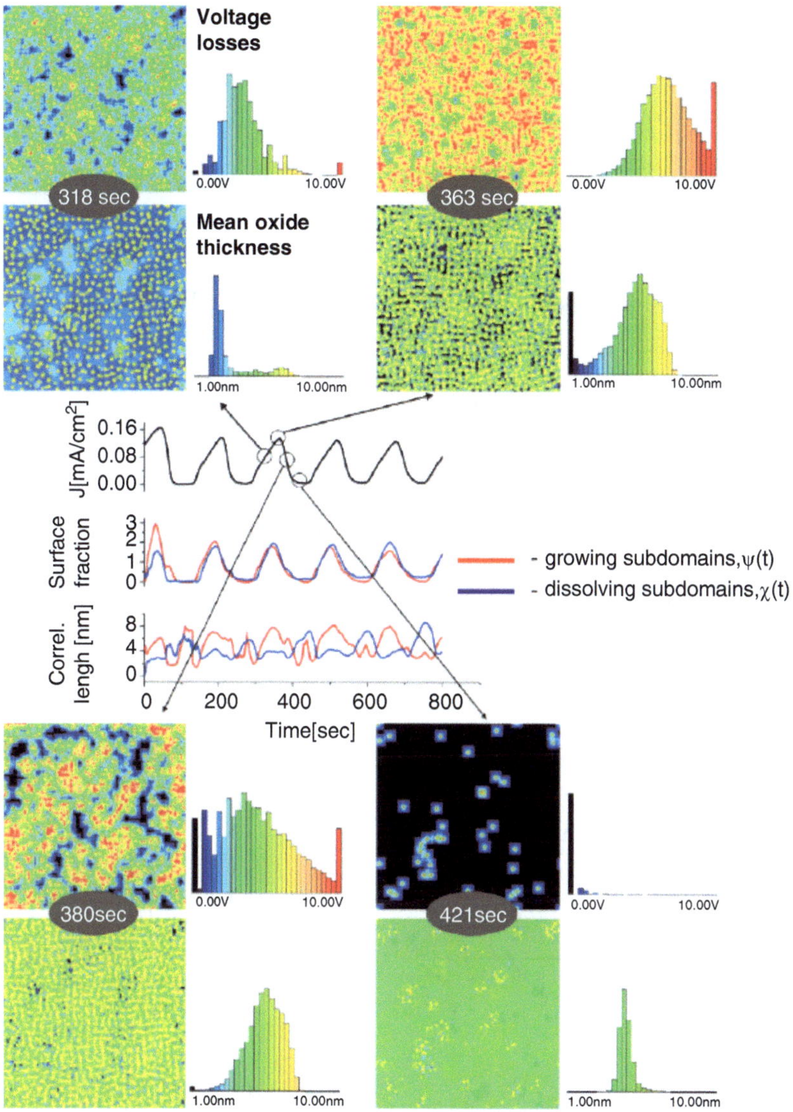

Fig. 4.15 Example of a fairly stable current oscillation at $U_{an} = 6$ V together with four screenshots of the oxide thickness and the voltage losses at the times indicated. The large bars in the histograms at the end of the scale show the sum of all values outside the range and are thus not "real." The time development of the four most important fit parameters for the AFAF are shown, too; for details refer to the text [14]. Reprinted from [14], Copyright 2007, with permission from Elsevier

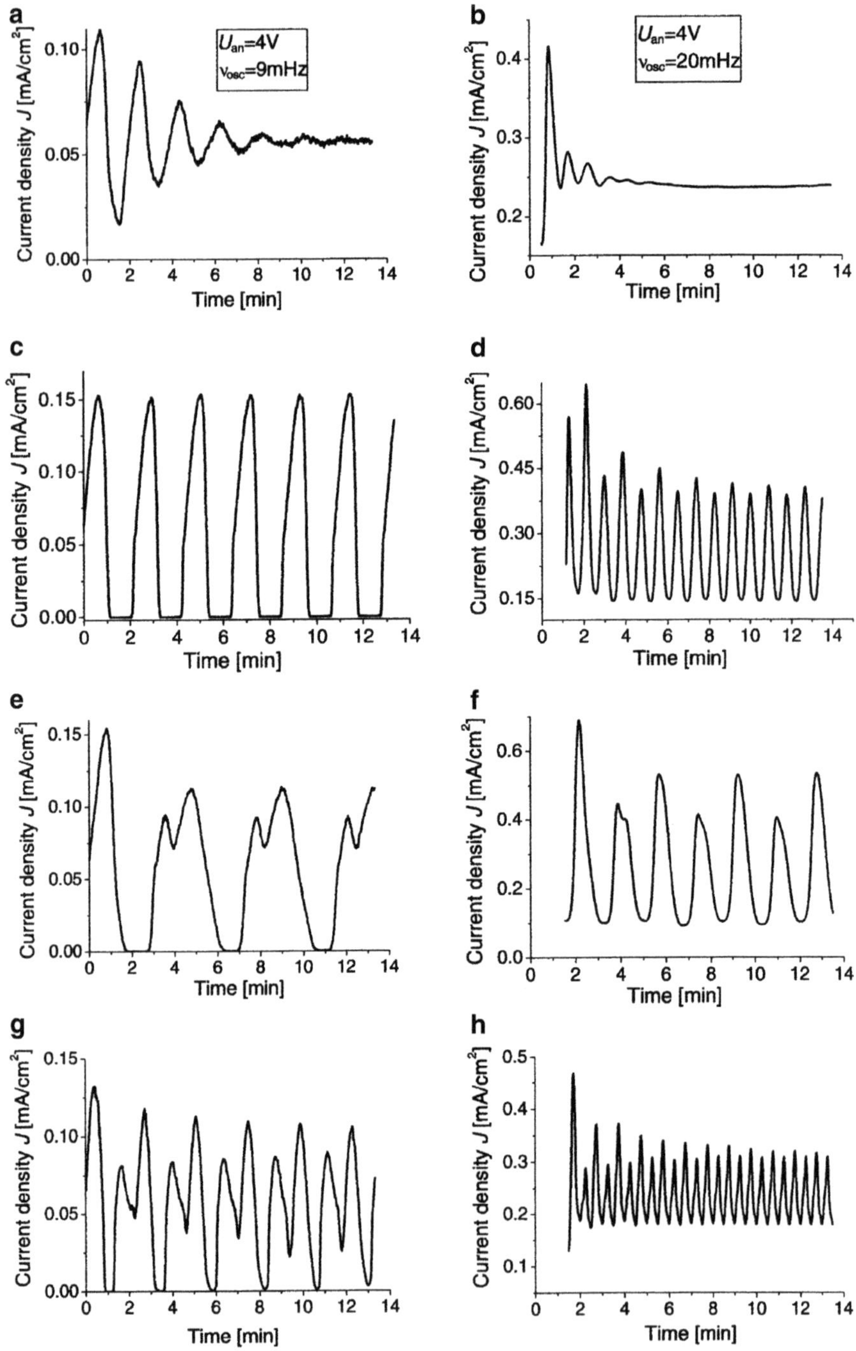

Fig. 4.16 Simulated (*left column*) and measured (*right column*) current density oscillation for forced oscillations and an average anodization voltage $U_{an} = 4$ V. Reference with zero amplitude, i.e., no modulation (**a**) and (**b**). Modulation frequency $\omega = 2\pi\nu_{osc}$ (**c**) and (**d**). Modulation frequency $\omega = \pi\nu_{osc}$ (**e**) and (**f**). Modulation frequency $\omega = 4\pi\nu_{osc}$ (**g**) and (**h**). Reproduced from [14], Copyright 2007, with permission from Elsevier

oxide layer thickness, temporal variations of not only the potential U, but also of the capacity C should be included: $j_{cap} = (dC/dt)U + (dU/dt)C$.

The CBM model of the Föll's group is in a way competitive to the approach proposed by the Lewerenz's group. Certain convergence of both approaches is not excluded, as, in the opinion of Föll et al. [21], the latter group in fact incorporated the "current bursts" in the recent version of the model [8]. The CBM approach is able to reproduce at least many different dynamic behaviors observed during the Si anodization in fluoride containing media, based on only one assumption—the probability functions for starting and stopping a single oscillator (CB) and many other not assumptions, but approximations [14]. However, it also has some limitations, two of physical and two of modeling nature. The physical ones are that (1) the Si surface must be covered with oxide at all times, and (2) all current produces oxide. The modeling limitations include: (1) low values of the HF concentration (low current densities) only and (2) the numerics involved in Monte Carlo program do not allow to simple scaling to larger sample areas. Foca et al. [14] give an illustrative example: a simulation run of a (1×1) μm^2 sample takes ca. 20 days which means that realistic sample sizes are out of reach. The last two limitations mean that the CBM model, in the version outlined here, cannot deal with large Si areas, covered with oxide layers experiencing stresses, as considered by Lewerenz's group. The authors declare thus further development of their CBM approach, aiming to incorporate new discoveries in the anodization of Si, particularly with respect to implementation of micro- and macrostructure of oxides, including self-organized single pore crystals. At this stage of research it is not yet clear, if CBM model can be applied also to the explanation of oscillations for other semiconductors, or if it remains valid only for silicon.

As a complement to the above models one should note that for slightly different system: anodization of Si in the mixture of HF and H_3PO_4, Parkhutik et al. have published several reports in which qualitatively different mechanism of oscillations was suggested, based on electrochemical experiments, supported by scanning electron microscopy (SEM), IR measurements and secondary ions mass spectroscopy [22–26]. In this mechanism, first the nanoporous oxide film grows on the Si surface; under galvanostatic conditions this causes the increase of the anodic electrode potential due to increase of the oxide thickness and consequent growth of its electrical resistance. Along with the thickening of the oxide film, the compressive stress at the oxide/silicon boundary is accumulating. This stress generates defects at SiO_2/Si interface which trigger out the local chemical attack on the Si wafer at a certain stage of the oxide growth. Isotropic etching of underlying Si wafer causes the peeling off the oxide layer (the electrode potential then decreases) and the Si surface is exposed to the growth of new portions of the oxide. Such a mechanism seems however to be specific for the Si in contact with HF mixed with phosphoric (or oxalic or sulfuric) acid. Also Parkhutik [27] has described the galvanostatic oscillations during the anodization of Si in aqueous solutions of sulfuric, phosphoric and oxalic acids, in the absence of HF, and interpreted the observed dynamics in terms of the growth and dissolution of silicon oxide layer

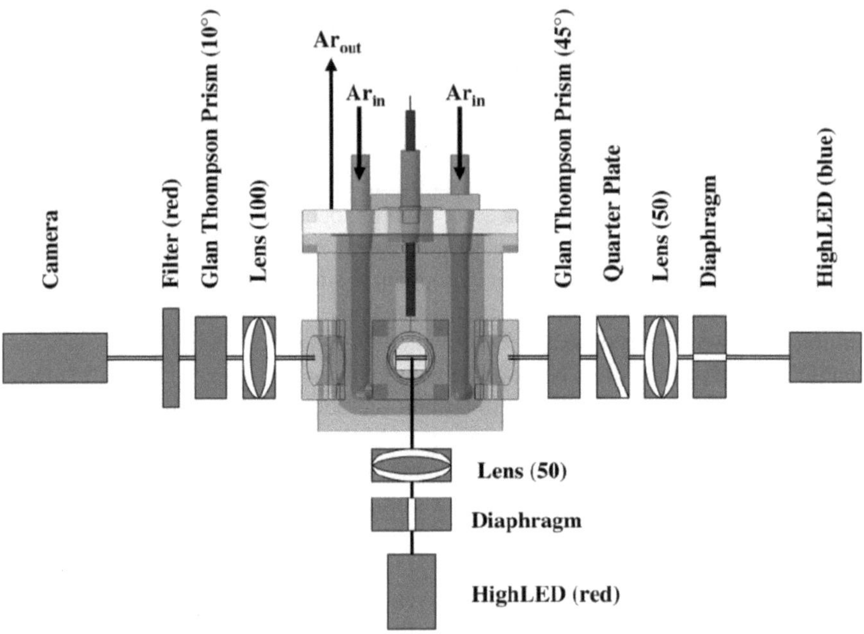

Fig. 4.17 Scheme of the electrochemical cell and the optical setup used for illumination and ellipsomicroscopic imaging. Reprinted with permission from [28] http://link.aps.org/abstract/PRL/v102/p194101. Copyright 2009 by American Physical Society

exhibiting porosity which causes the penetration of the acid anions into the oxide/semiconductor interphase.

As a last, recent experimental example of spatiotemporal self-organization during Si anodization in fluoride-containing media we shall describe here the application of contrast-enhanced optical microscopy, based on ellipsometry, for monitoring (in situ) of the rich spatiotemporal dynamics taking place at the semiconductor/electrolyte interface [28]. In these experiments, as the working electrode the *n*-Si (111), homogeneously illuminated with a red light emitting diode (630 nm) was used. As the counter and reference electrodes, the Pt wire bent into a ring and a saturated Hg/Hg_2SO_4 electrode, respectively, were employed. The experimental setup is sketched in Fig. 4.17.

Representative experimental data are collected in Fig. 4.18.

At appropriate (intermediate) illumination strength and voltages there were observed regular relaxation oscillations of the total current (Fig. 4.18f), together with sinusoidal, in-phase locked oscillations of the spatially averaged ellipsometric light intensity $<R>$ (Fig. 4.18e). In contrast to regular time series of these averaged quantities, snapshots of the electrode (Fig. 4.18a, b) have revealed that the oxide layers formed an intricate and time-dependent labirynthine pattern. In turn, Fig. 4.18c shows the temporal evolution along the 1D cut indicated in Fig. 4.18a. Based on that, one may think that uniform oscillation with the base frequency is superimposed on an irregular pattern that changes in time. Finally, Fig. 4.18d,

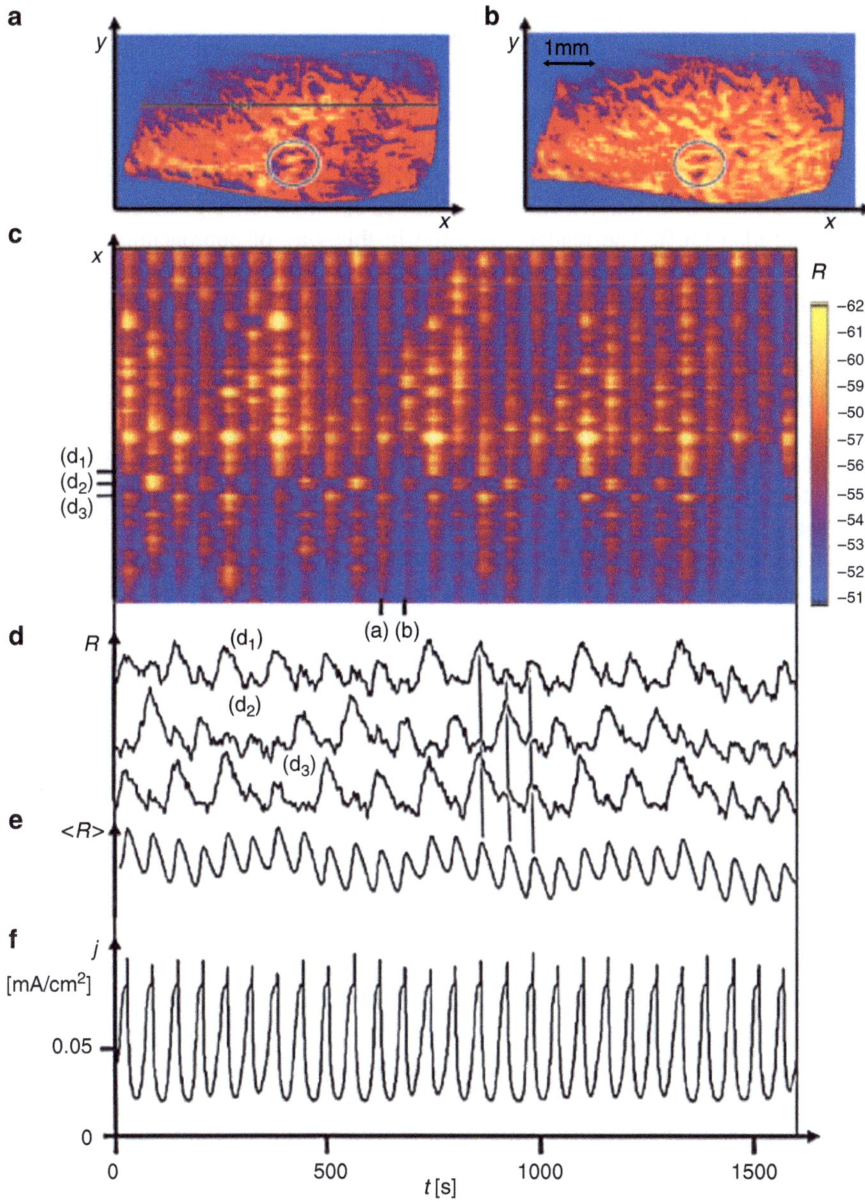

Fig. 4.18 Spatiotemporal data during n-Si electrodissolution at a voltage $U = 11.65$ V. (**a**, **b**) Ellipsomiscroscopic snapshots of the Si electrode taken at subsequent maxima of the average light intensity <*R*>. (**c**) Spatiotemporal evolution of $R(x, t)$ for the 1D cut indicated in (**a**). (**d**) Local time series of $R(t)$ for the three points indicated in (**c**). (**e**) Time evolution of <*R*>. (**f**) Time evolution of the total current. Reprinted with permission from [28] http://link.aps.org/abstract/PRL/v102/p194101. Copyright 2009 by American Physical Society

showing the local dynamics for the three points indicated in Fig. 4.18c, suggests that local oscillation amplitude evolves irregularly in time and is different for each oscillator shown. Nevertheless, the extrema of R for all individual oscillators are locked to the ones of the average signal $<R>$. The Fourier analysis allowed to determine the formation of unusual subharmonic cluster patterns during the oscillations. A theoretical model describing these experimental observations was based on a modified complex Ginzburg–Landau equation (cf., e.g., the review paper [29] and also [30]). The reader interested in this way of approaching nonlinear problems can also consult other works in that area. García-Morales and Krischer [31] have recently derived the nonlocal complex Ginzburg–Landau equation (NCGLE) valid for electrochemical systems with migration coupling.

Generally, modern studies of self-organization in semiconductors concern with their microporous structure. For silicon, the first important event in this area was the discovery of luminescent microporous Si [18, 32]. The porous Si, due to its various applications, is now industrially produced (cf., e.g., [33]). Nevertheless, the current or potential oscillations were usually confined to the electropolishing regime and the role of pore formation in this semiconductor was not invoked explicitly in early mechanisms of these instabilities. Although the original CBM was later extended to pore formation in Si, their simultaneous occurrence with the oscillations was not explicitly seen. It was even initially thought that for Si the voltage oscillations and pore formation were strictly separated phenomena [10]. Only later observations of pores in such III–V semiconductors, as InP and GaP, allowed to detect the synchronization between macroscopic oscillations and microscopic pore formation. In consequence, similar correlations were found also for Si [34]. The self-organized porous structure of III–V semiconductors and associated nonlinear dynamic phenomena are described in the following section.

4.1.3 The Outline New Model Involving Ohmic Potential Drops

Both previous models of oscillations during Si anodization were focused on the dynamic characteristics of an inhomogeneous, microscopic distribution of silicon oxide on the Si surface. They did not take into account the eventual role of ohmic potential drops (IR_s) which destabilize the state of the systems with negative differential resistance (NDR), leading to oscillatory and/or multistable regimes; many examples of such bifurcations were described in previous sections of this book. In this context it is useful to mention the very recent ellipsomicroscopic study of the anodic oxidation of p-type silicon in fluoride-containing electrolytes (NH_4F), reported by Miethe and Krischer [35]. A very intriguing conclusion of those studies is that *in contrast* to recent findings with n-type silicon electrode, for which cluster patterns were observed (thus also *in contrast* to both models, discussed above), all sustained oscillations in p-Si anodization proceeded *uniformly* on the entire electrode, accompanied by the oscillations of only the oxide layer thickness. Furthermore, also *in contrast* to n-type Si, these oscillations appeared to be only stable if a *sufficiently large serial ohmic resistance* was present in the electric circuit. The oscillations emerged then through a supercritical Hopf bifurcation. At larger ammonium fluoride concentrations also complex periodic, quasiperiodic, and deterministically chaotic oscillations were reported. In the opinion of Miethe and Krischer, their results constitute a strong premise toward treating the source of oscillations at the anodized *p*-Si electrode in a way analogous to that of the N-NDR oscillators, i.e., based on the interplay of positive and negative feedback loops in *p*-Si electrodissolution, created in the presence of ohmic drops. Furthermore, in spite of this serial resistance, the synchronization mechanism of the entire electrode surface is not considered to be a consequence of a global coupling, like in the interpretation of Chazalviel and Ozanam [36] who considered this global effect a principal source of synchronization of local oscillators, leading to sustained macroscopic current oscillations (cf. also Sect. 6.3.2, Vol. I). Instead, in the present interpretation the serial resistor is a necessary ingredient of only the basic oscillator, while the synchronization within the entire electrode surface is brought about by a synchronizing electrostatic coupling of neighboring sites which arises whenever local differences in the electric potential exist [35]. In conclusion of these new findings, it is necessary to consider an alternative, already third kind of models of instabilities in the anodization of (at least some) semiconductor electrodes, based this time essentially on the principles of nonlinear dynamics.

4.2 Self-organization in III–V Semiconductors

In Sect. 4.7.2 of volume I, the oscillatory reduction of H_2O_2 on GaAs semiconductor electrodes was analyzed. In this section, we describe the anodic processes involving this semiconductor. Van Meirhaeghe et al. [37] have reported current oscillations observed under potentiostatic conditions on an illuminated *n*-GaAs immersed in KCl + $Na_2B_4O_7$ solution (pH $\approx$ 10). These oscillations were clearly related to the potential region of a negative differential resistance, and occurred if the external resistance exceeded certain critical value and if the light intensity was also sufficiently high. With increasing external serial resistance, the shape of the oscillations changed from the regular single peak to the relaxation ones. When $Na_2B_4O_7$ was replaced by acetate or phthalate buffer, the oscillations did not set in, in spite of existence of NDR regions in the system characteristics. Thus, the latter systems constitute the examples of not very frequently met situation, where, in spite of NDR characteristics, oscillatory regime does not set in. At this early stage of understanding of electrochemical oscillations, their explanation was referred to the Degn model (outlined in Sect. 6.1.2.2, volume I), in which, besides the NDR region and external resistance, the role of concentration polarization near the electrode surface was indicated as the necessary condition for the onset of such instabilities. In the particular case of dissolution of GaAs electrode, if the potential is varied from relatively positive values, corresponding to the passive state of the electrode, toward more negative values, the region of NDR

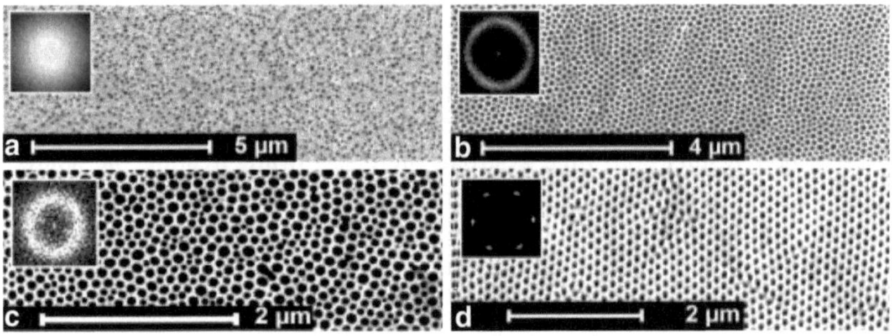

Fig. 4.19 Pores crystal formation in (100) InP as a function of etching conditions. The insets show Fourier transforms. (**a**) Nucleation layer of crysto pores; the structure is amorphous. (**b**, **c**) Pore structure of layers anodized at "low" (**b**) or "large" (**c**) voltages showing some degree of order, but no long-range order. (**d**) Pore structure of a layer obtained under optimized current–voltage conditions—a single crystal is obtained. The current pore structure in (111)B-oriented samples behaves similarly, except that the nucleation layer is not as easily visible. Reprinted from [38] with permission of John Wiley and Sons, Inc. Copyright 2003 Wiley-VCH GmbH & Co. KGaA

is reached, meaning the increase in the anodic current, associated with the dissolution process:

$$GaAs + 6h^+ + 8OH^- \rightarrow GaO_2{}^- + AsO_2{}^- + 4H_2O \quad (4.1)$$

The consumption of OH^- in this process causes its concentration polarization, in consequence of which the photocurrent decreases and thus also the ohmic potential drops. This shifts the electrode potential back to more positive values, at which repassivation occurs. The diffusion of OH^- ions from the bulk can now diminish the concentration polarization, the dissolution current increases and, due to increasing ohmic drops, the electrode potential shifts to less positive values, at which reactivation occurs. In fact, there is nothing qualitatively different in this oscillatory mechanism, compared to other typical N-NDR systems, only particular dissolution process is specific for the GaAs electrode in the given medium.

In recent decades, GaAs and other III–V semiconductors (as GaP and InP) attracted the interest of researchers due to the porous structures which is important for technical applications and is a source of various kinds of self-organization. The presently described III–V porous semiconductors, exhibiting efficient electroluminescent behavior and therefore used in optoelectronics, are produced by electrochemical etching ("anodization"), in which self-organized phenomena can be involved, as described in respective reviews or monographs, cf., e.g., [21, 38–40]. The self-organization manifests itself in the case of III–V semiconductors as the simultaneous occurrence of self-induced oscillations of pore-system properties in space and/or time.

The formation of spatial pore patterns was found for InP, provided that the voltage applied for etching in the potentiostatic experiment was optimally chosen, i.e., between certain minimum and maximum values, beyond which the long-range order is destroyed (see Fig. 4.19) [38, 41].

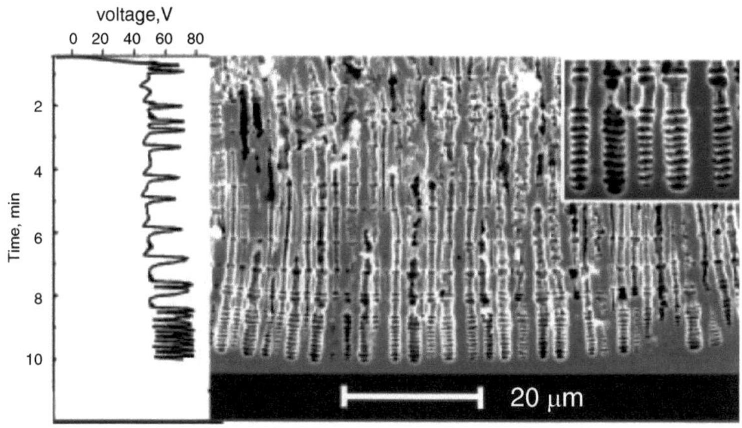

Fig. 4.20 Synchronized diameter oscillations in InP current pores and simultaneously occurring voltage oscillations. The inset shows an enlargement of the bottoms of the pores. Reprinted from [38] with permission of John Wiley and Sons, Inc. Copyright 2003 Wiley-VCH GmbH & Co. KGaA

Generally, the geometry of pore structures of semiconductors, i.e., their distinct intrinsic length scale, may result from: quantum wire effects (nanometer scale), avalanche break-through (10–100 nm scale) [42], space-charge region (10 nm–10 μm), Helmholtz layer thickness (3–5 Å), diffusion instabilities (micrometer to centimeter), treated in terms of linear stability analysis by Chazalviel et al. [43], and finally extrinsic scales defined by lithography [21]. It seems that for porous III–V semiconductors, this is the size of the space-charge region that determines the lattice constant in the crystals, as, e.g., for InP (but rather not for GaAs).

In turn, the spatiotemporal periodicity, meaning the pore-diameter oscillations, is always coupled to temporal current or voltage oscillations [44–46]. Macroscopic oscillations of current/voltage can appear if the individual microscopic oscillators are synchronized in an area comparable to the sample area. For III–V semiconductors, the correlations between oscillations and pore formation can depend on potentiostatic or galvanostatic conditions. For InP anodization under *galvanostatic* conditions there was found one-to-one correspondence between the voltage and diameter peaks (Fig. 4.20).

Under potentiostatic conditions, many individually synchronized domains could form, but with random phases between domains. The external current was then rather noisy (approximately constant) than oscillatory. It was however possible to achieve better (although rather temporal) synchronization of pores under potentiostatic conditions, if it is first reached under galvanostatic conditions and only then switched to potentiostatic ones providing the same (average) current density. The maintained synchronization during pore growth manifested then itself as the self-induced, quite regular current oscillations (cf. Fig. 4.21 which compares

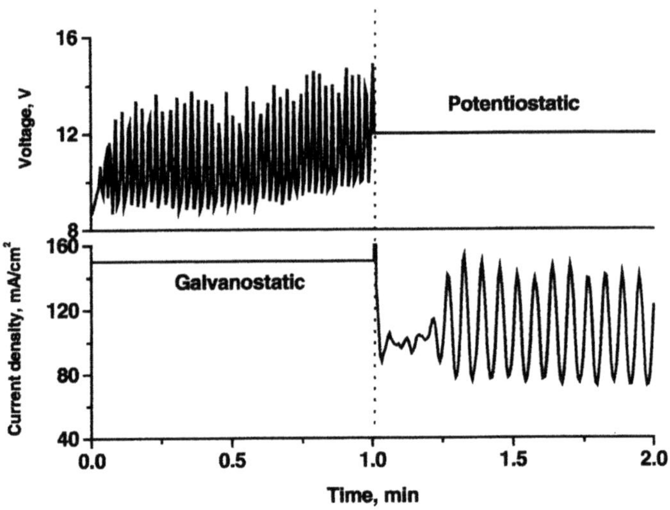

Fig. 4.21 Switching from galvanostatic to potentiostatic conditions after synchronization was obtained. Self-induced current oscillations were obtained for the first time in this way. Reprinted from [38] with permission of John Wiley and Sons, Inc. Copyright 2003 Wiley-VCH GmbH & Co. KGaA

two potentiostatic experiments, with the second preceded by the galvanostatic regime).

In order to understand, e.g., the potential oscillations in etching of III–V semiconductors, the following qualitative mechanism was proposed [21]. For such galvanostatic oscillations, the basic assumption, corresponding also to the CBM approach for Si dissolution, is that the current at the pore tip generally oscillates, while the diameter remains nearly constant, which means that the current density also oscillates. Each pore is described in terms of the equivalent circuit, as the resistor with the oscillating resistance $R(t)$, yielding its averaged value $<R>$. The total current flowing through all the pores is given by the voltage/current source applied to parallel connection of individual resistors. If the pores growth is unsynchronized, i.e., the phases of the oscillating resistances are uncorrelated (random), the total current will have certain average value $<I> = U/<R>$. When the growth of pores becomes synchronized due to the space charge interaction, the phase of pore currents becomes correlated and the total current starts to oscillate. Since the power supply maintains the total current, the only way to meet this condition is by oscillating the potential on (or at least most of) the semiconductor surface. In order to make the model, and corresponding simulated $E - t$ curves, more realistic, it would be necessary to ascribe to every pore certain capacity and take into account the flow of the corresponding capacitive current. The fact that the current varies in time in the form of the bursts, may have several reasons: ionic breakdown of oxides occurring at the very large field strength, when an oxide becomes thinner due to dissolution [43], the local breakdown of passivated areas, or the localized avalanche

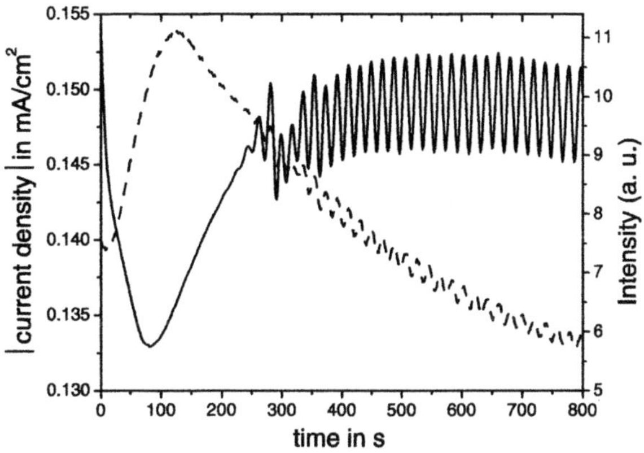

Fig. 4.22 Synchronous current (*solid line*) and electroluminescence (*dashed line*) oscillations during porous etching of *n*-GaP(100) potentiostatically at 40 V in 0.1 M HF solutions. Reprinted from [47], Copyright 2005, with permission from Elsevier

breakdown events [38]. The intrinsic time constants of such processes are the average duration of a current burst and the average time between two current bursts. Another important aspect of experimental detection of self-organization involving pores is that if synchronization does not manifest itself through regular macroscopic oscillations, the self-induced pore diameter oscillations may still occur, but that phenomenon is then only visible in SEM images.

With respect to electrochemical porous etching of *n*-GaP (100) in HBr or HF aqueous solution, the high intensity and oscillatory electroluminescence, correlating with the oscillations of current, was reported (Fig. 4.22). The orange–red light emission mechanism (especially strong for HBr medium) was ascribed as directly coupled to the electron transfer at the semiconductor–-electrolyte interface, in terms of the active etching of the pore front, associated with the electron-hole recombination as a source of light [47].

For GaAs and GaP, also additional self-organization processes were reported: the formation of pore domains, i.e., pores self-organized around a central structure (note that also for Si it was found the creation of domains in the form of "fractal" pores). If etching of (100) *n*-GaP is continued for a sufficient time, domains start to overlap or fuse, creating a cellular or "catacomb"-like porosity (Fig. 4.23) [48, 49].

In turn, electrochemical etching of (111) *n*-InP surface in an aqueous HCl solution can lead to a porous layer with a pillar structure characterized by quasi-isolated columns growing perpendicular to the initial surface and exhibiting an aspect ratio >100 [38, 50].

Finally, Föll et al. [38] have indicated that in the case of pore etching in III–V compounds, one observes an impressive, extreme case of self-organization: self-induced strong diameter oscillations occurring simultaneously with self-induced single-pore crystal formation, creating essentially a three-dimensional photonic crystal resembling that obtained extrinsically in Si [51].

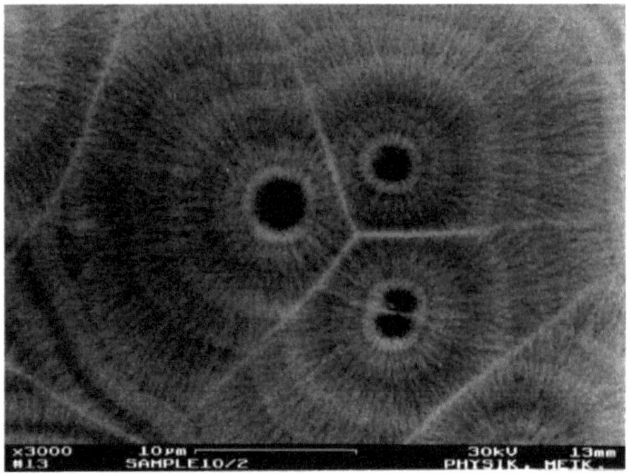

Fig. 4.23 SEM micrograph of the top surface of a GaP sample anodized in 0.5 m H_2SO_4 under temporary variations of the applied voltage from +5 to +15 V. Reprinted with permission from [48]. Copyright 1997 American Institute of Physics

4.3 Anodization of Ti and the Patterned TiO_2 Layers

4.3.1 Oscillations of Current During Anodization of Ti

Spatiotemporal self-organization in the anodic dissolution of metallic Ti is strictly related to the formation of semiconducting TiO_2 layer. Evidently, TiO_2 is one of the most investigated compounds in the materials science (more than 50,000 papers published over the past 40 years!), mainly due to its applications in photoelectrochemistry, such as photocatalysis or solar cell. Electrochemical oxidation of Ti surface, investigated for many years for different conditions, usually led to a compact oxide layer, but recent investigations revealed it porous structure and ability to form nanotubes.

Also, the formation of TiO_2 by anodization of Ti surface can be associated with the oscillations of the anodic current, accompanied with the formation of self-organized porous pattern of the oxide layer. The coexistence of these phenomena was reported in 2003 by Beranek et al. [52] for the case when titanium was anodized in the H_2SO_4 solution containing low concentration of HF (0.05–0.40 wt%) which strongly affected the amplitude of the oscillations. Obtaining of highly ordered porous TiO_2 layer (single pore diameter 140 nm, pore spacing 150 nm) required optimum experimental conditions. Electrochemical experiments were supported by scanning electron microscope (SEM) and X-ray photoelectron microscopy investigations which showed that the porous layer forms under a competition of TiO_2 formation and dissolution up to a limiting thickness of ~500 nm, and that the time for complete self-ordering for the investigated systems was of the order of several hours.

Very recent report on the current oscillations accompanying the Ti anodization under similar conditions was published by Warczak and Sadkowski [53]. The titanium electrode was first activated, under open circuit conditions, in 0.05 M H_2SO_4 containing small addition of NH_4F, with the moment of activation discerned by the rapid drop of an open circuit potential to very low value, lower than the potential of hydrogen gas evolution in the same electrolyte. Also, activation of Ti electrode could be reported by drop of its impedance for many orders of magnitude. The subsequent polarization of Ti anode in 1 M H_2SO_4 (or its mixture with ethylene glycol) was carried out in a two-electrode system, with the Pt electrodes as the counter and the virtual reference electrode. The oscillations were observed for only activated Ti electrodes and only if the polarization reached relatively high anodic voltages, at least 6 V (vs. the mercury/mercury sulfate electrode), up to 30 V (as a maximum voltage supplied by the potentiostat). The amplitude and frequency of the oscillations depended strongly on this voltage, but the intensive stirring of the solution did not visually change their shape. This suggested the origin of these instabilities located at the electrode/electrolyte interface rather than due to the transport processes. The authors suggested the correlation of the oscillations with, e.g., the ring-like patterns of TiO_2, reported among others by Mor et al. [54]. An addition of ethylene glycol was inspired by its beneficial effect on the growth of regular and long nanotubes on Ti [55, 56]. The effect of ethylene glycol on the oscillations was both trivial (increasing of ohmic resistance of the solution causing the anodic shift of oscillations) and nontrivial—the oscillations became more regular, compared to glycol-free samples, and their frequency decreased (Fig. 4.24). Indirectly, it suggests the correlation between the microscopic properties of porous TiO_2 layer and morphology of the oscillations. While this research will surely be continued, we shall further focus on the self-organized properties of TiO_2, formed under Ti anodization.

4.3.2 TiO_2 Nanotubes

The research in the area of TiO_2 fabrication was recently intensified by the finding of self-organized nanotubular oxide architectures which can be prepared by a simple, but optimized anodization of Ti metal surfaces. These nanotubular geometries provide large potential for enhanced and novel functional features, including enhanced photocatalytic performance. Zwilling et al. in 1999 have reported the formation of self-organized nanotubes during anodization of Ti in a fluoride-containing electrolyte [57], although they considered this surface structure as only "porous." Figure 4.25 illustrates schematically the effect of experimental conditions on the morphology of TiO_2 layers and indicates the possibilities of their doping.

In the course of further research, it was possible to detect the gradual transition from pores (*connected* oxide body around a pore) to tubes (*separated* oxide body around the pore) [58]. The conditions for obtaining the possibly high order of nanotubes have been adjusting for recent years, and recently almost ideally

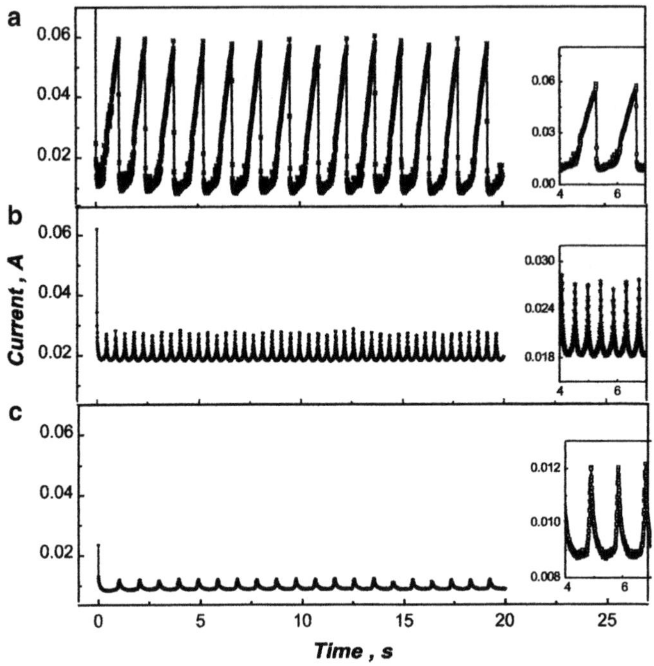

Fig. 4.24 Current oscillations recorded in (**a**) 1 M H_2SO_4, (**b**) 1:1 volume mixture of 1 M H_2SO_4 and ethylene glycol and (**c**) 1:5 volume mixture of 3 M H_2SO_4 and ethylene glycol. Anodizing at 20 V. Reproduced from [53], Copyright 2009, with permission from Elsevier

hexagonally ordered TiO_2 nanotube arrays several 100 μm in length, with tube diameters ranging from 10 to >200 nm, were obtained in the group of P. Schmuki. Furthermore, various modifications of the electrochemical parameters led to obtaining of new morphologies such as nanobamboo, nanolace, stacked layers, branched tubes, multiwalled tubes, and formation of amphiphilic TiO_2 nanotube stacks. Also, under specific conditions, self-ordering at two length scales could be obtained. TiO_2 nanotubes, also modified in various ways, find, among others, the following applications: (1) as photocatalysts, having a higher photocatalytic photoreactivity than a comparable nanoparticulate layer, (2) in the dye-sensitized solar cells, after the dye sensitization on TiO_2 (e.g., using ruthenium complex with bipyridyl, according to Gerischer and Tributsch [59]), with enhancement of efficiency, when the TiO_2 nanotubular layer replaces the TiO_2 nanoparticulate photoanode; (3) in electrochromic devices, due to such properties of TiO_2 also as nanotubes, which can also be lifted off from a metallic Ti substrate and transferred onto conducting glass; alternatively, complete anodization of thin Ti layers on conducting glass can be performed to construct effective transparent electrochromic devices. Other potential advantages and applications of TiO_2 nanotubes can be found in [58].

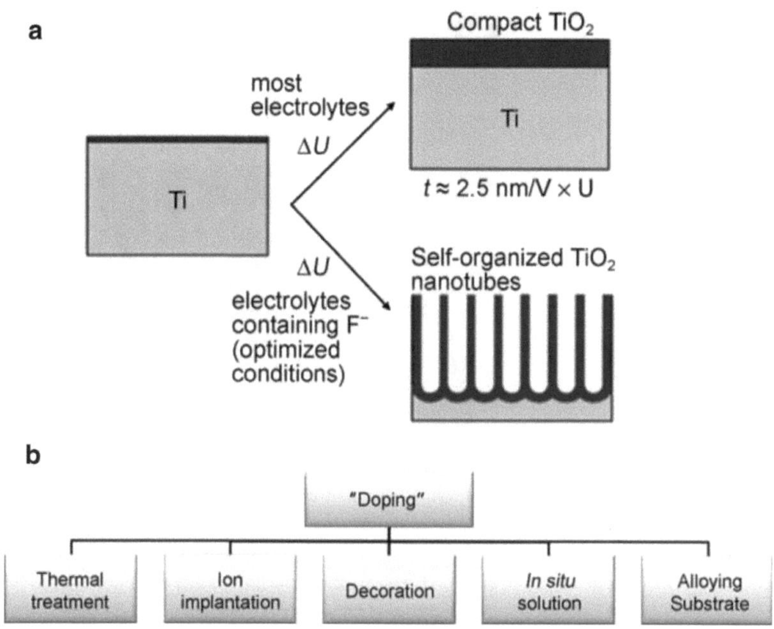

Fig. 4.25 (**a**) Schematic representation of the anodization process of Ti metal forming either a compact TiO_2 layer or (under specific optimized conditions) a self-organized TiO_2 nanotube layer and (**b**) several methods for doping TiO_2 layers. Reprinted from [58] with permission of John Wiley and Sons, Inc. Copyright 2010 Wiley-VCH GmbH & Co. KGaA

Figure 4.26 shows SEM images of TiO_2 nanoparticles and nanotubes grown on a Ti foil by optimized anodization.

The factors determining the morphology and properties of TiO_2 nanotubes during the controlled anodic oxidation of Ti substrate, in relation to other transition metal oxide nanostructures, are discussed in a very recent overview by Schmuki et al. [60]. One should also note that the electrochemical anodization is only one of the methods of producing the TiO_2 nanotubes.

4.3.3 TiO_2 Nanogrooves

Besides Ti anodization, the TiO_2 crystal as a substrate can also be a source of spatial self-organized patterns. Recently, Nakanishi et al. [61] have described the photoelectrochemical etching of n-TiO_2 (rutile) single crystals, in which there were obtained ordered arrays of nanogrooves with the spacing of ca. 300 nm over a macroscopically wide area of 0.5×0.5 cm at the surface without any use of imprints or templates. Moreover, the depth and location of nanogrooves could be regulated by irradiation with patterned light intensities which possibility is particularly important for the potential application of this procedure in nanotechnology of

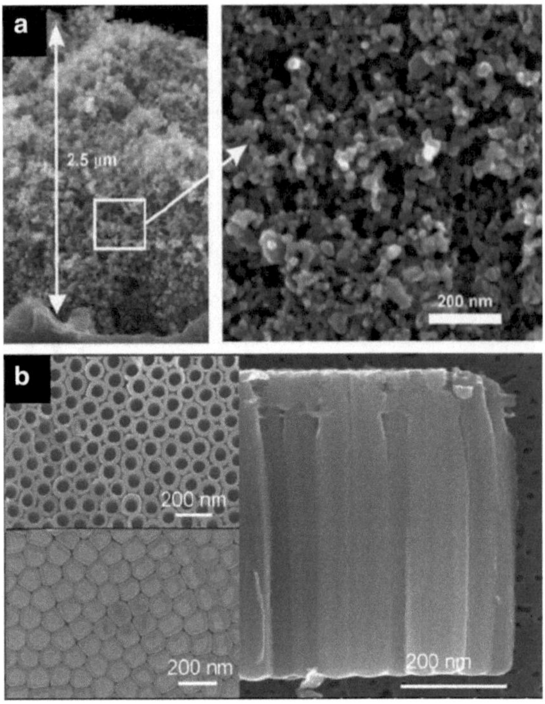

Fig. 4.26 Scanning electron microscopy (SEM) images of photoanodes consisting of (**a**) compacted TiO_2 nanoparticles and (**b**) self-organized TiO_2 nanotubes grown on a Ti foil by optimized anodization. Reprinted from [58] with permission of John Wiley and Sons, Inc. Copyright 2010 Wiley-VCH GmbH & Co. KGaA

materials of desired morphology. This work reflects the above-mentioned recent trend of increasing interest in the application of self-organization phenomena in producing the surface nanorelief structures at solid surfaces (cf., e.g., application of etching in vacuum [62, 63], buckling [64, 65], periodic precipitation [66], with practical applications [67, 68]).

Electrochemical etching (or photoetching) was used before [61] to produce ordered nanogrooves or holes at metal [69, 70] and semiconductor (cf., e.g., [38, 71, 72] surfaces. However, usually the patterns could be produced on very small areas, so in order to obtain a pattern on a macroscopic scale it was necessary to use appropriate external imprints or templates (as, e.g., in [38]). Also for electrochemical photoetching of n-TiO_2, the nanoholes and grooves were obtained before only on small areas [73, 74]. The presently described report is thus a continuation and improvement of these earlier works, achieved by manipulation with experimental conditions. The experimental setup is sketched in Fig. 4.27.

As the working electrode the single-crystal TiO_2 wafer (0.5 cm $\times$ 0.5 cm $\times$ 1.0 mm), with surfaces cut parallel to (110) face, was used. Prior to the electrochemical experiment, the wafer of TiO_2 was slightly reduced by heating at 550 °C for 3 h (such prolonged time was one of the crucial parameters for the quality of later obtained patterns), and cooled down under a hydrogen atmosphere to get n-type semiconductivity. The morphology of the surface was inspected with a high-resolution SEM and an AFM. Figure 4.28 shows the dependence of the photocurrent

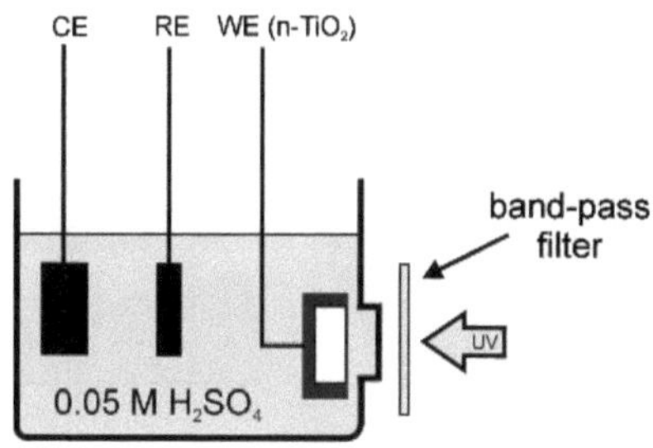

Fig. 4.27 Schematic illustration of the experimental setup for electrochemical photoetching of n-TiO_2. After [61]

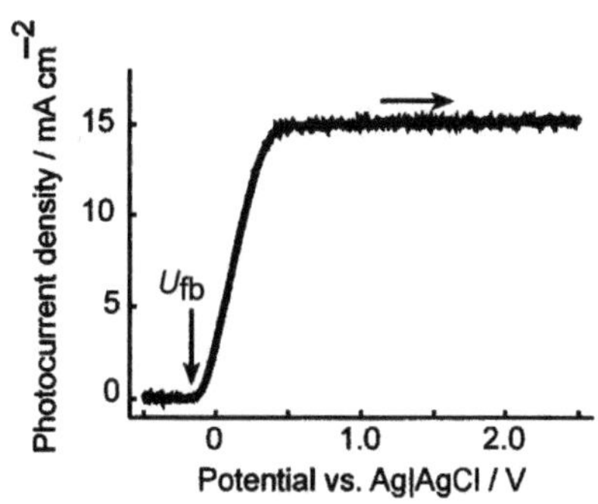

Fig. 4.28 The j_{pc} (photocurrent density) vs. U dependence for a (110)-cut n-TiO_2 (rutile) electrode in 0.05 M H_2SO_4. U_{fb}- flat-band potential. Reprinted with permission from [61]. Copyright 2007 American Chemical Society

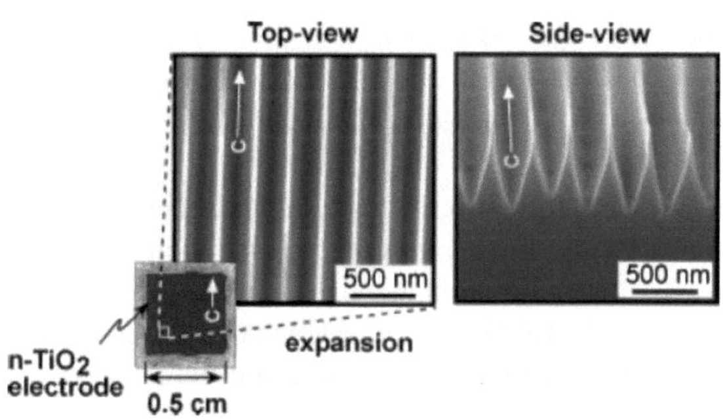

Fig. 4.29 SEM images of a part of the n-TiO_2 surface of 0.5 cm × 0.5 cm in size after the photoetching with the total charge passed $Q_p = 50$ C cm^{-2}. Reprinted with permission from [61]. Copyright 2007 American Chemical Society

density (j_{pc}) on the applied voltage U. The photoetching reaction proceeded with quantum efficiency of about 1% and competitively with oxygen photoevolution reaction. In turn, Fig. 4.29 shows the representative SEM images of a part on the n-TiO_2 surface, obtained after the photoetching at $U = 2.0$ V, under constant

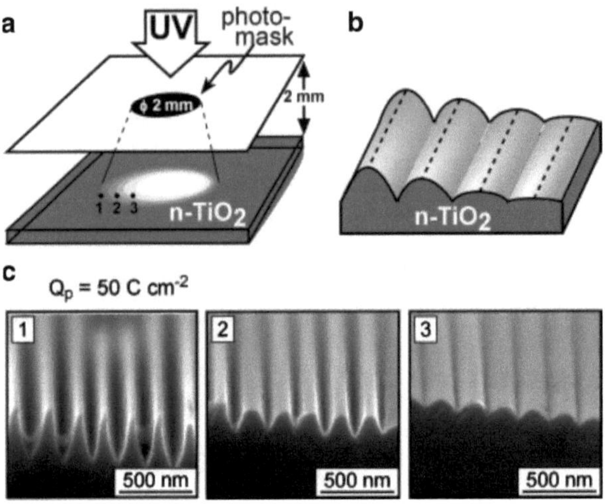

Fig. 4.30 (**a**) Schematic drawing of illumination with a varied intensity. (**b**) Schematic illustration of expected grooved array with varied groove depth. (**c**) SEM images for various local positions of the TiO_2 surface after photoetching with $Q_p = 50$ C m^{-2}. The numbers 1, 2 and 3 added on the SEM images mean that they were obtained at the positions marked by the same number on (**a**). Reprinted with permission from [61]. Copyright 2007 American Chemical Society

illumination at a constant intensity yielding $j = 15$ mA cm^{-2} (consumed in both parallel photoetching and oxygen evolution reactions). Since the same pictures were obtained for any other part of the surface, one could conclude that such highly ordered nanostructure was formed over entire macroscopic area of the electrode.

Further experiments showed that the irradiation with UV light with a spatial intensity gradient led to formation of nanogrooves with a variation of the depth. Figure 4.30 shows the results of the respective experiment when the UV-light incident onto the center of n-TiO_2 electrode was attenuated by inserting a photomask.

The mechanism of the formation of such highly ordered arrays of nanogrooves cannot be explained by the light interference since their shape and size are independent of the wavelength of illuminated light, but appear to be strongly dependent on the crystal faces and axes of TiO_2. Nakanishi et al. [61] have suggested the explanation in terms of dynamic self-organization via nonequilibrium kinetics of the surface instability, induced by autocatalytic photoetching (similar to the theory of formation of porous semiconductor). The schematic representation of this mechanism is visualized in Fig. 4.31.

According to this figure, the n-TiO_2 electrode, under anodic polarization, has upward band bending near the surface, and by the band bending the photogenerated holes in the valence band migrate to the surface and cause the photoetching. Let us assume that surface of n-TiO_2 is initially flat and apply a small sine curve perturbation on its shape (see again Fig. 4.31), which induces the modulation of the potential distribution (or the band bending) in the space charge layer of n-TiO_2.

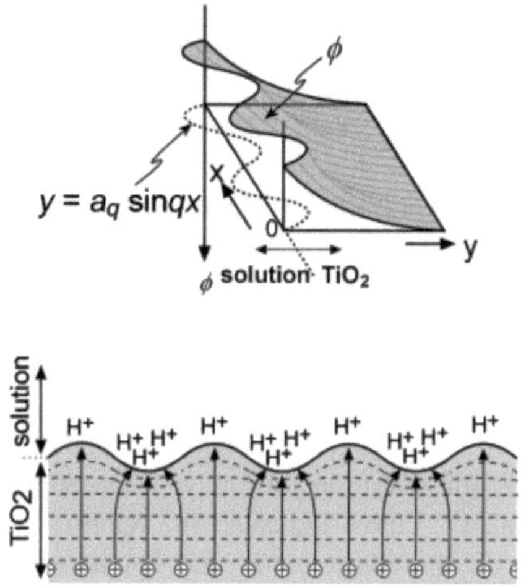

Fig. 4.31 Schematic models explaining the formation of an ordered nanogroove array. (**a**) The band bending near the n-TiO_2 surface; x, the coordinate along the n-TiO_2 surface; y, that toward the n-TiO_2 interior; ϕ, the electric potential; and $y = a_q \sin qx$, a curve representing fluctuation-induced surface hollows. (**b**) The distributions of ϕ in the n-TiO_2 crystal. *Dashed curves* in n-TiO_2: equi-potential surfaces. (*Circled plus*) Valence-band hole. Reprinted with permission from [61]. Copyright 2007 American Chemical Society

The key consequence is that the potential gradient in the TiO_2 increases near the bottom of the surface hollows, which accelerates the migration of photogenerated holes to the bottom of the surface hollows. This acceleration, in turn, induces enhancement of the photoetching reaction, which makes the surface hollows deeper. In other words, the autocatalytic photoetching occurs at the bottom of the surface hollows [61].

Figure 4.31 shows also an additional process that retards the photoetching reaction based on the following mechanism: the photogenerated holes migrating to the bottom of the surface hollows produce a high local concentration of H^+ ions in the electrolyte, due to the oxygen photoevolution reaction:

$$2H_2O + 4h^+ \rightarrow 4H^+ + O_2 \tag{4.2}$$

where h^+ refers to a photogenerated hole. The corresponding local decrease in pH causes the downward shift of the flat band potential U_{fb} of n-TiO_2 (for -0.059 V/pH, at 298 K). This in turn causes a decrease in the potential gradient in the space charge layer of n-TiO_2 near the bottom of the surface hollows and thus a decrease in the migration of photogenerated holes there, leading in this way to the retardation of the photoetching reaction. In terms of this explanation, the self-organized formation of nanogroove arrays with spatial periodicity becomes a consequence of a combination of this retardation process with the above-mentioned autocatalytic process [61]. Simple theoretical considerations led to the expression for the photoetching rate which indicated such wavenumber of the sinusoidal fluctuation which corresponded to its most enhanced growth and thus selected in the course of photoetching. It is useful to add that these considerations, involving, among others, the dependence of the pH decrease on the curvature of the surface hollows, correspond to the idea of the so-called Mullins-Sekerka instability [75], which was originally elaborated for the description of morphological stability of spherical particle undergoing growth controlled by diffusion of heat flow. In conclusion, the work by Nakanishi et al. [61] opens a novel effective route for formation of designed patterned nanorelief structures at semiconductor surfaces.

The reader interested in other materials that exhibit induced self-organized pore formation should inspect the literature concerning ZrO_2 [76] and Al_2O_3 [77]. Due to the regularity of pore structures of Al_2O_3 (having the size from 20 to 200 nm), its new applications as, among others, membranes for nanofiltration, quantum dots housing and substrates for microreactors, have been suggested. Also in a very recent report, Lee et al. [78] have indicated the relation between the modulated nanoporous structure of aluminum oxide and spontaneous current oscillations during the potentiostatic hard anodization of aluminum in 0.3 M $H_2C_2O_4$ medium, and the oscillatory mechanism is proposed. Theoretical analysis of nanoscale pore formation during aluminum anodization, performed in terms of linear stability analysis, was described by Thamida and Chang [79]. The extensive application of nanochannel alumina in the production of various nanomaterials (functional nanowires, nanotubes and membranes) is described in a recent specialized monograph by Ozin and Arsenault [80], which is also a recommendable source of information on other types of nanostructures. Note also that hexagonal patterns on anodized Al surface, of the convective origin, are described in Chap. 5.

4.4 Overview of Spatiotemporal Self-organization in Etched Semiconductors

To summarize the present state of knowledge, based on a very recent review by Föll et al. [21], one can identify the following kinds of self-organization phenomena at anodized semiconductor electrodes (see original and cited references for more details):

1. Oscillations of current/potential *in time*, explained in terms of CBM model, described in Sect. 4.1 (and, alternatively, in terms of approach by Lewerenz et al. or Krischer et al.), associated with spatiotemporal self-organization of anodized surfaces.
2. The formation of pore crystals and ordered pore domains (for more examples, see Fig. 4.32), meaning also oscillations of current *in space*; interestingly, such patterns were found not only for (100) n-InP, but also for (111) n-Si and even for (100) n-Ge anodized in HCl solution in water or water/DMSO mixtures [81, 82]. It is important to note that the pore crystal only forms with so-called current line or curro pores, that grow in the direction of current flow (more or less perpendicular to the surface) [21].
3. Oscillations of pore diameter and pore direction, illustrated by Fig. 4.33.

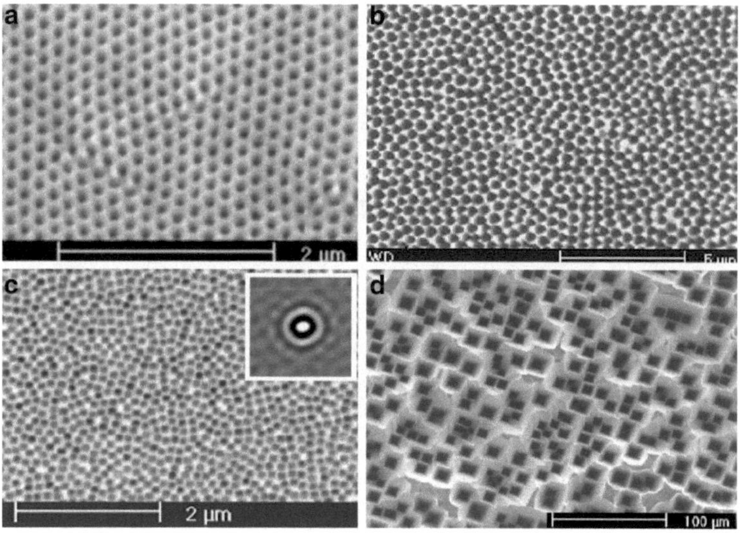

Fig. 4.32 Examples of current oscillations in space. (**a**) Single pore crystal in (100) n-type InP, (**b**) Short range order ("damped oscillation") in (111) n-type Si, (**c**) Frustrated crystal (see [21] for details) in (100) n-type Si. (**d**) Random pore distribution in (100) n-type Ge obtained in an HCl electrolyte (note the much larger scale). Reproduced from [21], Copyright 2009, with permission from Elsevier

Fig. 4.33 Self-induced pore diameter oscillations. (**a**) Crysto pores in GeAs; (**b**) Curro pores in InP; (**c**) Crysto pores in Si (TEM picture; insert shows details); (**d**) Curro pores in n-type Si. Reproduced from [21], Copyright 2009, with permission from Elsevier

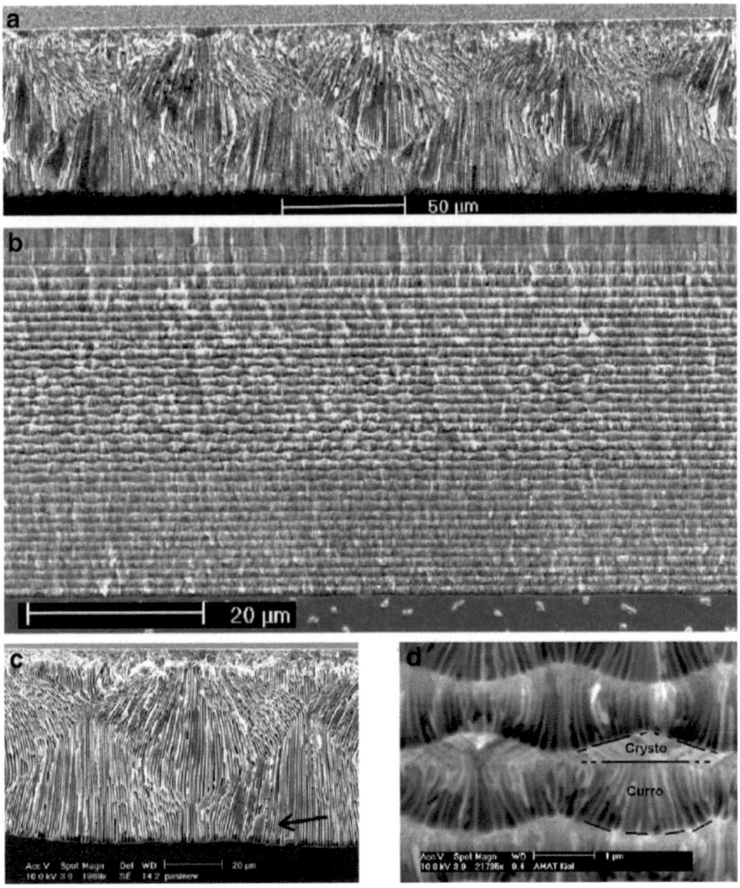

Fig. 4.34 Self-induced pore-bundle diameter oscillations. (**a**) Curro pores in (100) *n*-type InP, wavelength ~ 80 µm. (**b**) Externally induced periodic switch between crysto and curro pores with self-induced pore bundle oscillations in parts of the depth: wavelength ~ 2 µm. (**c**) Detail of (**a**). Note that individual pores still show some uncorrelated diameter oscillations; the *arrow* points to particularly well expressed ones; (**d**) Details of (**b**). Reproduced from [21], Copyright 2009, with permission from Elsevier

In this case, the oscillations in space of some pore geometry parameter, and oscillations in time of the current or voltage, do not necessarily appear together. As an example can serve the above-mentioned anodization of Ge in HCl solution [81], for which strong voltage oscillations during pore growth practically did not influence the pore geometry. As a relatively new phenomenon in the context of pore diameter oscillations, the "pore bundle" diameter oscillations or pore direction oscillations was indicated, so far reported only for *n*-InP (Fig. 4.34) [21].

4. The pore growth mode transitions is the concept introduced recently by the group of Föll, based on new findings in the course of experiments oriented on

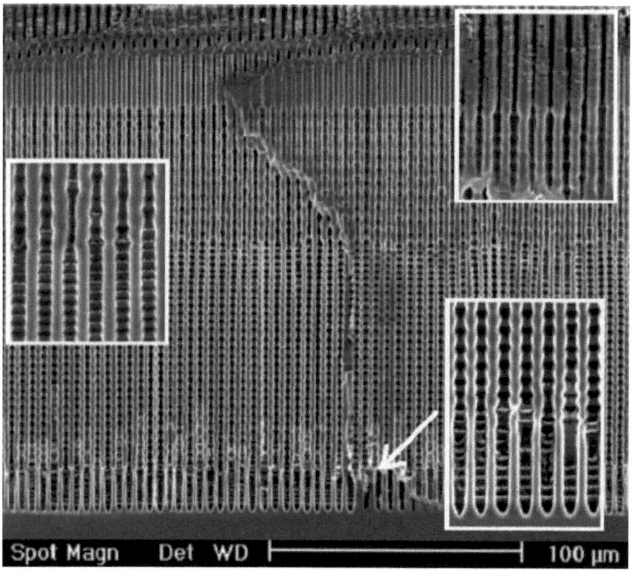

Fig. 4.35 Macropores in *n*-type Si growing with a 50% modulation of the current starting at a depth of about 25 μm. The insets illustrate the three growth mode changes encountered. Reproduced from [21], Copyright 2009, with permission from Elsevier

"fast etching" variant of *n*-Si anodization, performed for enhanced HF concentration and upon addition of carboxylic acids and/or carboxymethylcellulose sodium salt. In one of the experiments, the current was switched periodically at 2 min intervals between the nominal current and 50% of that value. It was then observed that variations of the pore diameter did *not* follow the current—upon the first reduction of the current, the pore diameters increased; and moreover, two growth mode changes occurred about halfway down the pores and before last ca. 20 μm (Fig. 4.35) [21].

5. The oscillation of the pore density with the pore depth in the crysto pores in (100) *n*-InP [83], which are considered to be a new expression of self-organization [21].

In conclusion, self-organization phenomena in semiconductor electrodes include various kinds of phenomena, which are however not yet completely understood. These problems remain thus a challenging task for the future investigations. As Föll et al. claim in their review [21], one of the main problems in this area is the lack of suitable experimental data. In search of them, the authors implemented the fast-Fourier-transform impedance spectroscopy (FFT IS). Accordingly, very recent report by Leisner et al. [84] contains relevant experimental studies of the growth mechanism of current-line-oriented pores in *n*-InP, considered the model system for electrochemical pore etching. FFT IS measurements were performed in situ during pore etching and supported by theoretical calculations. In this way pore growth parameters such as the space-charge region (SCR) width, the SCR potential, the capacitance at the pore tips, and the avalanche breakdown field strength were extracted. It was found that at the pore tips a constant field strength exists, which ensures that avalanche breakdown occurs, generating the necessary holes for the electrochemical dissolution of n-InP.

References

1. Grzanna J, Jungblut H, Lewerenz HJ (2000) A model for electrochemical oscillations at the Si/electrolyte contact. Part I. Theoretical development. J Electroanal Chem 486:181–189
2. Grzanna J, Jungblut H, Lewerenz HJ (2000) A model for electrochemical oscillations at the Si/electrolyte contact. Part II. Simulations and experimental results. J Electroanal Chem 486:190–203
3. Feller W (1971) An introduction to probability theory and its applications, vol II. Wiley, New York
4. Ozanam F, Blanchard N, Chazalviel JN (1993) Microscopic, self-oscillating domains at the silicon surface during its anodic dissolution in a fluoride electrolyte. Electrochim Acta 38:1627–1630
5. Gerischer H, Lübke M (1988) Electrolytic growth and dissolution of oxide layers on silicon in aqueous solutions of fluorides. Ber Bunsenges Phys Chem 92:573–577
6. Lewerenz HJ, Jungblut H, Rauscher S (2000) Surface analysis of the electropolishing layer on Si(111) in ammonium fluoride solution. Electrochim Acta 45:4615–4627
7. Lewerenz HJ, Skorupska K, Aggour M, Stempel T, Grzanna J (2009) Surface chemistry and electronics of semiconductor-nanosystem junctions I: metal-nanoemitter-based solar cells. J Solid State Electrochem 13:185–194
8. Grzanna J, Notz T, Lewerenz HJ (2008) Model for current oscillations at the Si/electrolyte contact: extension to spatial resolution. ECS Trans 16:173–180
9. Grzanna J, Notz T, Stempel T (2011) Nanopore morphology development during current oscillations at the Si/electrolyte contact. ECS Trans 33:127–135
10. Carstensen J, Prange R, Popkirov GS, Föll H (1998) A model for current oscillations in the Si-HF system based on a quantitative of current transients. Appl Phys A 67:459–467
11. Carstensen J, Prange R, Föll H (1999) A model for current-voltage oscillations at the silicon electrode and comparison with experimental results. J Electrochem Soc 146:1134–1140
12. Ozanam F, Chazalviel JN, Radi A, Etman M (1991) Current oscillations in the anodic dissolution of silicon in fluoride electrolytes. Ber Bunsenges Phys Chem 95:98–101
13. Chazalviel JN (1992) Ionic processes through the interfacial oxide in the anodic dissolution of silicon. Electrochim Acta 37:865–875
14. Foca E, Carstensen J, Föll H (2007) Modelling electrochemical current and potential oscillations at the Si electrode. J Electroanal Chem 603:175–202
15. Carstensen J, Christophersen M, Föll H (2000) Pore formation mechanism for the Si-HF system. Mater Sci Eng B69–70:23–28
16. Uhlir A (1956) Electrolytic shaping of germanium and silicon. Bell Syst Tech J 35:333–347
17. Uhlir A Jr, Uhlir (Williams) I (2005) Historical perspective on the discovery of silicon. Phys Stat Sol C 2:3185–3187
18. Canham LT (1990) Silicon quantum wire array fabrication by electrochemical and chemical dissolution of wafers. Appl Phys Lett 57:1046–1048
19. Lehmann V, Föll H (1990) Formation mechanism and properties of electrochemically etched trenches in n-type Silicon. J Electrochem Soc 137:653–659
20. Foca E, CarstensenJ FH (2005) Monte Carlo simulation of electrochemical oscillations in the electropolishing regime. Phys Stat Sol A 202:1524–1528
21. Föll H, Leisner M, Cojocaru A, Carstensen J (2009) Self-organization phenomena at semiconductor electrodes. Electrochim Acta 55:327–339

22. Parkhutik VP, Matveeva E (1999) Observation of new oscillatory phenomena during the electrochemical anodization of silicon. Electrochem Solid State Lett 2:371–374
23. Parkhutik V, Costa Gómez F, Moya Tarazona L, Fenollosa Esteve R (2000) Oscillatory kinetics of anodic oxidation of silicon – influence of the crystallographic orientation. Microelectron Reliab 40:795–798
24. Parkhutik V, Matveeva E, Perez R, Alamo J, Beltrán D (2000) Mechanism of large oscillations of anodic potential during anodization of silicon in H_3PO_4/HF solutions. Mater Sci Eng B69–70:553–558
25. Parkhutik V (2001) Silicon anodic oxides grown in the oscillatory anodisation regime – kinetics of growth, composition and electrical properties. Solid State Electron 45:1451–1463
26. Parkhutik V (2002) Chaos-order transitions at corroding silicon surface. Mater Sci Eng B 88:269–276
27. Parkhutik VP (1991) Kinetics, composition and mechanism of anodic oxide growth on silicon in water-containing electrolytes. Electrochim Acta 36:1611–1616
28. Miethe I, García-Morales V, Krischer K (2009) Irregular subharmonic cluster patterns in autonomous photoelectrochemical oscillator. Phys Rev Lett 102:194101-1–194101-4
29. Aranson IS, Kramer L (2002) The world of the complex Ginzburg-Landau equation. Rev Mod Phys 74:99–143
30. Kuramoto Y (1984) Chemical oscillations, waves and turbulence. Springer, Berlin
31. García-Morales V, Krischer K (2008) Nonlocal complex Ginzburg-Landau equation for electrochemical systems. Phys Rev Lett 100:054101-1–054101-4
32. Lehmann V, Gösele U (1991) Porous silicon formation: a quantum wire effect. Appl Phys Lett 58:856–858
33. Wehsporn RB, Schilling J (2002) Electrochemically prepared pore arrays for photonic-crystal applications. MRS Bull 26:623–626
34. Christophersen M, Langa S, Carstensen J, Tiginyanu IM, Föll H (2003) A comparison of pores in silicon and pores in III-V compound materials. Phys Stat Sol A 197:197–203
35. Miethe I, Krischer K (2012) Ellipsomicroscopic studies of the anodic oxidation of p-type silicon in fluoride containing electrolytes during current oscillations. J Electroanal Chem 666:1–10
36. Chazalviel JN, Ozanam F (2010) Current oscillations in the anodic dissolution of silicon: On the origin of a sustained oscillation on the macroscopic scale. Electrochim Acta 55:656–665
37. Van Meirhaeghe RL, Cardon F, Gomes WP (1979) Photocurrent oscillations at the *n*-GaAs/ electrolyte interface. Electrochim Acta 24:1047–1049
38. Föll H, Langa S, Carstensen J, Christophersen M, Tiginyanu IM (2003) Pores in III-V semiconductors. Adv Mater 15:183–198
39. Föll H, Carstensen J, Langa S, Christophersen M, Tiginyanu IM (2003) Porous III-V compound semiconductors: formation, properties and comparison to silicon. Phys Stat Sol A 197:61–70
40. Schmuki P, Lockwood DJ, Ogata Y, Isaacs HS (eds) (2001) Pits and pores II: formation, properties and significance for advanced materials. The Electrochemical Society, Phoenix, AR
41. Langa S, Christophersen M, Carstensen J, Tiginyanu IM, Föll H (2003) Single crystalline 2D porous arrays obtained by self organization in n-InP. Phys Stat Sol A 197:77–82
42. Lehmann V, Stengel R, Luigart A (2000) On the morphology and electrochemical formation mechanism of mesoporous silicon. Mater Sci Eng B69–70:11–22
43. Chazalviel JN, Wehrspohn RB, Ozanam F (2000) Electrochemical preparation of porous semiconductors: from phenomenology to understanding. Mater Sci Eng B69–70:1–10
44. Langa S, Carstensen J, Tiginyanu IM, Christophersen M, Föll H (2005) Selfordering in porous III-V compounds. In: Wehrspohn RB (ed) Ordered porous nanostructures and applications. Springer, Berlin, p 57
45. Langa S, Carstensen J, Tiginyanu IM, Christophersen M, Föll H (2001) Self-induced voltage oscillations during anodic etching of n-InP and possible applications for three-dimensional microstructures. Electrochem Solid State Lett 4:G50–G52

46. Harvey E, Buckley DN, Chu SNG (2002) Oscillatory behavior during the anodization of InP. Electrochem Solid State Lett 5:G22–G25
47. Wloka J, Lockwood DJ, Schmuki P (2005) High intensity and oscillatory electroluminescence observed during porous etching of GaP in HBr and HF electrolytes. Chem Phys Lett 414:47–50
48. Tiginyanu IM, Schwab C, Grob JJ, Prevot B, Hartnagel HL, Vogt A, Irmer G, Monecke J (1997) Ion implantation as a tool for controlling the morphology of porous gallium phosphide. Appl Phys Lett 71:3829–3831
49. Erne BH, Vanmaekelbergh D, Kelly JJ (1996) Morphology and strongly enhanced photoresponse of GaP electrodes made porous by anodic etching. J Electrochem Soc 143:305–314
50. Takizawa T, Arai S, Nakahara M (1994) Fabrication of vertical and uniform-size porous InP structure by electrochemical anodization. Jpn J Appl Phys 33:L643–L645
51. Schilling J, Müller F, Matthias S, Wehrspohn RB, Gösele U, Busch K (2001) Three-dimensional photonic crystal based on macroporous silicon with modulated pore diameter. Appl Phys Lett 78:1180–1182
52. Beranek B, Hildebrand H, Schmuki P (2003) Self-organized porous titanium oxide prepared in H_2SO_4/HF electrolytes. Electrochem Solid State Lett 6:B12–B14
53. Warczak M, Sadkowski A (2009) Oscillatory regime of titanium anodization under voltage control. Electrochem Commun 11:1733–1735
54. Mor GK, Varghese OK, Paulose M, Mukherjee N, Grimes CA (2003) Fabrication of tapered, conical-shaped titania nanotubes. J Mater Res 18:2588–2593
55. Macak JM, Schmuki P (2006) Anodic growth of self-organized anodic TiO_2 nanotubes in viscous electrolytes. Electrochim Acta 52:1258–1264
56. Prakasam HE, Shankar K, Paulose M, Varghese OK, Grimes CA (2007) A new benchmark for TiO_2 nanotube array growth by anodization. J Phys Chem C 111:7235–7241
57. Zwilling V, Darque-Ceretti E, Boutry-Forveille A, David D, Perrin MY, Aucouturier M (1999) Structure and physicochemistry of anodic oxide films on titanium and TA6V alloy. Surf Interface Anal 27:629–637
58. Nah YC, Paramavisam I, Schmuki P (2010) Doped TiO_2 and TiO_2 nanotubes: synthesis and applications. ChemPhysChem 11:2698–2713
59. Gerischer H, Tributsch H (1968) Elektrochemische Untersuchungen zur spektralen Sensibilisierung von ZnO - Einkristallen. Ber Bunsenges Phys Chem 72:437–445
60. Berger S, Hahn R, Roy P, Schmuki P (2010) Self-organized TiO_2 nanotubes: factors affecting their morphology and properties. Phys Stat Sol B 247:2424–2435
61. Nakanishi S, Tanaka T, Saji Y, Tsuji E, Fukushima S, Fukami K, Nagai T, Nakamura R, Imanishi A, Nakato Y (2007) Ordered nanogroove arrays on n-TiO_2 with a variation of the groove depth, formed by self-organized photoetching. J Phys Chem C 111:3934–3937
62. Zúñiga-Pérez J, Martínez-Tomás C, Muñoz-Sanjosé V, Munera C, Ocal C, Laügt M (2005) Faceting and structural anisotropy of nanopatterned CdO(110) layers. J Appl Phys 98:034311-1–034311-5
63. Sugawara A, Hembree GG, Sheinfein MR (1997) Self-organized mesoscopic magnetic structures. J Appl Phys 82:5662–5669
64. Bowden N, Brittain S, Evans AG, Hutchinson JW, Whitesides GM (1998) Spontaneous formation of ordered structures in thin films of metals supported on an elastomeric polymer. Nature 393:146–149
65. Ohzono T, Matsushita SI, Shimomoura M (2005) Coupling of wrinkle patterns to microsphere-array lithographic patterns. Soft Mater 1:227–230
66. Grzybowski BA, Bishop KJM, Campbell CJ, Fiałkowski M, Smoukov S (2005) Micro- and nanotechnology via reaction–diffusion. Soft Mater 1:114–128
67. Smoukov SK, Bitner A, Campbell CJ, Grzybowska KK, Grzybowski BA (2005) Nano- and microscopic surface wrinkles of linearly increasing heights prepared by periodic precipitation. J Am Chem Soc 127:17803–17807

68. Teranishi T, Sugawara A, Shimizu T, Miyake M (2002) Planar array of 1D gold nanoparticles on ridge-and-valley structured carbon. J Am Chem Soc 124:4210–4211
69. Masuda H, Fukuda K (1995) Ordered metal nanohole arrays made by a two-step replication of honeycomb structures of anodic alumina. Science 268:1466–1468
70. Yuzhakov VV, Chang HC, Miller AE (1997) Pattern formation during electropolishing. Phys Rev B 56:12608–12624
71. Schmuki P, Fraser J, Vitus CM, Graham MJ, Isaacs HS (1996) Initiation and formation of porous GaAs. J Electrochem Soc 143:3316–3322
72. Sugiura T, Itoh S, Ooi T, Yoshida T, Kuroda K, Minoura H (1999) Evolution of a skeleton structured TiO_2 surface consisting of grain boundaries. J Electroanal Chem 473:204–208
73. Nakato Y, Akanuma H, Shimizu J, Magari Y (1995) Photo-oxidation reaction of water on an n-TiO_2 electrode. Improvement in efficiency through formation of surface micropores by photo-etching in H_2SO_4. J Electroanal Chem 396:35–39
74. Tsujiko A, Kisumi T, Magari Y, Murakoshi K, Nakato Y (2000) Selective formation of nanoholes with (100)-face walls by photoetching of *n*-TiO_2 (rutile) electrodes, accompanied by increases in water-oxidation photocurrent. J Phys Chem B 104:4873–4879
75. Mullins WW, Sekerka TF (1963) Morphological stability of a particle growing by diffusion or heat flow. J Appl Phys 34:323–329
76. Berger S, Faltenbacher J, Bauer S, Schmuki P (2008) Enhanced self-ordering of anodic ZrO_2 nanotubes in inorganic and organic electrolytes using two-step anodization. Phys Stat Sol RRL 2:102–104
77. Jessensky O, Müller F, Gösele U (1998) Self-organized formation of hexagonal pore structures in anodic alumina. J Electrochem Soc 145:3735–3740
78. Lee W, Kim JC, Gösele U (2010) Spontaneous current oscillations during hard anodization of aluminum under potentiostatic conditions. Adv Funct Mater 20:21–27
79. Thamida SK, Chang HC (2002) Nanoscale pore formation dynamics during aluminum anodization. Chaos 12:240–251
80. Ozin GA, Arsenault AA (2005) Nanochemistry. A chemical approach to nanomaterials. RSC Publishing, Cambridge, CA
81. Fang C, Föll H, Carstensen J (2006) Electrochemical pore etching in germanium. J Electroanal Chem 589:259–288
82. Cheng F, Carstensen J, Föll H (2006) Electrochemical pore etching in Ge. Mater Sci Semicond Proc 9:694–700
83. Leisner M, Carstensen J, Cojocaru A, Föll H (2008) In-situ FFT impedance spectroscopy during the growth of n-type InP "crysto pores". ECS Trans 16:133–142
84. Leisner M, Carstensen J, Föll H (2010) Pores in *n*-type InP: a model system for electrochemical pore etching. Nanoscale Res Lett 5:1190–1194

Chapter 5
Convection as a Source of Self-Organization in Electrochemical Systems

5.1 Convection as a Self-Organized Phenomenon

5.1.1 The Navier-Stokes Equation

Convection is the macroscopic collective flow of particles of the fluids and gases, which can be caused by various driving forces. It is a nonlinear phenomenon which is mathematically described by the Navier–Stokes equation for the motion of viscous liquid:

$$\frac{\partial \boldsymbol{\vartheta}}{\partial t} + (\boldsymbol{\vartheta} \cdot \nabla)\boldsymbol{\vartheta} = \frac{1}{\rho}\left[-\nabla p + \eta \nabla^2 \boldsymbol{\vartheta} + \mathbf{F}\right] \tag{5.1}$$

where $\boldsymbol{\vartheta}$ denotes the vector of fluid velocity; ρ, the fluid density; η, its dynamic viscosity; p, pressure evolving in the fluid; and $\mathbf{F}$ is the volume driving force (N m^{-3}) for convection. For the cases described in this work, the pressures developing in the fluid are moderate and then the incompressibility of the fluid can be assumed:

$$\mathrm{div}\boldsymbol{\vartheta} = 0 \tag{5.2}$$

as the condition significantly simplifying numerical integration of Eq. (5.1), yielding the vector field of velocity $\boldsymbol{\vartheta}$.

The following driving forces $\mathbf{F}$ can operate as a source of convection:

1. ***Buoyancy forces***, caused by ***density gradients*** of the fluid in the gravitational field which gradients, in turn, can be induced by the thermal gradients or by the concentration gradients (also under isothermal conditions); the thermally induced convection is usually called the *Bénard–Rayleigh instability*.
2. ***Surface (interfacial) tension gradients***, which may be induced by thermal or concentration gradients, associated also with (electro)chemical processes at the interface; this type of convection is usually termed the *Bénard–Marangoni instability*.

M. Orlik, *Self-Organization in Electrochemical Systems II*,
Monographs in Electrochemistry, DOI 10.1007/978-3-642-27627-9_5,

3. ***Volume electric forces***, arising from the violation of the electroneutrality principle that may occur under specific conditions. The fluid motion of that type is called the *electrohydrodynamic* (EHD) convection. In physical systems, in order to induce the EHD convection, the high voltage (of the order of tenths kVs) is usually required for the nonconducting, nonelectroactive media, and then even the *unipolar* charge injection may occur. For electrochemical systems, the sufficiently high spatial charge in the solution may develop even for a few volt voltage if the thin layer of the electrolyzed solution contains only trace amount of the conducting species.

Irrespective of the particular nature of driving force, convection of all kinds may exhibit, under appropriate conditions, the *self-organized* nature: the motion of the fluid attains then the shape of cooperating rolls, forming also fingerprint-like shapes or hexagonal cells. Patterns of this type become examples of the spatial convective structures which are typical of one-component systems with thermal gradient. Often the terms: Bénard-Rayleigh and Bénard-Marangoni instabilities are referred to the formation of such self-organized *spatial* patterns. More complicated, multicomponent (including electrochemical) systems may exhibit also spatiotemporal patterns, with oscillatory variation of the pattern morphology, as well as of other system characteristics, as e.g., the flowing current, as a function of time. In view of these universalities, it is useful to discuss first the most general features of self-organization in convection.

5.1.2 Classical Bénard–Rayleigh Instability

The thermal convection or Bénard–Rayleigh instability in one-component system is well recognized and its basic description is relatively simple. In order to observe self-organization in thermal convection, usually one has to induce it in the sufficiently thin layer of the fluid, of a thickness ranging from tenths of micrometers up to a few millimeters. Apparently, the first historical experimental manifestation of self-organization in this convection was published already in 1900 by Bénard [1], who heated up the thin layer of the whale spermaceti oil. Above certain critical temperature gradient, the convective motion sets in and underwent self-organization into adjacent hexagonal cells (*Bénard cells*), visible from the top of the fluid layer (Fig. 5.1). One should mention that in this case the surface of the fluid was free, so the instabilities driven by density gradients could be accompanied by the gradients of the surface tension, leading to hexagonal structures.

Also more linear convective patterns, like that shown in Fig. 5.2, can emerge due to Bénard–Rayleigh instability.

The local buoyancy force density is defined as the product of the vector of the acceleration of gravity $\mathbf{g}$ and the local difference in the fluid density: $\mathbf{f} = \mathbf{g}\Delta\rho$. Furthermore, the free convection arises in the fluid if the vector of density gradient $\nabla\rho$ is not parallel to the vector of the acceleration of gravity $\mathbf{g}$ (i.e., when $\nabla\rho \times \mathbf{g} \neq 0$), since then the nonzero torque is created; physically it means

Fig. 5.1 (*Top*) Hexagonal convective cells (Bénard cells) in the thin layer of the fluid heated from below, from: Koschmieder EL (1994) Bénard convection. In: Adv Chem Phys, vol 26 (Eds Prigogine I, Rice SA), John Wiley & Sons, Inc. Hoboken, NJ, reproduced with permission; (*Bottom*) image of a single convective cell; Courtesy of Prof. M.G. Velarde from the Autonomous University of Madrid. Both images reproduced from [2]

non-uniform distribution of fluid density (i.e. also of temperature) along the horizontal coordinate. If both vectors are parallel, then the fluid may or may not set into convective motion, so this case has to be considered in terms of the fluid stability [4]. In other words, the fluctuation understood as the displacement of the small portion of the fluid may either vanish or undergo amplification, engaging the fluid in the convection on a macroscopic scale. For further discussion of the fluid stability under such conditions, one has to distinguish between the *horizontally* and *vertically* oriented systems which differ with the condition for the onset of convection.

In *horizontal* systems, shown in Fig. 5.3, the convection sets in *only above* certain critical temperature gradient. Below this critical value only the linear distribution of temperature develops, and the fluid remains quiescent. In other words, this steady state is stable, since all local fluctuations are damped.

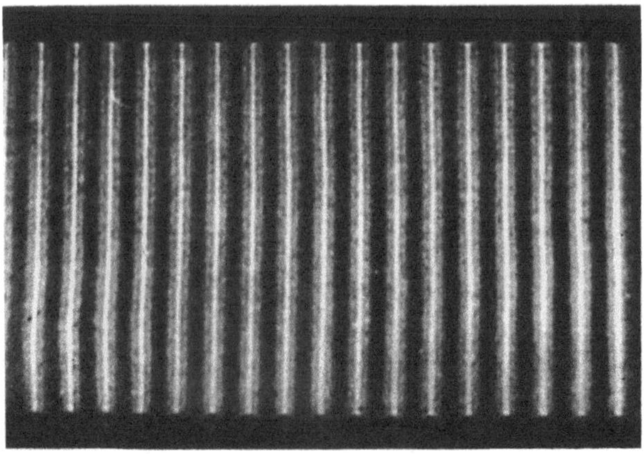

Fig. 5.2 Convective pattern of ordered rolls as alternative manifestation of Bénard–Rayleigh instability [3]. Reproduced with permissions of Hermann Publishers and Wiley-VCH Verlag GmbH & Co. KGaA. Copyright Hermann Publishers 1984

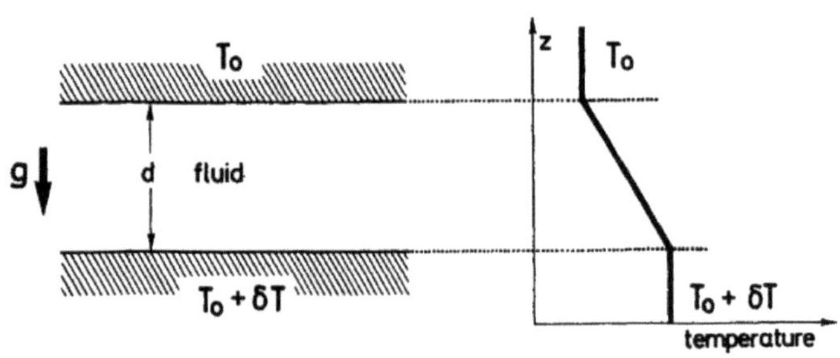

Fig. 5.3 Diagram of a Rayleigh–Bénard experiment. The fluid between two horizontal heat-conducting plates has—in the absence of convection—a constant temperature gradient (represented on the *right*). Its temperature goes from $T_0 + \delta T$ at the bottom of the layer to T_0 at the top [3]. Reproduced with permissions of Hermann Publishers and Wiley-VCH Verlag GmbH & Co. KGaA. Copyright Hermann Publishers 1984

Let us consider the local fluctuation – the displacement of the fluid in terms of schematic in Fig. 5.4: from H to H′ and from B to B′ which creates a precursor torque [3].

The time scale (characteristic time) of this displacement is the shorter, the higher the driving (buoyancy) force, and the longer, the higher the fluid viscosity:

$$\tau_{\text{disp}} \propto \frac{\eta}{\rho_0 g \alpha d(\Delta T)} \tag{5.3}$$

where the following parameters characterize the fluid: η, dynamic viscosity; α, expansion coefficient and ρ_0, mean density. Furthermore, d is the thickness of the fluid layer, g, is the gravitational acceleration and ΔT, is the temperature gradient.

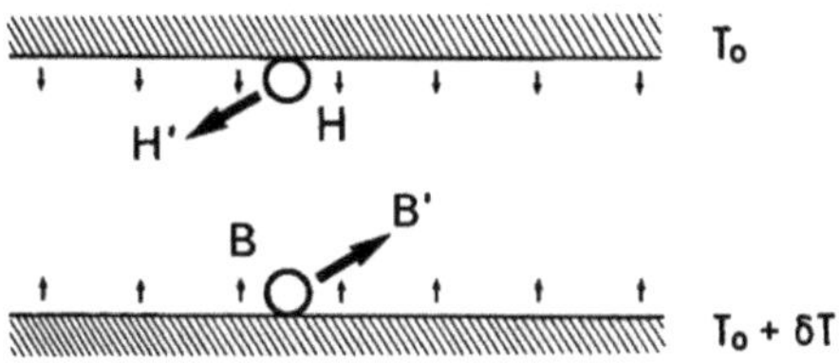

Fig. 5.4 Diagram of the generation of motion in the fluid. We consider two fluid elements H and B displaced to H′ and B′, respectively. This motion is damped or maintained according to the value of δT [3]. Reproduced with permissions of Hermann Publishers and Wiley-VCH Verlag GmbH & Co. KGaA. Copyright Hermann Publishers 1984

The source of the critical value of ΔT, only above which the fluid continues to displace, is the competitive diffusion of heat, caused by resulting *horizontal* temperature gradient between the zones H′ and B′; this process occurs in a time scale defined by:

$$\tau_{\mathrm{dif}} \propto \frac{d^2}{D_{\mathrm{T}}} \tag{5.4}$$

where D_{T} [m^2 s^{-1}] (often denoted also as κ or α) is the thermal diffusivity coefficient, defined in turn as:

$$D_{\mathrm{T}} = \frac{k}{\rho c_{\mathrm{P}}} \tag{5.5}$$

with k meaning the thermal conductivity coefficient [W m^{-1} K^{-1}], ρ is the fluid density [kg m^{-3}], and c_{P} is the specific heat capacity [J K^{-1} kg^{-1}] of the fluid. The physical sense of quantity (5.5) is to inform how quickly the given substance adjusts its temperature to that of their surroundings.

Obviously, if τ_{dif} is shorter than τ_{disp}, the fluctuation of the fluid motion is damped. Quantitatively, the necessary condition for the sustained convection is [3]:

$$\frac{\tau_{\mathrm{dif}}}{\tau_{\mathrm{disp}}} = \frac{\rho_0 g \alpha d^3 (\Delta T)}{\eta D_{\mathrm{T}}} = \frac{\rho_0 g \alpha d^4}{\eta D_{\mathrm{T}}} \left(\frac{\Delta T}{d}\right) \equiv Ra \geq Ra_{\mathrm{c}} \approx 1{,}708 \tag{5.6}$$

meaning that in the *horizontal* system the sustained convection sets in if the dimensionless Rayleigh number Ra exceeds it critical value Ra_{c} equal to ca. 1,708. In the thin-layer system is then a chance for the convection to self-organize into the counter-rotating, cooperating rolls, the cross-section of which is schematically shown in Fig. 5.5. This scheme indicates a very important quantitative aspect of the self-organized convection: that the characteristic size of the convection roll, in the ideal case having a circular cross-section, is equal (or, in practice, close) to the thickness of the fluid layer. This dependence can be used for the easy experimental confirmation of the convective origin of the given pattern, since then its characteristic size (wavelength) should appropriately vary with the thickness of the thin-layer system.

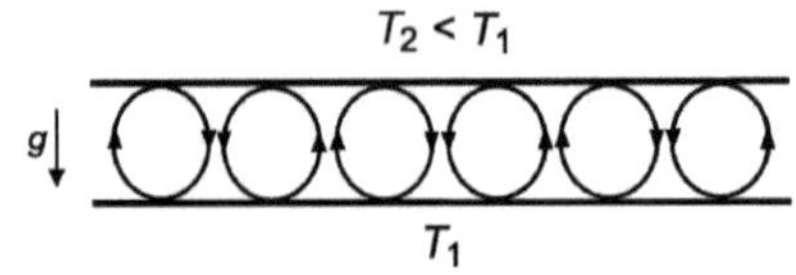

Fig. 5.5 General scheme of self-organization of the convective motion in the horizontal, thin-layer system, into the counter-rotating rolls of a diameter close to the thickness of the fluid layer. The fluid is heated from below ($T_1 > T_2$) and the arrow at symbol g indicates the direction of the gravitational field

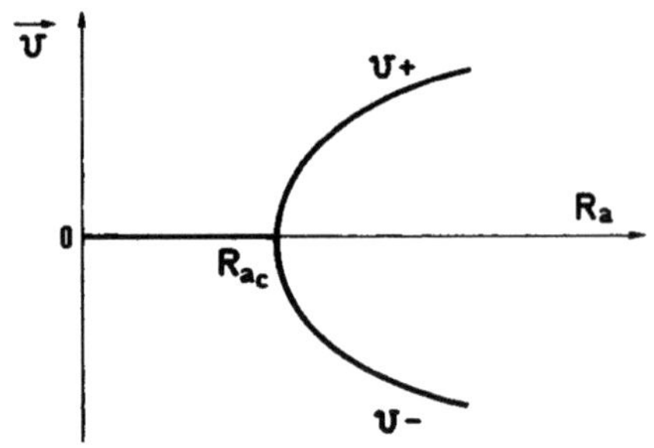

Fig. 5.6 Bifurcation diagram near the threshold Ra_c of the thermoconvective instability. In the experiment free of imperfections, the rolls engendered at Ra_c have equal probability of rotating in either direction. This is expressed by the existence of two branches denoted by v_+ and v_- [3]. Reproduced with permissions of Hermann Publishers and Wiley-VCH Verlag GmbH & Co. KGaA. Copyright Hermann Publishers 1984

Analysis of the Bénard–Rayleigh instability shows that the onset of convection has a *supercritical* nature: above the critical Rayleigh number the fluid velocity starts from zero and rises gradually. In other words, the transition to convective motion is "soft," contrary to the "hard" one if the velocity would jump immediately to large value at a bifurcation point, as in the subcritical case [3]. The schematic bifurcation diagram shown in Fig. 5.6 illustrates also the fact that in the perfect system the roles rotating in clockwise and counterclockwise direction appear with equal probability.

The formation of hexagonal patterns can be understood as the optimum way to fill up the (theoretically infinite) homogeneous space. For thermal convection, in every hexagonal cells the warmed fluid goes up along its inner area, cools down at the upper surface, and returns to the bottom along its walls.

Let us now consider the case of *vertically* oriented layer of the fluid (Fig. 5.7), for which the vector of the density gradient is not parallel to the vector of acceleration of gravity, so the condition for convection is always met. The initially created torque is not damped as the corresponding forces do not oppose each other, like in the horizontal system. Thus, contrary to the horizontal systems, the fluid will set into convective motion for any infinitesimal temperature gradient: it rises along the warmer boundary and falls along the colder boundary. This effect should be taken

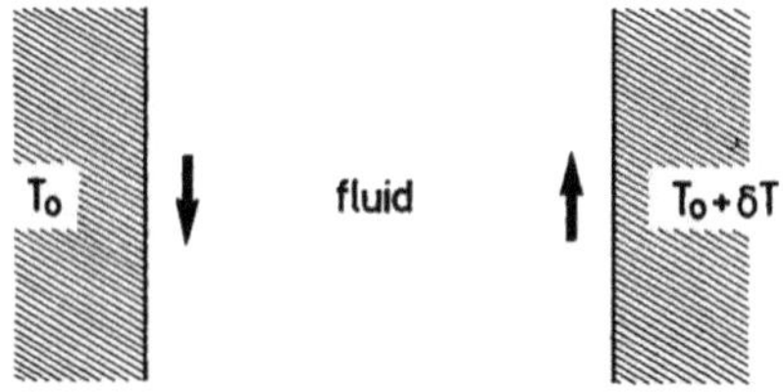

Fig. 5.7 The onset of convection for the vertically oriented layer occurs for any value of the horizontally imposed temperature gradient [3]. Reproduced with permissions of Hermann Publishers and Wiley-VCH Verlag GmbH & Co. KGaA. Copyright Hermann Publishers 1984

into account in the analysis of real systems as a possible source of a side effect, avoiding of which can be difficult.

Irrespective of the particular variant of the thermal convective system, one can characterize the onset of convection also from the energetic point of view. The conduction of heat is enhanced through convection which, however, means the transformation of a part of the inner energy of the system into the kinetic energy of this motion. In other words, the dissipation of the free energy is not total, since its part is stored in the kinetic energy of convection. The dissipative structure diminishes thus the entropy production.

The reader particularly interested in the detailed treatment of hydrodynamic (as well hydromagnetic) stability is advised to consult the classical monograph by Chandrasekhar [5].

5.2 Electrochemical Analogues of the Bénard–Rayleigh Instabilities

5.2.1 *The Cu|CuSO$_4$|Cu System*

Universalities in self-organized convection driven by density gradients are convincingly visualized through the comparison of the classical, nonisothermal systems discussed earlier and the isothermal systems, described in this section. Kawczyński and Baranowski [6–9] have constructed an electrolytic cell in which the convection set in due to fluid density gradients, caused by the concentration gradients instead of temperature gradients. The thin-layer cell consisted of two parallel copper electrodes, with $CuSO_4$ solution placed between them. Typical parameters of the system were: electrode size: 4 × 10 cm, interelectrode distance: $d = 0.6$–1.1 mm, $[CuSO_4] = 10^{-3}$ M. Upon passing the electric current, the dissolution of anode into Cu^{2+} ions and the simultaneous deposition of Cu^{2+} ions at the cathode occurred, associated with the migration of Cu^{2+} and $SO_4{}^{2-}$ ions in the solution. In the absence of hydrodynamic instabilities, the steady-state, linear concentration profile of $CuSO_4$ develops in the solution. Analogously to the thermal

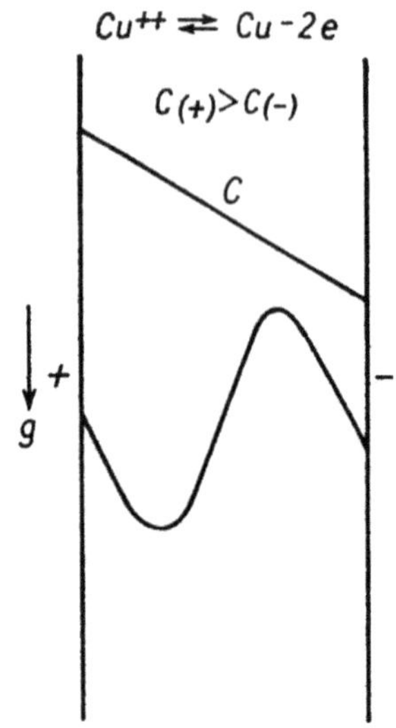

Fig. 5.8 Scheme of the vertical electrogravimetric setup for the electrochemically generated convection driven by density gradients. Reprinted from [9] with kind permission of Deutsche Akademie der Naturforscher Leopoldina—Nationale Akademie der Wissenschaften

convection, the flow of the electric current is now a source of free energy dissipation, through both the resistance of the electrode processes and the ohmic resistance of the electrolyte solution (the evolution of Joule heat). The resulting concentration gradient can then be considered the simplest dissipative structure in our system. Noteworthy this structure stores some energy—in fact, the concentration cell developed in the system can be a source of voltage after switching off the electrolysis.

Formal and morphological analogies with the thermal convection are very close. First, in the *vertically* oriented cell (see Fig. 5.8), the convection sets in for any concentration (density) gradient. Again, a part of inner system's energy is transformed into the kinetic energy of convection.

We shall devote now more attention to the *horizontally* oriented system and consider the galvanostatic and potentiostatic conditions of the electrolysis. If we want to avoid convection, the upper electrode must be the cathode, since then the solution of higher density develops in the vicinity of the bottom anode. When the top electrode is the anode, the denser fluid appears in the upper part of the system, which means, as in thermal convection, the nonequilibrium situation in the gravitational field and the solutions layers at the given electrodes have a tendency to exchange their places.

For *galvanostatic* control, the potential difference between the copper electrodes increases as a function of time due to developing concentration polarization—the higher voltage is then necessary to maintain the constant current. In the absence of convection, within a few minutes the steady-state voltage is reached, meaning that the linear steady-state concentration profile has developed. Figure 5.9 shows such an experimentally obtained characteristic for the case of the upper copper cathode (convection excluded).

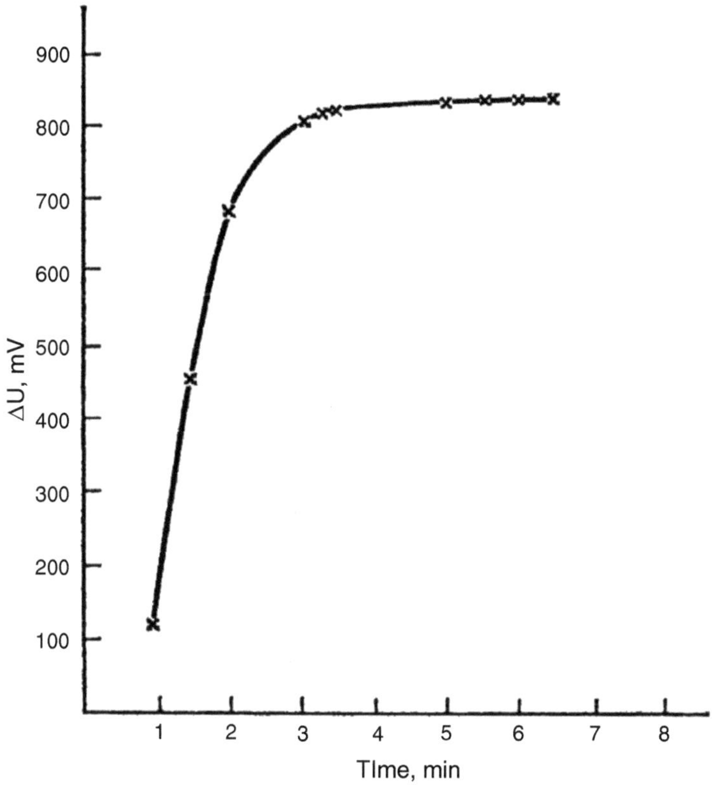

Fig. 5.9 Changes of the difference of the potential drop as a function of time in the case when the anode forms the bottom electrode. Reprinted from [6], Copyright 1972, with permission from Elsevier

On the other hand, if the top electrode is the anode, the voltage–time dependence is similar, *unless the electric current density exceeds critical value*, corresponding to the onset of convection—just like in the thermal system the critical temperature gradient is required. Due to convection the solution is stirred, the concentration polarization is thus minimized, and the potential difference appropriately decreases to the steady-state value with respect to both migration and convection (see Fig. 5.10).

Based on the value of this critical current I_{crit} one can calculate the critical concentration gradient of $CuSO_4$ at which the convection sets in [6, 10]:

$$\left(\frac{dc}{dz}\right)_{crit} = \frac{2c_0}{d}\frac{I_{crit}}{I_{lim}} \tag{5.7}$$

where z is the coordinate normal to the electrode surfaces; c_0, initial, uniform concentration of the $CuSO_4$ electrolyte; d, interelectrode distance; and I_{lim} is the limiting current that corresponds to the given experimental conditions.

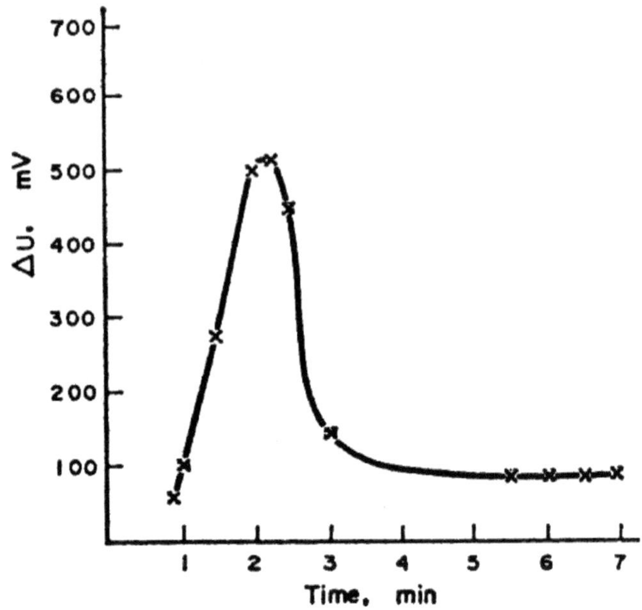

Fig. 5.10 Changes of the difference of potential drop as a function of time in the case when anode forms the top electrode (creation of an adverse density gradient). Reprinted from [6], Copyright 1972, with permission from Elsevier

In turn, under *potentiostatic* conditions [7], in the absence of convection the measured current continuously decreases, asymptotically tending to a steady-state value, corresponding to the linear $CuSO_4$ concentration profile. This occurs always if the top electrode is the cathode or, if it is the anode, only if the applied voltage does not exceed certain critical value, determining the critical concentration gradient (see Fig. 5.11, curve I). Above this critical value the convection sets in and it manifests itself as the transition from the monotonically decreasing current to its damped oscillations, finished with steady-state current of a value significantly higher than for the quiescent solution (Fig. 5.11, curve II). It is also noteworthy that under potentiostatic conditions the etching of the electrode was observed in the form of *hexagons* or *rolls*, with a characteristic size comparable with the interelectrode distance (2–3 mm).

It is thus clear that the isothermal electrochemical system, shown in Fig. 5.12, is an electrochemical analogon of the nonisothermal Bénard–Rayleigh system, with only different detailed source of the density gradient.

Furthermore, analogy with the thermal convection is also of quantitative nature. Kawczyński and Baranowski [8] have shown theoretically that the appropriately defined "electrochemical Rayleigh number":

$$Ra^* = \frac{g\beta(\partial\rho/\partial c)_T d^4}{\rho D \nu} \tag{5.8}$$

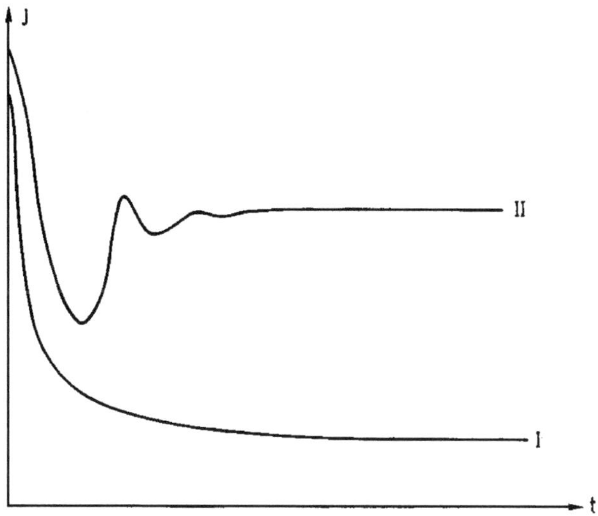

Fig. 5.11 The electrical current as a function of time measured at potentiostatic conditions (applied voltage $\Delta U = 0.2$ V). Distance between the electrodes: 2.11 mm, concentration of the copper sulfate solution: 2.6×10^{-3} M. Curve I: cathode forms the upper electrode (no convection). Curve II: anode forms the upper electrode (convection occurs). Reprinted from [7]

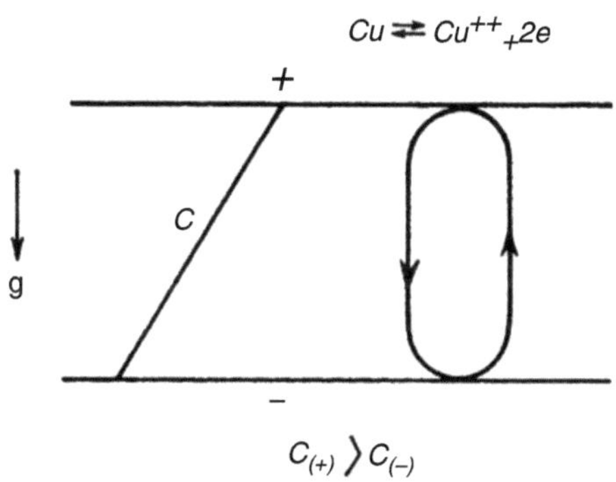

Fig. 5.12 Schematic of the electrochemical analogon of the Bénard–Rayleigh instability. Reprinted from [9] with kind permission of Deutsche Akademie der Naturforscher Leopoldina—Nationale Akademie der Wissenschaften

attains the same critical value of 1,708, at which the thermal convection sets in. In Eq. (5.8), β is the gradient of the $CuSO_4$ concentration c [replacing the temperature gradient in Eq. (5.6)], d is the interelectrode distance, D is the effective diffusion coefficient of $CuSO_4$, and other symbols have the same significance as for Eq. (5.6).

One should note that *experimentally* determined critical *Rayleigh* numbers Ra^* appeared to be lower than this theoretical value, both for galvanostatic [6] and potentiostatic [7] conditions of electrolysis. This is probably due to simplifications of the theoretical model. Several detailed sources of discrepancies, occurring in real electrochemical systems, were suggested. Nechiporuk and Elgurt [11] have shown that both the kinetics of electrode processes and various external influences might affect the onset of instability in this type of convective systems. After such corrections, meaning the appropriate modification of the boundary conditions, the concordance between the experimental and the theoretical critical *Rayleigh* numbers appeared to be better. Furthermore, Grigin and Shapovalov have suggested [12] that for the solution of the binary electrolyte, in addition to buoyancy forces, also electrostatic forces should be considered. More recently, Winkler et al. [13] have shown how the onset of convection in the Cu|$CuSO_4$|Cu electrolytic system might depend on the geometric parameters, like the ratio of the radius of the electrochemical cell to the interelectrode distance.

The electrolytic-convective system described above involved upper anode and lower cathode, since only in this arrangement the nonequilibrium density gradient in gravitational field can develop. However, theoretical considerations [14] predict that if the electrochemical characteristics of the process occurring in a thin-layer cell exhibit the negative differential resistance, e.g., due to potential-dependent adsorption of the inhibitor molecules adsorbed on the electrode surface (cf. Sect. 2.1.4, volume I), the *convective instability* may set in *also* for the *negative* values of the Rayleigh number. Formally this corresponds to the reversed density gradient, i. e., the copper cathode placed over the anode.

More recently, using a Michelson laser interferometer, Nakabayashi et al. [15] have described a very curious hydrodynamic pattern that appeared in such Cu| $CuSO_4$|Cu system, with upper Cu electrode as the anode. The observed vertical cross-sectional view of the convection pattern indicated just explicitly convective flow dividing the space of the fluid into roll cells, similar to those typical of Rayleigh–Bénard instability.

5.2.2 Oscillations in Electroformation of Ionic Liquids

A rather unusual example of the oscillations driven presumably by density gradients was observed by Caban et al. [16] during the electroformation of ionic liquids in undiluted nitromethane, during electrolysis at Pt microelectrodes, in the presence of tetrahexylammonium bromide (THABr) as a supporting electrolyte. Then on the surface of this electrode there is formed the microlayer of [$CH_3NO_2^{-}$] [THA^+]—the ionic liquid which is more viscous and less dense compared to pure corresponding nitroalkane. The gravitational rotated cell which allowed to determine above properties is shown in Fig. 5.13.

A priori unexpected, very reproducible oscillations of the current were reported at potentials more negative than −6 V (vs. Pt quasi-reference electrode), but only if the orientation of the microelectrode vs. the gravity vector was in the range 90°–180° (with the angle of 18° being an optimum value) (Fig. 5.14).

This suggested the role of gravitational removal of very small drops of ionic liquid from the depletion layer, which process should enhance the resistance and,

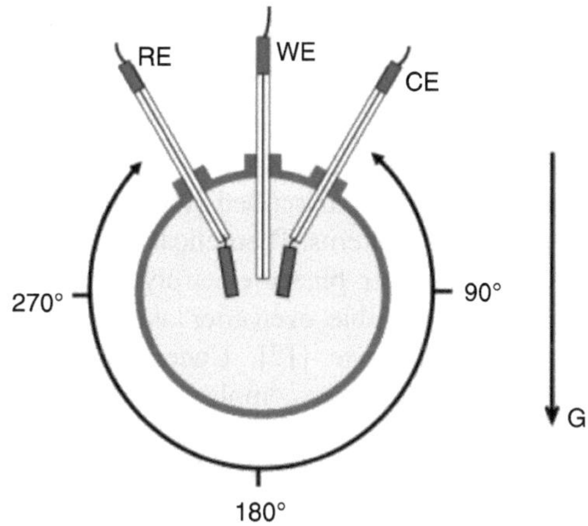

Fig. 5.13 Scheme of the gravitational rotated cell. Reprinted with permission from [16]. Copyright 2004 American Chemical Society

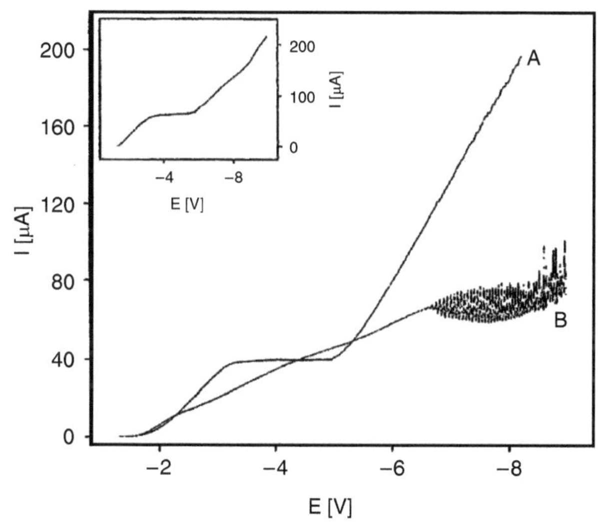

Fig. 5.14 Voltammetric responses of 25.4-μm-radius Pt microdisk in undiluted nitromethane. Supporting electrolyte was 5 mM THABr, scan rate = 50 mV/s; angle $\alpha = 0°$ (**a**) and 180° (**b**). Inset shows response for 0.02 M THABr, scan rate = 50 mV/s, $\alpha = 180°$, and $T = 21°C$. Reprinted with permission from [16]. Copyright 2004 American Chemical Society

due to increasing ohmic drops, shift the electrode potential to less negative values, at which current also drops. Next the potential, as well as the current are restored, and when the consecutive microdrop departs upward, the entire process repeats. Alternative, less probable (due to gravitational effect) explanation involves the increase of resistance due to electroreduction of counterion from the supporting electrolyte, followed by rebuilding of the ionic layer at decreased current.

5.3 Oscillatory Convective Instabilities Leading to Spatiotemporal Patterns

Convective instabilities may manifest themselves not only as the spatial, steady-state dissipative structures, as the ones described in Sects. 5.1 and 5.2, but also as the spatiotemporal (i.e., oscillatory) patterns. Theoretical considerations suggest that in the solution of $CuSO_4$ (or another binary electrolyte) the sustained *oscillatory convective instabilities* are not possible, even after taking into account the Coulomb forces developing in the electrolyte [12]. Under isothermal conditions such oscillations should be expected for more complex, i.e., multicomponent systems.

Recently, Grigin and Davydov [4] have analyzed a model electrochemical system, containing the solution of an electroactive species in the presence of an excess of supporting electrolyte (ensuring the electroneutrality of the solution and thus excluding substantial electric forces). In this model, the solution contained the salt with the electroactive cation and the electroinactive anion, the latter one being the same as for the supporting electrolyte (e.g., $AgNO_3$ + excess KNO_3 solution) placed between the horizontal silver electrodes. It was found that upon certain critical value of the current passing through the system, instabilities might arise. Theoretical stability analysis involved the temporal evolution of the perturbations of the fluid velocity, concentrations of species and electrode potential, proportional to $\exp(\lambda t)$ factor, where the eigenvalue λ can be a complex number. Its vanishing real part, upon varying system's control parameters, means reaching the stability boundary. Such analysis indicated the possibility of two convective instabilities: the Rayleigh–Bénard instability and oscillatory instability. The latter one sets in, if upon increasing faradaic current, its value critical for the oscillatory instability occurs before the critical value for the Rayleigh–Bénard instability. The oscillatory instability requires also the appropriate ratio between the volume buoyancy forces and the diffusion coefficients. In order to find the detailed conditions for that, additional assumptions had to be made. For example, in the simplest case, the exchange current density for the Ag^+/Ag redox process is assumed to be much higher than the limiting current, determined by the concentration of Ag^+ ions; this means that the effect of the finite kinetics of the electron transfer at the electrode/electrolyte interface may be considered negligible. For that case the conditions of instability were derived and expressed in terms of the numbers Ra_1 and Ra_2, where index "1" refers to Ag^+ ions, while index "2"—to K^+ ions. Ra_1 is a partial Rayleigh number for the electroactive Ag^+ ions [17]:

$$Ra_1 = \frac{gd^3c_{10}}{\rho \nu D_1}\frac{\partial \rho}{\partial c_1} \tag{5.9}$$

and Ra_2 is a partial Rayleigh number for indifferent K^+ cations:

$$Ra_2 = \frac{gd^3c_{10}}{\rho \nu D_1}\frac{\partial \rho}{\partial c_2} \tag{5.10}$$

where ν is the kinematic viscosity, c_{10} is the concentration of supporting electrolyte KNO_3 in the absence of current, c_1, c_2 denote the concentrations of Ag^+ and K^+, respectively, and other symbols have the same meaning as above.

In terms of these quantities, the condition for the spatial dissipative structures (or monotonic instability, or oscillation with zero frequency, $\omega = 0$) is [17]:

$$Ra_1 \geq 1,740 + \frac{1}{2}Ra_2 + \frac{D_1(D_3 - D_2)}{4D_2D_3}Ra_2 \tag{5.11}$$

while the condition for the spatiotemporal dissipative structures (oscillation frequency $\omega > 0$) is:

$$Ra_1 \geq 1,740\left(1 + \frac{2D_2D_3}{D_1(D_2 + D_3)}\right) + \frac{1}{2}Ra_2 \tag{5.12}$$

with the oscillation frequency given by:

$$\omega = 30\frac{\sqrt{D_1(D_2 + D_3)}}{d^2} \tag{5.13}$$

In Eqs. (5.11)–(5.13), D_1, D_2 and D_3 are the diffusion coefficients of Ag^+, K^+ and NO_3^- ions, respectively.

The oscillatory convective instabilities manifest themselves in the form of internal standing waves in the concentration of stratified liquid, characterized with certain frequency in the spatial period. Also the model electric current exhibited oscillations of low frequencies (0.0001–0.01 Hz), strongly dependent on the diffusion coefficients and the interelectrode distance. In [17], Grigin and Davydov have outlined the experimental verification of the predicted oscillatory convective instabilities. The conditions of such instabilities were later [18] discussed for more general cases. The neglected above effect of the limited electrode reaction rate at the electrode–solution interface, expressed through the exchange current density and the transfer coefficients, was considered and respective stability diagrams for the Rayleigh–Bénard and oscillatory convective instabilities in the (Ra_2, Ra_1) plane were constructed [18].

It is noteworthy that convective oscillatory instabilities in the multicomponent (more than binary) electrochemical systems, containing an excess of supporting electrolyte, should exhibit certain analogies with the thermal oscillatory convective instabilities arising also in multicomponent non-electrochemical systems, in which thermal diffusion (Soret effect) occurs. In the latter case, the buoyancy forces have two components: the first caused by temperature gradients, and the second one caused by the concentration gradient. For the thin-layer system containing binary liquid, with both temperature and concentration gradients existing throughout the solution, it was predicted that under such conditions not only steady-state Bénard cells should develop, but also oscillatory convective instabilities should emerge, as a result of the interplay of Bénard–Rayleigh type instabilities with the thermal diffusion flow. One should note that possible significant role of the Soret effect in the multicomponent systems with the Bénard–Rayleigh instability was suggested already by Prigogine. However, there are also obvious differences between the electrochemical and nonelectrochemical systems. Among others, different quantities determine the characteristic frequency of oscillations: in multicomponent electrochemical systems it is determined by the diffusion coefficients, while in thermal systems—by thermal diffusivity.[1]

Finally, an example of hydrodynamic instabilities causing electrochemical oscillations with pattern formation was described by Baier et al. [19]. During the potentiostatic oxidation of iodide ions in aqueous solution, the oscillatory variations of current were reported, associated with the formation of columns of descending liquid with periodic structure colored by iodine–starch complex. Such patterns were attributed to a periodically varying flow of solution in front of the electrode due to a density gradient. A simplified model of this instability was also formulated.

5.4 Bénard–Marangoni Instabilities

5.4.1 The Marangoni Number

The driving force for the Marangoni type of convective instability is the gradient of the interfacial (surface) tension, causing the motion of the solution surface which can undergo self-organization under appropriate conditions. Probably, the simplest example of such phenomena is the interaction of the surface of a thin layer of an alkaline (NaOH) solution of the bromocresol green indicator, placed in a Petri dish, with the absorbed HCl vapors [20]. The resulting convective vortices of the solution surface are easily monitored due to changes of the color of the indicator. However, the origin of these instabilities includes also the role of density gradients: the surface layer in which NaCl forms exhibits higher density than the lower portions and in this way the Rayleigh–Bénard instability also sets in, cooperating with the Marangoni instability. The role of density gradients can be easily proved by the fact that heating of the solution layer from the bottom enhances the evolution of the patterns. In fact, such cooperation of both kinds of instabilities can be quite a common case. Let us consider again the formation of the hexagonal convective cells (Fig. 5.1) that was above ascribed mainly to the Bénard–Rayleigh instability. In fact, this is probably a simplification and the detailed mechanism of the formation of these cells remains a subject of discussion. Rayleigh himself [21] has considered only buoyancy effects, while later Pearson has took into account only

[1] The reader interested in further developments of natural convective instabilities in electrochemical systems is advised to consult a more recent review : Volgin VM, Davydov AD (2006) Russ J Electrochem 42:567–608.

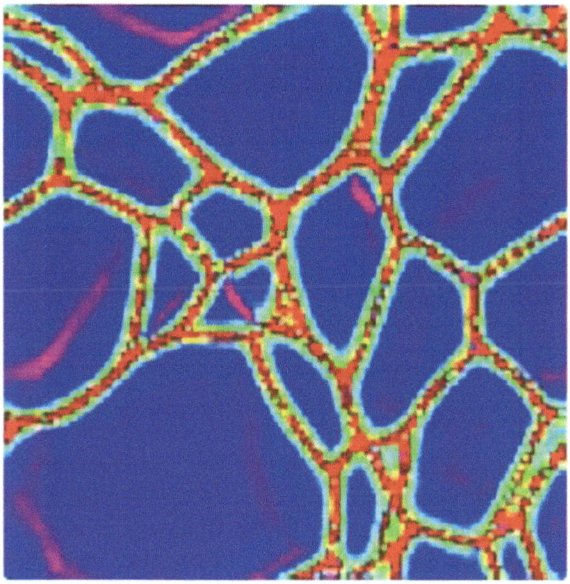

Fig. 5.15 Polygonal convection cells predicted from a model of Marangoni–Bénard convection at very high Marangoni number (proportional to the imposed temperature gradient). The model is derived for a layer of infinite depth, infinite Prandtl number, and periodic boundary conditions. The polygonal cell boundaries (*red*) are colder than their center (*blue*) and are compressed by the nonlinear advection of heat by the surface velocity field (from hot to cold). Source: http://www.ulb.ac.be/cenoliw3/nonlinear.html

the role of the surface tension gradients [22] which appear on the free surface of the fluid, as a consequence of temperature gradients. In turn, Nield has combined both approaches [23] and his theoretical considerations on the fluid stability led him to the conclusion that the hexagonal cells observed by Bénard in 1900 for the oil layer with a free surface were caused largely by surface tension gradients, with density gradients of a rather minor importance. Similar conclusions were drawn by Scriven and Sterling [24]. Of course, density gradients become the sole cause for convection if the fluid layer has no free surface.

Similarly as for the Bénard–Rayleigh instabilities, for the case when surface tension gradients are caused by temperature gradients, intercomparison of different systems is possible in terms of the dimensionless Marangoni number [23], the critical value of which separates the quiescent fluid from its convective regime:

$$Ma = \frac{(-\partial\sigma/\partial T)\beta d^2}{\rho\nu\kappa} \tag{5.14}$$

where σ is the surface tension, β is the temperature gradient, and other symbols have the same significance as for Eq. (5.6). For probably all typical liquids, the derivative $[-(\partial\sigma/\partial T)]$is greater than zero.

Figure 5.15 illustrates the model prediction of the pattern formed for the conditions of a very high Marangoni number.

In electrochemical systems, convective motion caused by the surface tension gradients was reported for various processes, occurring at both mercury and solid electrodes, including their electrodissolution and the electrodeposition of metallic phase on them.

5.4.2 Instabilities at the Mercury–Solution Interface

5.4.2.1 Polarographic Maxima and Cellular Convective Cells

Everybody familiar with fundamentals of polarography knows the phenomenon of polarographic maxima. Here we shall limit further considerations to the maxima of the first kind, caused by extra transport of electroactive species—the convection induced by surface tension gradients at the Hg/solution interface. Usually treated as nuisance, these maxima manifest itself as often significant enhancement of the current above the value predicted for only diffusion transport of reactants. We invoke this phenomenon here for two reasons.

First, because it may happen that in the potential range of polarographic maxima the current attains not only enhanced value, but also varies in time in an oscillatory manner. Already in 1973 Jakuszewski and Turowska [25] have analyzed the role of polarographic maxima in inducing electrochemical oscillations (as well as they discussed the significance of negative resistance in such instabilities, presenting thus one of the earliest analysis of oscillations of the NDR-type systems). The oscillations were found in the region of polarographic maximum associated with the electroreduction of 0.3 mM nitrobenzene in aqueous 0.05 M sulfuric acid medium. Later, Korolczuk and Matysik [26] have reported current oscillations within the potential range of polarographic maxima, e.g., for 0.5 mM $CrCl_3$ + 0.05 M NH_4Cl, at the potentials corresponding to the second wave of Cr(III) reduction (at -1.45 to -1.48 V vs. SCE). Oscillations of analogous nature were found also for the range of polarographic maxima associated with the electroreduction of Fe(II), Co(II), Zn(II), and Mn(II) in various, respective electrolytes (see [26] for details).

The second reason for invoking polarographic maxima is that the explanation of their origin is related to the hydrodynamic stability of the mercury/solution interface, i.e., it is a problem belonging to the field of nonlinear dynamics. Traditional explanation of polarographic maxima of the first kind is based on inhomogeneities in surface tension, but the detailed origin of these inhomogeneities was a subject of many discussions [27]. Moreover, in these traditional explanations the nonlinear nature of convection as the manifestation of instability in the system dynamics was not explicitly considered. For that reason one should invoke here the novel approach by Aogaki et al. [28, 29] who have elaborated the linear stability analysis of mercury electrode–solution interface under the conditions of the faradaic process. Closely related to that analysis was the experimental observation of

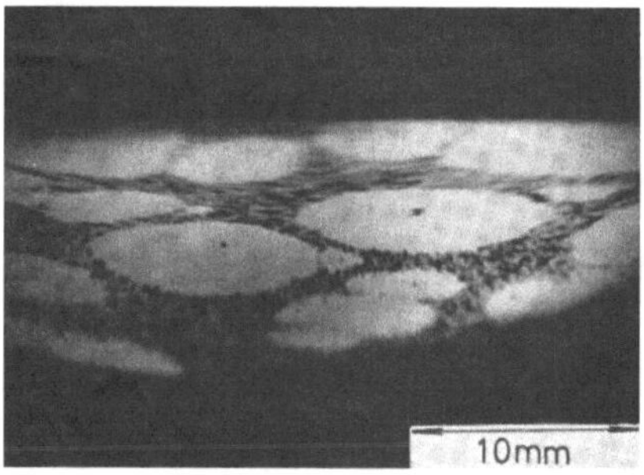

Fig. 5.16 Typical cellular convection pattern emerging at water/mercury interface after a short time of the reaction $Hg^{2+} + 2e \rightarrow Hg$ at the potential of −0.35 V (vs. the mercury - coated platinum wire as a reference electrode) in a solution with 1 mol m^{-3} $Hg(NO_3)_2$ + 10^2 mol m^{-3} KNO_3. Hg electrode diameter $\phi = 33.4$ mm. The patterns were visualized through the spatial distribution of the active carbon particles dispersed in the solutions as tracers. Reprinted from [30], Copyright 1985, with permission from Elsevier

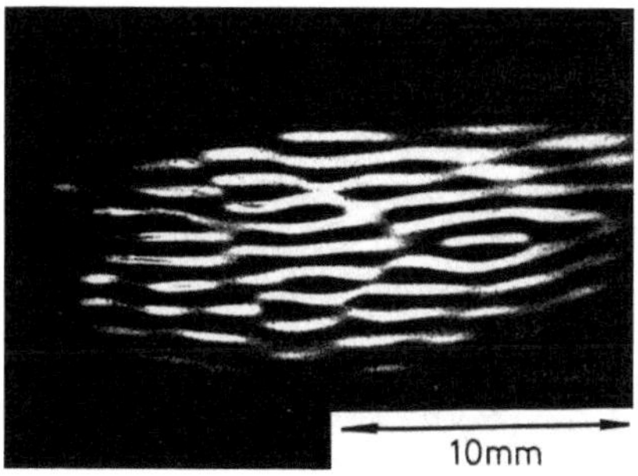

Fig. 5.17 Regular convex and concave patterns occurring at the Hg surface through resonance with the potential pulsating between −0.5 and 0 V with the frequency 167 Hz. Solution: 1×10^{-3} M $Hg(NO_3)_2$ + 0.10 M KNO_3. Reprinted from [31], Copyright 1985, with permission from Elsevier

the untypical and transient cellular *convective patterns* associated with the electroreduction of Hg(II) ions at the planar Hg electrode (Figs. 5.16 and 5.17) [30]. The system is electrochemically not well defined due to the fast surface synproportionation of Hg(II) and Hg(0) into Hg(I) species. Later studies performed by Gorzkowski et al. [32] have proved that this synproportionation contributed to the instability of the system, facilitating the onset of convection.

The transient nature of these patterns manifests itself in this way that within several seconds they turn into a random turbulent flow. In order to stabilize them, Makino et al. [31] have used the rectangular potential pulse for the electrochemical stimulation of the system. In this way, the growth of the fluctuations was repeatedly perturbed and one could even seek the conditions for resonance. The steady-state structures obtained in this way had, however, different morphology—they attained the shape of concave and convex patterns (Fig. 5.17).

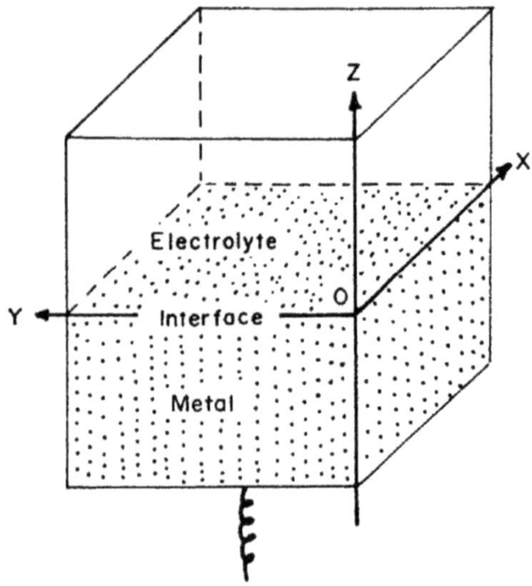

Fig. 5.18 Schematic configuration of liquid metal electrode and electrolyte solution; the solution contains the same metal ions as the electrode. Reprinted from [28], Copyright 1978, with permission from Elsevier

The hydrodynamic stability of such systems was also analyzed theoretically, for the model geometry shown in Fig. 5.18.

The linear stability analysis of this model system was performed. The motion of the fluid was described with the Navier–Stokes equation (5.1) with the driving force **F** defined through the surface tension gradient $\nabla\sigma$:

$$\frac{\partial \boldsymbol{\vartheta}}{\partial t} + (\boldsymbol{\vartheta} \cdot \nabla)\boldsymbol{\vartheta} = \frac{\mathbf{1}}{\rho}\left[-\nabla p_0 + \eta\nabla^2\boldsymbol{\vartheta} + \nabla \cdot (\nabla\sigma) \cdot \delta(z-0)\right] \quad (5.15)$$

where:

$$\nabla\sigma = \begin{bmatrix} \frac{\partial\sigma_0}{\partial x} & 0 & 0 \\ 0 & \frac{\partial\sigma_0}{\partial y} & 0 \\ 0 & 0 & 0 \end{bmatrix} \quad (5.16)$$

Furthermore, σ_0 denotes the interfacial (surface) tension, and, in this case ϑ and p_0 mean the *perturbations* of the fluid velocity and pressure, respectively. $\delta(z-0)$ is the delta function, i.e., it is equal to zero for all $z \neq 0$. Other symbols have the same significance as for Eq. (5.1). Also, the condition of the fluid incompressibility (5.2) was assumed.

The interested reader can find all details of the mathematical treatment in [28, 29], while here only most important points will be summarized. The flow of the (steady-state) faradaic current is associated with the (steady-state) concentration profile of the electroactive species developing along z axis, while along the x and y axes the concentration distribution is assumed initially homogeneous. Thus, the linearized concentration profile is described through the following dependence:

$$c_{ss}(z) = c_{ss}(z = 0) + \left(\frac{dc_{ss}}{dz}\right)_{z=0} \cdot z \tag{5.17}$$

The perturbation imposed on such initial distribution means the linear combination of the steady-state profile [Eq. (5.17)] with the perturbing component:

$$c = c_{ss} + \delta c \tag{5.18}$$

where the perturbation, characterized with wave numbers k_x and k_y, is defined as:

$$\delta c = \delta c_0(z) \exp\left[i\left(k_x x + k_y y\right) + pt\right] \tag{5.19}$$

Due to an excess of the supporting electrolyte, the resulting concentration c is ruled by the diffusion and convection only, and not by migration, so:

$$\frac{\partial c}{\partial t} + (\boldsymbol{\vartheta} \cdot \nabla)c = D\nabla^2 c \tag{5.20}$$

Concerning the boundary conditions, the hydrodynamic perturbations of the fluids: mercury phase and aqueous solution, due to viscosity, are assumed to decay for $z \to \pm\infty$. Also due to viscosity, there is no slip at the interface. Furthermore, the interface is assumed to remain flat under the infinitesimal fluid motion, due to the relatively high interfacial tension. Finally, the condition of conservation of momentum at the interface must be met.

Based on these assumptions, Aogaki et al. [29] have performed theoretical considerations which led them to the conditions under which infinitesimal perturbations can grow into macroscopic instabilities, giving rise to self-organized convective cells. Among others, the connection between the slope of the electrocapillary curve and the direction of the electrode process (reduction or oxidation) was proposed (see Fig. 5.19). For the electroreduction process, the theory predicts the convective instability within the potential range (II), limited with the equilibrium potential (here equal to zero) and the potential of zero charge, i.e., for the potential range, when the electrode surface is positively charged.

The feedback process involved in these instabilities was identified as the following sequence: (1) the streaming flow of the liquids is induced by the inhomogeneity of the interfacial tension; (2) this flow enhances the transport of the electroactive species near the interface which enhances the inhomogeneities of the interfacial tension; this "autocatalytic" rise of the intensity of convective motion eventually engages the entire interface and adjacent liquids into the macroscopic pattern of self-organized convective cells. Figure 5.20 shows the schematic profile of the flows near the interface.

In complement to classical explanations of the polarographic maxima, the present model shows how important is to consider not only the existence of tangential flows induced by interfacial tension gradients, but also whether such motion can be amplified or damped, in line with the principles of the stability analysis. Of course, the above conclusions should be valid not only for the

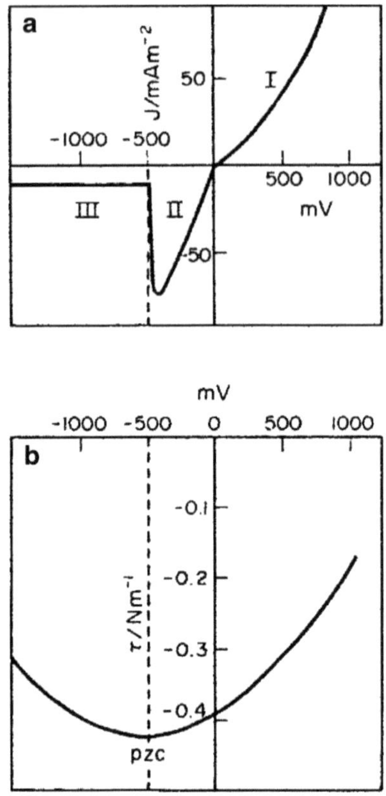

Fig. 5.19 Typical current–voltage curve (**a**) for a deposition–dissolution system of liquid metal and the corresponding electrocapillary curve (**b**); maximum wave appears only in the region II, satisfying the instability condition; $\tau = \sigma$ means surface tension. Reprinted from [28], Copyright 1978, with permission from Elsevier

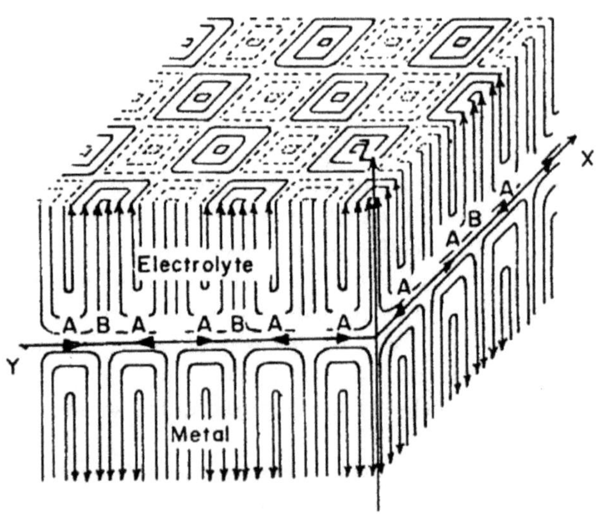

Fig. 5.20 Schematic profile of disturbing flow near the interface with a single wave length: cell patterns drawn by *solid* and *break lines* correspond to the upward and downward flows of the solution, respectively, and the interfacial motion is accelerated at A and retarded at B. Reprinted from [28], Copyright 1985, with permission from Elsevier

Hg–aqueous solution interface, but also, e.g., for the system: liquid lead electrode–fused KCl–LiCl system (at appropriately elevated temperature).

The Hg(II)/Hg electrochemical system discussed earlier occurs at such positive potentials and is characterized with such a high rate of the electron transfer ($k_s \approx 0.02$ cm s^{-1} [33]) that within the potential region important for the onset of hydrodynamic instabilities no kinetic, but only transport limitations affect the current. If these conditions are not met, other factors, such the finite rate of the electron transfer, have to be taken into account. Thus, following the papers by Aogaki et al. [28, 29], Nechiporuk and Elgurt have published [34] a more general theoretical analysis of the Marangoni instabilities associated with the electrodeposition of metals on liquid electrodes. In this model, the planar electrode–electrolyte interface is infinite in both x and y directions, but the thickness of the electrode and the solution layers are finite and equal to h_1 and $-h_2$, respectively, along the z axis. Moreover, external boundaries at these values are assumed to be solid phases. At $z = 0$, i.e., at the liquid electrode surface there occurs the electroreduction of metals ions, controlled by the rate of transport only, and the resulting metal atoms diffuse into the volume of the electrode. A simplifying assumption is that at the opposite boundary always the same, constant concentration of metal ions is kept, as there was an efficient source of decaying metal ions there. Finally, an excess of supporting electrolyte is assumed, leaving only diffusion as the significant transport of the electroactive ions. Based on equations analogous to those used by Aogaki, with appropriate boundary conditions, respective stability conditions were derived and expressed in terms of the appropriately modified Marangoni number. Generally, instability should occur for positive values of this number. It is noteworthy that this approach indicated the role of ohmic potential drops, not taken into account in the previous analysis. In the presence of ohmic drops it may happen that for the electroreduction process occurring even on the negative part of the electrocapillary curve, the Marangoni number still remains positive. Further analysis included the role of the finite rate of the electron transfer, including the potential distribution across the double layer. For galvanostatic conditions it was found, again for the electroreduction process, that the hydrodynamic instabilities occur in the range of the current from zero to the value forcing the polarization of the electrode to the potential of zero charge. In turn, for potentiostatic conditions, increasing ohmic potentials drops was found to be a factor destabilizing the system which should be stable for zero resistance. For given finite value of this resistance, the range of electrode potentials corresponding to instabilities increases with the increase in the thickness of the electrolyte layer and in the concentration of the electroactive ions.

5.4.2.2 The Three-Boundary Electrochemical Convective System

As a more complicated system with mercury electrode engaged in surface-tension driven instabilities, Jehring et al. [35] have described the current oscillations which spontaneously set in for the three-boundary system, in which two different solutions (liquids) with different surface tension remain in contact with the mercury electrode

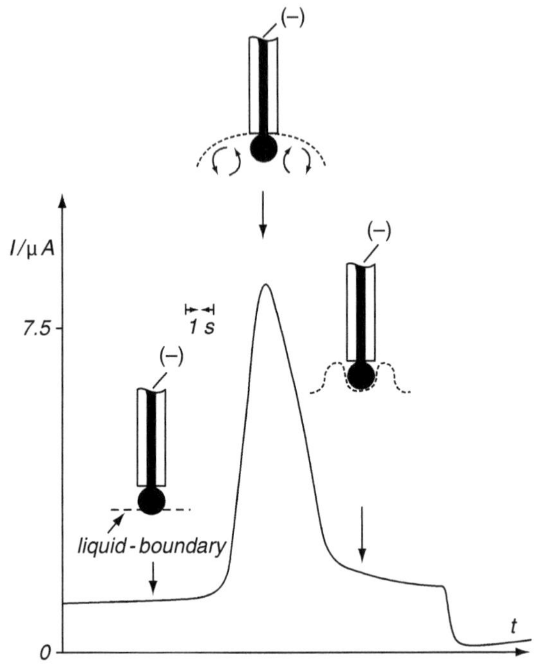

Fig. 5.21 Size and direction of liquid-streaming being a schematic representation of the source of convective oscillations in the three-phase boundary system: liquid-1–liquid-2–mercury electrode, for the case when the upper liquid–Hg phase boundary exhibits higher surface tension than the lower liquid-Hg phase boundary. Reprinted from [35], Copyright 1978, with permission from Elsevier

surface. “Liquid 1” was, e.g., the solution of 2 mg ml^{-1} acetylcholine chloride in 0.1 M NaCl and “liquid 2”—the solution of 6 M NaCl, and various types of mercury electrode (slowly dropping, hanging or sessile) were used. The electrode was externally polarized by the potentiostat working in a two-electrode mode. The oscillations (of the amplitude of the order of 1–20 μA and the period of the order of 0.05–0.1 Hz) started when the mercury surface contacted to the two-liquid boundary and simultaneously the onset of local convection was observed. A bit stochastic nature of these motions was the reason for the limited reproducibility of these oscillations. Schematic in Fig. 5.21 explains the origin of the oscillations for the case when the upper liquid–Hg interface exhibited higher surface tension than the other interface. The tangential motions drive the lower fluid up, until the mercury drop becomes fully surrounded by it. Then the driving force for convection ceases, until the original three phase boundaries are recovered, and the whole cycle then repeats. These processes are associated with the oscillations of the capacitive current, as upon changes of the surrounding liquid, the electrode surface is recharged. After addition of the electroactive species [e.g., thallium(I) ions] the oscillations are amplified, due to the additional contribution of the faradaic current, enhanced by the convective component.

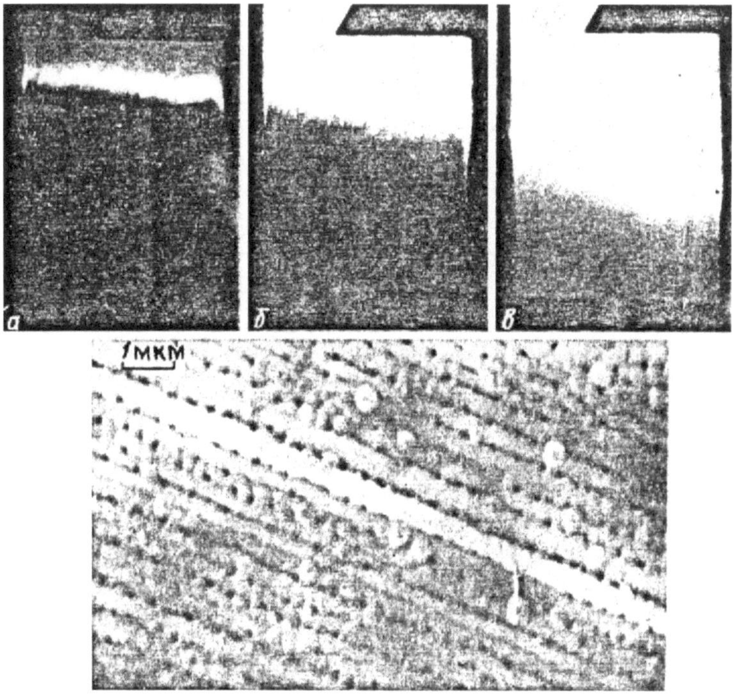

Fig. 5.22 (*Top*): The luminescent waves associated with the galvanostatic electrooxidation of AMg-2 aluminum alloys, in the form of the foil of the thickness 50–100 μm (if the thickness exceeded ca. 200 μm, the waves did not form). A front of 1.5–2 mm width forms at the edge, moves with the velocity 3–5 mm s^{-1} and decays at the opposite border. The next front appears after some time, not necessarily in the same place, or more than one fronts can be born and annihilate each other upon their meeting. It was found that these waves appeared only within the current range from 1.0 to 1.5 mA cm^{-2} and for temperatures below 80 °C. (*Bottom*) The periodic structure of the oxide layer, consisting of pores and cells, observed after passing of eight waves. The characteristic size of the pattern is 450 nm and increases with the potential difference between the working anode and the counter electrode. Reprinted from [36] with kind permission of Nauka Publishers, Moscow

5.4.3 *Instabilities at the Solid Electrode–Solution Interface*

5.4.3.1 Waves and Spatial Patterns in Aluminum Oxidation

The electrochemical instabilities of the Marangoni type are not limited to the liquid electrode–electrolyte interface, but may also occur in the case of the solid electrode. Tschernyshev [36] has observed the wave propagation accompanying the anodic oxidation of aluminum or its alloys in contact with the solution of boric acid in ethylene glycol with an addition of ammonia, under galvanostatic conditions. What is unusual, this process is accompanied with electrochemiluminescence (ECL) which for the AMg-2 alloy takes a form of luminescent waves (Fig. 5.22).

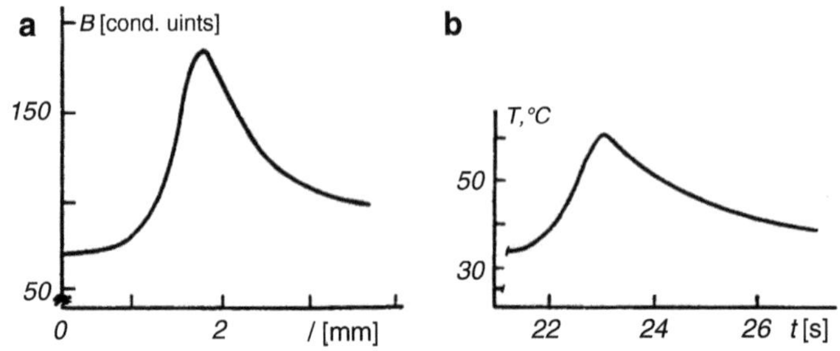

Fig. 5.23 (**a**) The spatial distribution of luminescence across the wave front and (**b**) associated variation of the temperature of a sample as a function of time, during the propagation of the wavefront. Reprinted from [36] with kind permission of Nauka Publishers, Moscow

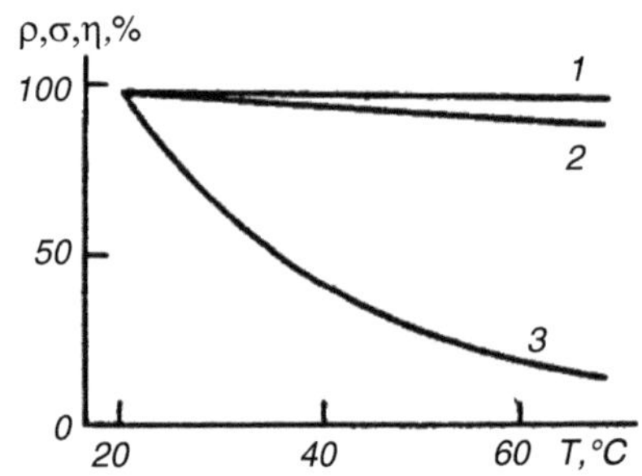

Fig. 5.24 Relative variations of the electrolyte density (1), interfacial tension (2), and viscosity (3) with temperature, for the experimentally applied temperature range. Reprinted from [36] with kind permission of Nauka Publishers, Moscow

The spatial distribution of luminescence and temporal variations of temperature of the metallic foil are shown in Fig. 5.23.

In the opinion of the author, these phenomena are of hydrodynamic nature and are due to the cooperating changes of the fluid density, interfacial tension and viscosity, caused by the variations of temperature, with viscosity exhibiting the most significant sensitivity (Fig. 5.24). Also, the role of high local electric potential, giving rise to electrokinetic transports inside the pores, can be considered. Furthermore, locally enhanced current density can lead to the thermal activation of the anodization process which means the enhanced growth of the oxide layer in these places.

Further examples of convective instabilities of Marangoni type, taking a form of convective cells, were reported and theoretically elaborated by Tenan et al. [37] for the electrooxidation of aluminum [38], and also for iron [39] electrodes, for which invoking of the role of convection in the formation of surface patterns seemed to be earlier not so common. For aluminum, Kleinke, Teschke, and Tenan [38] have described the formation of convective patterns on its surface upon anodic oxidation

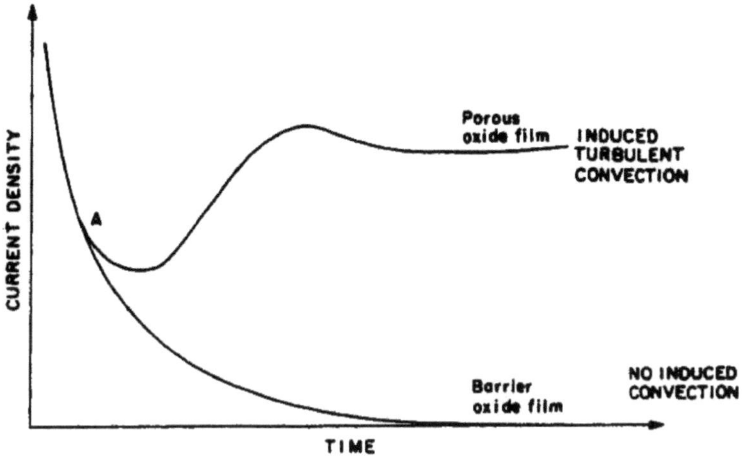

Fig. 5.25 Current density vs. time transient curves for the formation of barrier-type and porous-type anodic films on aluminum. Reproduced from [38] by permission of The Electrochemical Society

to the solutions of sulfuric, chromic, phosphoric, or oxalic acid. Under appropriate conditions the porous oxide layer was formed and it was crucial, for the sharp pattern to record, to remove the electrode from the solution at the first stages of the pore formation process [38]. The fundamental difference in current–time characteristics between the compact and the porous layer is shown in Fig. 5.25.

Contrary to the lower curve, the monotonically decreasing course of which is typical of progressing blocking of the electrode surface through the oxide film, the upper one indicates that after initial and relatively short formation of the compact layer, pores are formed and therefore the current can rise again, presumably with the contribution from convection.

The pore formation was thus reported for the aluminum anode (wire) to which a voltage step (as large as, e.g., 22 V) was applied; an initial sharp increase of the current was followed by its decrease and then again an increase (Fig. 5.26).

The explanation of this shape assumes that in the initially formed compact passivation layer (barrier), initial pores are formed [approximately at point (a)] which thinner the passivation barrier. The typical thickness of the passivation layer was estimated as, e.g., 23 μm, under typical conditions. Its exact composition is not known; one can speculate that initially $Al(OH)_3$ is formed which later transforms into hydrous aluminum oxide, but it is probably a simplification.

Experimentally it was found that well-developed patterns, revealed by scanning electron microscopy, are observed if the electrode is removed at either point (a) or (b) of the *I*–*E* characteristics. Typical patterns are shown in Fig. 5.27.

The cell size of the patterns (varying from ca. 20 to 50 μm for Al oxidized in H_2SO_4, at 25 °C and at 25 V) was found to be a decreasing function of the concentration of the given acid. Other experiments, involving formation of patterns on the cylindrically shaped (pencil-like) Al wire, for which the distribution of

Fig. 5.26 Current density vs. time curve for a 22 V step application to aluminum in 20% H_2SO_4 solution. The patterns were observed after removing the electrode from the solution at either (**a**) or (**b**) points of these characteristics. Reproduced from [38] by permission of The Electrochemical Society

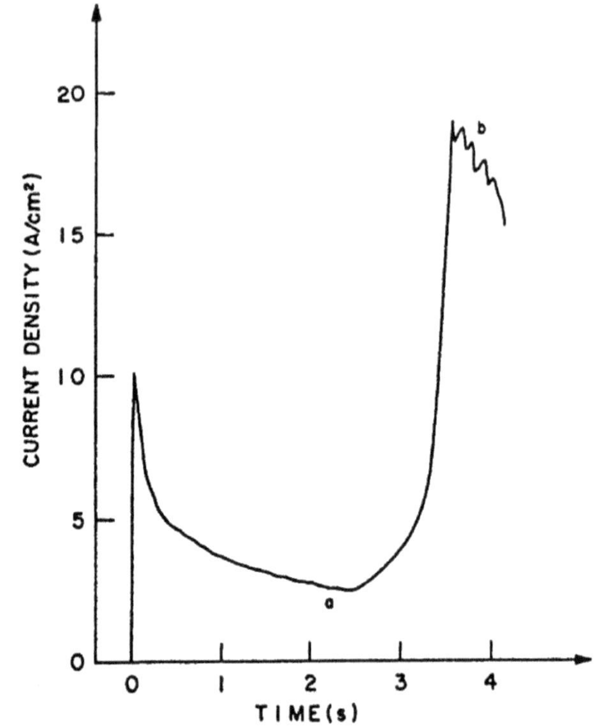

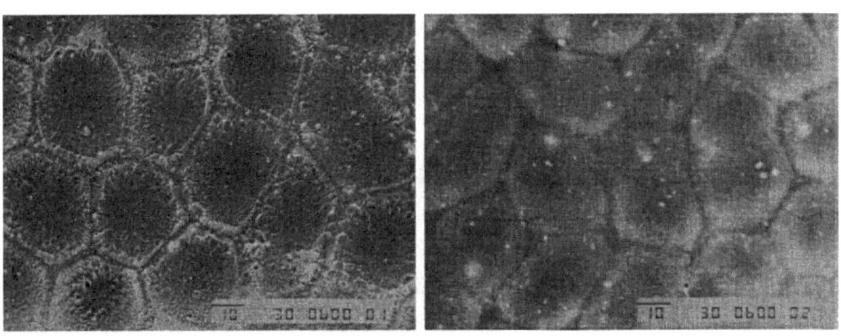

Fig. 5.27 Micrograph showing a patterned structure on an aluminum electrode in 20% H_2SO_4 anodized at 22 V (*left*: secondary electron imaging) and (*right*: backscattering electron imaging). Reproduced from [38] by permission of The Electrochemical Society

current density was inhomogeneous, allowed to find out that the cell size was also a decreasing function of the increasing current density. Furthermore, wettability measurements [38] have led to the estimation of the $\partial\sigma/\partial c$, i.e., of the dependence of the interfacial tension on the solute concentration. The formation of patterns like those shown in Fig. 5.27 was finally ascribed to the solute concentration inhomogeneities at the film–solution interface, causing local variations in the

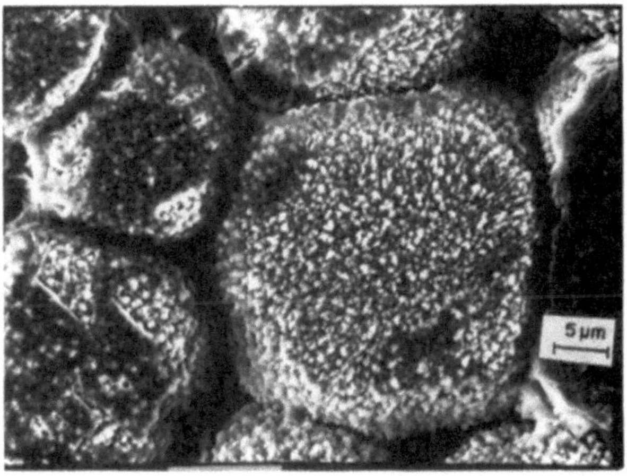

Fig. 5.28 SEM photograph of a pattern observed for a polished iron electrode, in 1 M H_2SO_4, switched from −650 to −100 mV vs. SCE. Reprinted with permission from [39]. Copyright 1990 American Chemical Society

film–solution interfacial tension. Weak capillary forces induce then the electrolyte convection in the vicinity of the electrode surface and determine the patterns which appear when a more viscous layer is formed over the electrode surface. In more detail, the driving force for the formation of the patterns is the combination of the electric current density and the gradient $(\partial\sigma/\partial c)$, included in the dimensionless Marangoni number (see below).

5.4.3.2 Patterns in Iron Oxidation

Teschke, Tenan et al. [39] have described pattern formation on the iron electrode in contact with 1 M or 0.5 M H_2SO_4 solution, during its anodization to potentials corresponding either to complete or partial passivation, including the transient range of the oscillations of current (Figs. 5.28–5.30). These patterns should be compared with the current–potential characteristics shown in Fig. 5.31, in order to recognize the passivation regions.

Since these patterns were associated with visible intensive convective motion of the iron(II) sulfate layer growing in the solution adjacent to the iron surface, it was assumed that gradients of interfacial tension appearing in the liquid iron(II) sulfate film are a direct cause for that phenomenon. These gradients are in turn caused by the microscopic fluctuations in the local current distribution which induce the inhomogeneous concentration distribution of iron(II) sulfate. The characteristic size of the convective cells is assumed to depend on the distance along which the concentration gradient of $FeSO_4$ develops. Analogous mechanism was proposed for the formation of other patterns in the same system, at the edges of the electrode, when the local nonuniformities in current/concentration distribution are even higher.

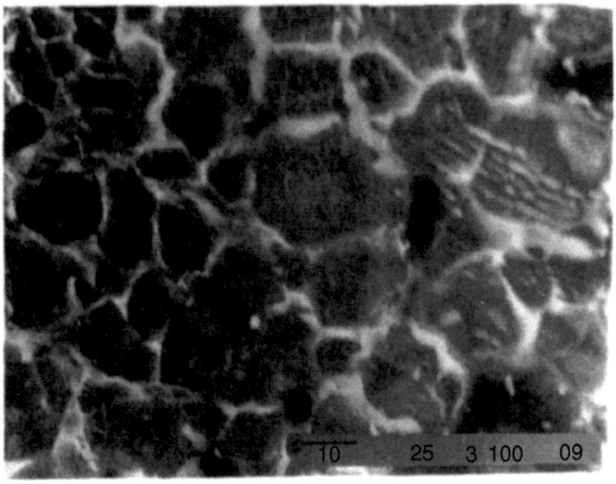

Fig. 5.29 SEM micrograph observed for an iron electrode in 1 M H_2SO_4, switched from −650 to +250 mV vs. SCE after one current spike. It is possible to observe the iron electrode grain structure below the iron sulfate pattern. Reprinted with permission from [39]. Copyright 1990 American Chemical Society

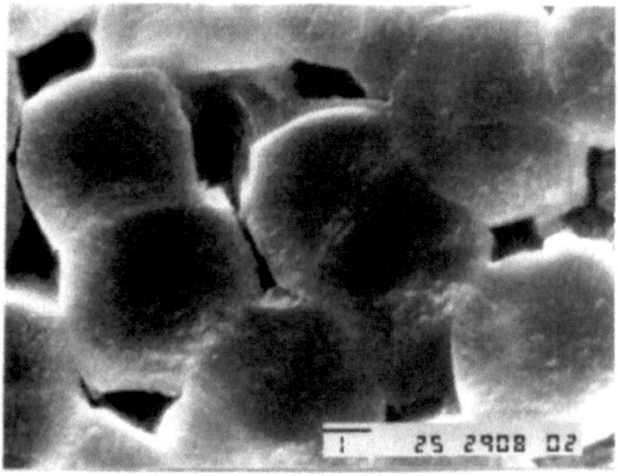

Fig. 5.30 Micrograph observed for an iron electrode in 1 M H_2SO_4 switched from −650 mV to +500 mV vs. SCE. Reprinted with permission from [39]. Copyright 1990 American Chemical Society

5.4.3.3 The Pore Formation Theory

In order to explain the experimentally observed patterns, as shown above in this section, Tenan et al. [37] have elaborated the theory which will be briefly outlined below. The initial course of the electrochemical process is quite simple: the bare surface of a metal electrode, externally polarized to sufficient anodic potential in

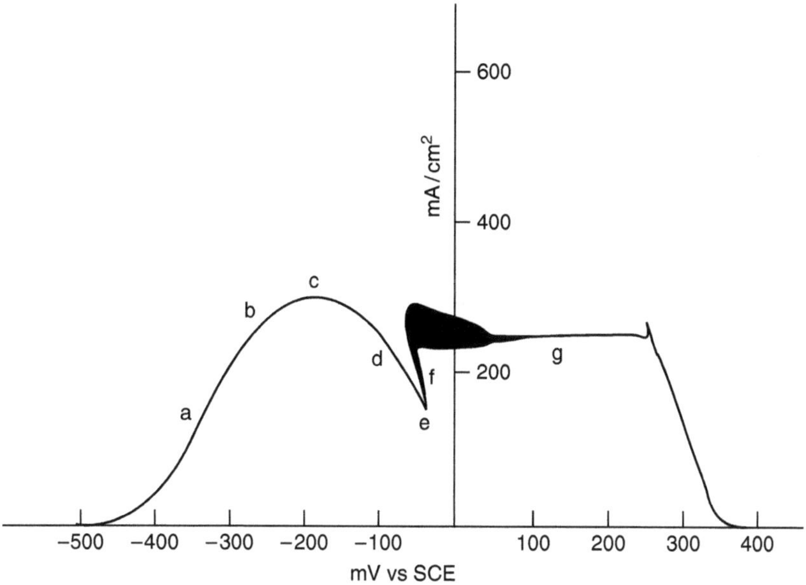

Fig. 5.31 Typical current vs. potential (in mV vs. SCE) curve obtained with a polarization device which transfers a negative resistance, for a polished iron electrode in 1 M H_2SO_4 (scan rate 5 mV/s). Reprinted with permission from [39]. Copyright 1990 American Chemical Society

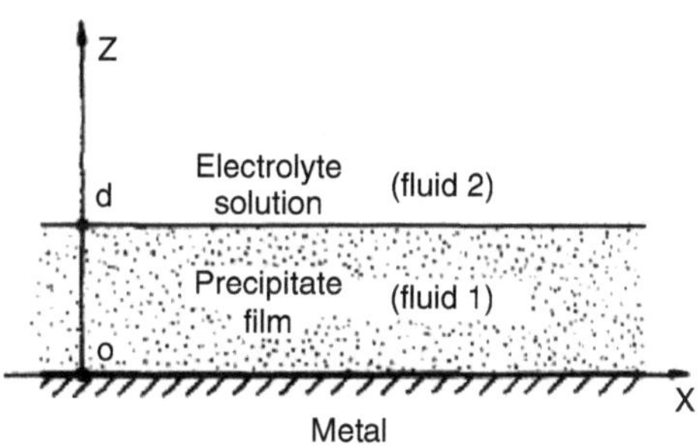

Fig. 5.32 The model scheme of the two-fluid system configuration used in the pore formation theory. Reprinted with permission from [37]. Copyright 1990 American Chemical Society

contact with the acid solution, undergoes oxidation. As a consequence, the concentration of metal ions in the electrolyte increases and when it exceeds an appropriate supersaturation value, the salt/hydroxo salt/hydrous oxide precipitates which forms a film adhering to the metal surface (Fig. 5.32).

Further dissolution of metal electrode is accompanied with diffusion of ions through this film and the adjacent solution of the electrolyte. The role of ionic migration is considered negligible [37]. In the absence of any other effects, it is expected that under galvanostatic conditions the steady state will eventually develop in this system.

According to the proposed mechanism, it is further assumed that the fluctuations in the concentration distribution may cause inhomogeneities of the interfacial tension at both interfaces, of which the fluid 1–fluid 2 interface may be set into convective motion, if counteracting viscous forces will not damp it. This means that the model should prove that under given conditions these (always present in real systems) fluctuations will be either enhanced or damped, proportionally to $\exp(\beta t)$ term, in dependence on the positive or negative sign of the (real part of) β coefficient. The full model means thus the construction of the dynamical, electrochemical diffusion-convective system, the stability of which will be analyzed with the linear stability theory. The mathematical treatment involves: (1) the Navier–Stokes equation (5.1), in which the driving force for convection neglects gravitational effects on the fluid flow, (2) the diffusion equation (the Fick's law with added convection), and also (3) the continuity Eq. (5.2). The reader interested in the details of the mathematical treatment of the problem is referred to the original paper [37], while below only most important conclusions will be collected.

The conditions for the convective instability of the model system from Fig. 5.32 can be expressed in terms of the appropriately defined Marangoni number (denoted here as N_{Ma}) which shows the balance between the driving force for convection, proportional to the surface tension gradient, and the dissipative counteraction due to viscosity:

$$N_{\mathrm{Ma}} = \frac{i_{\mathrm{a}}}{nF} \frac{\partial \sigma}{\partial c_2} \frac{d^2}{D_2^2 \eta_2} \tag{5.21}$$

In this definition of Ma, i_{a} means the constant anodic current density, σ is the interfacial tension at the fluid 1–fluid 2 interface, while c and D mean the concentration and the diffusion coefficients of metal ions, respectively, in fluid 2 of dynamic viscosity η_2. Determination of the stability of the given system requires the knowledge of viscosities η and densities ρ of both fluid 1 and fluid 2, as well as of diffusion coefficients D of metal ions in both phases. The following exemplary ratios were used: η_1/η_2 varying from 1 to 10 and D_1/D_2 varying from 0.1 to 1.2, while $\rho_1/\rho_2 = 2$. Note that the gradient $\partial\sigma/\partial c_2$ can be either positive or negative, both sufficiently positive and negative values meaning the onset of convection, with the region of stability (quiescent fluids) placed between these borders. Figure 5.33 shows these regions in the coordinate system: Marangoni number vs. the product of the thickness of the precipitate layer (d) and the inverse of the perturbation length α.

In turn, Fig. 5.34 shows schematically the direction of flows forming the convective cells for both signs of $\partial\sigma/\partial c_2$ derivatives.

The direction of flows depends on that whether the interfacial tension in place A is greater or lower than in place C. Based on this scheme one could suppose that the direction of flow is the only difference in the dynamics, caused by the reversion of the sign of $\partial\sigma/\partial c_2$. However, the conditions of stability for positive and negative concentration gradient are different. Let us come back to Fig. 5.33 which shows that for positive Marangoni numbers the system becomes unstable with respect to perturbations of relatively great lengths (i.e., for $\alpha d < 1$), while for negative Marangoni numbers—instability can appear for perturbations of relatively small

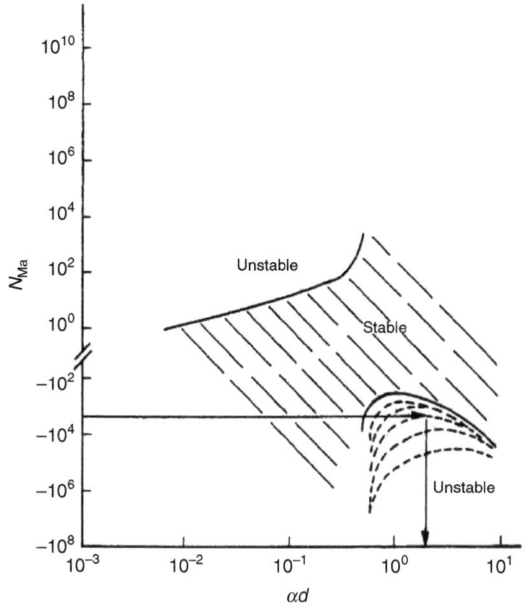

Fig. 5.33 Plot of the Marangoni number N_{Ma} (driving force/dissipative force) as a function of the product of the *inverse* of the perturbation length α, with the film thickness d. *Solid lines* are typically neutral stability curves that separate the stable from the unstable regions. *Dashed lines* represent the solutions of the characteristic equation for real positive values of the growth constant β. The *arrows* indicate schematically the procedure for obtaining the perturbation length of the fastest growing perturbation from a given Marangoni number (see [37]) for details. Reprinted with permission from [37]. Copyright 1990 American Chemical Society

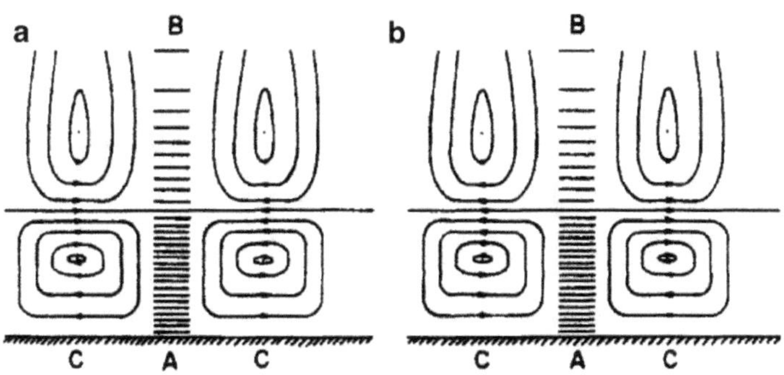

Fig. 5.34 Flow patterns (schematic) initiated in a region where the concentration has increased (column AB). Flow pattern for (**a**) $\partial\sigma/\partial c_2>0$ and (**b**) $\partial\sigma/\partial c_2<0$. Reprinted with permission from [37]. Copyright 1990 American Chemical Society

lengths (i.e., for $\alpha d > 1$). Further analysis of the effects of varying parameter values shows that also the physical sources of stability are different: in systems with positive $\partial\sigma/\partial c_2$ the stability is achieved due to low interfacial tension at the fluid 1–fluid 2 interface, while for negative $\partial\sigma/\partial c_2$ the stability is caused by relatively large surface viscous dissipation.

Fig. 5.35 Typical transverse cross-section of the perturbated concentration profile for a hexagonal pattern, calculated at a distance $z = 0.9d$ from the electrode surface (cf. Fig. 5.33). Reproduced from [38] by permission of The Electrochemical Society

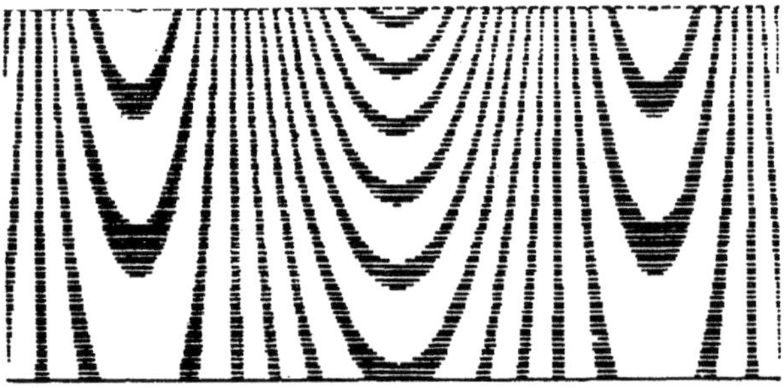

Fig. 5.36 Typical longitudinal cross-section of the perturbated concentration profile for a hexagonal pattern. Reproduced from [38] by permission of The Electrochemical Society

Based on the Marangoni number it is further possible to predict the length of the dominant perturbation λ, meaning the perturbation for which the growth rate is the greatest [i.e., for which β in $\exp(\beta t)$ term has the largest real positive value]. In Fig. 5.33 dashed lines indicate solutions for the real positive values of β, from which one determines the corresponding value of αd; assuming further that the perturbation length is defined as the hexagon diagonal, one can derive that $\lambda/d = 8\pi/(3\alpha d)$ and plot λ/d as a function of the Marangoni number.

Schematic representation of the concentration profiles, as well as of the fluid velocity field, corresponding to the hexagonal convective patterns, is shown in Figs. 5.35–5.37.

Fig. 5.37 Typical transverse cross-section of the fluid velocity inside the formed film. Reproduced from [38] by permission of The Electrochemical Society

Such a velocity distribution of the fluid means that it will drag ions and other particles with different velocities. This may cause local aggregation of precipitate particles and their enhanced deposition on the surface of the solid electrode, which attains morphology revealing the geometry of convective flows, visible even after emerging from the solution for the purpose of the electronic microscopic analysis.

In appropriate experiments it appeared difficult to observe such ideally regular hexagonal (or square) patterns, like those predicted by the theory, since the real pictures exhibited some random irregularities. Nevertheless, the basic predictions of the model appear to be confirmed. Taking as an example the iron–1 M H_2SO_4 system, a comparison of the experiment and theory was performed [37]. For that purpose the solution density (ρ_2 = 1.26 g cm^{-3}) and viscosity (η_2 = 2.7 cP), the iron(II) sulfate solubility ($c_{sat} = 1.5 \times 10^{-3}$ mol cm^{-3}), and the diffusion coefficient of Fe^{2+} ions $D_2 = 2.5 \times 10^{-6}$ cm^2 s^{-1} were used. The supersaturation ratio $s = c_{precipitation}/c_{sat}$ was taken as equal to 2.2, and the thickness of the precipitate layer (generally dependent on the current density at the anode i_a) was assumed as equal to $d = 8.3$ μm, for $i_a = 500$ mA cm^{-2}. Furthermore, based on wettability data, the $\partial\sigma/\partial c_2$ was found as equal to -1.3×10^4 dyn cm^2 mol^{-1}. Other parameters remained unknown, so they had to be assumed; among them: average film–electrolyte interfacial tension $\sigma^0 = 10$ dyn cm^{-1}, the fluid density ratio $\rho_1/\rho_2 = 2$, diffusion coefficient ratio $D_1/D_2 = 0.05$, and viscosity ratio $\eta_1/\eta_2 = 10$. Calculations led to the result that the fastest growing perturbation in the case of hexagonal cell patterns would have a perturbation length (i.e., the hexagon diagonal) ranging from 15 to 32 μm, in dependence on surface dilational and shear viscosities, the latter value being very close to the experimentally determined cell diagonal $\lambda_{exp} = 34$ μm.

The reader interested in other type of patterns, formed due to interplay of the metal anodic dissolution and convection caused by the rotation of the iron (stainless steel) disk electrode, will find their brief characteristics in Sect. 5.10.

5.5 Bistability Caused by Potential-Dependent Convection

5.5.1 Experimental Realization

Apart from the pattern formation on the mercury surface during the Hg^{2+} electrolysis, the Hg(II)/Hg electrolytic system described in Sect. 5.4.2 [28–30] is quite interesting also because it exhibits an unusual type of the NDR region in the *I–E* characteristics, i.e., this region caused by the potential dependence of the convection rate. Such a source of N-NDR region was not mentioned in Chapter 2 of volume I, where only the decrease of the electron transfer rate due to three effects was considered: the decrease of the surface concentration of the reactant due to double-layer effect, the decrease of the surface concentration of adsorbed electrocatalyst, the increase of surface concentration of an inhibitor, including the passivation effect.

Experiments oriented on the studies of instabilities caused by the convection-driven N-NDR region were recently performed by Gorzkowski et al. [32] in the electrolytic cell shown in Fig. 5.38, containing mercury both as a cathodic and anodic material. Such construction allowed to keep the composition of the electrolyzed solution at the quasi steady-state level even for prolonged electrolytic times, since the amount of Hg(II) consumed at the cathode was balanced by the equal amount of Hg(II) ions produced as a result of Hg anode dissolution.

In this system, within a large potential range of the electrode reaction: Hg(II) + 2e → Hg, the faradaic current is largely controlled by the self-induced convective inflow of Hg(II) ions from the solution bulk to the electrode surface. The rate of this convection is in turn strongly dependent on the imposed electrode potential: it passes through the maximum, and when it decreases, the region of the N-NDR develops on the current–potential characteristics (Fig. 5.39) (cf. also [30, 40]).

As for other N-NDR systems, the destabilization of the steady states belonging to the negative slope of the *I–E* curve requires appropriate ohmic potential drops, realized by inserting the adjustable external ohmic resistance in the electric circuit of the working electrode. It is noteworthy that systematic experimental studies for varying serial resistance R_s and externally applied voltage U treated as the bifurcation parameters have revealed only bistability (Fig. 5.40), but no sustained oscillations of the faradaic current were reported.

5.5.2 Linear Stability Analysis

One can pose a question, whether oscillations are indeed impossible in this system, or do they occur, but in a such tiny bifurcation parameter region that it is very difficult to find them experimentally. Intuitively, one can suppose that the

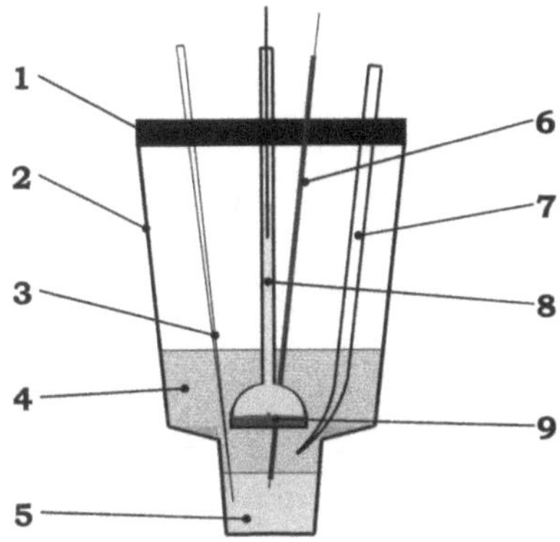

Fig. 5.38 Scheme of the electrolytic cell ("the electrochemical flow reactor") designed for recording of bistability and oscillations accompanying the electrode processes of mercury ions at the mercury electrode (1) teflon leakproof cover of the cell; (2) electrolytic cell; (3) movable argon inlet, allowing passing of the gas either through the mercury layer (5), or the studied solution (4), or above its surface; (5) mercury layer of an upper surface area 3.30 $\pm$ 0.15 cm^2 (in dependence of the layer thickness); (6) steel contact to Hg (in glass rod); (7) salt-bridge (filled up with 0.1 M HNO_3) connected to the reference electrode, (8) counter electrode composed of mercury layer in the tube, separated from the solution through the sintered glass (Schott funnel). Reprinted from [32], Copyright 2008, with permission from Elsevier

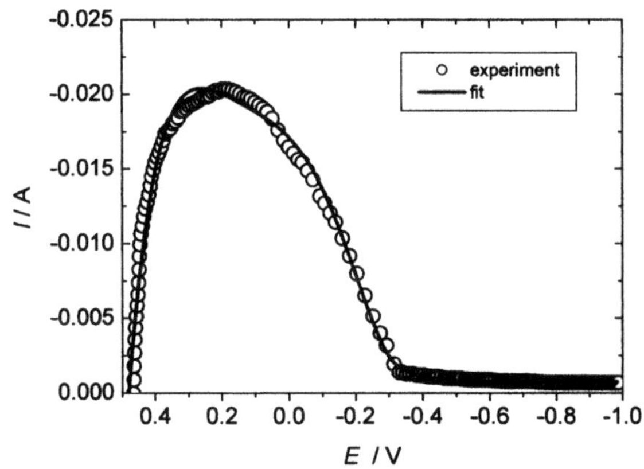

Fig. 5.39 Experimental (open circle) and fitted (line) quasi-steady state *I–E* characteristics of the Hg(II) electroreduction at the flat Hg electrode (A = 3.3 cm^2) under conditions when the convection sets in at positive potentials and passes through the maximum as a function of increasing negative electrode potential. Original composition of the solution: 7 mM $Hg(NO_3)_2$ + 0.1 M HNO_3. The experimental dependence was corrected for ohmic potential drops (R_u = 10 Ω). Reprinted From [40], Copyright 2008, with permission from Elsevier

oscillations are not possible, since the origin of NDR is only the potential dependence of the transport rate with the surface concentration of the reactant fixed at zero value within the entire NDR region. The surface concentration is thus not a variable which could (relatively slowly) drive the system away from the steady state. Similarly, every voltammetric peak, recorded at, e.g., the hanging or static Hg electrode exhibits the region of negative resistance (following the peak value), but it

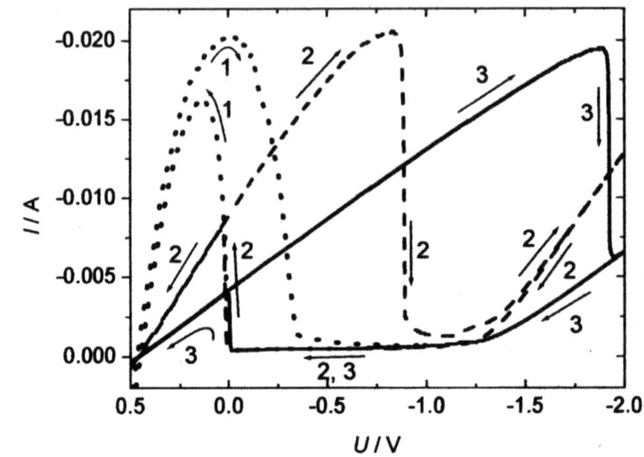

Fig. 5.40 Hysteresis (bistability) in the I–U characteristics upon the cyclic changes (100 mV s^{-1}) of the external voltage U during the electroreduction of Hg(II) at the flat Hg electrode, for different total serial resistances of the circuit (1) $R_u = 10\ \Omega$, $R_s = 0$ (no externally applied resistance), (2) $(R_u + R_s) = 50\ \Omega$, (3) $(R_u + R_s) = 100\ \Omega$. Curve (1) shows a distinct NDR region which in curves (2, 3) turns into the bistable (hysteresis) regions, indicating the destabilization of the steady states from curve (1) through the saddle–node bifurcations. High currents correspond to self-induced convection, whereas low currents correspond to quiescent mercury surface, thus it is the bistability between the convection and the convection-less regime. Reprinted from [32], Copyright 2008, with permission from Elsevier

is again the negative slope caused by the decrease of the diffusion rate with time, associated with linearly changing potential, with the surface concentration of the reactive species being then fixed again at a practically constant, almost zero value. In other words, when the N-NDR comes from the characteristics of the transport, and not of the electron-transfer interfacial kinetics, with the concentration at electrode surface fixed, there is only one dynamical variable operating—the interfacial potential drop E. The (continuous) system with one degree of freedom is not able to oscillate. The above intuitive explanation will now be confirmed in terms of stability analysis which should thus indicate the saddle–node bifurcations, leading to bistability and exclude the Hopf bifurcations, leading to the oscillations. The model dynamical system was constructed according to the following assumptions:

1. The faradaic current of the Hg(II) electroreduction is approximated in terms of the concept of the linear (Nernst) diffusion layer of a thickness δ:

$$I_f(E) = -nFAD_{ox}\left(\frac{\partial c_{ox}}{\partial x}\right)_0 \approx -nFAD_{ox}\left(\frac{c_{ox}^0 - c_{ox}(0,t)}{\delta}\right)_0 \tag{5.22}$$

with n—number of electrons exchanged, F—Faraday constant, A—electrode surface area, D_{ox}—diffusion coefficient of the reducible Ox (i.e., Hg(II)) species, with its concentration equal to c_{ox}^0 in the solution bulk and equal to $c_{ox}(0, t)$ at the electrode surface.

2. As the electrode surface area A is constant and the electrode material does not undergo refreshment (as it would be the case for the streaming Hg electrode), the capacitive current flows only if the electrode undergoes recharging:

$$I_c = C_d A \left(\frac{dE}{dt} \right) \tag{5.23}$$

where C_d is the differential capacity of the Hg electrode (per unit area) in contact with the studied solution, for simplification assumed further as being a constant (average) value.

3. Since the studied potential range of the Hg(II) electroreduction was largely more negative than the formal potential of the Hg(II)/Hg couple, the concentration of Hg(II) at the electrode surface was set to zero: $c_{ox}(0, t) \approx 0$. In consequence, Eq. (5.22) simplifies to the form:

$$I_f(E) \approx -nFAD_{ox} \frac{c_{ox}^0}{\delta} \tag{5.24}$$

   where the minus sign indicates the cathodic current.

4. In terms of Eq. (5.24), the variations of the faradaic current as a function of the electrode potential, like $I = f(E)$ shown in Fig. 5.39, can be explained in terms of variation of only the Nernst diffusion layer δ with E, according to schematic in Fig. 5.41.
5. Since for the generation of the oscillations at least two first-order differential equations are necessary, the effective electrode potential E and the thickness of the diffusion layer δ were chosen as the corresponding *two dynamic variables*. Based on the charge conservation principle and a quantitative balance of a transport of Hg(II) in the diffusion layer, the following equations describing the dynamics of the electrochemical systems were derived [40]:

$$\frac{dE}{dt} = \frac{U - E}{C_d A R_s} + \frac{nFD_{ox} c_{ox}^0}{C_d \delta} \equiv f(E, \delta) \tag{5.25}$$

$$\frac{d\delta}{dt} = \frac{2D_{ox}}{\delta} - 2v(E) \equiv g(E, \delta) \tag{5.26}$$

6. The steady-state current means the steady-state thickness of the Nernst diffusion layer; this is achieved when the diffusion flow of the Hg(II) ions toward the electrode surface is exactly balanced by the convective flow of Hg(II) ions from the solution bulk, i.e., if the diffusion and convective fluxes are equal: $D_{ox}(\partial c_{ox}/\partial x)_{ss} = c_{ox} v_{ss}$. In terms of the present model, this condition simplifies to:

$$D_{ox} \left(\frac{c_{ox}^0}{\delta} \right)_{ss} = c_{ox}^0 v_{ss} \tag{5.27}$$

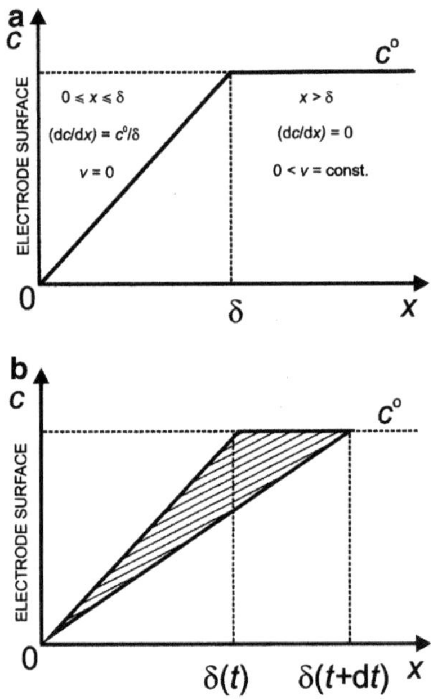

Fig. 5.41 (**a**) The linearized concentration profile for the conditions where convection of the velocity v affects the thickness of the Nernst diffusion layer (δ) and thus the faradaic current, with zero surface concentration of Ox species imposed [$c_{ox}(x = 0, t) = 0$]. For simplicity the component of the fluid velocity, normal to the electrode surface, was assumed constant throughout the solution outside the Nernst layer; (**b**) schematic picture of the variations of the thickness of the linearized Nernst diffusion layer, explaining the variation of the current I. Reprinted from [40], Copyright 2008, with permission from Elsevier

Based on Eqs. (5.27) and (5.24) the net convection velocity can be calculated from the value of the steady-state current at a given potential:

$$v_{ss}(E) = \frac{-I_{f,ss}(E)}{nFAc^0_{ox}} \tag{5.28}$$

with the expression for the corresponding steady-state thickness of the Nernst diffusion layer:

$$\delta_{ss}(E) = \frac{D_{ox}}{v_{ss}} = \frac{-nFAD_{ox}c^0_{ox}}{I_{f,ss}(E)} \tag{5.29}$$

Under such assumptions, the stability analysis was performed [40] for Eqs. (5.25) and (5.26) linearized around the steady state:

$$\begin{aligned}\frac{\mathrm{d}E}{\mathrm{d}t} &= \left(\frac{\partial f}{\partial E}\right)_{\mathrm{ss}}(E - E_{\mathrm{ss}}) + \left(\frac{\partial f}{\partial \delta}\right)_{\mathrm{ss}}(\delta - \delta_{\mathrm{ss}}) \\ &= \left(\frac{-1}{C_{\mathrm{d}}AR_{\mathrm{s}}}\right)_{\mathrm{ss}}(E - E_{\mathrm{ss}}) - \left(\frac{nFD_{\mathrm{ox}}c_{\mathrm{ox}}^{0}}{C_{\mathrm{d}}\delta^{2}}\right)_{\mathrm{ss}}(\delta - \delta_{\mathrm{ss}})\end{aligned} \tag{5.30}$$

$$\begin{aligned}\frac{\mathrm{d}\delta}{\mathrm{d}t} &= \left(\frac{\partial g}{\partial E}\right)_{\mathrm{ss}}(E - E_{\mathrm{ss}}) + \left(\frac{\partial g}{\partial \delta}\right)_{\mathrm{ss}}(\delta - \delta_{\mathrm{ss}}) \\ &= -2\left(\frac{dv}{dE}\right)_{\mathrm{ss}}(E - E_{\mathrm{ss}}) - \left(\frac{2D_{\mathrm{ox}}}{\delta^{2}}\right)_{\mathrm{ss}}(\delta - \delta_{\mathrm{ss}})\end{aligned} \tag{5.31}$$

which generated the corresponding Jacobian matrix:

$$\mathbf{J} = \begin{bmatrix} \left(\frac{\partial f}{\partial E}\right)_{\mathrm{ss}} & \left(\frac{\partial f}{\partial \delta}\right)_{\mathrm{ss}} \\ \left(\frac{\partial g}{\partial E}\right)_{\mathrm{ss}} & \left(\frac{\partial g}{\partial \delta}\right)_{\mathrm{ss}} \end{bmatrix} = \begin{bmatrix} \frac{-1}{C_{\mathrm{d}}AR_{\mathrm{s}}} & -\left(\frac{nFDc_{\mathrm{ox}}^{0}}{C_{\mathrm{d}}\delta^{2}}\right)_{\mathrm{ss}} \\ -2\left(\frac{\mathrm{d}v}{\mathrm{d}E}\right)_{\mathrm{ss}} & -\left(\frac{2D_{\mathrm{ox}}}{\delta^{2}}\right)_{\mathrm{ss}} \end{bmatrix} \tag{5.32}$$

The crucial for the diagnosis of bifurcations, the trace and the determinant of **J** are given by:

$$\mathrm{Tr}(\mathbf{J}) = \left[\frac{-1}{C_{\mathrm{d}}AR_{\mathrm{s}}} - \left(\frac{2D_{\mathrm{ox}}}{\delta^{2}}\right)_{\mathrm{ss}}\right] \tag{5.33}$$

$$\begin{aligned}\mathrm{Det}(\mathbf{J}) &= \left[\frac{1}{C_{\mathrm{d}}AR_{\mathrm{s}}}\right]\left[\frac{2D_{\mathrm{ox}}}{\delta^{2}}\right]_{\mathrm{ss}} - 2\left(\frac{nFD_{\mathrm{ox}}c_{\mathrm{ox}}^{0}}{C_{\mathrm{d}}\delta^{2}}\right)_{\mathrm{ss}}\left(\frac{\mathrm{d}v}{\mathrm{d}E}\right)_{\mathrm{ss}} \\ &= \left[\frac{1}{R_{\mathrm{s}}} + \left(\frac{\mathrm{d}I}{\mathrm{d}E}\right)_{\mathrm{ss}}\right]\left[\frac{2D_{\mathrm{ox}}}{C_{\mathrm{d}}A\delta^{2}}\right]_{\mathrm{ss}}\end{aligned} \tag{5.34}$$

As indicated in Chap. 2 of volume I, the saddle–node bifurcation occurs when the Det changes its sign, i.e., for $\mathrm{Det}(\mathbf{J}) = 0$. The Hopf bifurcation requires that the conditions: $\mathrm{Tr}(\mathbf{J}) = 0$ with $\mathrm{Det}(\mathbf{J}) > 0$ are met [let us remember that for $\mathrm{Tr}(\mathbf{J}) < 0$ all perturbations would exponentially decay, while for $\mathrm{Tr}(\mathbf{J}) > 0$ they would exponentially increase, which evolution in nonlinear systems would end up with a limit cycle, corresponding to sustained oscillations].

From Eq. (5.33) it follows that in terms of our simple model the condition $\mathrm{Tr}(\mathbf{J}) = 0$ is *never met* in this case, since $\mathrm{Tr}(\mathbf{J})$ is always negative, so all perturbations of the steady state are damped. Thus, the oscillations originating through the Hopf bifurcation are *not* possible. The latter conclusion is in line with the experimentally observed lack of oscillations and above intuitive predictions. Formally, this situation occurs because the a_{11} element of the Jacobi matrix (5.32), being the $\partial/\partial E[\mathrm{d}E/\mathrm{d}t]$ expression, is always negative, so the electrode potential E in this system is *not* an autocatalytic variable, contrary to the systems in which the current is controlled not

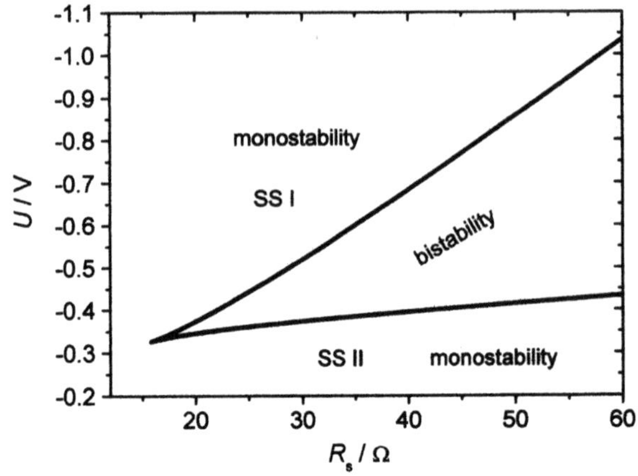

Fig. 5.42 Bifurcation diagram of Eqs. (5.25) and (5.26) showing the points of saddle–node bifurcations for the range of (U, R_s) parameters used in the experiments. There are no points of the Hopf bifurcations present. Reprinted from [40], Copyright 2008, with permission from Elsevier

only by the rate of transport, but also by rate of the electron-transfer step at the electrode–solution interface (see 2.2.4, volume I).

On the contrary, the determinant may change its sign if the condition $1/R_s + (dI/dE)_{ss} = 0$ is met, and if this sum is negative, we enter the region of existence of saddle points. Since the serial resistance R_s can only be positive, the slope $(dI/dE)_{ss}$ must be negative, i.e., the saddle–node bifurcation occurs for the states belonging to the N-NDR region of the I–E characteristics shown in Fig. 5.47. In this way, we derive (and confirm) the well-known condition for the saddle–node bifurcation and resulting bistability in the one-dimensional dynamical system (see Sect. 2.2.1.1, volume I):

$$R_s \geq -\left(\frac{dE}{dI}\right)_{ss} \tag{5.35}$$

The theoretical bifurcation diagram constructed from the condition $\mathrm{Det}(\mathbf{J}) = 0$ is shown in Fig. 5.42, while Fig. 5.43 shows the exemplary nullclines of Eqs. (5.25) and (5.26), corresponding to the bistable behavior.

Thus, starting from the two-dimensional model we end up with in fact the one-dimensional case. Thereby, the electrochemical cell from Fig. 5.38 could be replaced with, e.g., the tunnel diode of the same I–E characteristics and the resulting bifurcation diagram would be also the same as in Fig. 5.42. For the sake of certain generalization, one can indicate that the system described in this section is a rare example of the N-NDR system, the dynamical characteristics of which do not allow oscillatory behavior through the Hopf bifurcation. In the literature [41] there was discussed also another, more complicated mechanism of the same dynamical features, but so far it seems to remain a hypothetical model (see Sect. 3.6.2, volume I).

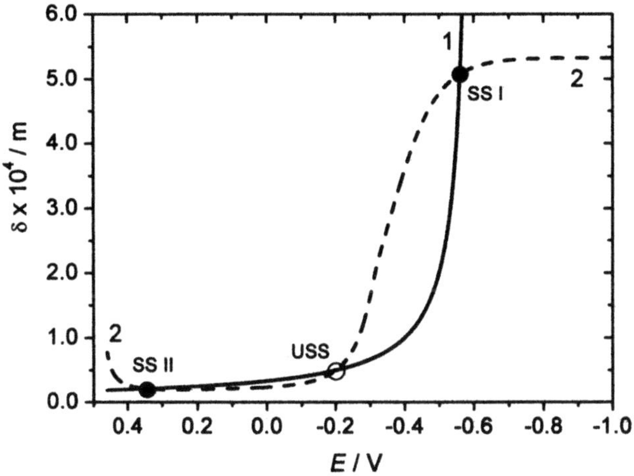

Fig. 5.43 Nullclines of Eqs. (5.25) and (5.26), corresponding to the condition $dE/dt = 0$ (curve 1) and $d\delta/dt = 0$ (curve 2) for $U = -0.6$ V and $R_s = 50\ \Omega$ (bistable regime). *Filled circles* denote the stable steady states (SS I, SS II), while the *hollow circle*—the unstable steady state (USS). Reprinted from [40], Copyright 2008, with permission from Elsevier

Finally, one should also note that if the Hg(I) solution in the experimental setup from Fig. 5.38 is enriched with sodium chloride, so that the calomel layer is formed, the system turns to the galvanostatic oscillator [32]. The oscillations of the electrode potential are then associated with periodical bursts of convection that disturbs the calomel layer which in quiescent states cover the mercury surface.

5.6 Self-Induced Convection in the Processes at Hg Electrodes in Nonaqueous Media

The electroreduction of Hg ions at Hg electrodes, described in previous sections, is obviously not the only process accompanied with a self-induced convection of the mercury electrode surface, as the polarographic maxima were found as phenomena accompanying many electrode reactions [27, 42], under appropriate experimental conditions. Also, such convective instabilities are not reserved for only aqueous media. In this section, we describe the convection-induced current oscillations recorded at the hanging mercury drop electrodes (HMDE). These oscillations will be further abbreviated as CVCAO, meaning the "Cyclic Voltammetric Anodic Current Oscillations" [43]. The reactants involved in the electrode processes are molecular oxygen, nitrobenzene (NB), 1,4-dinitrobenzene (DNB), benzoquinone (BQ), 2,3,5,6-tetramethylbenzoquinone (TMBQ), benzophenone (BP), azobenzene (AB), 2,1,3-benzothiadiazole (BTD), 7,7,8,8-tetracyanoquinodimethane (TMBQ), methyl viologen dichloride (MV^{2+}), and tris(2,2′-bipirydine)ruthenium(II) dichloride ($Ru(bpy)_3{}^{2+}$) in dimethyl sulfoxide (DMSO) solutions containing 0.1 M tetraethylammonium perchlorate (TEAP). In this case, the common property of the Hg–0.1 M TEAP/DMSO interface was the potential of zero charge, determined as −0.27 V vs. Ag|AgCl|NaCl(sat.) reference electrode. The currents oscillations,

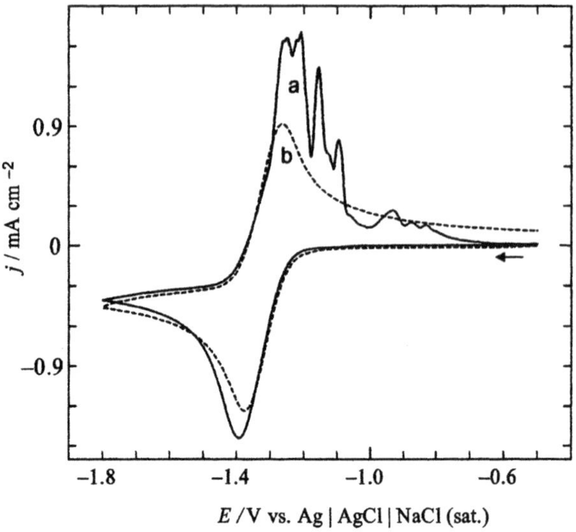

Fig. 5.44 Comparison of current oscillations reported on the anodic peak of the cyclic voltammogram for the BTD/BTD$^{\bullet-}$couple, recorded at HMDE (curve a), compared with the smooth course reported for the glassy carbon (GC) electrode (curve b). Composition of the sample: 5.0 mM BTD + 0.1 TEAP (DMSO). Scan rate 0.1 V s^{-1}. The formal potential of the BTD/BTD$^{\bullet-}$ couple was estimated as ca. −1.32 V, i.e., more negative than pzc = −0.27 V. Reprinted with permission from [43]. Copyright 2004 American Chemical Society

i.e., CVCAO, were found only for those redox couples who have the formal potentials more negative than pzc, i.e., for: $BP^0/BP^{\bullet-}$, $O_2{}^0/O_2{}^{\bullet-}$, $AB^0/AB^{\bullet-}$, $Ru(bpy)_3^{2+}/Ru(bpy)_3^{+}$, $BTD^0/BTD^{\bullet-}$, $NB^0/NB^{\bullet-}$, $DNB^0/DNB^{\bullet-}$, $DNB^{\bullet-}/DNB^{2-}$, $TMBQ^0/TMBQ^{\bullet-}$, $MV^{2+}/MV^{\bullet+}$, and $BQ^{\bullet-}/BQ^{2-}$, and also if the concentrations of these redox species and of TEAP were appropriately adjusted.

Typical experimental manifestation of CVCAO effect is shown in Fig. 5.44, for the electrode processes of 2,1,3-benzothiadiazole. In the first, *cathodic* scan BTD is reduced to the $BTD^{\bullet-}$ radical which causes pink color of the solution around the electrode and in the backward, *anodic* scan the enhancement and oscillations of the current are associated with the visible, vigorous convection of the colored solution.

Thanks to electrochromic features of the BDT + e → $BDT^{\bullet-}$ process the self-induced motion of the liquid near the electrode surface could have been visualized by using a CCD-color video camera [44] (Fig. 5.45). For that purpose also a hemispherical mercury pool electrode (HSMPE) was employed.

Comparative experiments were made also for the mercury pool electrode of the circular cross-section, which revealed various modes of the solution motion: upward, downward, clockwise and anticlockwise, depending on particular experimental conditions. In the opinion of the authors [44], the observed hydrodynamic instabilities can be understood in terms of the theory of polarographic maxima of the first kind, elaborated by Aogaki et al. [28], who have derived the correlation between the direction of the electron flow at the interface and the slope of the

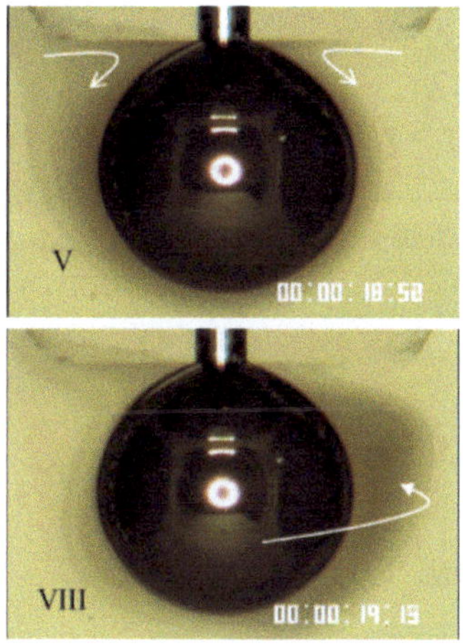

Fig. 5.45 Exemplary images of the HMDE with adjacent solution recorded during the electrooxidation of *pink colored* $BDT^{\bullet -}$ radical; image (V) corresponds roughly to the anodic peak current of the voltammogram, while image (VIII) was recorded for the more anodic potential, around the half of the descending anodic peak. *Arrows* indicate the direction of the motion of the solution. Simultaneously, the vigorous shaking of the HMDE sphere was observed. See original reference for the full set of images I–X. Reprinted in part with permission from [44]. Copyright 2006 American Chemical Society

electrocapillary curve at a given potential (cf. Sect. 5.4.2). The schematic correlation of the course of the electrocapillary curve with the motion of mercury surface is shown in Fig. 5.46, where surface tension is here denoted as γ. Evidently, these studies give deeper insight into the nature of polarographic maxima of the first kind, treated in terms of nonlinear dynamics.

5.7 The Old and New Versions of the Beating Mercury Heart

In this section, we describe the last example of convective motions driven by interfacial tension gradients, i.e., of Marangoni type. We now come back to mercury/aqueous solution interface, in order to describe the simple, but visually very effective experiments with "beating mercury heart," abbreviated further as BMH and surprisingly still unfamiliar to some chemists. As Najdoski et al. [45] have indicated, it is not easy to identify unambiguously a discoverer of that electrochemical dynamical system: the credit could go either to Ermann or to Carl Adolf Paalzow (1858). Avnir [46] has ascribed the first report about BMH to Gabriel Lipmann (1873), who however was inspired by Wilhelm Kühne, professor of physiology at Heidelberg University, who apparently learnt about that from Paalzow.

Anyway, in its simplest version, particularly useful for the lecture demonstrations, a large drop of mercury ($\phi = 1$–2 cm) is placed in the watch glass and fully covered with the acidic (H_2SO_4) solution of $KMnO_4$ or $K_2Cr_2O_7$ as the oxidant. Then the side area of the drop is carefully touched only with the steel wire through the solution (see Fig. 5.47). At this moment the drop spontaneously starts pulsating, in spite of the

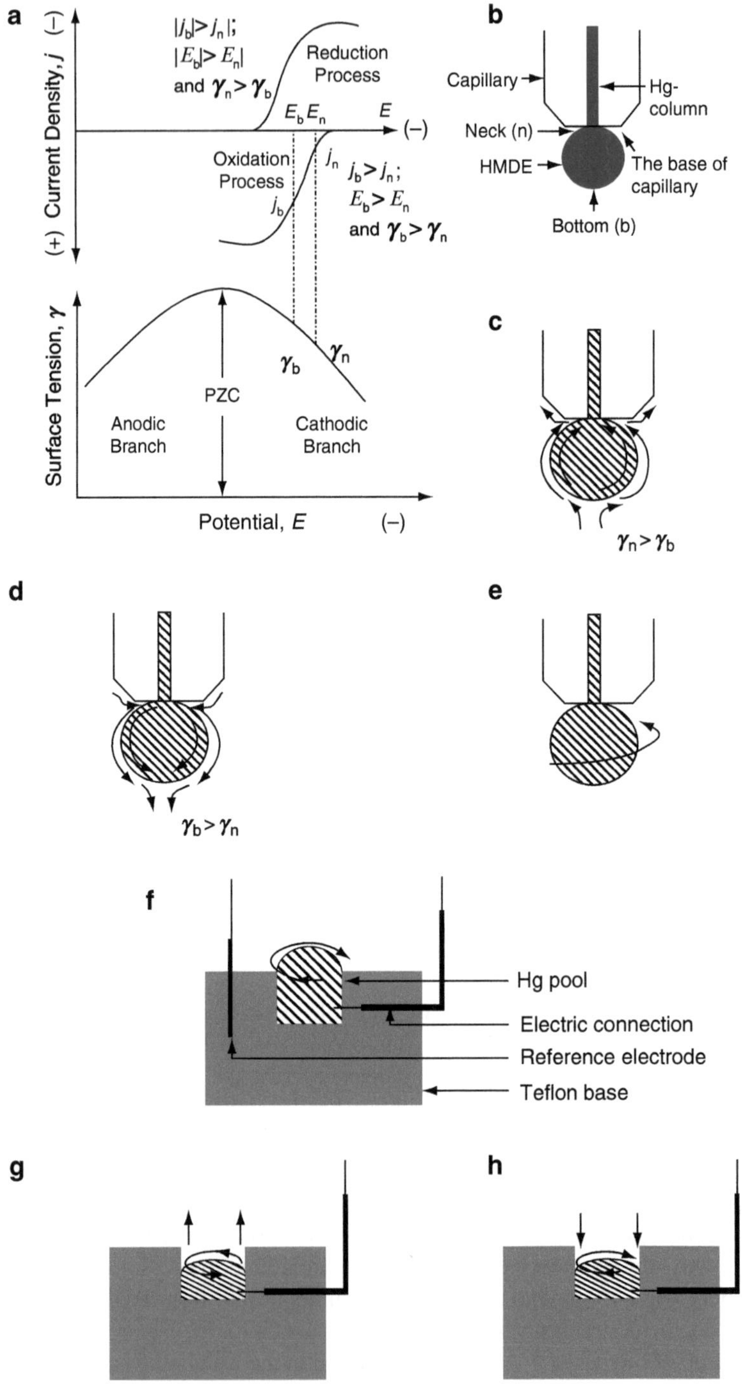

Fig. 5.46 Schematic representations of (**a**) an electrocapillary curve (ECC) and electrochemical processes (i.e., reduction and oxidation), (**b**) a cross-section of HMDE, (**c**) upward and (**d**) downward streaming of the HMDE surface and its adjoining solution, (**e**) a circular motion in the solution

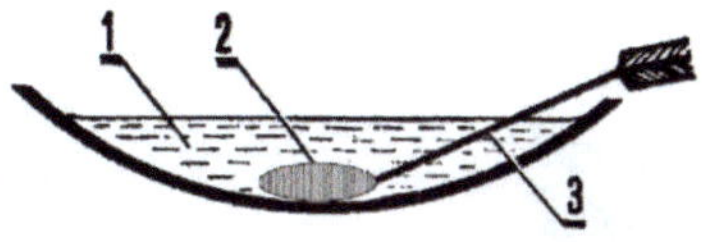

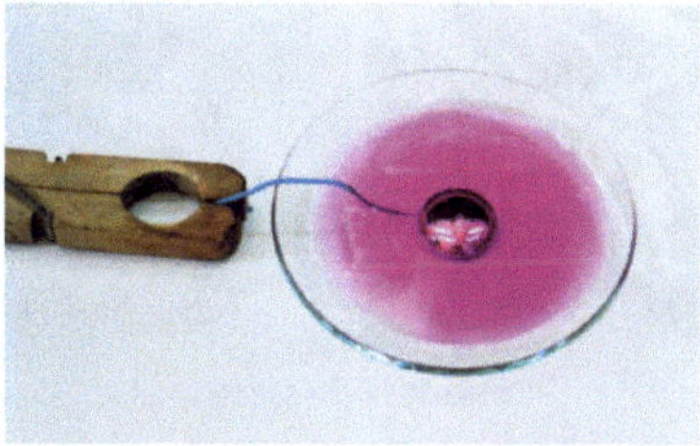

Fig. 5.47 (*Top*) Typical schematic experimental setup for the classical "beating mercury heart" (1) acidified (H_2SO_4) solution of $KMnO_4$ or $K_2Cr_2O_7$, (2) mercury drop, (3) steel wire; (*bottom*) corresponding simple experimental realization, with $KMnO_4$ solution as the oxidizing medium

immobilized position of the wire. The shape of the drop changes from the flattened to the more spherical shape, indicating thus significant periodical changes of the interfacial tension in the oscillatory cycle.

Typical modes of those pulsations are shown in Fig. 5.48, but more complex (e.g., pentagonal shapes) are also possible.

The mechanism of these phenomena is related to, shown in Fig. 5.49, the dependence of the interfacial tension at the mercury–solution interface on the surface charge of the metal, known as the electrocapillary curve (cf. also Sect. 5.4.2). This dependence illustrates that the shape of the Hg drop is closest to spherical one at the potential of zero charge, corresponding to the maximum interfacial tension at the Hg/electrolyte solution interface. Depending on the particular construction of the BMH, in its pulsations either the negative or positive branch of the electrocapillary curve is engaged (see below).

Initially, when the steel wire does not yet touch mercury, its surface undergoes certain oxidation to Hg_2^{2+} ions which adsorb on the Hg surface, also in the form of Hg_2SO_4 film, both factors causing the decrease in the interfacial tension and thus flattening of the drop. At this state its side area should touch the immobilized steel wire. The metallic iron reduces then $Hg_2{}^{2+}$ ions to metallic mercury and thus the adsorbed Hg_2SO_4 film disappears; the interfacial tension abruptly increases,

Fig. 5.46 (continued) adjoining to the HMDE surface, (**f**) a sectional view of a hemispherical mercury pool electrode (HSMPE), and (**g** and **h**) a sectional view of the partially filled HSMPE. The potential of zero charge (PZC) and the potentials negative and positive than the PZC on the ECC are indicated. The relationships among current density (j), potential (E), and surface tension (γ) with respect to the electrochemical processes at potentials negative than PZC are also shown. The *arrows* shown by the curvatures in the panels (**e–h**) indicate the direction of circular motion, and *upward* and *downward arrows* shown in panels (**g**) and (**h**) represent the directions of the expansion and shrinking of the mercury pool during the reduction and reoxidation processes, respectively. Reprinted with permission from [44]. Copyright 2006 American Chemical Society

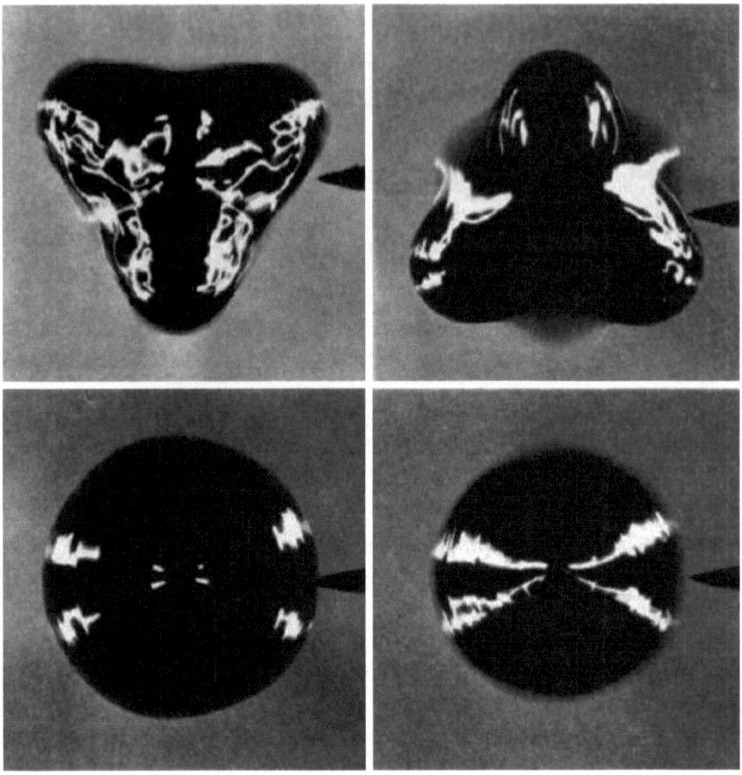

Fig. 5.48 Typical modes of the electrochemical-mechanical oscillator from Fig. 5.47; the *top row* indicates pulsation resembling "beating mercury heart."; the *bottom row* shows concentric pulsating modes. Reprinted with permission from [46]. Copyright 1989 American Chemical Society

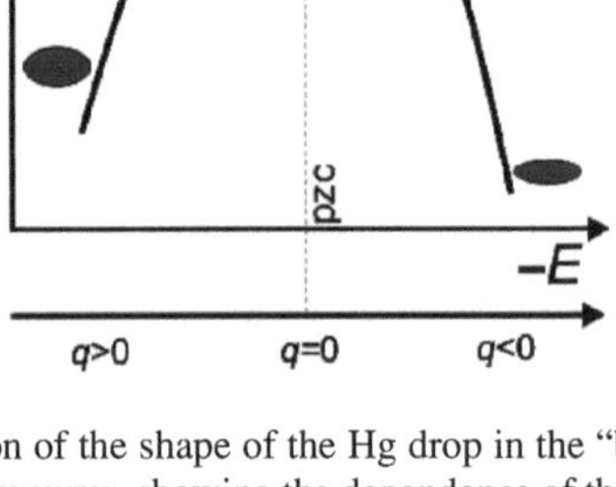

Fig. 5.49 Schematic correlation of the shape of the Hg drop in the "beating mercury heart" with the course of the electrocapillary curve, showing the dependence of the interfacial tension σ on the metal surface charge (q), with σ attaining maximum at the potential of zero charge (pzc)

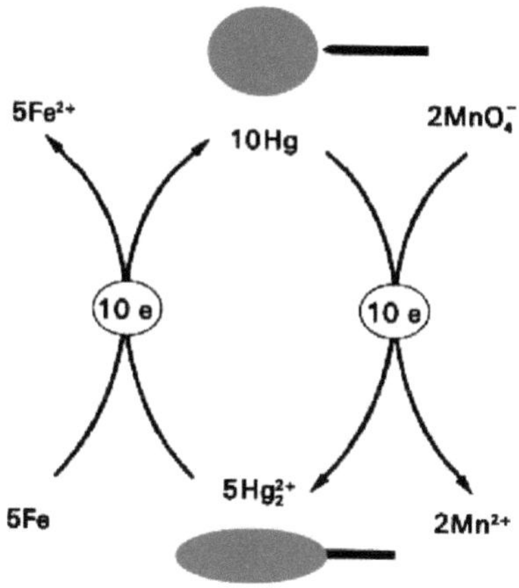

Fig. 5.50 Correlation of the shape of Hg drop in/without side contact with steel wire with the electrochemical processes at the Hg–Fe–$KMnO_4$(aq) interfaces. Based on scheme from [46]

making the shape of Hg drop more spherical, what causes the loss of contact of its surface with the steel wire. Then the surface oxidation of Hg occurs again and the whole cycle shown in Fig. 5.50 repeats periodically.

The time scale of the processes shown in Fig. 5.50 is strongly dependent also on the hydrodynamic properties of the Hg drop subjected not only to surface tension gradients, but also to gravitational forces. Extensive theoretical treatment of this problem would require solution of the Navier–Stokes equation (5.1) with appropriate boundary conditions but such an exact analysis, quite complex, at least to my knowledge was not performed so far. But even if it is done one day, it should only confirm (and deepen) the mechanism of the phenomenon, the essential electrochemical-hydrodynamic construction of which is already understood. Regarding the electrochemical characteristics, one can add that the redox processes occurring in "mercury heart" can be summarized as the oxidation of metallic iron with $KMnO_4$ (or $K_2Cr_2O_7$), with the Hg/Hg_2^{2+} couple serving as a mediator. In such a highly oxidizing medium, Fe^{2+} ions are further oxidized to Fe^{3+}.

Various modifications of the "beating mercury heart" were later proposed. For example, the mercury drop was polarized externally with the inert (Pt) wire, with respect to appropriate reference electrode, using the conventional potentiostat working in the two-electrode mode. Then the electrons reducing Hg_2^{2+} ions come not from the metal phase of Fe, but from an external source of electricity. Some new demonstrations of BMH involving either classical setup or the use of external power supply have been recently described by Najdoski et al. [45]. Figure 5.51 shows other modifications in which the metal rod has to touch the mercury layer from the top, thus differently than in the case shown in Fig. 5.47 when such a configuration would make sustained oscillations impossible. The detailed proposed mechanism of pulsations in the arrangements from Fig. 5.51 can be found in [47, 48].

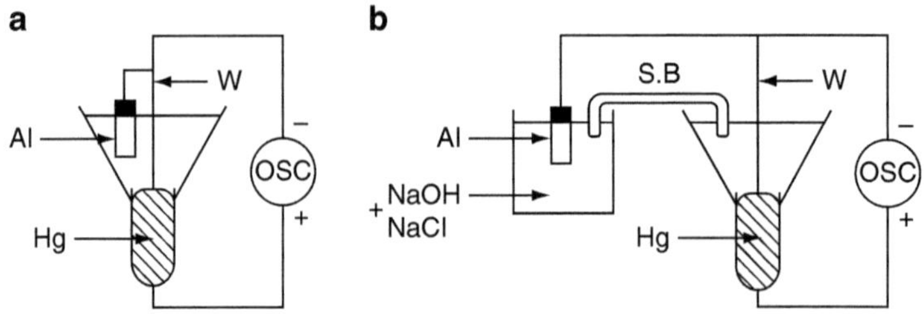

Fig. 5.51 Modified versions of "beating mercury heart": the apparatus in which oscillations were studied. In this geometry, the mercury meniscus is convex and the center of the meniscus is initially set even with the top of the tube. In (**a**), the corroding electrode (Al) is in the same vessel as the mercury. In (**b**), the corroding metal is located in a separate container connected to the solution above the mercury by 0.1 M NaCl salt bridge (SB). The polarity of the oscilloscope (OSC) is as indicated. Reprinted from [47]

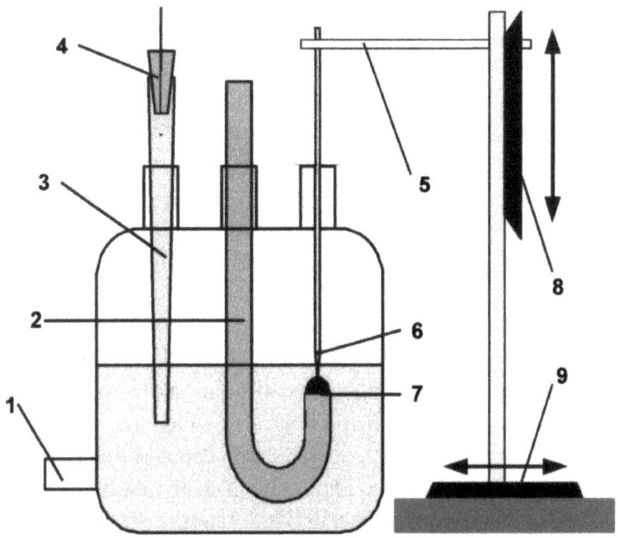

Fig. 5.52 Construction of the convective "pulsating mercury heart" based on the electroreduction of Hg^{2+} on mercury hemisphere. (1) Argon inlet (outlet not visualized), (2) the glass base for the mercury drop ($\phi = 3.9$ mm), (3) salt bridge filled up with 0.1 M HNO_3, (4) Ag|AgCl reference electrode, (5) connector between the sharp-tip steel or tungsten rod (6) and the system of the stepping motors; (7) mercury drop; (8, 9) movable elements driven by the stepping motors (not shown) in vertical (8) and horizontal (9) directions, with the resolution of ± 1 μm, being a computer-controlled mechanical module of the SECM equipment. For simplification neither the counter Pt electrode, fixed close to mercury drop, nor the stepping motor, moving the rod in horizontal direction, perpendicular to the plane of the page, is visualized. Reprinted from [32], Copyright 2008, with permission from Elsevier

Later such studies were undertaken also by Kim et al. [49] who have performed detailed analysis of oscillation mechanism for both the iron-triggered and aluminum-triggered cases by probing the electrode potential, electrocapillarity, and the shape of the mercury surface. One of the conclusions was that the range of potential oscillation is on positive side of the electrocapillary maximum of Hg in the case of Fe-triggered oscillations in acidic solutions, and on the negative side in the case of Al-triggered oscillations in basic solutions (this explains the different positions of the wires with respect to Hg surface in each case). Also, contrary to typical views in the literature, and also cited above, the authors claim that the ultimate cause of the Hg oscillation is the fluctuating surface charge density without involving a film formation on the mercury surface or oxidation of mercury. The surface charge density changes alternately in the positive direction by an electrochemical reduction and in the negative direction by contact with the corroding metals. This fluctuating charge leads to fluctuating surface tension of mercury on one side of the electrocapillary maximum, which in turn induces the oscillation of the mercury body. Thus, in opinion of Kim et al., the addition of dichromate (or permanganate) salt, usually prescribed for the demonstration experiment is not a requisite for realizing the oscillation. In fact, my personal experience with the beating mercury heart in the version shown in Fig. 5.48 proves that it oscillates even if after some time the permanganate solution becomes completely colorless. One can suppose that low, invisible amount of $KMnO_4$ is still present in the solution and oxidizes the Hg surface, but in view of conclusions drawn by Kim et al., in fact, the surface oxidation of Hg may appear not absolutely necessary.

Another variant of pulsating mercury drop was recently described by Gorzkowski et al. [32] in which at the hemispherical mercury drop contacted with the $Hg(NO_3)_2$ or $Hg_2(NO_3)_2$ solution, upon external polarization the electroreduction of Hg(II) or Hg(I) ions to metallic mercury occurred, associated with violent convection at the Hg/solution interface (Fig. 5.53); in fact, it was the modification of the idea developed by Aogaki et al. [28–30], described in Sect. 5.4.2. For the appropriate position of the iron or tungsten rod tip, the high-amplitude current oscillations were reported (Fig. 5.54), explained as reflecting periodical bursts of convection, separated by quiescent periods when, due to convection-driven deformation of the Hg drop shape, its connection with the solid metal rod was broken. The amplitude of these oscillations was so high that it could not be explained in terms of diffusion-only control of the electroreduction of Hg ions and thus the contribution from convection had to be involved in the explanation.

Furthermore, it is noteworthy that "beating hearts" based on another liquid metals can be constructed. Ealy and Niewahner [50] have described the "gallium beating heart" which obviously required temperature of the metal, and thus also of the solution, enhanced up to 50 °C. The principal scheme of the experiments resembles that of the classical BMH: 15 g of gallium is placed in the Petri dish and 50 mL of 6 M H_2SO_4 is added and the dish is placed on a hot plate. When the gallium melts, 3–4 mL of 0.1 M $K_2Cr_2O_7$ is added to the dish. Under these conditions the gallium layer may form a dull, flat puddle which, upon contact with the iron nail, turns into the shiny quasi-spherical shape. If the nail is positioned carefully so that its contact with gallium occurs only when forms a dull puddle, the spontaneous oscillations may set in. Chemical processes are also analogous to those occurring in the beating mercury heart: the dichromate(VI) ions oxidize the surface

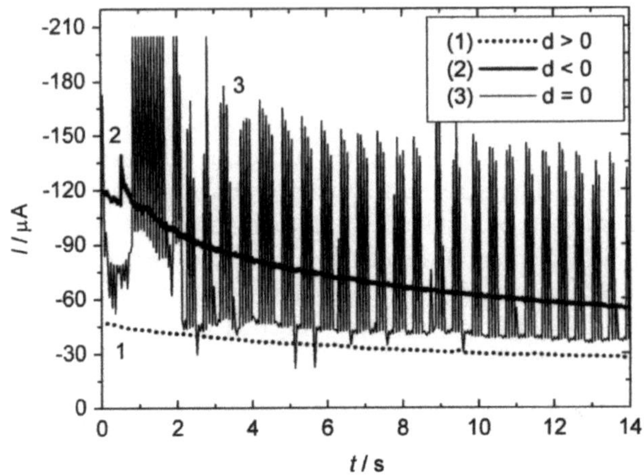

Fig. 5.53 Current–time dependences, observed in the cell from Fig. 5.52, filled up with 0.001 M $Hg(NO_3)_2$ + 0.1 M HNO_3, for different distances d between the steel rod and a mercury drop (curve 1: $d > 0$, rod without contact with Hg drop; curve 2: $d < 0$, rod immersed in Hg drop, curve 3: $d = 0$, rod in contact with the mercury surface). Applied potential = −0.40 V vs. Ag|AgCl. Only for $d = 0$ the oscillations are observed and high values of the current suggest the contribution of convective motion to the transport; for $d > 0$ the diffusion-controlled current flowing at the steel rod surface is observed, for $d < 0$ the nonoscillatory current flows through the surface of both steel rod and Hg hemisphere. Reprinted from [32], Copyright 2008, with permission from Elsevier

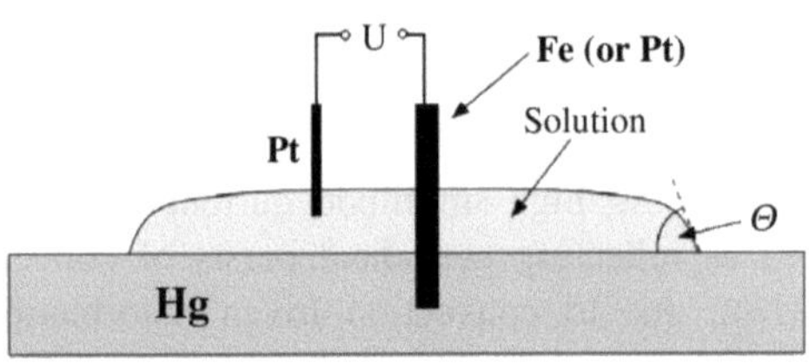

Fig. 5.54 Scheme of the experimental setup. The angle Θ at the curve shared by three interfaces is given by the Young's equation: $\sigma_{MA} - \sigma_{MS} = \sigma_{SA} \cos \Theta$, with σ's meaning the surface tensions at the mercury–air (MA), mercury–solution (MS), and solution–air (SA) interfaces. Reprinted from [51], Copyright 2007, with permission from Elsevier

atoms of gallium to Ga^{3+} ions which form the surface film of $Ga_2(SO_4)_3$ precipitate, decreasing the surface tension of Ga. Metallic iron oxidizes to Fe^{3+} ions, reducing simultaneously Ga^{3+} ions to Ga, which means destruction of the $Ga_2(SO_4)_3$ precipitate and the surface tension increases.

Finally, an interesting "inverted" construction of the system related to BMH was recently published by Maiworm et al. [51]. In their setup it is not the mercury that beats, but a drop of the potassium dichromate (1.16 mM)–sulfuric acid (7 M) solution on a surface of mercury placed in a dish of 15 cm diameter. An iron stick is put vertically through the drop at its center and remains constantly immersed in the mercury (Fig. 5.54).

Measurements of the voltage U exhibited its spontaneous oscillations, accompanied with visually observed mechanical oscillations of a drop of oxidant solution on a Hg surface. As the cause for the work of this electrochemically driven "motor" are considered the changes of the surface tension at the drop's bottom.

5.8 Convective Instabilities Caused by Gas Evolution Reactions in Electrode Processes

Besides the hydrodynamic instability of the Hg–solution interface due to surface tension gradients, also evolution of gaseous hydrogen due to the H^+ electroreduction, or of oxygen due to water electrooxidation, may cause the convection which affects the course of the electrode processes. For example, Kirowa-Eisner et al. [52] have reported current maxima of the first kind accompanying the electroreduction of H^+ ions at the Hg electrode. In consequence, if there occur also other parallel electrode processes, the presence of hydrogen-generated convection manifests itself by the distortion of all the currents from their diffusion-controlled values. Recently, it was found that if the hydrogen evolution occurs simultaneously with the electroreduction of Cd^{2+}, Cu^{2+}, or Eu^{3+} at the Hg electrode, the nonadditivity of the faradaic currents of the component processes is observed which can be explained just in terms of distortion of the concentration distributions of the electroactive species through convection associated with hydrogen evolution [53].

5.8.1 Oscillations in the Electrode Reactions of Anions at Solid Electrodes

The hydrogen evolution can induce convection also for the processes occurring at the solid electrodes and cause even electrochemical oscillations under appropriate conditions. Here we enter the subject of interesting literature discussion on the origin of such oscillations in particular systems: whether they indeed are caused by convection or only by the presence of the so-called additional current carrier, with the eventual convection meaningless for instabilities, as Strasser et al. [54] have proposed for the interpretation of iodate(V) reduction (see Sects. 3.5 and 4.2, volume I). Drawing the attention of the reader to this discussion, we present here the original interpretation of the authors of relevant papers, based on the idea of gas-induced convection.

For the electroreduction of $Fe(CN)_6^{3-}$ in 1 M NaOH on a stationary Pt electrode, Li et al. [55] have shown that under potentiostatic conditions the system remained stable, but high-amplitude oscillations of the electrode potential set in when the imposed current exceeded the limiting current (Fig. 5.55).

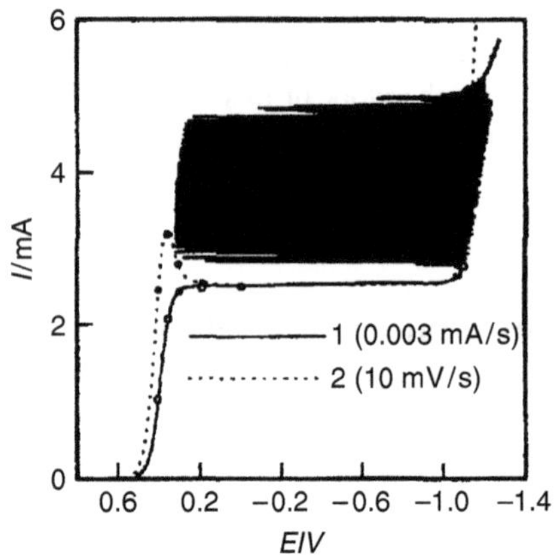

Fig. 5.55 Voltammograms of 0.8 M $Fe(CN)_6^{3-}$ in 1 M NaOH recorded on a Pt electrode under (1) current and (2) potential sweep, with the indicated scan rates, respectively. *Hollow circles* indicate the potentials for which the impedance spectra were taken. Potentials were measured with respect to the Hg(l)|HgO(s)|1 M NaOH reference electrode. Reprinted from [55], Copyright 1997, with permission from Elsevier

Analysis of Fig. 5.55 shows that (1) the potentiostatic electroreduction of $Fe(CN)_6^{3-}$ ions is followed by the electroreduction of H^+ to gaseous hydrogen at potentials more negative than −1.0 V and (2) the amplitude of oscillations switches between the potentials close to the beginning of the plateau of the $Fe(CN)_6^{3-}$ electroreduction and the potentials covering also the hydrogen evolution. Thus, it is suggested that hydrogen evolution must be considered in the mechanism of those oscillations. The proposed mechanism consists of the following steps. As long as the imposed current is lower than its limiting value, the electrode potential attains the stable value, belonging to the *I–E* characteristics. When the current exceeds the limiting value, the surface concentration of $Fe(CN)_6^{3-}$ drops to zero during the transition time τ, determined by the Sand equation [56]:

$$I\tau^{1/2}/c_{\text{ox}}^{0} = nFAD_{\text{ox}}^{1/2}\pi^{1/2}/2 \tag{5.36}$$

and the electrode potential moves to more negative values:

$$E = E_{\tau/4} + \frac{RT}{nF}\ln\frac{\tau^{1/2} - t^{1/2}}{t^{1/2}} \tag{5.37}$$

until hydrogen evolution occurs which ensures the flow of the imposed current. However, the negative feedback sets now: the convection generated by nucleation, growth, and movement of hydrogen bubbles causes the replenishment of the $Fe(CN)_6^{3-}$ concentration in the preelectrode layer and therefore further hydrogen evolution is no longer necessary to keep the imposed current, so the electrode

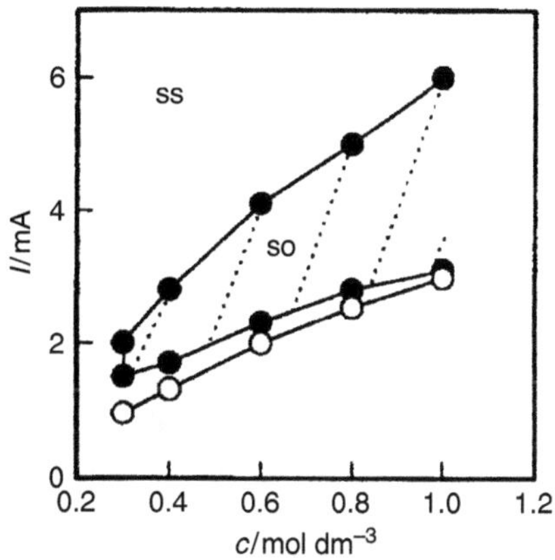

Fig. 5.56 Measured boundaries (*filled circle*) in the parameter plane: applied current vs. $Fe(CN)_6^{3-}$ concentration that separate the region for which sustained oscillations were observed (labeled so) from the region for which only stable stationary states were found (labeled ss), and relationship of the limiting current with $Fe(CN)_6^{3-}$ concentration (*hollow circle*). Reprinted from [55], Copyright 1997, with permission from Elsevier

potential returns to more positive values and the convection stops. This completes the oscillatory cycle which repeats when the surface $Fe(CN)_6^{3-}$ concentration drops to zero again. When the imposed current is as high as twice the limiting value, the hydrogen evolution becomes so predominant, that the convective replenishment of $Fe(CN)_6^{3-}$ concentration is not sufficient to switch the system to the *I–E* characteristics of $Fe(CN)_6^{3-}$ electroreduction. A competitive steady state, determined by the hydrogen evolution reaction, becomes now stable.

Systematic studies of the regions of occurrence/absence of oscillations in this system led to the bifurcation diagram shown in Fig. 5.56.

This diagram does not reveal the richness of the oscillatory modes. For example, if—for the fixed $Fe(CN)_6^{3-}$ concentration—the imposed current is gradually increased, the following sequence was reported: chaotic → quasi-periodic → periodic → quasi-periodic → damped oscillations (see Fig. 5.57).

Later Li et al. [57] have also shown that both galvanostatic and potentiostatic (the latter ones in the presence of ohmic potential drops) oscillations occur for the electroreduction of $Fe(CN)_6^{3-}$ and other anions (IO_3^-, $S_2O_8^{2-}$) if coupled with the hydrogen evolution (Fig. 5.58).

The qualitative mechanism of these current oscillations includes the following positive and negative feedbacks. Within the oscillatory peak the rise of current is caused by the positive feedback, when initially applied negative potential is sufficiently high to cause the electroreduction of both the anions and the hydrogen cations; the molecular hydrogen formed causes the convection which, as an extra transport, supplies additional portions of anions toward the electrode surface and thus makes the total current even higher. This increase of current is, however,

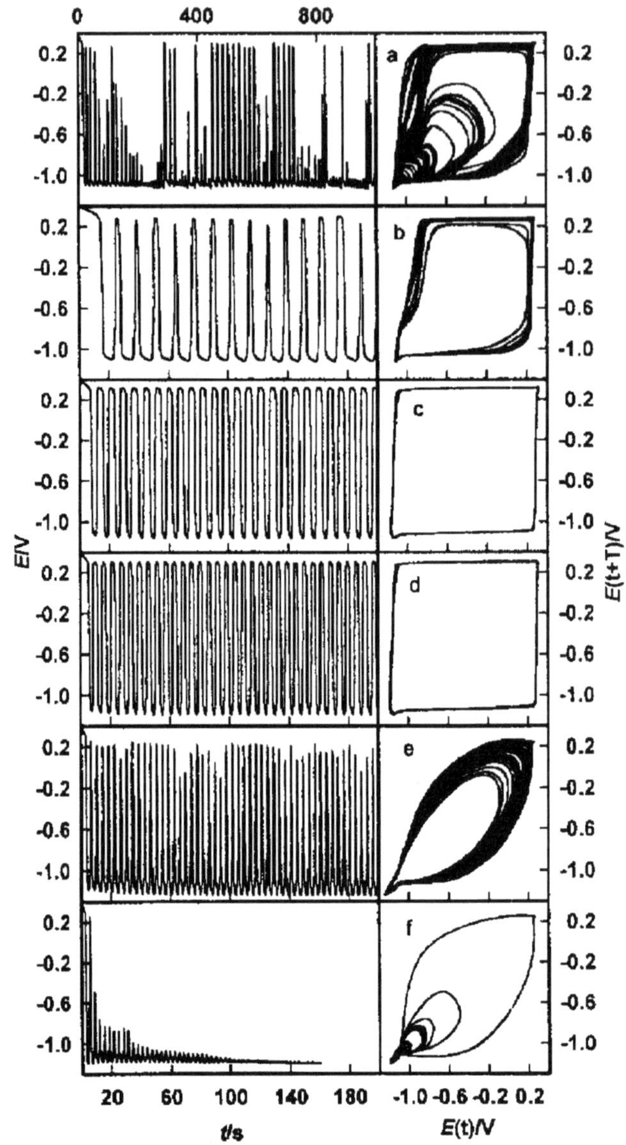

Fig. 5.57 *Left*: time series of potential oscillations for 0.8 M $Fe(CN)_6^{3-}$ reduction in 1 M NaOH solution at different applied currents (mA) (**a**) 2.8, (**b**) 2.9, (**c**) 3.5, (**d**) 4, (**e**) 4.8, (**f**) 5. The oscillations are represented both by the $E - t$ sequences and the corresponding phase portraits, constructed from the time series by time delay technique, with the delay times T (s) equal to (**a**) 0.8, (**b**) 0.8, (**c**) 0.6, (**d**) 0.5, (**e**) 0.2, (**f**) 0.2. Reprinted from [55], Copyright 1997, with permission from Elsevier

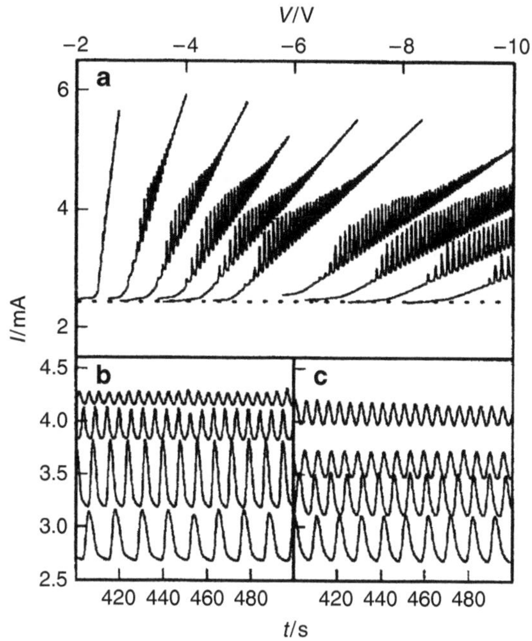

Fig. 5.58 Current oscillations for 0.8 M $Fe(CN)_6^{3-}$ reduction on platinum in a 1 M NaOH solution. (**a**) Current–voltage curves. Voltage scanning at 10 mV s^{-1} with the external series resistance R_e (Ω): 0; 200; 400; 600; 800; 1,000; 1,400; 1,800; 2,200; 2,600 (from left to right). The resistance of the solution (R_s) is 48 Ω. The *dotted line* represents the limiting current plateau (2.44 mA). (**b**) Time series at different voltage (V) 8.5; 8; 7; 6.2 (from top to bottom) while fixing the R_e at 1,400 Ω. (**c**) time series at different R_e: (Ω): 600; 700; 800; 950 (from top to bottom) while fixing the voltage at 5 V. Reprinted from [57], Copyright 1997, with permission from Elsevier

followed by its decrease caused by the negative feedback: due to increasing ohmic drops *IR* the effective interfacial potential drop of the working electrode decreases to less negative values, so the hydrogen evolution (and thus the convection) stops. The lack of convection lowers the flux of the anions and the current decreases. Then again the positive feedback starts to operate: in the presence of low ohmic drops the interfacial potential drop *E* returns to such negative values that the hydrogen evolution can start again and the whole cycle repeats.

Further studies have revealed that not only the electroreduction, but also the electrooxidation process can become a source of quite analogous oscillations. Galvanostatic oscillations were found for the electrooxidation of $Fe(CN)_6^{4-}$, when the imposed anodic current exceeded the limiting current of $Fe(CN)_6^{4-}$ and then oscillation amplitude temporarily reached the range of evolution of gaseous oxygen (Fig. 5.59). In turn, potentiostatic oscillations were reported for various values of an external resistance (see Fig. 5.60).

These observations suggested certain universality in the discussed oscillatory mechanism which, for the Ox + ne = Red process presumably required evolution

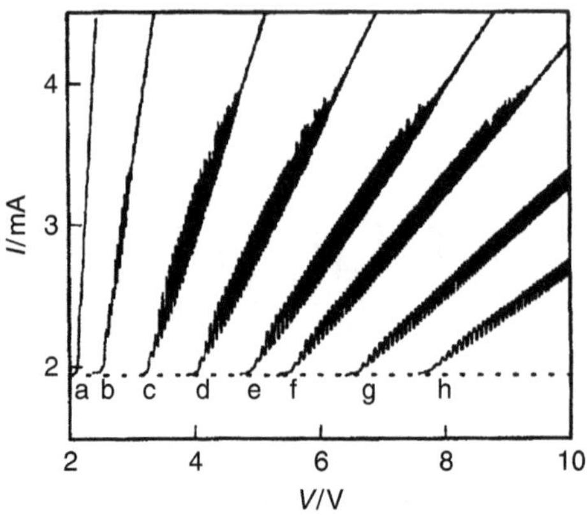

Fig. 5.59 Anode current–voltage curve in a solution of 0.6 M $Fe(CN)_6^{4-}$ + 1 M NaOH by voltage scan at 10 mV s^{-1} with different external resistances (Ω) (**a**) 0; (**b**) 200; (**c**) 600; (**d**) 1,000; (**e**) 1,400; (**f**) 1,800; (**g**) 2,400; (**h**) 3,000. The resistance of the solution R_s was 55 Ω. The *dotted line* represents the limiting current plateau (1.94 mA). Reprinted from [57], Copyright 1997, with permission from Elsevier

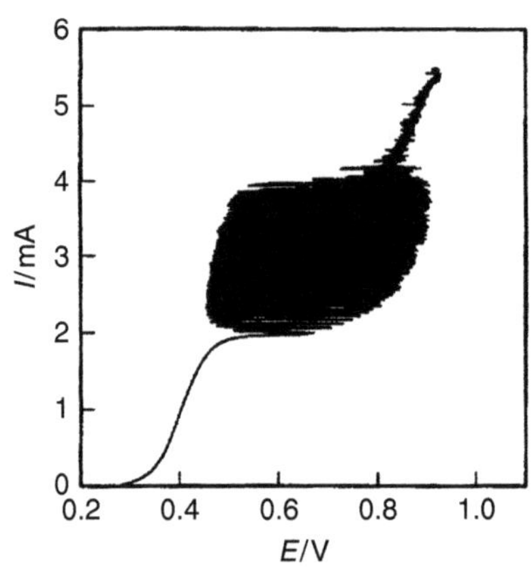

Fig. 5.60 A typical voltammogram for the oxidation of 0.6 M $Fe(CN)_6^{4-}$ (in 1 M NaOH) on a Pt electrode, under current scanning (0.003 mA s^{-1}). Reprinted from [57], Copyright 1997, with permission from Elsevier

of gas at the potentials more negative (for reduction) or more positive (for oxidation) than the formal potential of Ox/Red couple. As it follows from Sect. 3.4 of volume I, galvanostatic oscillations can be associated with the hidden negative resistance (HN-NDR type oscillator), meaning that the negative impedance does not manifest itself on the dc (zero ac frequency) curves but becomes detectable under conditions of the impedance measurements, for moderate ac frequencies. In fact, for the $Fe(CN)_6^{3-}$ electroreduction the reported spectra revealed the negative

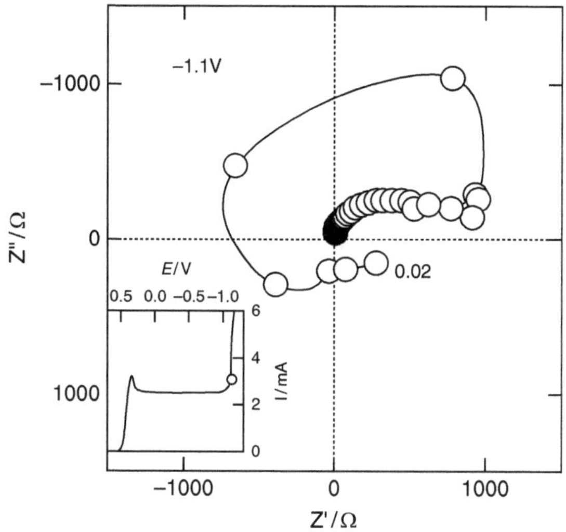

Fig. 5.61 Impedance diagram at −1.1 V for 0.8 M $Fe(CN)_6^{3-}$ reduction in a 1 M NaOH solution on a platinum electrode. Frequency range: 1,000–0.02 Hz. The inserted plot is the polarization curve for the same system obtained by potential scan at 10 mV s^{-1}, and the location where the impedance spectra were taken, was indicated with *circle* on the curve. Reprinted from [57], Copyright 1997, with permission from Elsevier

impedance for the range of nonzero frequencies, when the electrode potential entered the region of hydrogen evolution, with the slope dI/dE being positive under dc conditions (Fig. 5.61). This negative impedance generally seems to be caused by the contribution from convection caused by hydrogen bubbles, but the detailed mechanism of its formation seems to be not well recognized.

Again, as indicated earlier, the reader is encouraged to become aware of the interesting discussion on the proposed oscillatory mechanism outlined in Sect. 3.5 of volume I and discussed more extensively in Sect. 4.2 of volume I.

Finally, regarding again the galvanostatic reduction of $Fe(CN)_6^{3-}$, the oscillations of the Pt electrode potential during this process were recently used by Wilson et al. [58] for construction of the dual-electrode setup, with the enzyme electrode as the other one. Electrochemical oscillations in $Fe(CN)_6^{3-}$ reduction on Pt induced then sympathetic current oscillations at the enzyme electrode, although the detailed mechanism of this coupling remains a subject of controversy (the convection caused by bubbling hydrogen or local pH variations being ones of the factors considered).

5.8.2 *Convection-Driven Oscillations in Methanol Oxidation*

Nonconvective, oscillatory electrooxidation of methanol and its electrochemical mechanism was described in Sect. 5.6 of volume I. In addition to that analysis, we describe here this process carried out under conditions suggesting the contribution

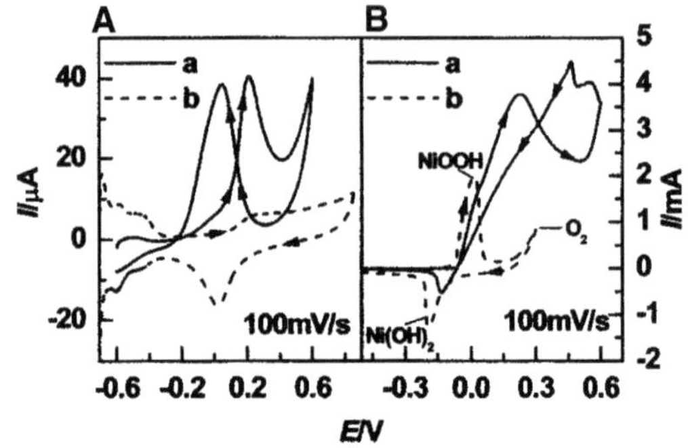

Fig. 5.62 Cyclic voltammograms on (**A**) Pt and (**B**) NNHF electrode for (a) 1 M CH_3OH in (**A**) 1 M $HClO_4$ or (**B**) 1 M NaOH. The *dashed lines* of (b) are CVs for the base solutions. Reprinted from [59], by permission of The Royal Society of Chemistry

from convection, caused by parallel oxygen evolution. Huang et al. [59, 60] have reported the oscillatory oxidation of methanol, in an alkaline medium, at the nanostructured nickel hydroxide film (NNHF) electrode, with the average particle size of 100 nm. These studies involved electrochemical measurements coupled with in situ temporal- and spatial-resolved Raman spectroscopy. Figure 5.62 shows the comparison of CVs for methanol at a Pt (left) and NNHF (right) electrodes. The voltammograms exhibit a crossing loop, which course is considered by Li et al. [61, 62] the criterion of occurrence of electrochemical oscillations (cf. also Fig. 4.20, volume I).

According to Fig. 5.63, for the NNHF electrode the oscillations are observed under galvanostatic conditions. Similarly as for the electrooxidation of anions, described earlier, the oscillations occur only if the imposed current exceeds the limiting value, and the potential amplitude of the oscillations reaches, at its most positive values, the foot of the oxygen evolution wave.

The amplitude of these potential oscillations was compared with the redox equilibria of nickel compounds. For mechanistic consideration it was also important that in situ Raman studies did *not* indicate the presence of CO_{ads} (contrary to Pt electrode), but instead the vibrational bands typical of NiOOH, formate, and carbonate groups were detected. In conclusion, it was suggested that the redox couple $Ni(OH)_2$/NiOOH acts as the mediator for the electrocatalytic oxidation of methanol, in which formate and carbonate ions are the intermediates (recent studies suggest that the formates are the intermediates also for Pt electrode). The network of corresponding processes is shown in Fig. 5.64. In addition, since NiOOH is a black compound, whereas $Ni(OH)_2$ is light green, the potential-dependent interconversion of these compounds can be observed visually, being the example of electrochromism.

From the quantitative point of view, considering higher peak current of methanol of oxidation at NNHF electrode, compared to Pt (cf. Fig. 5.62), one concludes that NNHF, allowing for different reaction mechanism, exhibits better than Pt electrocatalytic properties for methanol. Finally, the shape of cyclic

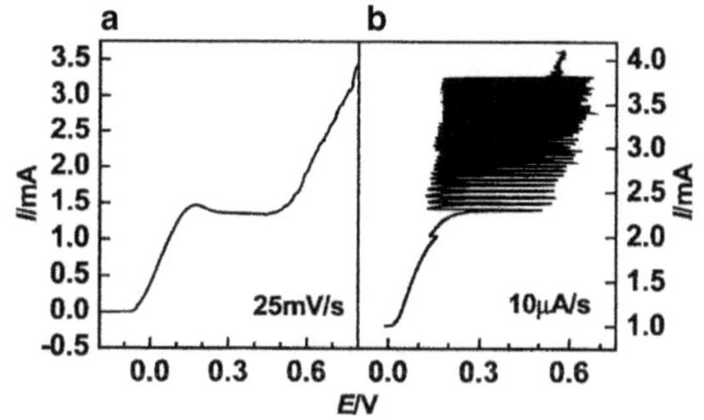

Fig. 5.63 Current–potential curves measured by (**a**) potential and (**b**) current scanning for 1 M CH_3OH oxidation in 1 M NaOH on the NNHF electrode. Reprinted from [59], with permission from Royal Society of Chemistry

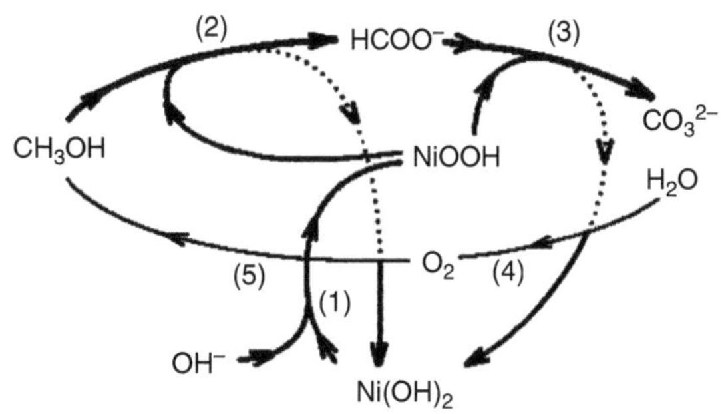

Fig. 5.64 A mechanistic network diagram for the oscillatory electrocatalytic oxidation of methanol on the NNHF electrode. Reprinted from [60] with kind permission from Springer Science + Business Media

voltammogram, in which the current in the backward scan exhibits enhanced value, is explained in terms of the extra contribution to the mass transfer rate from convection generated by the oxygen evolution. In this sense, the mechanism of the oscillations is in fact analogous to that proposed above for the reduction of IO_3^-, $Fe(CN)_6^{3-}$, or $S_2O_8^{2-}$ (with hydrogen evolution) or the oxidation of $Fe(CN)_6^{3-}$(with oxygen evolution) under analogous conditions. When the imposed current exceeds the limiting value, the surface concentration of methanol drops to zero, but then the electrode is charged to more positive potentials, until the current of oxygen evolution, added to that of methanol oxidation, satisfies the imposed current. The evolution of oxygen bubbles causes the convection which replenishes the diffusion layer with additional portions of methanol. The electrode potential moves then to less anodic values, but this stops the oxygen evolution and convection; in consequence the electrode potential has to increase again in order to cause oxygen evolution, etc. One should add that during the oscillations of the electrode potential, the *periodic* evolution of gaseous oxygen was also observed.

5.8.3 Convection-Driven Oscillations and Dendritic Formation During Metal Deposition

The formation of dendritic patterns during metal deposition, assuming the absence of convection, but predominating role of diffusion as the source of such patterns, was described in Sect. 2.11. Here, as the complement to that material, there are described other examples of such patterns, the creation of which was accompanied by the convection caused by parallel hydrogen evolution. This is case of the deposition of tin, one of these metals for which electrochemical oscillations can be associated with both electrooxidation and electroreduction processes. Here we consider cathodic oscillations. Piron et al. [63] have described damped oscillations of the potential of the Sn electrode immersed in the alkaline (KOH) solution of SnO, occurring under galvanostatic conditions, for the cathodic current density higher than the limiting current density (Fig. 5.65).

Only during the formation of the first peak the evolution of gaseous hydrogen is observed, since only then the electrode attains the negative potential sufficiently high for both reduction of the water molecules and such supersaturation of H_2 in the solution that H_2 bubbles can form. Simultaneous in situ observation of the electrode surface indicated its increasing roughness. Scanning electron micrographs (SEM) revealed the correlation of oscillations with the pattern morphology on the Sn surface, for which transient formation of dendritic structures was observed (Fig. 5.66). This also means that in the course of the oscillations the real, effective surface area is constantly increasing, up to significant degree of roughness attained at the end of the oscillations.

Figure 5.66 shows that at the first peak, i.e., point (b) in Fig. 5.65, many small crystal particles form on the surface (image A). Later, during the oscillatory course, i.e., for points (c) and (d) in Fig. 5.65, dendritic growth takes place (images b, c) and finally, when oscillations cease [point (e) in Fig. 5.65], the whisker-like structures, providing a very large surface area, develop [63].

Another important experimental observation was also the decay of oscillations if the solution was stirred. Based on these data, the authors have proposed the following mechanism of the oscillations: the relatively high imposed constant current density causes the electroreduction of a significant portion of $HSnO_2^-$ ions in the preelectrode layer of the solution; accordingly, the electrode potential moves initially to such negative potentials, that the electroreduction of water molecules becomes possible. Both the evolved hydrogen bubbles *and the density gradient* caused by the resulting concentration gradient of Sn(II) species induce the convective mixing of the solution which enhances the transport of $HSnO_2^-$ ions from the solution bulk to the interface. The electrode potential then moves back to less negative values, at which hydrogen evolution is stopped. As it does not happen again, one can suppose that the oscillating electrode potential does not reach a sufficiently negative potential any more. The convection is, however, continuously operating due to the concentration gradient of $HSnO_2{}^-$ species, and therefore oscillations of the electrode potential persist for some time yet. With increasing roughness of the electrode surface, the effective current density continuously decreases until it reaches so low values that the depletion of the preelectrode

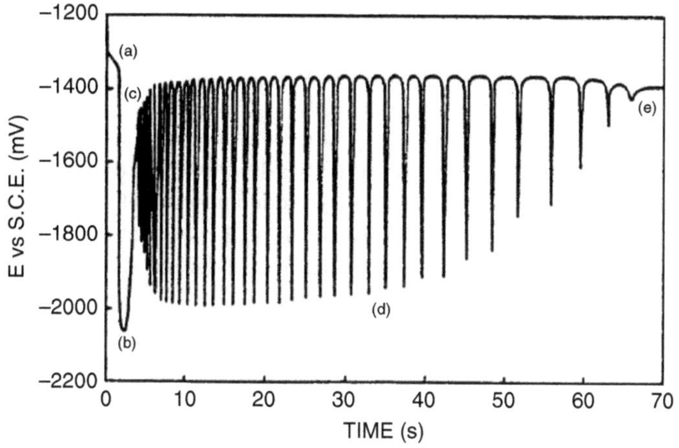

Fig. 5.65 Potential oscillation of the tin electrode in the solution of 0.23 M SnO + 6.25 M KOH at a cathodic current 800 mA m^{-2} (anode–Pt plate, potential of Sn electrode measured vs. SCE). Reproduced from [63] by permission of The Electrochemical Society

Fig. 5.66 Scanning electron micrographs of the tin electrode surfaces at different moments during the potential oscillation at 800 A m^{-2}, to compare with a, b, c, d times of the $E - t$ course in Fig. 5.65; (**a** ≡ top left)—at the time (b) in Fig. 5.65 (magnification 1,500 times); (**b** ≡ top right)—at the time (c) in Fig. 5.65 (200 times); (**c** ≡ bottom left)—at the time (d) in Fig. 5.65; (**d** ≡ bottom right)—at the time (e) in Fig. 5.65 (1,500 times). Reproduced from [63] by permission of The Electrochemical Society

layer with $HSnO_2^-$ is so low, that solution density gradient appears to be insufficient for the onset of convection which then ceases, and so the oscillations. Thus, in terms of the above interpretation this system should be considered a kind of convection-driven electrochemical oscillator.

The morphology of Sn deposits is in this way not unambiguously explained. Therefore, more recently, Tada et al. [64] have described the tuning of the spacing and thickness of this metal latticeworks by modulation of self-organized potential oscillations in Sn electrodeposition. This tuning of the spacing and thickness of Sn was achieved by the modulation of the oscillations, induced by changing the applied current density (j_{ap}) and the concentration of Sn(II) ions (c_{Sn}). Figure 5.67 shows the correlation between the phase of the potential oscillations and the growth of Sn pattern.

Figure 5.67 should be compared with the scheme of formation of Sn deposit shown in Fig. 5.68.

At point (1) in Fig. 5.67, due to passive layer covering the Sn electrode surface, its potential is so negative (to meet the imposed current density), that even hydrogen evolution occurs. Under such conditions, in course of diffusion of spherical geometry toward the peaked part of the deposit surface, the autocatalytic growth of crystals starts, resulting in the formation of sharp needles (stage 2). This increases the roughness, i.e., the effective surface of the electrode, so the current density decreases and the potential then moves to less negative values, which shift continues until the autocatalytic growth of crystals stops, i.e., when the diffusion-controlled conditions of Sn deposition disappear (stage 3). Now, a charge-transfer controlled deposition of Sn occurs, which leads to cuboids surrounded with the thermodynamically stable (110) and (011) faces. At this stage, however, the Sn surface is gradually oxidized with time, since the electrode potential is now more positive than the redox potential for Sn oxidation. This process takes also an autocatalytic character, due to partial formation of Sn oxide, with resultant $Sn^{\delta+}$ atoms inducing positive polarization of adjacent surface Sn atoms, which in turn causes an easier attack of negatively charged OH^- species on them. The rapid passivation of the surface causes, through rapid decrease in effective surface area, and thus a rapid increase in effective current density, the rapid shift of the potential toward negative values, corresponding to stage 4 which is equivalent to stage 1, so the cycle is closed. Further considerations, concerning the role of Sn(II) concentration and applied current, the reader can find in the original reference.

Similar phenomena associated with the oscillatory zinc electrodeposition were studied, among others, by St-Pierre and Piron [65, 66]. Oscillations were observed also under galvanostatic conditions, for Zn electrode immersed into the alkaline solution of Zn(II) salt [presumably mainly tetrahydroxozincate(II) ions], when the electrode process occurs:

$$Zn(OH)_4^{2-}(aq) + 2e \rightarrow Zn(s) + 4OH^-(aq) \tag{5.38}$$

If the imposed cathodic current density is high enough, the electroreduction of Zn(II) ions at the interface is faster than their diffusion transport from the bulk, so the concentration of Zn(II) near the electrode surface drops, causing an increase in the negative potential of Zn electrode, until the hydrogen evolution occurs. The

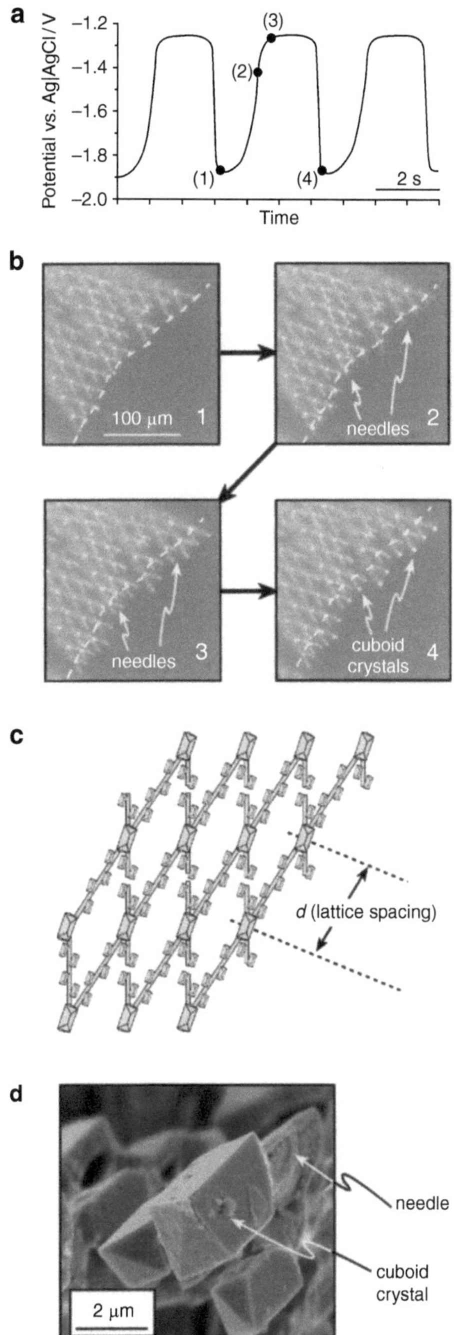

Fig. 5.67 (**a**) A potential oscillation at $j_{ap} = -36\ \mathrm{mA\ cm^{-2}}$ in 0.2 M Sn(II) + 4.0 M NaOH. (**b**) Optical microscopic (OM) images of Sn deposits taken at various stages of the potential oscillation. The number, (1)–(4), added on the OM images mean that they were obtained at stages of the potential oscillation marked by the same numbers in (**a**). (**c**) A schematic illustration of the latticework in (**b**). (**d**) A scanning electron microscopic (SEM) image of cuboid crystals formed at the stage of the positive-side potential of the oscillation. Reprinted from [64], Copyright 2005, with permission from Elsevier

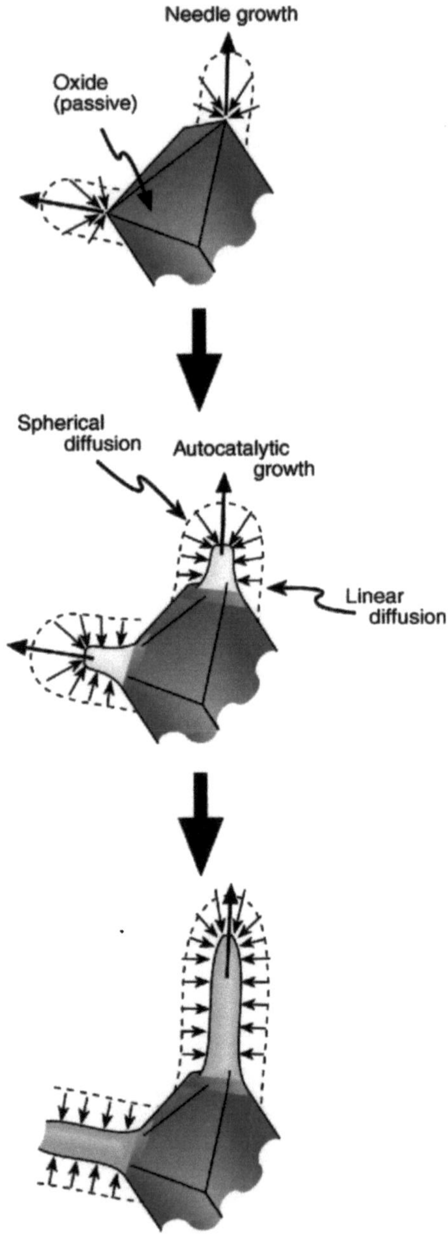

Fig. 5.68 Schematic drawings of Sn deposits in the course of the transition from the negative- to positive-side potential of the oscillation. Reprinted from [64], Copyright 2005, with permission from Elsevier

hydrogen bubbles cause the mixing of the solution and this additional convective transport accelerates the transport of Zn(II) species from the solution bulk to the interface. An increase in this concentration causes the electrode potential to return to less negative values, the hydrogen evolution then stops, and the increase of concentration polarization repeats again [65]. Different regions, of which the

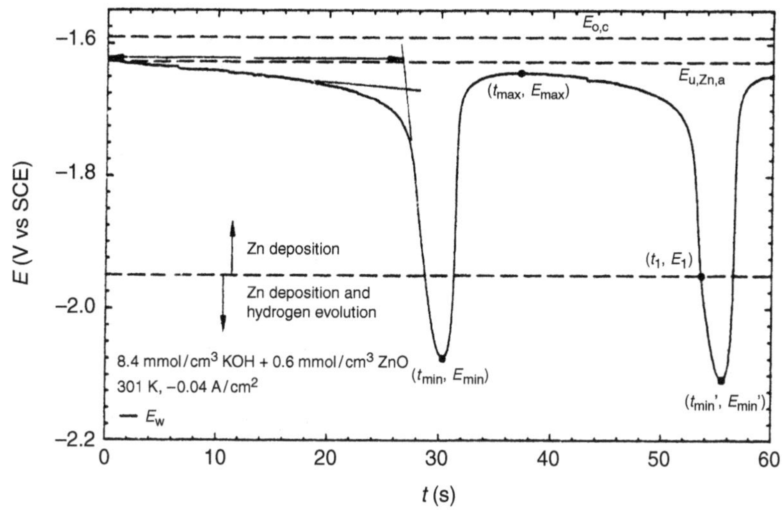

Fig. 5.69 Oscillatory variations of the Zn electrode potential and characteristic points used to define variables. Reproduced from [66] by permission of The Electrochemical Society

oscillatory course is composed, are shown in Fig. 5.69. Noteworthy, the nucleation of hydrogen bubbles occurs only when the concentration of dissolved hydrogen exceeds several times the thermodynamically determined saturation value. Thus, the bubble formation occurs later than the electroreduction of H^+ ions to H_2 begins, roughly at the minima of the oscillating potential shown in Fig. 5.68 [66], thus similarly as for Sn electro-deposition.

The mechanism of oscillations for both Sn and Zn electrodeposition is thus quite similar and therefore can be summarized in a way, suggesting the search of other systems of that type [67]: (1) the limiting current density for the deposition of metal must be observed before the hydrogen evolution reaction becomes significant; (2) the imposed current density must exceed the limiting value for the metal deposition, allowing thus the electrode potential to reach the hydrogen evolution process upon increasing concentration polarization of metal ions in the solution. Finally, when the electrode surface becomes substantially rough (due to powdery or dendritic deposits) and thus its real surface area increases, the local current density eventually drops below the limiting value and then oscillations cease since the electrode potential no longer attains the values negative enough to cause hydrogen evolution and the onset of convection. Usually, due to this effect, oscillations cease after several cycles.

One should add that, similarly as in the case of Sn deposition, also for Zn it was found a significant correlation between the oscillatory regime (both under galvanostatic and potentiostatic conditions) and dendritic structure of deposits on the electrode surface. According to Suter and Wong [68], the more orderly the dendrites, the larger the amplitude of oscillations. It was further suggested [69] that the oscillations of the concentration field, and in consequence, of the current density, are caused by a competition between the faceted dendritic growth and the diffusion of electrolyte toward the growing dendrite. From the point of view of nonlinear dynamics, it is also noteworthy that formation of various surface deposits in the thin-layer Zn|$ZnSO_4$|Zn system may cause such transitions, like period-doubling bifurcations leading from relaxation, periodic oscillations to deterministic chaos [70, 71].

5.9 The Self-Organized Electrohydrodynamic Convection

5.9.1 Principles of the Electrohydrodynamic (EHD) Convection

The driving force (expressed here as volume force in [N m^{-3}]) for the EHD convection originates from the interaction of the electric field **E** with the excess space charge density q_{ex}, carried by the fluid:

$$\mathbf{F}_{ex} = q_{ex}\mathbf{E} \tag{5.39}$$

This nonzero space charge density (i.e., deviation from electroneutrality) requires nonlinear spatial distribution of the electric field, according to the *Poisson's* equation:

$$q_{ex} = \varepsilon_r \varepsilon_0 \text{ div } \mathbf{E} = -\varepsilon_r \varepsilon_0 \nabla^2 \varphi \tag{5.40}$$

where ε_r is the relative permittivity of the fluid, ε_0 is the permittivity of the vacuum, and φ is the electric potential. The EHD convection sets in when the fluid becomes unstable under the influence of this force. One should realize that even relatively small deviations from electroneutrality can produce quite significant electric forces and the convection becomes then the way to decrease the system's free energy. Obviously, the volume force $\mathbf{F}_{ex}$ defined in this way can be inserted to the Navier–Stokes equation (5.1) and the corresponding vector field of the fluid velocity can be computed.

Generally, all charge carriers can move due to diffusion, migration, and convection, so the total flux of every carrier s of a charge $z_s e$ includes contributions from all these kinds of transport, respectively:

$$\mathbf{f}_s = -D_s \nabla c - \frac{z_s F D_s}{RT} c_s \nabla \varphi + c_s \boldsymbol{\vartheta} \tag{5.41}$$

In turn, the total faradaic current, summarizing the contributions from all charge carriers, is related to the individual fluxes in a following way:

$$\begin{aligned}\mathbf{j} = F\sum_{s} z_s\mathbf{f}_s &= -F\sum_{s} z_s D_s \nabla c_s \\ &\quad -\frac{F^2}{RT}\left(\sum_{s} z_s^2 D_s c_s\right)\nabla\varphi + F\left(\sum_{s} z_s c_s\right)\boldsymbol{\vartheta} \\ &= -\sum_{s} D_s \nabla q_s - \left(\sum_{s} u_s q_s\right)\nabla\varphi + q_{ex}\boldsymbol{\vartheta}\end{aligned} \tag{5.42}$$

where $u_s = FD_s|z_s|/RT$ is the ionic mobility of species s, according to this Einstein - Smoluchowski equation. The spatial distribution of the faradaic current density meets the principle of the charge conservation:

$$\operatorname{div}\mathbf{j} = -\frac{\partial q_{ex}}{\partial t} \tag{5.43}$$

In classical, low-voltage electrochemistry, particularly in the presence of high excess of indifferent supporting electrolyte, the excess charge q_{ex} is negligible outside the region of the double layer, so the motion of the fluid does not contribute to the conductivity of the system; also the condition (5.43) reduces to zero divergence of the current density vector. But for the case of the high-voltage EHD, considered below for physical systems, the fluid may exhibit so significant deviations from electroneutrality ($q_{ex} \neq 0$), that the relative contribution of the $q_{ex}\boldsymbol{\vartheta}$ term to the overall current density in Eq. (5.42) can become quite remarkable.

The EHD convection can occur in various fluid systems, under very different conditions, particularly with respect to the magnitude of voltage applied between the electrodes. It is thus reasonable to distinguish between the "high voltage," "medium voltage," and "low voltage" EHD, corresponding to the voltages varying from the tenths of kilovolts to only a few volts. The high-voltage version of this phenomenon, when the voltage of ca. 20 kV or more is applied to the thin (a few mm thick) layer of dielectric fluid in order to inject charge carriers (electrons or ions from the ion-exchange membranes) is a subject of interest of physicists rather than of chemists. Nevertheless, the basic knowledge about this kind of phenomena is quite instructive. According to Felici [72, 73], the EHD convection was first reported already in 1838, when M. Faraday has described his own experiment with the flow of current through the dense insulating fluid which then set into a violent motion.

In the simplest version of the high-voltage EHD convection, charge carriers of only one sign are injected from one side of the thin-layer fluid system. This is the so-called *unipolar charge injection* which may cause the fluid to move and self-organize into various convective patterns (rolls, cells, etc.) of a morphology resembling that of classical Bénard–Rayleigh convection, although now also under rather isothermal conditions. The fact that fluid motion takes a form of the self-organized rolls under EHD conditions means the generation of the "vorticity" (vortex vector) $\boldsymbol{\omega}$ due to the nonzero $(\nabla q_{ex}) \times \mathbf{E}$ vector [73]. If a strong positive feedback between $\boldsymbol{\omega}$ and $(\nabla q_{ex}) \times \mathbf{E}$ exists, the rotational motion of the fluid is accelerated to the level limited by dissipative (viscous) interactions.

The simplest experimental manifestation of the onset of EHD convection is based on the measurement of electric current flowing through the thin layer of the fluid under applied voltage. For the case of the above mentioned unipolar charge injection, in the *absence* of EHD convection this current is still relatively low, being the pure *space-charge-limited* current (SCLC) [73, 74]. Its value is defined by the Langmuir-Child's law:

$$j_{\mathrm{sclc}} = \frac{9}{8}\varepsilon_{\mathrm{r}}\varepsilon_0 u_{\mathrm{s}}\frac{U^2}{d^3} \tag{5.44}$$

where U is the voltage applied to the thin layer of a fluid of a thickness d and u_s is the ionic mobility of (positive or negative) injected charge carriers (s). In terms of this equation, this is exclusively a migration current or, in other words, the Ohm's law is met. The onset of convection obviously accelerates the motion of charge carriers and then the current exceeds, even considerably, the value predicted by Eq. (5.44). Besides direct optical observation of the fluid, this enhancement of the current can be one of the premises for the occurrence of additional, convective transport in the system.

Visual observations allowed to report cooperating convective rolls in several experimental systems. One of the examples includes the thin layer of silicon fluid ($d \leq 0.5$ mm) exposed to an electron beam [75]. The injected electrons became trapped by the liquid molecules (or by impurities) to form negative ions and this created an internal space charge and associated electric field. Below the critical current density the convection was not observed and the pure SCLC [cf. Eq. (5.44)] was reported. The onset of convection occurred only above certain critical value of current, what is analogous to the critical temperature gradient necessary to induce density-driven convection in the gravitational field. For the sake of generalization, Schneider and Watson [74] have defined the electrical analogue of the *Rayleigh* number:

$$Ra_{\mathrm{e}} = \frac{\varepsilon_{\mathrm{r}}\varepsilon_0 U}{u_{\mathrm{s}}\eta} \tag{5.45}$$

The critical value of Ra_e, above which the EHD convection sets in, was theoretically estimated as close to 99 and this assessment was confirmed experimentally [75]. Also other "electric" analogues of dimensionless numbers, characterizing the convection, were defined for the case of EHD phenomena, like, e.g., electric *Nusselt* number, electric *Prandtl* number, electric *Reynolds* number [76].

The convection induced by unipolar charge injection was experimentally studied also for the following systems (fluid/injected ions): methanol/H^+, chlorobenzene/Cl^-, ethanol/H^+ or Cl^-, nitrobenzene/Cl^-, propylene carbonate/Cl^-, nitrobenzene/K^+ [77], and pyralene/Cl^- [76]. In the latter case, for the thin layer ($d = 0.3$ mm) of the pyralene 1460 there were observed hexagonal convective cells at overcritical value $U = 1.1U_c$ and slightly elongated (quasi-cellular) motion at $U = 2U_c$ where

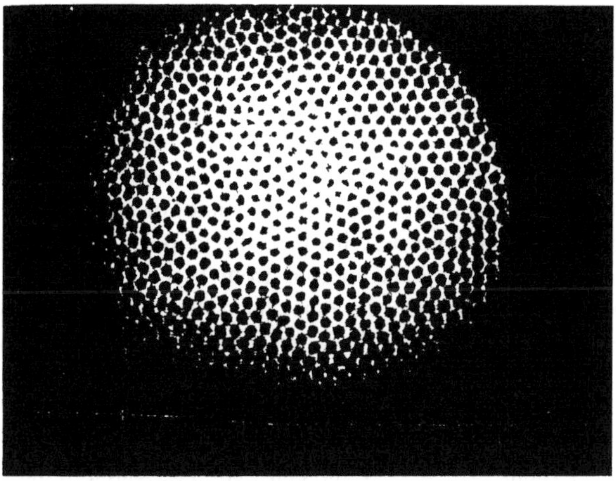

Fig. 5.70 Schlieren photograph of the motion pattern in Pyralene 1460 under strong unipolar injection ($U = 60$ V, thickness of the fluid layer $d = 0.3$ mm). Reprinted from [78], copyright 1978, with permission from Elsevier

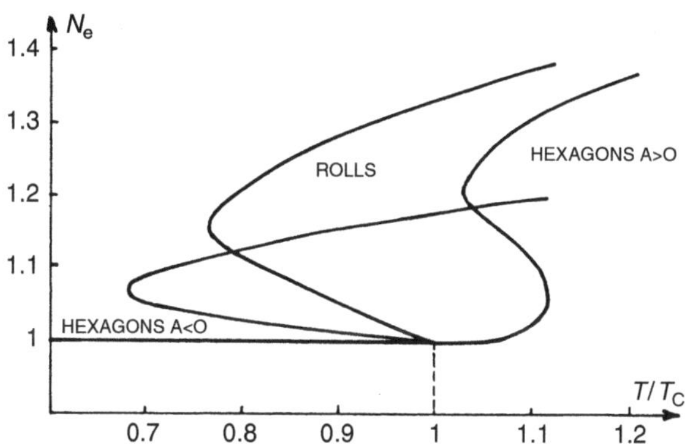

Fig. 5.71 The stability diagram revealing possible EHD patterns in the simple model of the unipolar charge injection to the thin layer of dielectric fluid. The electric Nusselt number (expressing the ratio of the convective current to the current in convection-less regime) is plotted against the critical stability parameter. Reprinted from [78], copyright 1978, with permission from Elsevier

$U_c = 50$ V means the critical voltage for the onset of EHD convection (Fig. 5.70). These studies have shown that not always satisfactory agreement is observed between theoretical and experimental critical values of dimensionless numbers, indicating the onset of convection. According to Atten [79] one of the reasons for that discrepancy is that theoretical stability criteria are defined for the stationary state while the experiments are usually carried out under transient conditions. Furthermore, the derivation of such numbers is theoretically derived based on the linear stability analysis while the phenomenon of self-organized EHD convection is evidently nonlinear. Therefore, according to Atten and Lacroix [78] a *nonlinear* stability analysis of such processes would be necessary which approach would explain better the formation of convective rolls and hexagonal cells. Figure 5.71 shows the diagram of possible patterns.

In the case of *bipolar* (or ambipolar) charge injection, two ion-exchange membranes (e.g., with mobile Na^+ and Cl^- ions, respectively) are applied as the opposite electrodes [80]. Although essential principles of the flow of current and of EHD onset are similar to those described for the unipolar injection, the system is more complicated, since the ions injected from two sides eventually meet and recombine in the fluid phase. On the other hand, the bipolar injection makes the system more symmetrical and since the larger space of the liquid is now subjected to driving electric forces, the convection sets in at lower critical voltages that are given by the new critical value (=46) of the electrical *Rayleigh* number (5.45). Experimental results obtained for pyralene 1460/(Na^+, Cl^-) system [81] have shown that just above the critical voltage $U_c = 25$ V the convection appeared in the form of "two-dimensional," or fingerprint-like rolls (considered the modes preferable for such systems of a higher symmetry) which turned into hexagonal "honeycombs" at voltages approximately doubling the above critical value. A theoretical nonlinear stability study [81] of such phenomena leads to conclusions which are quite well concordant with the experiments. In particular, contrary to the unipolar charge injection, for the critical voltage U_c (1) there is no discontinuous increase of current (only the increase of the slope of the *I–U* relationship is observed), (2) the hysteresis of the current as a function of the voltage scan is *not* present, and (3) the fluid motion indeed attains the shape of the rolls. Finally, the convection induced by *bipolar* charge injection is much more similar to the classical *Rayleigh–Bénard* stability problem than the fluid motion caused by the unipolar injection.

5.9.2 EHD Convection in Liquid Crystals

When the electric field develops in the thin layer of the nematic liquid crystal, the convection exhibiting self-organization into various patterns can set in. Such phenomenon was first reported by Williams [82, 83] for a thin layer ($d = 10$–200 μm) of nematic liquid phase (*p*-azoxyanisole) exposed to electric field of $E =$ 1,000–3,500 V cm^{-1}. As in the case of the density-driven convection in the horizontal thin-layer system (Sect. 5.1), the fluid motion was reported only above the critical electric field intensity (e.g., for $E \approx 1{,}000$ V cm^{-1}). This motion took the form of regular patterns, consisting of long parallel regions, called now the *Williams domains.* These patterns are observed for all temperatures within the range between the melting point of p-azoxyanisole (117°C) and the point of phase transition to the isotropic liquid phase (134°C). Noteworthy, such patterns were observed for both dc and ac ($f = 1$ kHz) electric fields. An important premise for the convective origin of these patterns is the dependence of the domain size on the thickness of layer of the liquid crystal, like in the case of Bénard–Rayleigh

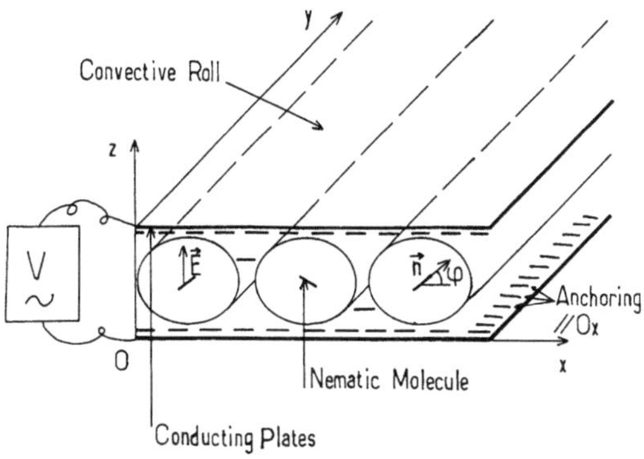

Fig. 5.72 Schematic representation of the experimental setup for the observation of Williams domains in the thin layer of the liquid crystal upon externally applied electric fields. The size of the cross-section of convective rolls adapts to the layer thickness [84]. The conducting plates can be made of glass with the transparent tin oxide or indium oxide conductive layer. Reprinted from [84] with kind permission from Springer Science + Business Media

instability. Typical experimental arrangement and schematic view of the convective rolls are shown in Fig. 5.72.

The optical observation of the Williams domains is possible due to differences in refractive index of the regions in which ordering of rod-like molecules of the nematic phase occurs, when the electric field is applied. Typical shapes of Williams domains and other types of EHD patterns are shown in Fig. 5.73.

By varying the amplitude and frequency of ac voltage one can observe also more complex patterns, including varicose-like and two-dimensional (rectangular), and even transition to chaotic structures. This means the existence of further analogies between the convection induced in liquid crystals and analogous phenomena in other self-organizing convective systems with different driving forces.

The Williams domains were later analyzed theoretically, among others by *Helfrich* [85] who has elaborated the model of these phenomena, based on the idea of a space charge combined with the anisotropy of the liquid crystal conductivity. Under typical conditions, the application of ac, instead of dc voltage, prevents the continuous electrolytic decomposition of the liquid crystal at the interface. The electric conductivity is so low that significant current cannot flow through the system, but the slight deviations of the liquid phase from electroneutrality (nonzero space charge) may develop. On the other hand, for the formation of a space charge, large enough to create a conduction-induced torque, a concentration of ions higher than certain minimum value is necessary. The latter remark means that formation of the Williams domains is partly the electrochemical problem and that the convection in liquid crystals may be dependent on the composition of the particular liquid crystal phase which is not always precisely known, as small amount of ionic species may come from various impurities. The electroconvection in liquid crystals is still quite intensively studied (see, e.g., [86, 87]).

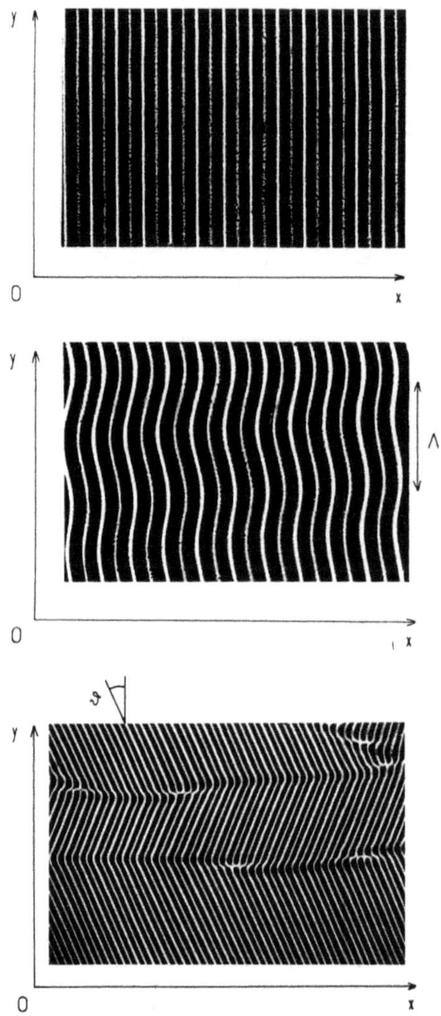

Fig. 5.73 Various types of EHD patterns in the liquid crystal observed upon increasing voltage with constant ac frequency 60 Hz: (*top*) classical Williams domains, (*middle*) undulatory patterns; (*bottom*) zig-zag patterns. Thickness of the liquid crystal layer $d = 50$ μm, lateral cell dimensions 2×3 cm. The initial molecular alignment of the nematic denoted by the unit vector **n** (the director) is parallel to x direction. The electric field is applied across the sample along z direction. The first manifestation of EHD pattern yields ca. 600 rolls, and the diameter of each roll = 3 cm/ 600 = 50 μm is well concordant with the layer thickness. Reprinted from [84] with kind permission from Springer Science + Business Media

5.9.3 EHD Convection in Colloidal Systems

The EHD convection can occur also in colloidal systems, when the current is conducted through charged micelles moving in an electric field (electrokinetic phenomenon). The effect of illumination on the motion of electrophoretic particles, leading to the term "photoelectrophoresis," was discovered by Tulagin [88] for particles of organic dyestuffs, with some practical applications to color display system.

For the understanding of the phenomena described below, it is necessary to invoke basic scheme of the interaction of photons with semiconductor. Absorption of light in the semiconductor causes the formation of the electron–hole pair (e^-, h^+) with the electron in the conduction band and the holes in the valence band. The holes may oxidize the species in the solution or accept the electrons from the cathode, while photoinduced electrons may reduce dissolved species or be accepted by the anode in the electrochemical experiment. If there is no kinetic balance between these oxidation and reduction processes, the semiconductor particle acquires a respective net electric charge. Thus, the measurements of the electrophoretic mobility of such particles allow to detect both the net surface charge and its eventual changes due to photoinitiated redox reactions. The relation between the electrophoretic mobility u_E of the particles and their zeta potential ζ, in the medium of relative permittivity ε_r and dynamic viscosity η, can be expressed using Smoluchowski equation:

$$u_E = \frac{v_E}{E} = \frac{\varepsilon_0 \varepsilon_r \zeta}{\eta} \tag{5.46}$$

where v_E is the particle velocity and E is the electric field intensity.

The electrophoresis of colloidal semiconductors—illuminated TiO_2 particles dispersed in aqueous media of various pH—was described, among others, by Boxell and Kelsall [89, 90]. Earlier the properties of dispersed TiO_2 particles under illumination were studied among others by Bard et al. [91, 92].

Regarding the self-organized motion of fluid, accompanying with these phenomena, one should invoke a series of papers by Takahashi et al. [93–97] who have described the photoelectrophoresis of TiO_2 (rutile-type) particles of a 1-μm diameter, suspended in organic liquid (1:1 mixture of benzene and olive oil). The experimental setup consisted of the thin layer of colloidal suspension placed between two transparent electrodes, with the interelectrode distance of the order of tenths of micrometers [95]. Alkyl chains are bounded to the surface of TiO_2 particles, which in the dark acquire thus a negative ξ potential and therefore, when the electric field is applied, they migrate to the anode and remain there at rest. But if the anode is transparent and its upper surface is illuminated, the TiO_2 particles undergo there photoexcitation, with the formation of the electron–hole pair. This is followed by the oxidation process (positive charge injection) at the anode, as a result of which the net charge (or zeta potential) of TiO_2 particles changes to positive values and the particles migrate toward the cathode. Injection of the electrons from the cathode causes the particles to recover negative charges and migrate back to anode, etc. This movement of colloidal particles was accompanied by a macroscopic motion of the surrounding liquid which exhibited self-organization into convective patterns [94, 95], resembling those of the classical Bénard–Rayleigh instability (Fig. 5.74).

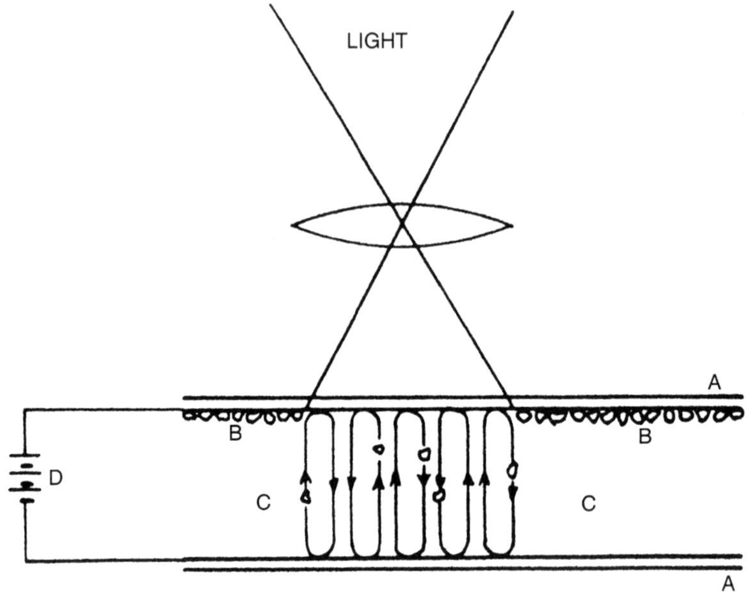

Fig. 5.74 Movement of TiO_2 particles under photoelectrophoresis: (A) transparent electrodes, (B) TiO_2 particles, (C) organic liquid containing dyes, (D) electric source. Reprinted with permission from [95]. Copyright 1979 American Chemical Society

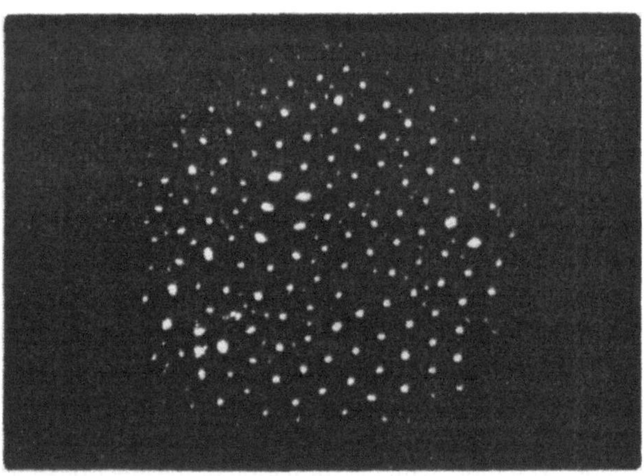

Fig. 5.75 Hexagonal convective pattern due to electrohydrodynamic instability induced by photoelectrophoresis of TiO_2 particles. The regular array of *white spots* represents the positions of streams from the anode to the cathode. Reprinted with permission from [95]. Copyright 1979 American Chemical Society

Further analogies with Bénard–Rayleigh instability concern with the critical voltage at which this convection sets in. Strictly speaking, upon increasing the voltage applied between the electrodes two thresholds were observed. The first one is a voltage $U_0 \approx 80$ V which is critical for the charge transfer between the anode

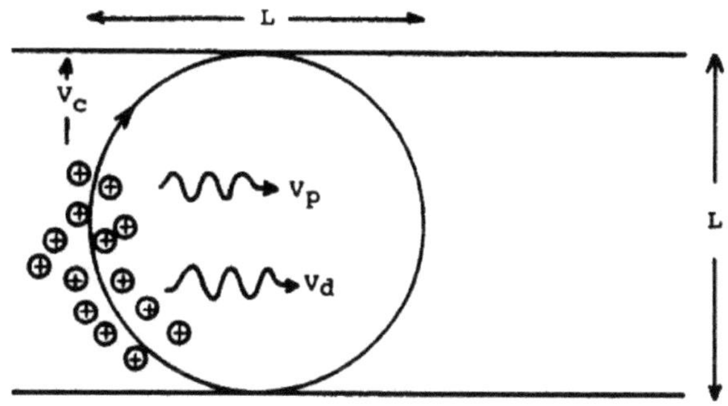

Fig. 5.76 Schematic representation of the cross-section of a single convective roll, generated by the movement of positively charged colloidal TiO_2 particles, dragging the fluid. *Symbols* denote the velocities: v_c—of convection normal to the electrode surface, v_P—of migration, and v_d—of diffusion parallel to the electrode surfaces. Reprinted with permission from [95]. Copyright 1979 American Chemical Society

and the photoexcited negatively charged particles. When the voltage was higher than U_0 but still lower than the next critical value U_c (≈200 V), i.e., when the fluid still remained quiescent, the relatively small current flew due to only migration of charged particles. Only at $U > U_c$ the convective circular flow set in and its *hexagonal* arrangement was observed through the transparent anode (Fig. 5.75).

Furthermore, the characteristic size of this pattern (the distance between the neighboring spots in Fig. 5.75) was of the same order as the thickness of the suspension layer, which observation only confirms the convective origin of the reported phenomena (cf. Fig. 5.5). Analogously to the thermal convection (Sect. 5.1), the source of threshold for the onset of the fluid motion can be understood as the result of the competition between the convective motion driven by charge particles in the direction normal to the electrode surfaces and the destroying effect of diffusion and migration of particles along the direction parallel to the electrodes. In terms of notation used in Fig. 5.76, the self-organized convective flows are maintained provided that the vertical convection velocity v_c is sufficiently high, compared to horizontal components of the migration v_p and diffusion v_d velocities.

Accordingly, the following dimensionless numbers, leading to quantitative criteria of EHD stability of the system considered, were derived [95]:

$$R = \frac{v_c}{v_p} \propto \frac{\varepsilon_r V}{\mu \eta} \tag{5.47}$$

$$R' = \frac{v_c}{v_d} \propto \frac{nqVL^2}{D_0 \eta} \tag{5.48}$$

where ε_r is the relative electric permittivity of the benzene–oil medium (ca. 3), $V = EL$, i.e., the product of the electric field intensity E and the thickness of the suspension layer L; μ, mobility of a charged colloidal particle; η, dynamic viscosity of the benzene–oil medium; n, number density of particles carrying electric charge q; and D_0 is the diffusion coefficient of those particles. In order to compute the critical values of those numbers, Takahashi et al. [95] have determined the mobility, the charge, and the diffusion coefficients of TiO_2 particles in the given medium and calculated the following critical values of R and R':

$$R = 115 \tag{5.49}$$

$$R' = 2.25 \times 10^{11} \tag{5.50}$$

The convection appears to be much faster than diffusion which thus plays relatively insignificant role in destruction of the fluid motion. Thus, this is the migration of particles along the vector of the electric field, produced by convective perturbation, which decides whether this perturbation will decay or will be amplified toward the macroscopic self-organized fluid motion.

In the simplest version, the stability analysis of the Navier–Stokes equation applied to this system discussed the role of the dimensionless number (5.47) in the prediction of the evolution of the convective fluid perturbation, yielding the critical value 115, quite close to 99 determined earlier by Schneider and Watson [74], but for the unipolar charge injection which is obviously not the case of the present colloidal system. Also, one can pose the question, if the system studied meets (or not) the electroneutrality condition. Therefore, Takahashi et al. [97] have extended later their analysis for the bipolar charge injection, with the emphasis put on the thermodynamic characteristics of the system analyzed. The photoelectrophoretic system was described in more detail: in the dark the TiO_2 particles of 1 μm diameter carried a negative charge of −48e, which under constant illumination, at the anode, underwent recharging, exhibiting the positive charge of 3,500e which caused the reversed migration of TiO_2 toward the cathode; there the particle recovered negative charge, being now equal to −660e, and could migrate back to anode. In consequence, in the model, the cyclic photoelectrophoretic movement between the cathode and the anode involved a particle that could exist in two states: $q_+ = 3{,}500e$ and $q_- = -660e$. Because of difference in the absolute values of these charges, the migration velocities of positively and negatively charged particles were different, with $v_+ > v_-$, the symbols (+) and (−) meaning the velocities of the opposite directions of motion of cations and anions, respectively. Furthermore, in the authors' interpretation the quiescent fluid in its steady state was characterized with no space charge in the bulk, i.e., the electroneutrality condition was met there: $q_+n_+ + q_-n_- = 0$. When the sufficiently large voltage was applied, the charge was injected at respective electrodes, and the charged fluid moved in the electric field: in the region of uniformly distributed charge density ρ_+ the fluid moves with the velocity vector $+v$, in the region of uniformly distributed charge density ρ_- with the velocity $-v$, while in between the quiescent fluid remains electroneutral (cf. Fig. 5.77).

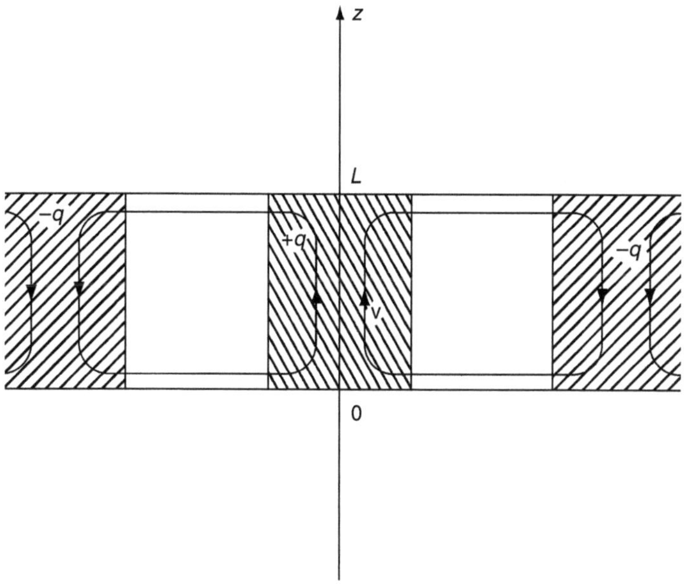

Fig. 5.77 Model of convective unit cell, sketching the motion of a charged ($+q$, $-q$) fluid in the zones of oppositely directed streams, separated with the areas of a quiescent, electroneutral fluid. Based on [97]

The authors have further analyzed the geometry of the hexagonal cell in terms of the Wigner–Seitz's unit cell[1] and derived the characteristic size (lattice constant) of the most favorable hexagonal convective cell as being proportional to the interelectrode distance L (the thickness of fluid layer), according to the dependence:

$$a = 2 \times 3^{-1/4} L \cong 1.520L \tag{5.51}$$

which also indicates that the cross-section of the convective roll is not circular but slightly flattened.

Finally, the new dimensionless number describing the hydrodynamic stability of the system considered was derived:

[1] The (primitive) Wigner–Seitz cell—the geometrical construction used mainly in crystallography, for the description of the crystal lattice: such a cell around given lattice point means the locus of points in space which are closer to that lattice point than to any other lattice points. In two-dimensional lattice, the Wigner–Seitz cell is formed by perpendicular bisectors of every line connecting the given lattice point with the neighboring lattice points. This formalism can be applied also to description of hexagonal convective cells, as they, just like Wigner-Seitz cells, fill all area (or space) without leaving any gaps.

$$R = \frac{\varepsilon_r^2 \varepsilon_0^2 V^2}{nq\mu\eta L^2} \tag{5.52}$$

and its critical value, corresponding to the onset of the photoelectrohydrodynamic convection, was estimated as close to eight. In this equation, n means the number density of particles carrying charge q and exhibiting mobility μ in the medium of dynamic viscosity η [97].

5.9.4 *The Low-Voltage EHD Convective Luminescent Patterns*

5.9.4.1 Experimental Setup and Typical Patterns

In this section we describe the self-organized convection, visualized by the patterns of electrochemiluminescence (ECL) emitted from the thin-layer cell, in which the interelectrode distance varied from ca. 10 to 200 μm. The luminescent patterns were observed during the dc, potentiostatic or potentiodynamic electrolysis of the molecules of rubrene (5,6,11,12-tetraphenyl-naphthacene), dissolved in 1,2-dimethoxyethane. In the following description, rubrene will be abbreviated by R and 1,2-dimethoxyethane by 1,2-DME.

The principle of observed phenomena is the following: if the proper voltage ($U = 3$–6 V) is applied between the electrodes, rubrene molecules are being reduced at the cathode and oxidized at the anode. The resulting ionic radicals (R^+, R^-) diffuse and migrate toward the respective opposite electrodes, i.e., toward each other. When they meet in the solution, they recombine, restoring electroneutral, but energetically excited rubrene molecules R*, which return to the ground electronic state with the emission of orange light [98], visible with a naked eye through the transparent electrode(s). As such electrode, a glass plate on which the thin layer of semiconducting ITO (indium-tin oxide) was sputtered, served. In the absence of convection, and for the homogeneous electrode surfaces the observed luminescence should be uniformly distributed. However, if the fluid is set into intensive convective motion, the luminescence becomes patterned. This is because the convective flows of fluid change spatial distribution of rubrene ions, so that R^+ and R^- recombine with different rates in various places; in practice it usually means that luminescence is visible only in selected places, determined by the spatial geometry of convective flows. Hence, rubrene molecules play here a double role: (1) of the electroactive substance and (2) of the "light indicator" of the self-organized convection. These convective patterns of luminescence, involving *quasi*-hexagonal arrangements of bright points on the dark background, were discovered by Köstlin and Schaper [99], in search of new types of electrodisplays. Schaper [100] has also suggested the application of such a thin-layer ECL cell for the detection system in high-pressure liquid chromatography systems. The same group of researchers also outlined the mechanism of the generation of convection,

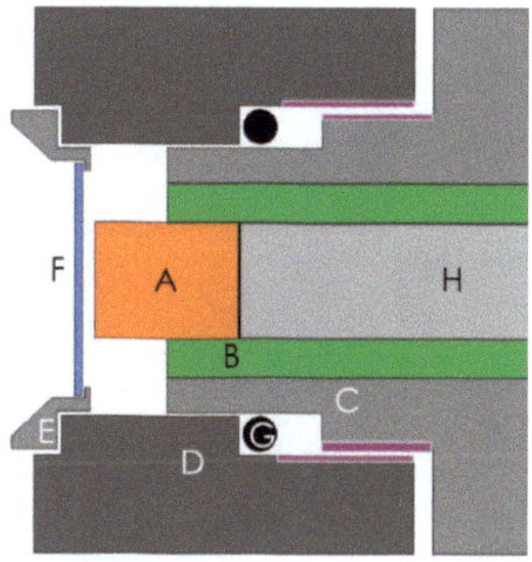

Fig. 5.78 Simplified cross-section of the thin-layer electrochemical cell: A: copper (gold, platinum) electrode, B: KEL-F insulating layer; C: metallic (bronze and steel) movable block; D: metallic (bronze) fixed block connected with part C by the thread; E: steel ring for fixing the glass electrode; F: transparent electrode (glass plate coated with ITO); G: sealing O-ring; H: copper rod (for copper electrode A and H denote a single copper rod). For clarity, several O-rings, threads, and supply and outlet lines for the solution and the inert gas (argon) are omitted. Reprinted with permission from [108]; color version. Copyright 1998 American Chemical Society

suggesting its EHD nature [101–103]. Similar phenomena were described later by Jaguiro et al. [104–107] who have extended the set of observed luminescent convective structures for more elongated patterns, resembling fingerprints (similar to those known for the thermal convection) [107]. Concerning the mechanism of these phenomena, all the above authors postulated that the flow of current occurred in the presence of only negligible amount of ionic contaminants and thus the appearance of charge carriers in the solution might be explained in terms of the predominant "unipolar charge injection" (see Sect. 5.9.1). In this particular case, this practically unipolar injection would be realized through the predominant production of rubrene anions at the cathode, with only modest formation of rubrene cations at the anode. However, if this were the case, in the absence of convection the current density should be close to the value predicted by (Eq. 5.44), being then of the order of ca. 1 nA cm^{-2}, i.e., six orders of magnitude lower than the experimentally observed current density (ca. 1 mA cm^{-2}). Such difference cannot be explained by the enhancement of the solution conductivity through the convective transport of the space charge of the current due to convection [cf. term $q_{ex}\boldsymbol{\vartheta}$ in Eq. (5.42)], since then the unrealistic value of the fluid velocity, reaching ca. 600 m s^{-1} would have to be assumed. Thus, these early suggestions on the mechanism of the flow of current and the source of the space charge required reconsideration.

More systematic experimental and mechanistic studies of these phenomena, including numerical modeling, were later performed by Orlik et al. [108–112]. Here these works will be briefly summarized. Figure 5.78 shows the schematic cross-section of the thin-layer electrolytic cell which allows to change smoothly the interelectrode distance. The magnitude of distance was verified based on capacity measurements of the capacitor formed by the parallel planar electrodes A and F.

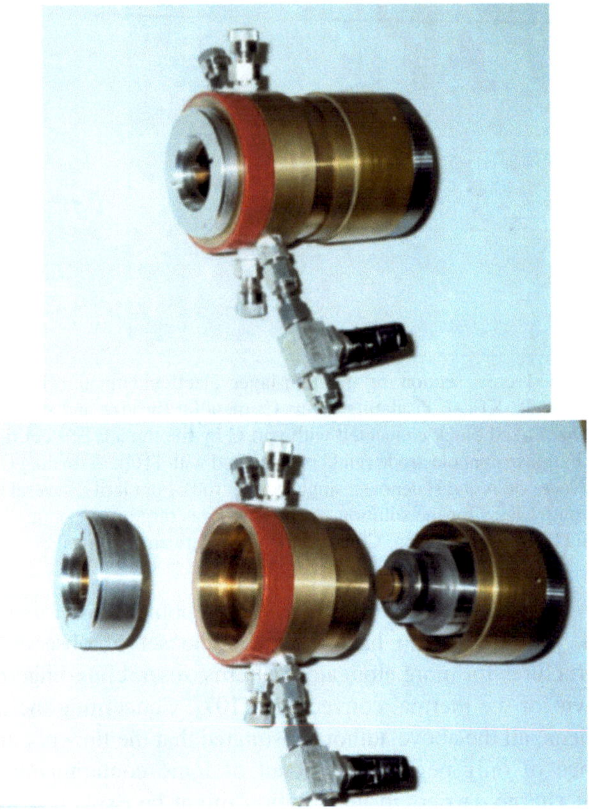

Fig. 5.79 The thin-layer electrolytic cell for the observation of the luminescent convective patterns accompanying the electroreduction of rubrene in 1,2-dimethoxyethane, and corresponding to the construction principles shown in Fig. 5.78. *Top picture*: assembled cell; *bottom picture*: cell disassembled into parts containing the transparent glass electrode (*left*) and the cylindrical metal (copper) electrode (*right*)

The realistic view of the cell, constructed according to this scheme is shown in Fig. 5.79.

Prior to the experiments both rubrene and the solvent were very carefully purified, particularly from traces of oxygen and water that could act as the scavengers of the formed rubrene radicals. The luminescent patterns were observed with the CCD camera fixed in front of the transparent electrode (**F**) and connected to the S-VHS recorder. In every experiment the cell was filled up with the saturated (ca. 4×10^{-3} mol dm^{-3}) solution of rubrene in 1,2-DME. However, only negligible current was then measured and no visible luminescence was reported. At this stage of studies it was thus very important to note, that for the appearance of luminescence the current density must be as high as 0.1–2 mA cm^{-2}, but this required an enhancement of the solution conductivity through *addition* to a sample of a (relatively low) amount of the supporting electrolyte: 5×10^{-6} to

1.2×10^{-5} mol dm^{-3} tetrabutylammonium hexafluorophosphate (abbreviated further by [TBA$^+$][HFP$^-$]). Obviously, the current was then conducted not only by rubrene ions produced at the electrodes, but also by the ions of this supporting electrolyte. Evidently, the solutions used in earlier experiments, described in the literature cited earlier, had to contain ionic contaminants at the concentration level precisely unknown, but comparable with now *intentionally* added amount of [TBA$^+$][HFP$^-$]. In consequence, the original idea of the unipolar charge injection of rubrene anions as the predominant charge carriers had to be replaced by a practically symmetrical *bipolar* injection of rubrene cations and anions at the respective electrode/electrolyte interfaces (i.e., like in typical, low-voltage electrochemical systems). In conclusion, the presence of supporting electrolyte in the thin-layer system appeared to be necessary for *two* reasons: (1) it ensured the sufficient electric conductivity of the solution and (2) its ions were necessary for the formation of electric double layers at the electrodes, within which the sufficiently large potential drop had to develop for the low-voltage electrolysis to occur. The role of the first factor was experimentally proved by the increase of the intensity of luminescence with increasing current, depending on the concentration of [TBA$^+$] [HFP$^-$] (Fig. 5.80a). In turn, the role of the second factor was clearly visualized by the correlation of the increasing region of luminescence with the rise of current flowing in the subsequent potential cycles (Fig. 5.80b).

Obviously the way in which the latter conclusion was drawn, requires a more detailed explanation. The effect shown in Fig. 5.80b can be observed only for such a thin-layer cell in which the interelectrode space remains in contact with the surrounding solution (like in the construction shown in Figs. 5.78 and 5.79). When the voltage is applied, the ions of supporting electrolyte from the solution of supporting electrolyte are consumed to the buildup of the double layers at both electrodes. The initial concentration of the electrolyte is, however, so low that the bulk of the solution in the interelectrode space becomes largely depleted with the TBA$^+$ and HFP$^-$ ions. The ohmic resistance of the solution thus significantly increases and thus the sufficiently large current of rubrene electrolysis cannot flow; also, the potential drop within the double layer could become yet insufficient for the occurrence of the electron transfer between the rubrene molecules and the electrodes. But simultaneously this decay of the electrolyte in the interelectrode space causes its concentration gradient with respect to the large volume of intact, surrounding solution. In this way at the edges of the electrodes, there induces the diffusion of missing electrolyte into the interelectrode space. The progress of the concentration front of supporting electrolyte as a function of time is matched by the gradual increase of zone of orange luminescence and simultaneous increase of faradaic current (see again Fig. 5.80b).

The process of recovering the initial electrolyte concentration in the interelectrode space can be accelerated by the cyclic decrease and increase of the interelectrode distance which facilitates the exchange of the solutions between the interelectrode space and its surrounding. This "conditioning" of the sample was used in order to obtain possibly uniform luminescence distribution over the entire electrode surface. Such procedure is better than a simple increase of the initial

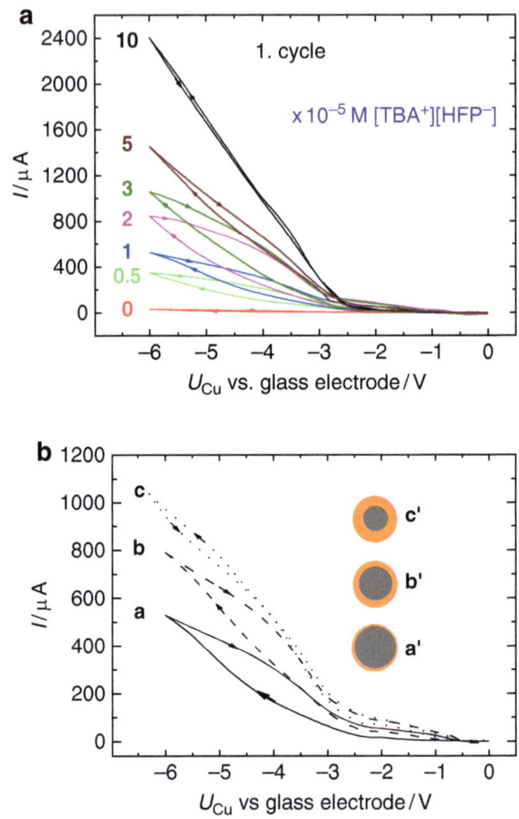

Fig. 5.80 Current–voltage (*I*/*U*) characteristics of the electrochemical thin-layer cell ($d = 64$ μm) filled with a saturated solution of rubrene in 1,2-dimethoxyethane, for a potential sweep rate 50 mV s^{-1}. (**a**) First current–voltage cyclic curve for different concentrations of [TBA$^+$][HFP$^-$]: (1) 0, (2) 5×10^{-6}, (3) 1×10^{-5}, (4) 2×10^{-5}, (5) 3×10^{-5}, (6) 5.2×10^{-5}, (7) 9.7×10^{-5} M; (**b**) a–c: three subsequent current–voltage cyclic curves for 1.2×10^{-5} M [TBA$^+$][HFP$^-$]; a′–c′: the luminescent (*orange*) areas of the (*dark in gray zones*) *circular electrode surface* observed simultaneously with the *I*–*U* curves a–c. Reprinted with permission from [108]; color version. Copyright 1998 American Chemical Society

concentration of the supporting electrolyte in the solution, since for too high concentrations of [TBA$^+$][HFP$^-$], when the coulombic forces are expected to be less significant, the convection was found to decay (the patterns of light turned into almost homogeneous layer of luminescence), which fact, on the other hand, corroborated the earlier suggestion on the EHD nature of this convection. The "conditioning" allowed one to find the interval of [TBA$^+$][HFP$^-$] concentration, optimum for the observation of luminescent patterns.

Typical examples of the patterns of luminescence, observed for different initial bulk concentrations of supporting electrolyte and applied voltages, are shown in Fig. 5.81. This figure shows also a general tendency in the morphology of patterns: transition from slowly rotating "tennis balls" at low voltage, through "fingerprints"

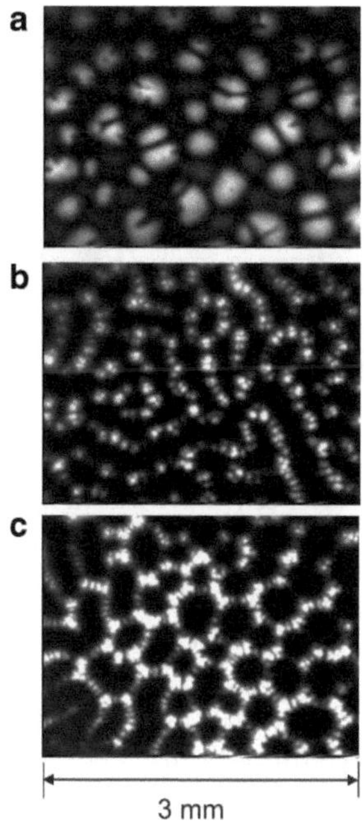

Fig. 5.81 Typical convective luminescent patterns, visible as bright (*orange*) areas, through the transparent electrode in the thin-layer electrolytic cell from Figs. 5.77–5.78, for the interelectrode distance $d = 175$ μm and initial concentration of $[TBA^+][HFP^-] = 7 \times 10^{-6}$ mol dm^{-3}. The copper electrode was polarized negatively with the voltage: (**a**) 3.0 V; (**b**) 3.5 V, (**c**) 4.0 V vs. the copper counterelectrode. Hexagonal convective cells predominate for relatively high voltages (up to 6 V) and relatively low concentrations of the supporting electrolyte ($\leq 7 \times 10^{-6}$ mol dm^{-3}), while they appear difficult to observe at $[TBA^+][HFP^-] > 1.2 \times 10^{-5}$ mol dm^{-3}. Reprinted with permission from [110]. Copyright 1999 American Chemical Society

at intermediate voltages, to quasi-hexagonal cells at highest voltages applied. In addition, for given voltage, with increasing concentration of supporting electrolyte, hexagons transformed back into "fingerprints," so hexagons evidently were favored when the electric field in the solution was relatively high.

One should emphasize that the electroconvective origin of patterns is additionally confirmed by the direct observation of the fluid motion (with intensity increasing with rising voltage), in the areas of bright orange luminescence.

A more quantitative description of those patterns included studies of the position of areas of luminescence in the interelectrode space and the dependence on the characteristic size (wavelength) of the patterns on the varying interelectrode distance (i.e., on the thickness of the solution layer). For relatively low concentrations of supporting electrolyte (like for the patterns shown in Fig. 5.81), the luminescence was found to occur in the solution near the cathode surface, while for higher $[TBA^+]$ $[HFP^-]$ concentrations the luminescence at both cathode and anode surfaces was discovered. In this case, when observed through the front transparent electrode, the luminescent spots, existing close to the anode, were located between the spots formed close to the cathode. The distance between the closest traces of

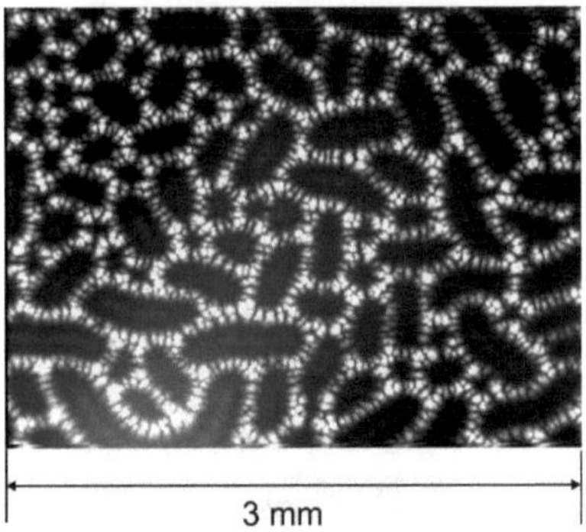

Fig. 5.82 Exemplary spatially chaotic convective pattern appearing in the thin-layer cell filled with the saturated solution of rubrene in 1,2-dimethoxyethane, with 1.2×10^{-3} M $[TBA^+][HFP^-]$ as a supporting electrolyte. A voltage $U = 4$ V was applied between the platinum cathode and the anode-glass plate covered by transparent ITO layer. Interelectrode distance $d = 90$ μm. Reprinted with permission from [109]. Copyright 1998 American Chemical Society

luminescence at the same electrode (d_{lum}) was in most cases equal approximately to doubled interelectrode distance ($d_{lum} \approx 2d$), so the distance between the neighboring places of cathodic and anodic luminescence observed through the transparent electrode was close to d. For better clarity of this description, consult schematic Fig. 5.84 in further part of this text. These dependences were not so straightforward for such experimental conditions, for which the *spatially chaotic* arrangement of cellular convection was reported [109] (Fig. 5.82).

More recent experiments performed by Dini et al. [113] have proved that analogous luminescent patterns could be observed in the same experimental arrangement, but for the electrolysis of 9,10-diphenylanthracene, instead of rubrene. Also Jaguiro [114] has more recently returned to the idea of application of such ECL thin layer cells as potential displays, elaborating the experimental conditions and construction details, favoring their extremely long-time (even 5-year-long) stability.

In search of the mechanism of the formation of these convective patterns, it was necessary to confirm its EHD origin in terms of possibly electrochemically realistic quantitative model. Such a model involved first the definition and calculation of the driving force for the EHD convection and, in the next step, modeling the fluid motion subject to such driving force, made by numerical integration of the Navier–Stokes equation. Such computations indicated the spontaneous self-organization of the fluid motion into convective rolls.

5.9.4.2 Shape of Convective Streams and Spatial Distribution of Luminescence

The qualitative scheme of the transport of all charge carriers within the single convective cells is shown in Fig. 5.83.

In turn, the spatial distribution of luminescent spots in the interelectrode space can be understood if one assumes that visible luminescence occurs only in those

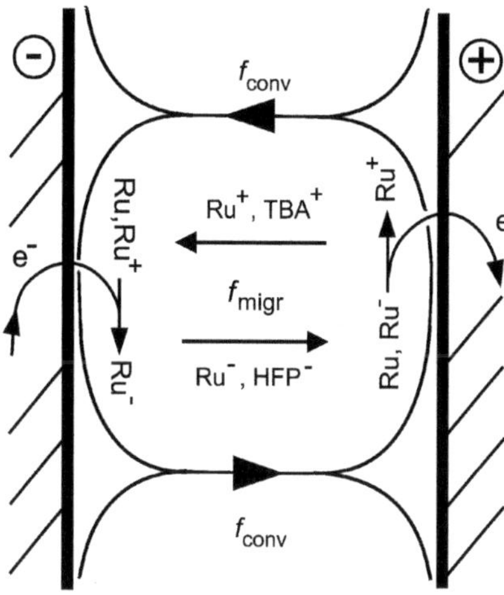

Fig. 5.83 Schematic picture of electrochemical processes occurring in a single convective roll of the EHD thin-layer cell. Electrode reactions of rubrene at the interfaces are accompanied with convection (f_{conv}) and migration (f_{migr}) of ions of rubrene and the supporting electrolyte. For simplification, diffusion fluxes were not indicated. Recombination between R^+ and R^- in the front of both electrodes leads to chemiluminescence. Reprinted with permission from [108]. Copyright 1998 American Chemical Society

places at the given electrode, which are hit by the most intensive stream of fluid carrying rubrene counterions formed at the opposite electrode. This is schematically shown in Fig. 5.84.

It is further noteworthy that R^+, R^-, and TBA^+ ions are expected to have a similar ionic mobility while a much smaller TFP^- anion should have an appropriately higher mobility, assessed as greater for the factor of about 4.4. This difference may explain why, contrary to the simplified Fig. 5.84, the distribution of luminescence in real system is not symmetrical with respect to the middle plane of the interelectrode space. Particularly for relatively low concentration of a supporting electrolyte (~10^{-6} M), the intensity of luminescence at the cathode is usually greater than at the anode. With the supporting electrolyte concentration increasing to ca. 10^{-5} M, the intensity of the luminescence at the anode becomes comparable to that at the cathode, but the morphology of both sets of luminescent spots is however not identical.

The scheme of convective rolls shown in Fig. 5.84 explains the straight-line fragments of the "fingerprint-like" luminescent patterns. In order to understand the formation of hexagonal luminescent convective cells, the distribution of streams and places of luminescence, shown in Fig. 5.85, was proposed [111].

This scheme shows that the hexagonal arrangement of luminescence at one electrode (being the cathode in most experiments) should be associated with a single central spot of light at the anode. This spot is, however, difficult to detect in most experiments (cf. Fig. 5.81c) due to the above-mentioned significant asymmetry in the intensity of luminescence, usually much weaker, and often even invisible

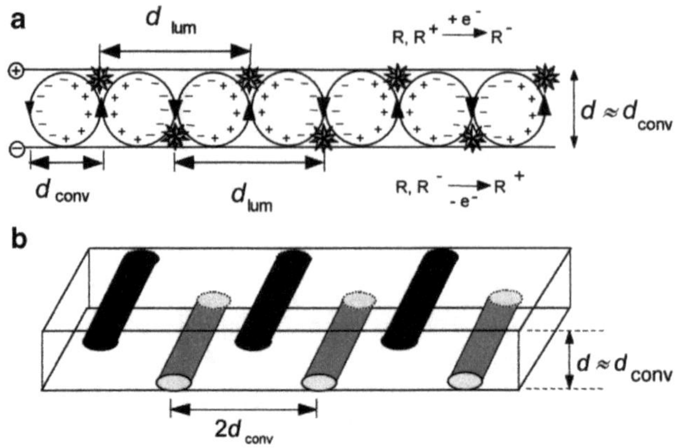

Fig. 5.84 Mechanism for generation of linear patterns of luminescence exhibiting the spatial period (d_{lum}) doubling the interelectrode distance (d): (**a**) visible luminescence (*asterisks*) occurs where the convective streams carrying rubrene ions formed at one electrode approach the other electrode (cf. Fig. 5.83); (**b**) three-dimensional distribution of cathodic–anodic luminescence; the traces in the cathodic and anodic layers are mutually shifted for the value comparable with the interelectrode distance d in the plane parallel to the electrodes. Reprinted with permission from [108]. Copyright 1998 American Chemical Society

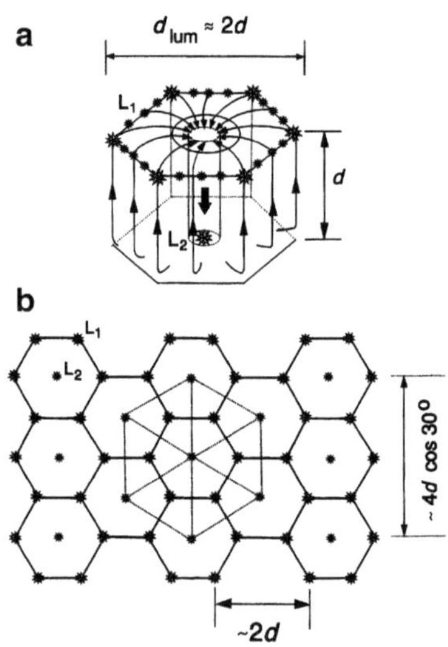

Fig. 5.85 (**a**) Schematic shapes of the fluid flows, forming a single convective cell, manifesting itself as an hexagonal arrangement of luminescence at one electrode, with a predicted central spot at the opposite electrode; (**b**) corresponding expected schematic view of the luminescent solution, observed through the front transparent electrode [111], to compare with Fig. 5.86. Reproduced by permission of PCCP Owner Societies

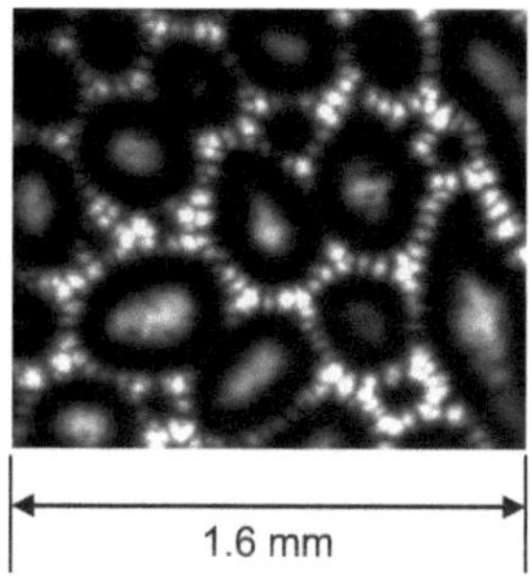

Fig. 5.86 Experimentally obtained set of full luminescent cathodic hexagonal cells, with the central spots at the anode. Parameters of the experiment: Cu cathode, glass/ITO anode, $U = 4$ V, $d = 90$ μm, $[TBA^+][HFP^-] = 1.2 \times 10^{-5}$ mol dm^{-3}. Picture size: 1.6 × 1.3 mm [111]. Reproduced by permission of The PCCP Owner Societies

at the anode. Only for some, carefully chosen conditions, it is possible to report the full picture of hexagonal cells (Fig. 5.86), which clearly confirms the schematic imagination suggested by Fig. 5.85.

5.9.4.3 Modeling the Driving Force for the EHD Convection

According to the idea of EHD convection, the driving force is assumed to be of the electrostatic origin, i.e., to come from the interaction of the space charge of the fluid with the externally applied electric field. The existence of nonzero space charge means the emergence of local, subtle deviations of the solution from electroneutrality, sufficient to produce significant force acting on the fluid. The source of this situation is the electrolysis of rubrene molecules at the respective electrodes, associated with the diffusion and migration of all particles in the solution. At this step of calculations the convection was not yet considered, and the following detailed assumptions formed the basis for numerical calculations [109]:

1. The progress of diffusion and migration of all dissolved species occurs only along the x co-ordinate normal to the electrodes surfaces (i.e., the model of the convection-less system was yet one-dimensional and shows only the *principle* of calculations of the driving force, later extended to two spatial dimensions).
2. Due to the low concentration of a supporting electrolyte, much lower than the concentration of the electroactive rubrene species, the electric current, flowing through the thin-layer cell, is controlled by the actual, relatively large ohmic solution resistance R_s which is *not* constant during the electrolysis due to continuous changes in the concentrations and spatial distribution of ions in the interelectrode space.
3. The spatial distribution of the electric field $E(x, t)$ is related to the magnitude of the actual faradaic current $I(t)$ and to the distribution of the electric conductivity

determined by the presence of ions of the concentrations c_s, charges $z_s e$, and diffusion coefficients D_s:

$$E(x,t) = \frac{RT}{F^2 \sum_s z_s^2 D_s c_s} \left[\frac{I(t)}{A} + F \sum_s z_s D_s \frac{\partial c_s}{\partial x} \right] \tag{5.53}$$

where A is the surface of the parallel electrodes; R, gas constant; T, temperature, and F is Faraday constant.

4. The nonlinear course of $E(x, t)$ determines the spatial distribution of the excess charge $q_{ex}(x, t)$, according to the Poisson equation (5.40). The product of $q_{ex}(x, t)$ and $E(x, t)$ calculated in this way defines the local volume electric force density exerted on the fluid.
5. Due to reversibility of the redox reactions of rubrene, the surface concentrations of molecules and ions of rubrene are interrelated through the Nernst equations (for the cathode at $x = 0$ and the anode at $x = d$):

$$E_{cath} = E_1^0 + \frac{RT}{F} \ln \frac{c_R(0,t)}{c_{R^-}(0,t)} = E_2^0 + \frac{RT}{F} \ln \frac{c_{R^+}(0,t)}{c_R(0,t)} \tag{5.54}$$

$$E_{anod} = E_2^0 + \frac{RT}{F} \ln \frac{c_{R^+}(d,t)}{c_R(d,t)} = E_1^0 + \frac{RT}{F} \ln \frac{c_R(d,t)}{c_{R^-}(d,t)} \tag{5.55}$$

6. In the two-electrode system, the total voltage U, decreased for the ohmic potential drops in the solution, distributes into such individual electrode potentials, that the same faradaic current flows through the cathode and the anode; this is the case if:

$$E_{cath} = \left[\frac{E_1^0 + E_2^0}{2} \right] - \left[\frac{U - |I| R_s}{2} \right] \tag{5.56}$$

$$E_{anod} = \left[\frac{E_1^0 + E_2^0}{2} \right] + \left[\frac{U - |I| R_s}{2} \right] \tag{5.57}$$

7. The net diffusion and migration fluxes (f) of the molecules and ions of rubrene at the electrode surfaces ($x = 0$, $x = d$) are equal to zero (no rubrene adsorption occurs on the electrode surfaces):

$$f_R(0,t) + f_{R^+}(0,t) + f_{R^-}(0,t) = 0 \tag{5.58}$$

$$f_R(d,t) + f_{R^+}(d,t) + f_{R^-}(d,t) = 0 \tag{5.59}$$

8. The faradaic current $I_f(t)$ is related to these fluxes in the following way:

$$I_f(t) = FA[f_R(0,t) + 2f_{R^+}(0,t)] = -FA[f_{R^-}(0,t) - f_{R^+}(0,t)]$$
$$= -FA[f_R(d,t) + 2f_{R^-}(d,t)] = FA[f_{R^+}(d,t) - f_{R^-}(d,t)] \quad (5.60)$$

   Simultaneously, since the components of the supporting electrolyte (denoted below as Cat ≡ TBA^+ and An ≡ HFP^-, respectively) are nonelectroactive, their individual *surface* fluxes are equal to zero.
9. The transport of all species (rubrene and supporting electrolyte) is described by the system of partial differential equations:

$$\frac{\partial c_s}{\partial t} = -\frac{\partial f_s}{\partial x} + T_{\mathrm{chem,s}} \quad (5.61)$$

   In Eq. (5.61), $T_{\mathrm{chem,s}}$ is a term describing the kinetic contribution from the chemical recombination of rubrene ions:

$$T_{\mathrm{chem,s}} = \begin{cases} -k_r c_{R^+} c_{R^-} & s = R^+, R^- \\ +2k_r c_{R^+} c_{R^-} & s = R^0 \\ 0 & s = \mathrm{Cat, An} \end{cases} \quad (5.62)$$

with k_r being the rate constant of this recombination. For the ions of supporting electrolyte (Cat, An) any interaction was neglected, which means also an assumption of their insignificant ion pairing, mainly due to low concentration of the supporting electrolyte, since the low dielectric permittivity of the 1,2-DME solvent ($\varepsilon_r = 2.238$ at 20 °C [115]) rather favors inter-ionic association.

Based on the above assumptions, the differential equations (5.61) were numerically integrated using the explicit finite differences method [116, 117], yielding thus data for the calculations of faradaic current, and the distribution of both excess charge and the electric field in the solution. Details of these calculations are given in [109]. Here will be summarized only representative results, limited to the realistic case of higher diffusion coefficient (and thus ionic mobility) of the anions of a supporting electrolyte, compared to assumed equal diffusion coefficients of other ions present in the system. The integration of *partial* differential equations for all the species ensures the calculation of their, varying in time, realistic concentration profiles, the solution resistance, and the corresponding electric current. In course of these calculations, due to a low concentration of supporting electrolyte, lower than rubrene concentration, the current is controlled by the relatively high solution resistance and thus only a small fraction of rubrene molecules is oxidized and reduced at respective electrodes.

Figures 5.87 and 5.88 show the model development of concentration profiles of all ionic species and corresponding spatial distribution of the electric field, and the

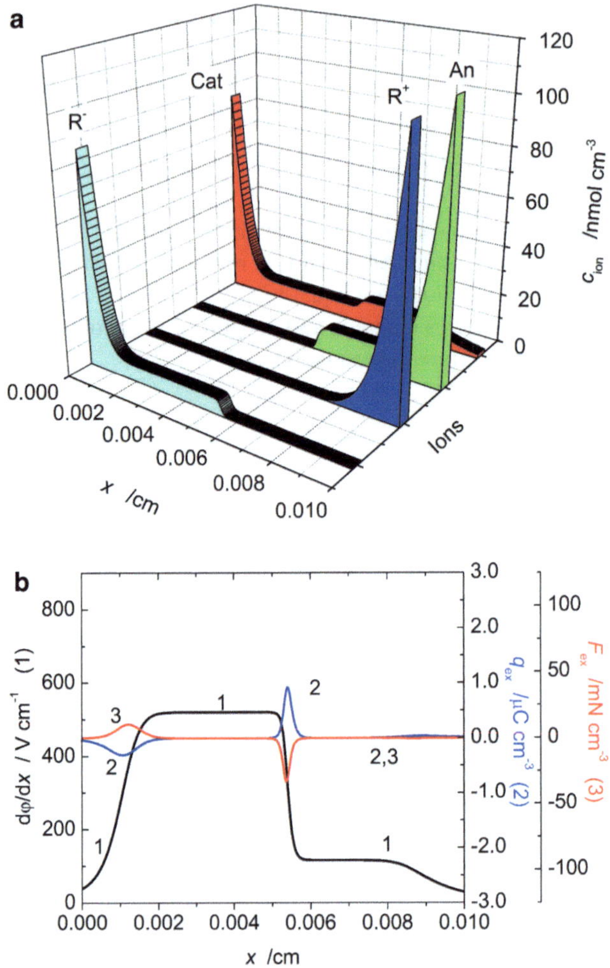

Fig. 5.87 (**a**) Simulated interelectrode distribution of rubrene anions R^-, rubrene cations R^+, anions (An) and cations (Cat) of the supporting electrolyte before meeting of the fluxes of R^+ and R^-, leading to their recombination and emission of chemiluminescence; the spatial distribution of rubrene molecules (R) is not visualized; (**b**) the corresponding distribution of: (1) gradient of the electric potential, (2) local uncompensated charge q_{ex}, and (3) driving force density F_{ex}. Equal diffusion coefficients of rubrene species and cations of a supporting electrolyte: $D_{R+} = D_{R-} = D_{Cat} = 6.2 \times 10^{-6}$ cm^2 s^{-1} and a higher diffusion coefficient for anions of supporting electrolyte, $D_{An} = 2.8 \times 10^{-5}$ cm^2 s^{-1} were assumed. Model time = 30 ms after beginning of the electrolysis caused by potential difference $U = 5$ V; actual current $I = 0.148$ mA, solution resistance $R_s = 18.19$ kΩ. Cathode is placed at $x = 0$, anode at $x = d = 0.01$ cm. Model sample: 4×10^{-3} mol dm^{-3} rubrene + 1×10^{-5} mol dm^{-3} [TBA$^+$][HFP$^-$] in 1,2-DME. Reprinted with permission from [109] (color version). Copyright 1998 American Chemical Society

driving force for the EHD convection. Figure 5.87 corresponds to the electrolysis time when diffusion profiles of rubrene ions did not meet yet, while 5.88 shows what happens to the concentration of ions and the local distribution of the electric field, when R^+ and R^- ions just meet and recombine.

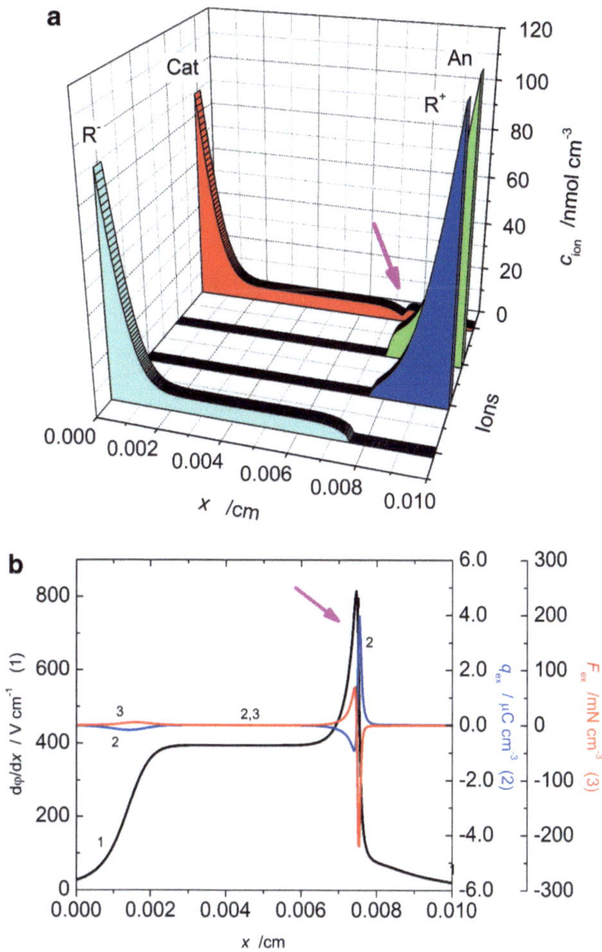

Fig. 5.88 Legend as for Fig. 5.87, for model time = 50 ms, actual current $I = 0.113$ mA, solution resistance $R_s = 23.91$ kΩ. The *arrows* indicate the zone in which recombination of rubrene ions occurs, giving rise to substantial local perturbation of the distribution of the electric field and the development of huge local electric forces. Reprinted with permission from [109] (color version). Copyright 1998 American Chemical Society

Analysis of Fig. 5.88 shows how the recombination of rubrene ions into neutral rubrene molecules lead to a local decay of a part of charge carriers. In consequence, the highly nonlinear course of the electric field develops in this region which is associated with the local formation of relatively high electric forces. It means that besides the global self-organized motion of the fluid also the secondary effect—its intensive local motion, in the close vicinity to the places in which rubrene ions recombine, may occur. This may explain the "discrete" structure of convective flows which look like they are composed of many splitted substreams [109].

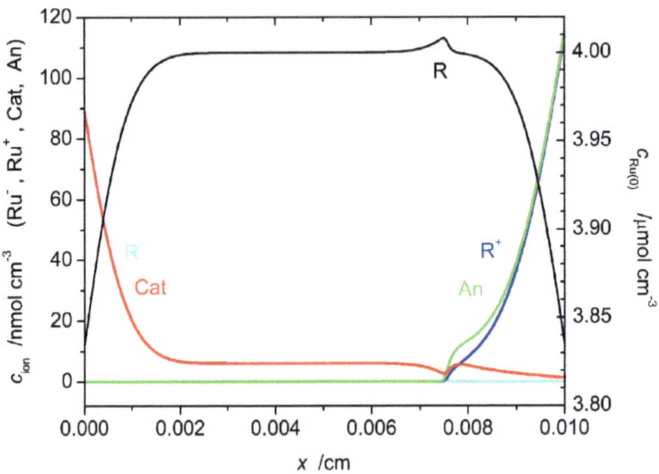

Fig. 5.89 Comparison of concentration profiles of all ionic species and rubrene molecules (R) for conditions when rubrene ion radicals meet and recombine in the diffusion-controlled process, for the conditions corresponding to Fig. 5.88 (i.e., for model time $t = 50$ ms). Reprinted with permission from [109] (color version). Copyright 1998 American Chemical Society

For comparison with Fig. 5.88, Fig. 5.89 collects the concentration profiles of all species, showing in addition the local enhancement of the concentration of rubrene molecules within the zone of rubrene ions recombination.

The electric forces calculated in a way presented above exert a pressure on liquid electrolyte from the cathodic region toward the anode and vice versa. In terms of this realistic model, it is possible to estimate the magnitude of these volume forces as equal to from a few to a few tenths of mN cm^{-3}. One should note that they are similar to (or even higher than) buoyancy forces which are able to cause the self-organized thermal convection. Thus, at least the magnitude of presently calculated electric forces should be sufficient to cause a pressure resulting in the electroconvective motion of the fluid. However, the question remains, whether these forces can make this motion self-organized into cooperating rolls or cells. In order to prove that, further calculations were needed that should involve integration of the Navier–Stokes equation (5.1) in at least *two-dimensional space*, since in one-dimensional space the electric force is exactly balanced by the pressure gradient, i.e. convection does not set in.

5.9.4.4 Modeling the Self-Organized EHD Convection

For the simulation of the convective rolls, the model interelectrode space was extended to two dimensions (employing three-dimensional space would require enormously long computation time). Accordingly, the volume force for the EHD convection is now realistically defined as:

$$\mathbf{F}_{\text{ex}} = q_{\text{ex}}\mathbf{E} = -q_{\text{ex}}\nabla\varphi = \varepsilon_{\text{r}}\varepsilon_0\left(\frac{\partial^2\varphi}{\partial x^2} + \frac{\partial^2\varphi}{\partial y^2}\right)\nabla\varphi \tag{5.63}$$

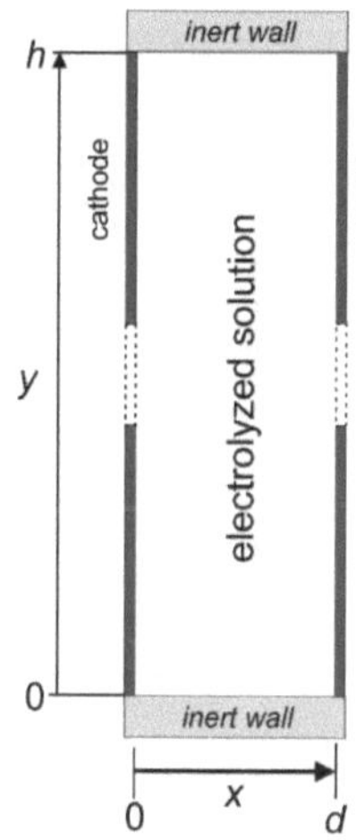

Fig. 5.90 Schematic setup of the modeled electrochemical-convective thin-layer cell; the electrolyzed solution of rubrene and supporting electrolyte in 1,2-DME is placed between the vertical electrodes of a length h and bounded from the top and the bottom by inert, nontransparent walls. Interlectrode distance $d = 100\ \mu\text{m}$; the electrode height $h \geq 4d$. Reprinted with permission from [110]. Copyright 1999 American Chemical Society

The procedure of numerical integration of the *Navier–Stokes* equation (5.1) with the electric driving force formed due to the rubrene electrolysis involved two coupled calculation modules: the "electrochemical" and "hydrodynamic" one. The electrochemical module is constructed analogously to the concept described in the previous section, extended now to two-dimensional space and with the convective transport added:

$$\mathbf{f}_s = -D_s \nabla c_s - \frac{z_s F D_s}{RT} c_s \nabla \varphi + c_s \boldsymbol{\vartheta} \tag{5.64}$$

where c_s, D_s are the concentration and diffusion coefficient of species s; z_s, ionic charges: $z_{R^+} = z_{Cat} = +1$, $z_{R^-} = z_{An} = -1$, $z_R = 0$; φ, electric potential; and $\boldsymbol{\vartheta} = [\mathrm{u}, \mathrm{v}]$ is the vector of fluid velocity.

In the hydrodynamic module, the *Navier–Stokes* equation for the two-dimensional space is solved. Figures 5.90–5.92 show schematically the construction of the model system and the interrelation between these computation modules. The electrochemical module requires relatively fine space grid which, if repeated in the hydrodynamic module, would cause unreasonably long integration time of the Navier–Stokes equation. Therefore, for the latter case a coarse space grid was applied and components of the vector of fluid velocity were projected onto the fine "electrochemical grid" using bilinear interpolation.

In terms of these two space discretization schemes, at every time step Δt of the numerical integration, in the electrochemical module the progress of diffusion, migration, and convection for all dissolved particles and the associated distribution

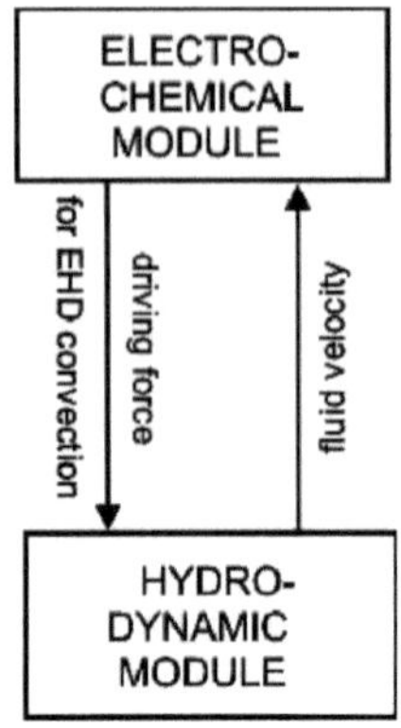

Fig. 5.91 Scheme of the coupling between the electrochemical (transport of particles) and hydrodynamic (fluid motion) modules of the numerical procedure, simulating the onset of EHD convection in the thin-layer electrolytic cell. Reprinted with permission from [110]. Copyright 1999 American Chemical Society

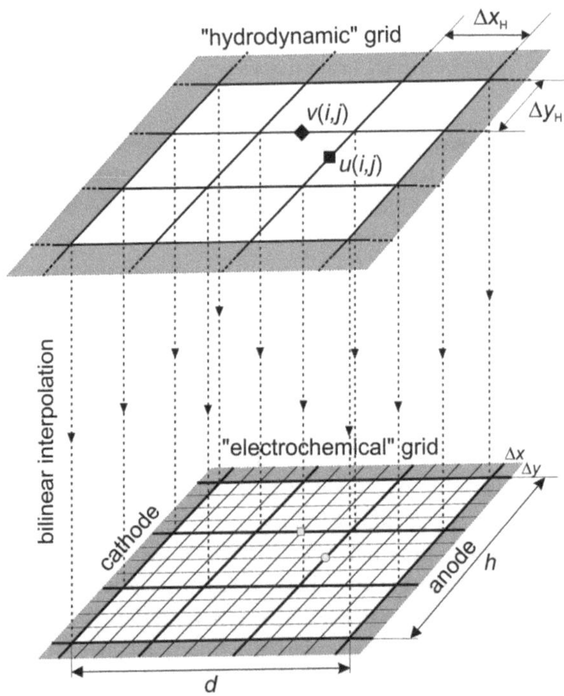

Fig. 5.92 Scheme of superposition of two space grids of different resolution used for the integration of (1) Navier–Stokes equation (*upper layer*) and (2) equations of electrochemical transports (*lower layer*). The *x*- and *y*-components of fluid velocities are first calculated at the edges of the cells of the "hydrodynamic grid" and then interpolated onto the edges of the cells of the "electrochemical" grid. Reprinted with permission from [110]. Copyright 1999 American Chemical Society

of the electric driving force were calculated. Using this driving force, in the hydrodynamic module the corresponding distribution of the fluid velocity was determined, introduced back to electrochemical module and used there for the calculation of convective contribution to the fluxes of all species. It was assumed that the fluid was quiescent initially and permanently at all the four boundaries of the model system (*no-slip* boundary conditions [118]). Detailed analytical expressions and corresponding finite differences notation were described in [110].

Application of appropriate voltage between the electrodes initiates the electrolysis of rubrene and development of all concentration profiles along the x-direction. As long as the system remains homogeneous along the electrode surface (y-direction), no convection sets in. In order to induce it, a small fluctuation of the solution ohmic resistance was introduced across the solution at $y = h/2$. The resulting fluid motion took a shape of a pair of counter-rotating convective rolls which *spontaneously*, upon further calculations, turned into set of neighboring cooperating rolls that eventually covered almost the entire space of the model system (Figs. 5.93 and 5.94) [112].

Figures 5.93 and 5.94 prove that the electric force, defined by Eq. (5.63), can indeed give rise to EHD convection that undergoes *spontaneous* self-organization in a universal way, depicted schematically in Fig. 5.5. One should comment that analogous self-organized motion, although of simpler spatial distribution, one obtains also for equal diffusion coefficients of all species, so the system is then ideally symmetrical with respect to the transport properties of rubrene species and supporting electrolyte components. Also in this model the uncompensated electric charge is initially distributed homogeneously in the solution and this distribution is distorted only by fluctuation that undergoes enhancement to macroscopic fluid motion.

The presented numerical model was designed to reproduce possibly accurately the electrochemical and hydrodynamic properties of the system, in order to avoid artifacts that could arise in a more simplified approach. Therefore, the numerical results seem to be a convincing proof of the EHD origin of the low-voltage convection in the thin-layer cell, leading to various luminescent patterns. However, in this way only the initial stage of the convection evolution (i.e., relatively low velocities) could be modeled, since with (presumably autocatalytically) increasing velocity of convection, the calculations became numerically unstable. Stability is kept if the model is strongly simplified in this way that the calculation of the actual fluid velocity from the driving force determined in the "electrochemical module" is replaced by the, fixed in time, such a spatial distribution of the model force that generates the counter-rotating convective cells. If the electrode length (h) is divided into N spatial elements Δy (indexed with j variable) and the interelectrode distance (d) is divided into M spatial elements Δx (indexed with i variable), the integer h/d ratio defines the number of neighboring virtual square chambers, in which separate convective cells will develop. In the following equal spatial steps, Δx and Δy will be assumed. The chambers are indexed along the y axis with the integer variable k ranging from 0 to h/d. Then one can define the following distribution of the model driving force:

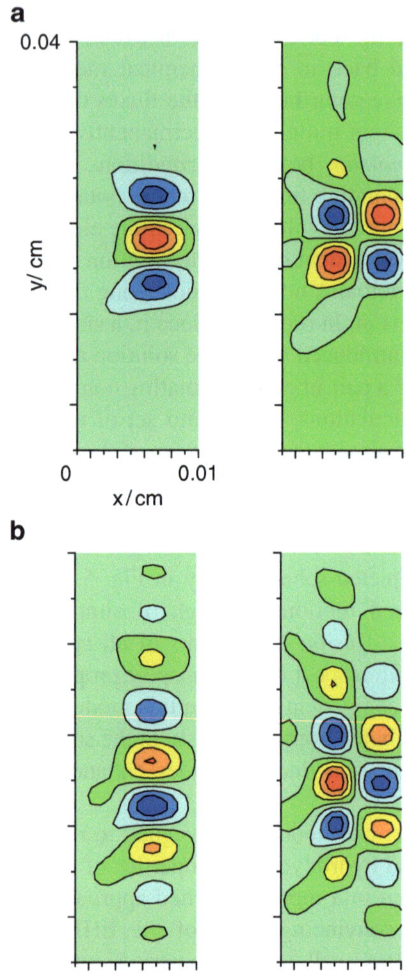

Fig. 5.93 Progress of self-organization in the evolution of EHD convection as a function of time, shown in terms of (left column) the x-component and (right column) the y-component of a fluid velocity. (**a**): $t = 208$ ms, x-component ranging from 2×10^{-8} m s^{-1} (*red*) to -1.5×10^{-8} m s^{-1} (*dark blue*), y-component ranging from 1.2×10^{-8} m s^{-1} (*red*) to -1.3×10^{-8} m s^{-1} (*dark blue*). (**b**): $t = 583$ ms, x-component ranging from 2×10^{-7} m s^{-1} (*red*) to -1.7×10^{-7} m s^{-1} (*dark blue*), y-component ranging from 1.2×10^{-7} m s^{-1} (*red*) to -1.1×10^{-7} m s^{-1} (*dark blue*). The diffusion coefficient of the anion of the supporting electrolyte $D_{An} = 2.8 \times 10^{-5}$ cm^2 s^{-1}, the diffusion coefficients of all other species equal to 6.2×10^{-6} cm^2 s^{-1}. The 1% local decrease of the solution resistance at $y = h/2$ was introduced at time $t_{fluct} = 17.0$ ms. $U = 3.6$ V. Reprinted from [112], Copyright 2000, with permission from Elsevier

For $k = 0, \ldots, h/d$; for $j = k \times M$; for $i = 1, \ldots, M - 1$:

$$F_{ex}(i,j) = \begin{cases} -\Phi & \text{if } k \bmod 2 = 0 \\ +\Phi & \text{if } k \bmod 2 \neq 0 \end{cases} \tag{5.65}$$

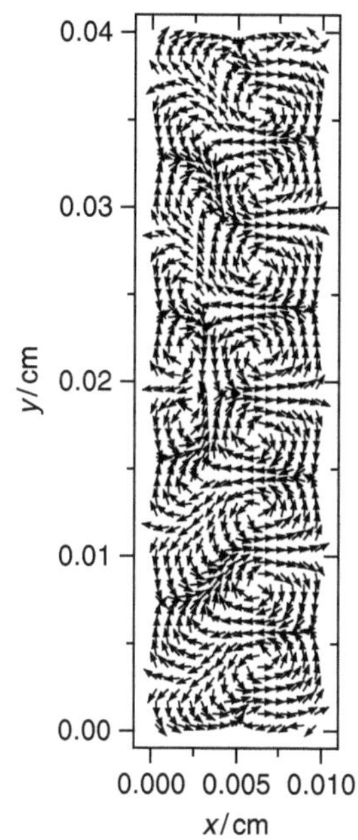

Fig. 5.94 The vector plot of the velocity of convection corresponding to Fig. 5.93b. A slight irregularity of motion is caused by the violent agitation of fluid. Reprinted from [112], Copyright 2000, with permission from Elsevier

meaning that the vectors of the force, normal to the electrode surfaces, change their directions (from toward the anode to toward the cathode) along the walls of the subsequent virtual chambers (Fig. 5.95a). Using this model approach, one can vary the magnitude of Φ in Eq. (5.65) in a relatively wide interval, including such values, for which the generated convection velocity appears to be sufficient for distinct separation of R^+ and R^- ions in the circulating streams. Based on that distribution of concentrations of R^+ and R^- ions one can determine the spatial distribution of the luminescence, with its intensity proportional to the local concentration product: $[R^+]\times[R^-]$. Figure 5.95b shows that if convection velocity is high enough, the areas of strongest luminescence appear indeed in the zones close to the electrode surface, with the cathodic and anodic luminescence occurring alternately, exactly as it was suggested earlier (see Fig. 5.84): maximum concentration product $[R^+]\times[R^-]$ occurs in those places close to the electrode surface, which are hit by the stream of rubrene counterions, carried from the opposite electrode by the convective stream. Of course, due to assumed simplified distribution of the driving force, this proof has limited, at most semiquantitative nature.

Nevertheless, the following simple reasoning confirms the validity of above model calculations. If experimentally observed convective structures exhibit distinct dark and light areas (as in Figs. 5.81, 5.82, 5.86), this means that for dark places the convective flux of rubrene ions, and of associated counterions of a background electrolyte, along the y-direction (parallel to electrode surfaces) should overcome their diffusion and migration flux along the x-direction (perpendicular to electrode surfaces). In terms of Eq. (5.64), the appropriate *minimum* y-component

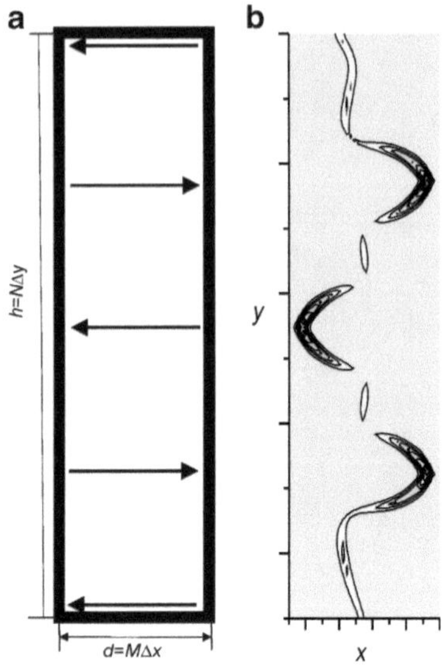

Fig. 5.95 (**a**) Model spatial distribution of the driving force, causing cellular convection [cf. Eq. (5.65)]. (**b**) Model distribution of luminescence coming from recombination of rubrene cations and anions in the presence of cellular convection. Simulation parameters of the electrochemical processes: external voltage $U = 4$ V, formal potentials of R/R^- and R/R^+ couples: $E_1^0 = -1.13$ V, $E_2^0 = +1.37$ V, respectively. Initial bulk concentration of rubrene $c_{Ru}^0 = 4 \times 10^{-3}$ M, concentration of inert supporting 1:1 electrolyte $= 1 \times 10^{-5}$ M. Interelectrode distance $d = 100$ μm, electrode height $h = 400$ μm. For simplicity, equal diffusion coefficients of all species were assumed: $D = 6.2 \times 10^{-6}$ cm^2 s^{-1}. *Red color* denotes highest intensity of luminescence. Picture (**b**) reproduced from [111] by permission of the PCCP Owner Societies

of the velocity of convection v^*, necessary to overcome the migration and diffusion along x-direction, must meet the following condition:

$$v \geq \left| \frac{-D_s\left(\frac{\partial c_s}{\partial x}\right) - \frac{z_s F D_s c_s}{RT}\left(\frac{\partial \varphi}{\partial x}\right)}{c_s} \right| \equiv v^* \qquad (5.66)$$

For simplification, justified by relatively high electric fields in the solution, we assume further that the *migration* contribution (second term in the numerator) is predominant over diffusion for the rubrene ions. Then, using a priori given or computed typical numerical values: $z_s = 1$, $(\partial\varphi/\partial x) = 2 \times 10^2$ V cm^{-1}, $D_s = 6.2 \times 10^{-6}$ cm^2 s^{-1}, $c_s = 10^{-7}$ mol cm^{-3}, and $T = 298$ K, one obtains the *minimum* velocity v^*, necessary for the separation of luminescent areas as equal to 0.048 cm s^{-1}. The simulations showed that for maximum y-component of velocity exceeding this value (e.g., ~0.08 cm s^{-1}), distinct regions with and without luminescence indeed develop, while for maximum y-component of velocity lower than v^* (e.g., ~7×10^{-4} cm s^{-1}) such a separation did not occur [111].

Concluding, self-organized convective patterns described in this section, compared to other manifestations of self-organized convection, included in this chapter, clearly show the universalities characterizing this phenomenon—structures of very similar morphology can be obtained for driving forces of completely different nature. In electrochemical systems, the convection can be induced "indirectly"—by producing either the appropriate gradients (isothermal or not) of density/surface tension, or "directly"—by the exertion of coulombic force on the fluid into which electric charge is injected. Also combinations of different driving forces in the same system are possible. An interesting aspect is also that some of the patterns that develop on the surfaces of solid electrodes can be explained only in terms of self-organized convection. The self-organized convection in electrochemical systems was also described in review papers by Ertl [119] and Orlik [120].

5.9.5 Electroconvection in Membrane Systems

The membrane oscillators are described in Chap. 6. Here, as the introduction to those systems, we mention here the specific case: a possible role of EHD convection in the characteristics of some membranes. Rubinstein and Maletzki [121] have described the electrochemical characteristics and performed stability analysis of the highly permselective, electrically inhomogeneous cation-exchange membrane (C-membrane), often applied in electrodialysis. The shape of the experimentally reported current–voltage dependence exhibited, above certain threshold voltage, the current higher than the limiting value—predicted by the classical theory of concentration polarization, applied to the membrane system. This enhanced current corresponds to region III in Fig. 5.96.

For the explanation of this "overlimiting conductance" of the studied system, Rubinstein and Maletzki [121] have employed the idea of the EHD convection as a corresponding extra transport. As above, this convection originates from the interaction of the electric field with the space charge in the fluid phase. In more detail, the bulk of the solution was assumed to be practically electroneutral, with the small excess charge developing only in the boundary layer of a thickness on the length scale of tens of micrometers, adjacent to membrane. Furthermore, the analysis showed that there could develop the concentration distribution of the electrolyte, which excluded the possibility for the fluid to keep its mechanical equilibrium, if membranes are inhomogeneous. The onset of convection causes the mixing of the solution which diminishes the concentration polarization and, in consequence, enhances the current.

Similar type of membrane electroconvective systems was analyzed later by Vorotyntsev et al. [122] who have performed the relevant stability analysis. The system consisted of the binary electrolyte solution placed between two membranes which were assumed to exhibit infinitely low resistance, to be permeable for anions (counterions), and to exhibit zero permeability for the cations (co-ions) of the binary electrolyte which is present also on external sides of both membranes. The "electrokinetic instability" of the fluid, involving the electro- and capillary osmotic slipping at the membrane (electrode)–solution interface, could occur when the electric current

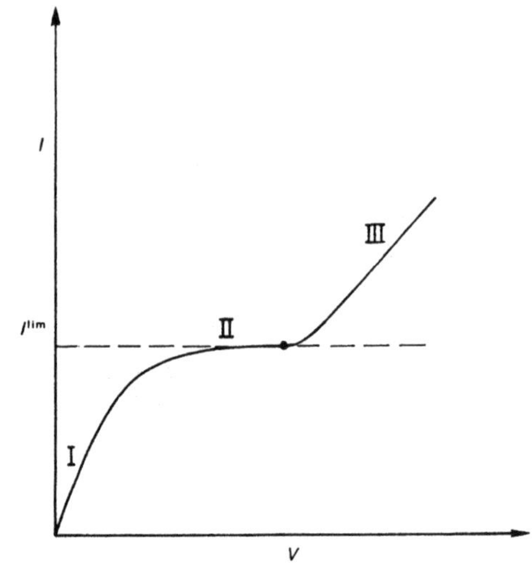

Fig. 5.96 Typical schematic current–voltage characteristics of a cation-exchange membrane. Reproduced from [121] by permission of the Royal Society of Chemistry. Classical theory of concentration polarization (without convection) explains the regions I and II, i.e., the asymptotic approaching of the limiting value, denoted with *dashed line*. Starting from the *black spot* the overlimiting, noisy (smoothed on the picture) current develops (region III), which indicates the extra contribution from EHD convection to the system's conductivity

approached its limiting value (corresponding to linear concentration distribution in the cell), provided that the tangential gradients of the electrolyte concentration and/or the potential were close to the interface. In other words, periodical flows could occur without coulombic forces in the bulk phase which was here considered to be quasi-electroneutral. For the onset of the periodically structured hydrodynamic instability, also additional condition had to be met: the diffusion coefficient of co-ions had to be sufficiently low, of the order of 10^{-6} cm^2 s^{-1} compared to ca. 10^{-5} cm^2 s^{-1} for typical ions in apueous solutions. Simultaneously, the authors discussed the difference between their approach and the model by Grigin [123] who considered the existence of the space charge in the bulk of the cell in the absence of convection. In terms of this alternative Grigin's approach, the *difference* in diffusion coefficients of ions was crucial for the generation of the lateral electric field, generated by the formation of lateral concentration gradients and thus—for the onset of hydrodynamic instability. Both approaches discuss however different types of instability.

5.10 Interaction of Spatial Pattern Formation and Forced Convection

Although the content of this chapter is essentially devoted to self-organized convection, at the end it is useful to note also the studies in which the pattern formation was observed in the presence of forced convection, caused by rotation of

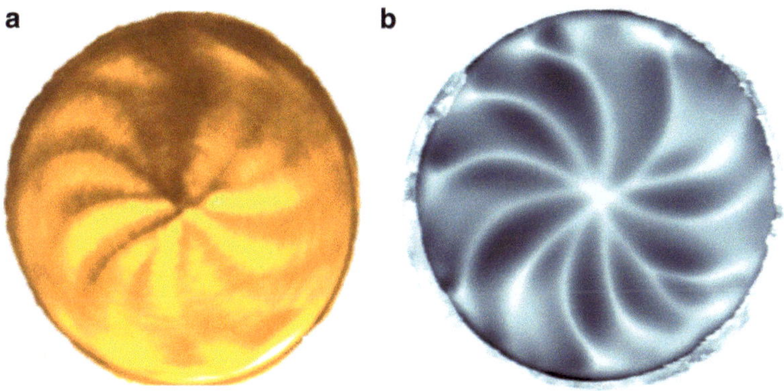

Fig. 5.97 (**a**) Vortices in the solution beneath the electrode surface (from CCD picture); (**b**) electrode surface after 60 min etching [from UBM (laser focus) scan]; (**a**) and (**b**) is the same electrode, with rotational speed of 10 rpm. Reproduced with permission from [125]. Copyright 2002 World Scientific

the disk electrode. Also under such conditions the free convection can set in due to formed density gradients and contribute to the emerging instabilities. Such studies of the coupling between chemical and hydrodynamic instabilities during the electrochemical dissolution of metals were reported by Baune et al. [124–126]. The iron (stainless steel) disk electrode, either stationary or rotating, was galvanostically dissolved in concentrated (3.5 M) iron(III) chloride solution. For the case of slowly rotating (0–25 rpm) steel electrode in an unstirred solution, visible convection vortices in the solution beneath the electrode were observed which correlated with the etched patterns on its surface. In fact, especially for such slow rotation rates, the free convection, driven by density gradients, can effectively influence the morphology of etched patterns. It was found that in those parts of the electrode in which the solution had a lower density during the dissolution process, the steel electrode was etched more strongly. This spatial distribution has also an additional dynamic aspect: in gravitational field, these lower density parts of fluid mean its regions directed upwards (toward the electrode), as they are initiated by sinking of the higher density solution. The faster dissolution of the electrode surface exposed to incoming solution of lower density is understandable, since ascending fluid is almost a fresh bulk solution, i.e., it is more reactive as a dissolution reagent. In turn, when this solution becomes used after the dissolution, due to convection vortices generated in the solution it always moves to the regions beneath the electrode in which it is sinking. These dynamic effects provoke an intensification of the surface instability and lead to the generation of etched patterns in the surface [124].

Figure 5.97 clearly shows that the streams of the solution of lower density [lighter areas in part (a)] correlate with the zones of more advanced etching of the electrode surface, visualized in part (b).

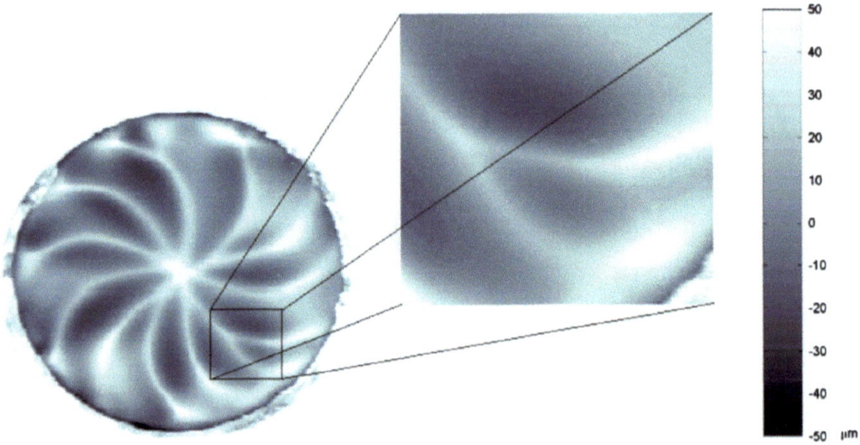

Fig. 5.98 Enlargement of the spatial bifurcation of a pattern at the electrode etched for 60 min at a rotational speed of 10 rpm. Reproduced with permission from [125]. Copyright 2002 World Scientific

Interestingly, the rotational speed of the disk affected not only the shape of the stationary spiral-like patterns in a rather trivial way, but also at certain value caused the change of the direction of their curvature (the convex pattern changed into the concave one). Furthermore, for the rotational speed of ca. 10 rpm, the spiral-like patterns exhibited splitting of the spirals, a phenomenon classified as a spatial bifurcation (Fig. 5.98).

Furthermore, together with the pattern formation the oscillations of the electrode potential were reported. They were interpreted as a new type of such instabilities as caused by a specific mechanism involving the interaction between hydrodynamic vortex patterns in the boundary layer and the electrochemical dissolution process, in addition coupled via the topographically structured surface of the electrode.

Finally, at *fast* rotational speed (2,000–6,000 rpm) the spiral-like etched patterns were found to exhibit invariant curvature, even under varying experimental conditions. Quantitatively, the invariant logarithmic shape of those spirals was characterized with the aspect ratio of $1/\sqrt{2}$ for the radial and tangential velocity [126]. The works described in this section contribute to understanding of the coupling between electrochemical processes and hydrodynamic flow at dissolving disk electrodes.

References

1. Bénard H (1901) Les tourbillons cellulaires dans une nappe liquide transportant de la chaleur par convection en régime permanent. Ann Chem Phys 23:62–144
2. Briggs J (1992) Fractals. The patterns of chaos. Simon and Schuster, New York (German edition: (1993) Chaos. Neue Expeditionen in fraktale Welten. Carl Hanser Verlag, p 97)

3. Bergé P, Pomeau Y, Vidal C (1984) Order within chaos. Towards a deterministic approach to turbulence. Wiley, New York
4. Grigin AP, Davydov AD (1998) Oscillatory convective instability in electrochemical systems. J Electroanal Chem 450:7–12
5. Chandrasekhar S (1961) Hydrodynamic and hydromagnetic stability. Oxford University Press, New York
6. Baranowski B, Kawczyński AL (1972) Experimental determination of the critical Rayleigh number in electrolyte solutions with concentration polarization. Electrochim Acta 17:695–699
7. Baranowski B (1980) The electrochemical analogon of the Bénard instability studied at isothermal and potentiostatic conditions. J Non Equilib Thermodyn 5:67–72
8. Baranowski B, Kawczyński AL (1970) Hydrodynamic stability in liquid electrochemical systems with concentration polarization. Rocz Chem (Ann Soc Chim Polonorum) 44:2447–2459
9. Baranowski B (1989) Dissipative Strukturen als ein energetisches Problem. Nova acta Leopoldina NF 60:91–107
10. Baranowski B, Fuliński A (1958) Theory of electrogravimetric separation in two-component electrolyte solution. Rocz Chem 32:303–313 (in Polish)
11. Nechiporuk VV, Elgurt IL (1991) Effects of electrochemical process conditions on the hydrodynamic instability of systems with concentration polarization. Electrochim Acta 36:321–323
12. Grigin AP, Shapovalov AP (1987) Izv Akad Nauk SSSR Ser MZhG 5:8
13. Winkler I, Plevan I, Nechiporuk V (1996) Region of occurrence of convective cells in the Cu^{2+}/Cu electrochemical system. Electrochim Acta 41:2743–2745
14. Nechiporuk VV, Elgurt IL, Andrushevskiy NM (1991) Hydrodynamic instability of electrochemical systems with concentration polarization and negative effective resistance. Electrochim Acta 36:143–146
15. Nakabayashi S, Yanagida M, Uosaki K (1996) Dissipation structure of electrochemical hydrodynamic convection. J Phys Chem 100:714–717
16. Caban K, Donten M, Stojek Z (2004) Electroformation of microlayers of ionic liquids in undiluted nitromethane and its homologues. Unusual oscillations behind the range of limiting steady-state current. J Phys Chem B 108:1153–1159
17. Grigin AP, Davydov AD (1998) Natural convection in electrochemical systems. Russ J Electrochem 34:1111–1136
18. Aleksandrov RS, Grigin AP, Davydov AD (2001) Oscillatory convective instability in a ternary electrolyte: effect of the electrode reaction rate. Russ J Electrochem 37:5–11
19. Baier G, Kummer U, Sahle S (1999) An electrochemically induced oscillatory instability. J Phys Chem A 103:33–37
20. Bowers PG, Soltzberg LJ (1989) Chemical generation and visualization of hydrodynamic instability: an extremely simple demonstration of self-organization. J Chem Educ 66:210
21. Lord Rayleigh (1916) On the convection currents in a horizontal layer of fluid when the higher temperature is on the under side. Phil Mag 32:529–546
22. Pearson JRA (1958) On convection cells induced by surface tension. J Fluid Mech 4:489–500
23. Nield DA (1964) Surface tension and buoyancy effects in cellular convection. J Fluid Mech 19:341–352
24. Scriven LE, Sternling CV (1964) On cellular convection driven by surface-tension gradients: effects of mean surface tension and surface viscosity. J Fluid Mech 19:321–340
25. Jakuszewski B, Turowska M (1973) Voltamperometric minima and electrochemical oscillations. J Electroanal Chem 46:399–410
26. Korolczuk M, Matysik J (1985) Current oscillations within the range of polarographic maximum potentials. J Electroanal Chem 195:277–281
27. Galus Z (1976) Fundamentals of electrochemical analysis. PWN, Warsaw, pp 459–469

28. Aogaki R, Kitazawa K, Fueki K, Mukaibo T (1978) Theory of polarographic maximum current – I. Conditions for the onset of hydrodynamic instability in a liquid metal electrode system. Electrochim Acta 23:867–874
29. Aogaki R, Kitazawa K, Fueki K, Mukaibo T (1978) Theory of polarographic maximum current – II. Growth or decay rate of the electrochemical and hydrodynamic instability. Electrochim Acta 23:875–880
30. Makino T, Morioka K, Aogaki R (1985) Occurrence of cellular convective flow accompanying the polarographic maximum value of the first kind. J Electroanal Chem 190:261–265
31. Makino T, Morioka K, Aogaki R (1985) Regular convection occurring at the interface by the resonance with pulsated potential in mercury electrode system. J Electroanal Chem 190:267–270
32. Gorzkowski MT, Jurczakowski R, Orlik M (2008) Electrochemical oscillations and bistability in the redox processes of mercury ions, coupled with the self-induced convection of Hg surface. J Electroanal Chem 815:135–144
33. Wrona PK, Galus Z (1982) Mercury. In: Bard AJ (ed) Encyclopedia of electrochemistry. Dekker, New York, pp 2–227
34. Nechiporuk VV, Elgurt IL (1991) Principles of pattern formation caused by Marangoni effects in electrochemical systems. Ukrainian Khim Zh 57:500–507 (in Russian)
35. Jehring H, Huyen NV, Horn E (1978) Current-oscillations caused by differences in surface tension at the mercury/liquid interface. J Electroanal Chem 88:265–268
36. Tschernyshev VV (1990) Autowave processes at anodic oxidation of aluminum. Elektrokhimiya 26:847–850 (in Russian)
37. Tenan MA, Teschke O, Kleinke MU, Galembeck F (1990) Surface tension effects on cell pattern formation during anodic metal dissolution. Langmuir 6:1640–1646
38. Kleinke MU, Teschke O, Tenan MA (1991) Pattern formation on aluminum electrodes. J Electrochem Soc 138:2763–2770
39. Teschke O, Kleincke MU, Galembeck F, Tenan MA (1990) Pattern formation on iron electrodes in sulfuric acid solutions. Langmuir 6:829–833
40. Orlik M, Gorzkowski MT (2008) The simple model and linear stability analysis of the electrode process with the NDR region caused by the potential-dependent convective transport. J Electroanal Chem 617:64–70
41. Koper MTM, Sluyters JH (1994) Instabilities and oscillations in simple models of electrocatalytic surface reactions. J Electroanal Chem 371:149–159
42. Milner GWC (1957) Principles and applications of polarography. Longman, London
43. Mominul Islam Md, Okajima T, Ohsaka T (2004) Current oscillatory phenomena based on redox reactions at a hanging mercury drop electrode (HMDE) in dimethyl sulfoxide. J Phys Chem B 108:19425–19431
44. Mominul Islam Md, Okajima T, Ohsaka T (2006) Eccentric phenomena at liquid mercury electrode/solution interface: upward, downoward, and circular motions. J Phys Chem B 110:8619–8625
45. Najdoski M, Mirceski V, Petruševski DS, Rosenberg D (2007) Mercury beating heart: modifications to the classical demonstration. J Chem Educ 84:1292–1295
46. Avnir D (1989) Chemically induced pulsations of interfaces: the mercury beating heart. J Chem Educ 66:211–212
47. Lin SW, Keizer J, Rock PA, Stenschke H (1974) On the mechanism of oscillations in the "beating mercury heart". Proc Natl Acad Sci USA 71:4477–4481
48. Keizer J, Rock PA, Lin SW (1979) Analysis of the oscillations in "beating mercury heart" systems. J Am Chem Soc 101:5637–5649
49. Kim CW, Yeo IH, Pail WK (1996) Mechanism of the mercury beating heart: an experimental study of the electrochemical-mechanical oscillator. Electrochim Acta 41:2829–2836
50. Ealy JL, Niewahner J (1993) Gallium beating heart. J Chem Educ 70:491–492

51. Maiworm A, Schmick M, Markus M (2007) Transduction of electrochemical into mechanical oscillations. Phys Lett A 361:488–492
52. Kirowa-Eisner E, Schwarz M, Rosenblum M, Gileadi E (1995) Temperature dependence of the transfer coefficient for the hydrogen evolution reaction on the DME. J Electroanal Chem 381:29–37
53. Wiśniewski A, Czerwiński WK, Paklepa P, Wrona PK, Orlik M (2008) The convection-driven non-additivity of the faradaic currents for the parallel reduction of some metal and hydrogen cations at the mercury electrode. J Electroanal Chem 623:120–128
54. Strasser P, Lübke M, Eickes C, Eiswirth M (1999) Modeling galvanostatic potential oscillations in the electrocatalytic iodate reduction system. J Electroanal Chem 462:19–33
55. Li Z, Cai J, Zhou S (1997) Potential oscillations during the reduction of $Fe(CN)_6^{3-}$ ions with convection feedback. J Electroanal Chem 432:111–116
56. Bard AJ, Faulkner L (2001) Electrochemical methods. Fundamentals and applications. Wiley, New York
57. Li Z, Cai J, Zhou S (1997) Current oscillations in the reduction or oxidation of some anions involving convection mass transfer. J Electroanal Chem 436:195–201
58. Wilson BP, Yliniemi K, Murtomäki L, Kontturi K (2009) Sympathetic current oscillations at an enzyme electrode induced by potential oscillations at a Pt surface. Electrochem Commun 11:2328–2331
59. Huang W, Li Z, Peng Y, Niu Z (2004) Transition of oscillatory mechanism for methanol electro-oxidation on nano-structured nickel hydroxide film (NNHF) electrode. Chem Commun 12:1380–1381
60. Huang W, Li ZL, Peng YD, Chen S, Zheng JF, Niu ZJ (2005) Oscillatory electrocatalytic oxidation of methanol on an $Ni(OH)_2$ film electrode. J Solid State Electrochem 9:284–289
61. Li ZL, Yu Y, Liao H, Yao SZ (2000) A universal topology in nonlinear electrochemical systems. Chem Lett 29:330–331
62. Li ZL, Ren B, Niu ZJ, Xiao XM, Zeng Y, Tiang ZQ (2002) On the criteria of instability for electrochemical systems. Chin J Chem 20:657–662
63. Piron DL, Nagatsugawa I, Fan C (1991) Cathodic potential oscillation of Sn(II)/Sn electrodes in KOH solution under constant current conditions. J Electrochem Soc 138:3296–3299
64. Tada T, Fukami K, Nakanishi S, Yamasaki H, Fukushima S, Nagai T, Sakai S, Nakato Y (2005) Tuning of the spacing and thickness of metal latticeworks by modulation of self-organized potential oscillations in tin (Sn) electrodeposition. Electrochim Acta 50:5050–5055
65. St-Pierre J, Piron DL (1987) A model for the potential oscillations of the zinc electrode polarized cathodically in an alkaline medium. J Electrochem Soc 134:1689–1694
66. St-Pierre J, Piron DL (1990) Mechanism of cathodic potential oscillations of the zinc electrode in alkaline solutions. J Electrochem Soc 137:2491–2498
67. Hudson JL, Tsotsis TT (1994) Electrochemical reaction dynamics: a review. Chem Eng Sci 49:1493–1572
68. Suter RM, Wong PZ (1989) Nonlinear oscillations in electrochemical growth of Zn dendrites. Phys Rev B 39:4536–4540
69. Wang M, Ming NB (1992) Concentration field oscillation in front of a dendrite tip in electrochemical deposition. Phys Rev A 45:2493–2498
70. Argoul F, Arneodo A (1990) Experimental evidence for deterministic chaos in electrochemical deposition. J Phys France 51:2477–2487
71. Argoul F, Huth J, Merzeau P, Arneodo A, Swinney HL (1993) Experimental evidence for homoclinic chaos in an electrochemical growth process. Physica D 62:170–185
72. Felici N (1969) Phénomènes hydro et aérodynamiques dans la conduction des diélectriques fluids. Rev Gen Electr 78:717–734

73. Felici N (1972) D.C. conduction in liquid dielectrics (Part II) electrohydrodynamic phenomena. Direct Curr 2:147–165
74. Schneider JM, Watson PK (1970) Electrohydrodynamic stability of space-charge-limited currents in dielectric liquids. I. Theoretical study. Phys Fluids 13:1948–1954
75. Watson PK, Schneider JM, Till HR (1970) Electrohydrodynamic stability of space-charge-limited currents in dielectric liquids. II. Experimental study. Phys Fluids 13:1955–1961
76. Lacroix JC, Atten P, Hopfinger EJ (1975) Electro-convection in a dielectric liquid layer subjected to unipolar charge injection. J Fluid Mech 69:539–563
77. Hopfinger EJ, Gosse JP (1971) Charge transport by self-generated turbulence in insulating liquid submitted to unipolar injection. Phys Fluids 14:1671–1682
78. Atten P, Lacroix JC (1978) Electrohydrodynamic stability of liquids subjected to unipolar injection: non-linear phenomena. J Electrostat 5:439–452
79. Atten P (1974) Electrohydrodynamic stability of dielectric liquids during transient regime of space-charge-limited injection. Phys Fluids 17:1822–1827
80. Felici N, Lacroix JC (1978) Electroconvection in insulating liquids with special reference to uni- and bi-polar injection: a review of the research work at the CNRS Laboratory for Electrostatics, Grenoble 1969–1976. J Electrostat 5:135–144
81. Lacroix JC, Atten P (1978) Double injection with recombination: EHD linear and non linear stability study. J Electrostat 5:453–461
82. Williams R (1963) Liquid crystals in an electric field. Nature 199:273–274
83. Williams R (1963) Domains in liquid crystals. J Chem Phys 39:384–388
84. Joets A, Ribotta R (1984) Electro-hydro-dynamical convective structures and transitions to chaos in a liquid crystal. In: Wesfreid JE, Zaleski S (eds) Cellular structures in instabilities, Lecture notes in physics 210. Springer, Berlin, pp 294–306
85. Helfrich W (1969) Conduction-induced alignment of nematic liquid crystals: basic model and stability considerations. J Chem Phys 51:4092–4105
86. Buka Á, Dressel B, Otowski W, Camara K, Toth-Katona T, Kramer L, Lindau J, Pelzl G, Pesch W (2002) Electroconvection in nematic liquid crystals with positive dielectric and negative conductive anisotropy. Phys Rev E 66:051713-1–051713-8
87. Buka Á (1989) Pattern formation in liquid crystals. Phys Scr T25:114–117
88. Tulagin V (1969) Imaging method based on photoelectrophoresis. J Opt Soc Am 59:328–329
89. Boxell C, Kelsall GH (1991) Photoelectrophoresis of colloidal semiconductors. Part 1. The technique and its applications. J Chem Soc Faraday Trans 87:3537–3545
90. Boxell C, Kelsall GH (1991) Photoelectrophoresis of colloidal semiconductors. Part 2. Transient experiments on TiO_2 particles. J Chem Soc Faraday Trans 87:3547–3556
91. Frank SN, Bard AJ (1977) Heterogeneous photocatalytic oxidation of cyanide ion in aqueous solutions at titanium dioxide powder. J Am Chem Soc 99:303–304
92. Kraeutler B, Bard AJ (1978) Heterogeneous photocatalytic preparation of supported catalysts. Photodeposition of platinum on titanium dioxide powder and other substrates. J Am Chem Soc 100:4317–4318
93. Toyoshima Y, Takahashi A, Nozaki H, Iida T, Sukigara M (1977) Photogr Sci Eng 21:29
94. Takahashi A, Toyoshima Y, Sukigara M (1977) Proceedings of the Tokyo symposium'77 on photo and electroimaging, p 295
95. Takahashi A, Alkawa Y, Toyoshima Y, Sukigara M (1979) Photoelectrophoresis and photoelectrohydrodynamic instability of TiO_2 particle suspension systems. J Phys Chem 83:2854–2858
96. Aikawa Y, Takahashi A, Sukigara M (1982) Charge-transport in photoslurry electrode cell based on TiO_2 particle suspension displaying highly organized convection. J Colloid Interface Sci 89:588–590
97. Aikawa Y, Shimoda H, Kobayashi K, Takahashi A, Takeda S, Sakata T, Sukigara M (1982) Electrohydrodynamic dissipative structure in unipolar and bipolar injection cases. Bull Chem Soc Jpn 55:654–659

98. Faulkner L, Bard AJ (1977) Techniques of electrogenerated chemiluminescence. In: Bard AJ (ed) Electroanalytical chemistry, vol 10. Dekker, New York, p 1
99. Köstlin H, Schaper H (1980) Electrochemiluminescence by dc of rubrene displaying highly organized electrohydrodynamic convection. Phys Lett 76A:455–458
100. Schaper H (1981) A novel hybrid detection system based on electrochemiluminescence in electrolyte-free solutions. J Electroanal Chem 128:335–342
101. Schaper H, Köstlin H, Schnedler E (1982) New aspects of D-C electrochemiluminescence. J Electrochem Soc 129:1289–1294
102. Schaper H, Schnedler E (1982) Electrolyte-free electrochemiluminescence in thin-layer cells. Current efficiency studies. J Phys Chem 86:4380–4385
103. Schaper H, Schnedler E (1982) On the charge distribution in thin-layer cells under electrostatic conditions. Application to electrolyte-free electrochemiluminescence. J Electroanal Chem 137:39–49
104. Jaguiro P (1996) Electrochemiluminescence observation and measurements of electrohydrodynamic streams. Proc SPIE Int Soc Opt Eng 2729:131–138
105. Jaguiro P, Smirnov A (1995) Electrochemiluminescence display technology. Proc SPIE Int Soc Opt Eng 2373:401–408
106. Jaguiro PV, Zhivnov VA (1987) Investigation of electrochemiluminescence in cells with submicrometric interelectrode distances. Izv Akad Nauk SSSR Ser Phys 51:591–595 (in Russian)
107. Solodkov VT, Jaguiro PV, Zhirnov VA (1990) Application of electrochemiluminescence for investigation of conduction of liquid dielectrics. Elektron Obrab Mater 152:34–36 (in Russian)
108. Orlik M, Rosenmund J, Doblhofer K, Ertl G (1998) Electrochemical formation of luminescent convective patterns in thin-layer cells. J Phys Chem B 102:1397–1403
109. Orlik M, Doblhofer K, Ertl G (1998) On the mechanism of electrohydrodynamic convection in thin-layer electrolytic cells. J Phys Chem B 102:6367–6374
110. Orlik M (1999) On the onset of self-organizing electrohydrodynamic convection in the thin-layer electrolytic cells. J Phys Chem B 103:6629–6642
111. Orlik M (1999) On the geometrical shape of self-organizing electro-hydrodynamic convective flows generated in thin-layer electrolytic cells. Phys Chem Chem Phys 1:5359–5367
112. Orlik M (2000) On the self-organization of the convective motion of fluid, driven by electric forces in the thin-layer electrolytic cells. Electrochem Commun 2:522–526
113. Dini D, Doblhofer K, Ertl G (2000) Nucleation of solution convection channels as the first step in electro-hydrodynamic pattern formation. Phys Chem Chem Phys 2:1183–1186
114. Jaguiro P (2001) Long time stability of electrochemiluminescence organic light emitting structures. In: Proceedings of the 10th international SID symposium "Advanced display technologies". Minsk, Belarus, pp 116–119
115. CRC (1983–1984) CRC handbook of chemistry and physics, 64th edn. CRC, Boca Raton, FL
116. Feldberg SW (1969) Digital simulation: a general method for solving electrochemical diffusion-kinetic problems. In: Bard AJ, Rubinstein I (eds) Electroanalytical chemistry, vol 3. Dekker, New York, pp 199–296
117. Britz D (2005) Digital simulation in electrochemistry, 3rd edn. Springer, Berlin
118. Griebel M, Dornseifer Th, Neunhoffer T (1995) Numerische Simulation in der Strömungsmechanik. Eine praxisorientierte Einführung, Vieweg, Braunschweig/Wiesbaden
119. Ertl G (1998) Pattern formation at electrode surfaces. Electrochim Acta 43:2743–2750
120. Orlik M (2002) The fluid convection as the self-organised phenomenon in electrochemical systems. Recent Res Dev Phys Chem 6:495–525, Transworld Research Network, Trivandlum, Kerala, India
121. Rubinstein I, Maletzki F (1985) Electroconvection at an electrically inhomogeneous permselective membrane surface. J Chem Soc Faraday Trans 87:2079–2087
122. Zholkovskii EK, Vorotyntsev MA, Staude E (1996) Electrokinetic instability of solution in a plane-parallel electrochemical cell. J Colloid Interfac Sci 181:28–33

123. Grigin AB (1985) Coulombic convective instability of a binary electrolyte in a cell with flat parallel electrodes. Elektrokhimiya 21:52–56 (in Russian)
124. Baune M (2002) Coupling of chemical and hydrodynamic instabilities at the electrochemical dissolution of metals. Ph.D. thesis, Universität Bremen, Germany
125. Baune M, Plath J (2002) Chemically induced hydrodynamic pattern formation: slowly rotating disk electrode under dissolving conditions and genesis of spatial bifurcation. Int J Bifurcat Chaos 12:2209–2217
126. Baune M, Breunig-Lyriti V, Plath PJ (2002) Invariant hydrodynamic pattern formation: fast rotating disk electrode under dissolution conditions. Int J Bifurcat Chaos 12:2835–2845

Chapter 6
Liquid Membrane and Other Membrane Oscillators

6.1 Dynamics of the Liquid–Liquid Interface and Liquid Membrane Systems

In complement to the hydrodynamic instabilities driven by surface tension gradients, enclosed in Sects. 5.4–5.7, we describe here the dynamics of the liquid–liquid interface, formed by two fluids of very limited miscibility, e.g., by water and 2-nitropropane. First reports on this type of dynamical systems have been published by Dupeyrat and Nakache [1, 2], and later analogous studies were continued also by Yoshikawa et al. [3–9].

A representative example of that type of the systems is the water–2-nitropropane system [3]. The key point is that each phase should contain the dissolved species that is significantly better soluble in the other phase. In this way the initial state of such system is far from equilibrium and the natural trend to decrease the free Gibbs energy is the flow of each dissolved species to the other phase. In the example described here, the aqueous phase contains initially the hexadecyltrimethylammonium bromide (or cetyltrimethylammonium bromide, CTAB) the large organic CTA^+ cation of which exhibits, due to its hydrophobicity, higher affinity to the organic phase. In turn, the phase of 2-nitropropane contains initially dissolved picric acid (HA), better soluble and also partly dissociating in water ($pK_a \approx 0.2$).

When these two phases are brought into contact, with the organic layer placed over the aqueous layer, the oppositely directed flows of CTAB and HPi begin. This transport is associated with subtle oscillations of various characteristics: pH of aqueous solution, interfacial potential drop (Fig. 6.1), and interfacial tension. The latter phenomena are easily detectable as visible oscillations of the meniscus at the interface of both liquids. It is interesting that under such non-equilibrium conditions the presence of surfactant (CTAB) does not depress the surface tension gradients, but on the contrary, it is responsible for their periodic temporal variations.

Yoshikawa and Matsubara [3] have proposed the mechanism of these oscillatory phenomena. First, one should note that CTA^+ cations in both phases form micelles above certain critical CTA^+ concentrations. Those micelles exhibit inverted

M. Orlik, *Self-Organization in Electrochemical Systems II*,
Monographs in Electrochemistry, DOI 10.1007/978-3-642-27627-9_6,

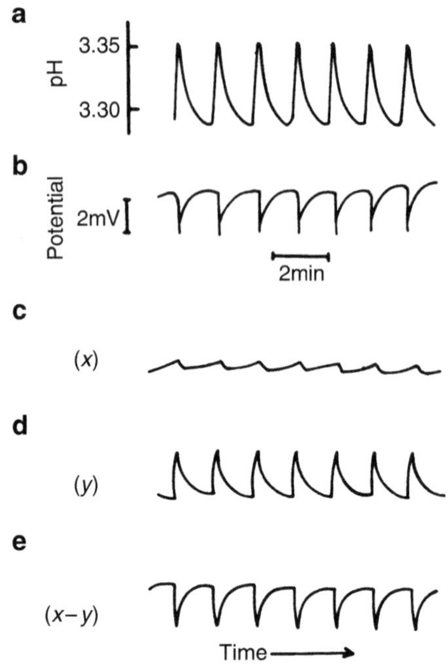

Fig. 6.1 Temporal oscillations of (**a**) pH in the aqueous phase, (**b**) electrical potential between the aqueous and organic phases, and (**c–e**) the results of computer simulation. x and y denote concentration of picrate anion (A^-) and hexadecyltrimethylammonium ion (CTA^+) near the interface, respectively, and $[x-y]$ is thus the difference between the concentrations of A^- and CTA^+. Reprinted with permission from [3]. Copyright 1983 American Chemical Society

orientation of chains: in water, hydrophilic ammonium groups are exposed to the surrounding water molecules, with aliphatic chains in the inner part, while in 2-nitropropane those chains are orientated to the organic phase. The detailed mechanism involves the following three steps, which are also schematically visualized in Fig. 6.2 [3].

Step I. CTA^+ ions, present as micelles in the aqueous phase, move toward the interface and attain such a state that hydrophilic ammonium group and the alkyl chain are orientated in the aqueous and organic phases, respectively. Simultaneously, the molecules of picric acid move toward the interface and dissolve in the aqueous phase, increasing the concentration of picrate anion (A^-) near the interface.

Step II. Simultaneously, the concentrations of CTA^+ cations and picrate anions (A^-) increase gradually, and CTA^+ ions form a monolayer structure on the interface. In both steps I and II, the driving force is mainly caused by the hydrophobic interactions between detergent and picrate.

Step III. When the concentration of CTA^+ reaches a critical value, CTA^+ ions suddenly transfer the interface toward the organic phase accompanied with the formation of the inverted micelles (this means the "cooperative" movements of CTA^+ ions). In this step an additional driving force arises from the formation of micelles in the organic phase. When the concentration of CTA^+ ions at the interface decreases to a lower critical value, the system returns to Step I.

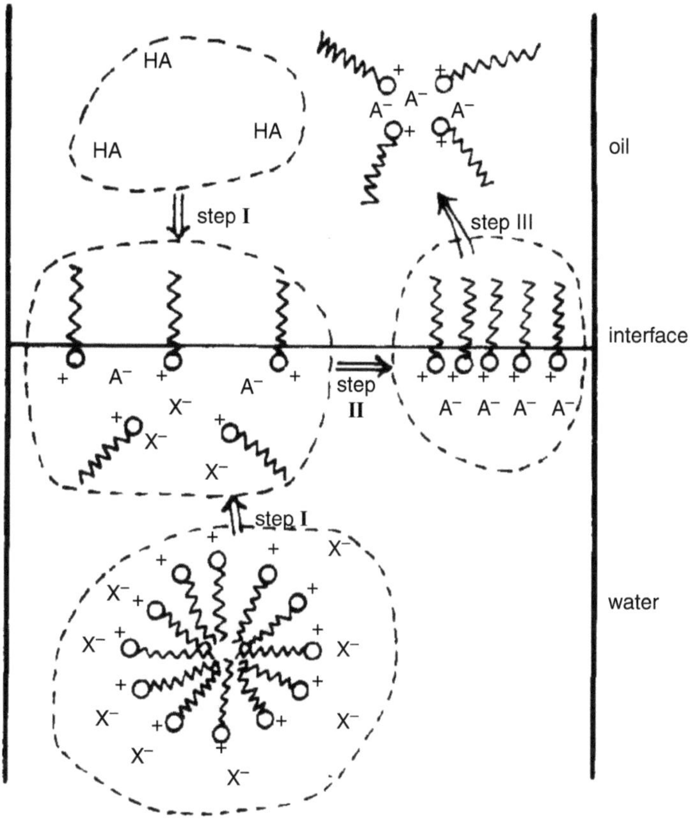

Fig. 6.2 Schematic representation of the mechanism of the oscillation in the liquid/liquid system. Reprinted with permission from [3]. Copyright 1983 American Chemical Society

These steps repeat cyclically until the distribution of the CTAB and HA concentrations at both sides of the interface approaches their equilibrium values.

The simple mathematical model was found to reproduce the oscillations of concentrations of picrate anions A^- (denoted as x) and CTA^+ cations (denoted as y), respectively, near the interface:

1. For $y_{\min} \rightarrow y \rightarrow y_{\max}$ (when y is increasing, steps I and II):

$$\frac{dx}{dt} = \alpha(y - x) \tag{6.1}$$

(diffusion of picric acid from the bulk organic phase towards the interface)

$$\frac{dy}{dt} = c - \beta(y - x) \tag{6.2}$$

(diffusion of CTAB from the bulk aqueous phase toward the interface)

2. For $y_{max} \rightarrow y \rightarrow y_{min}$ (when y is decreasing, step III)

$$\frac{dx}{dt} = \alpha(y - x) - \frac{\gamma y^n}{\theta + y^n} \tag{6.3}$$

$$\frac{dy}{dt} = c - \beta(y - x) - \frac{\delta y^n}{\theta + y^n} \tag{6.4}$$

with the last two terms in Eqs. (6.3) and (6.4) expressing the cooperative behavior between CTA^+ ions in step III. Parameters of numerical calculations, results of which are shown in Fig. 6.1c–e, were the following: $y_{min} = 0.4$, $y_{max} = 0.8$, $\alpha = 0.3$, $\beta = 2$, $c = 0.25$, $\gamma = 2$, $\delta = 30$, $\theta = 2$, and $n = 3$. In terms of this model, the variations of interfacial potential drop can be correlated with the difference in the concentrations of A^- and CTA^+ ions, i.e., with $(x-y)$. In turn, variations of $[A^-]$ (x) were found to correlate well with the variations of pH of aqueous phase, although also other factors contribute to this value in a real system.

6.2 Liquid Membrane Systems as the Model Chemoreceptors

6.2.1 Construction and Oscillatory Dynamics of the Liquid Membrane System

Systems of the type described above can be extended for one more aqueous phase and then the middle organic layer becomes a liquid membrane. Such membranes are known for their practical applications in metal separation technology, waste water treatment [10], with possible water purification from detergents [11]. Also, liquid membranes can be considered the simplest chemical models of biomembranes. Here we shall focus on the nonlinear dynamical properties of such systems and their potential practical applications. For example, the oscillators of this type were used by the Yoshikawa's group to construct the model systems exhibiting the molecular recognition by appropriate variation of the oscillatory characteristics upon addition of various compounds. The particular experimental system, based on the U-shaped cell, is sketched in Fig. 6.3 [5, 9].

In such system the oscillations were induced after certain initial period, provided some amount of ethanol (or another aliphatic or aromatic alcohol) was added. Figure 6.4 proves that the response depended strongly on particular alcohol. The detailed composition of every phase is given in the caption.

It was further found that the characteristics of the oscillations depended also on the concentration of given alcohol. This effect was qualitatively explained in terms of the incorporation of alcohol molecules into the micelles of CTA^+ ions and in this way influencing their structure at the interface, one of the key factors for the oscillations. The presence of alcohol could also influence the rate of migration of

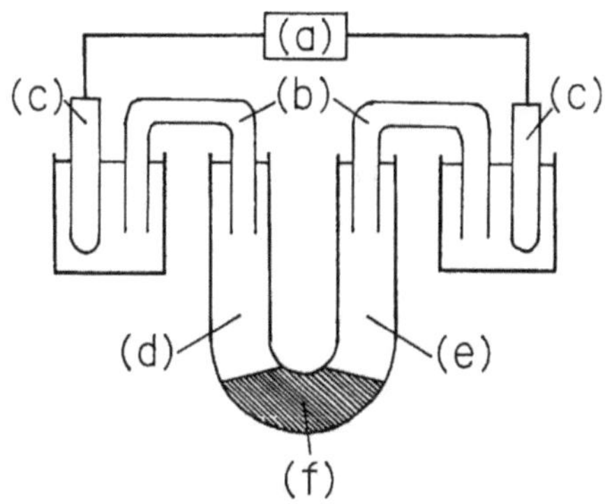

Fig. 6.3 Diagram of the U-shaped experimental apparatus for studying liquid membrane oscillator: (**a**) millivoltmeter, (**b**) salt bridge (3 M KCl), (**c**) Ag/AgCl electrode, (**d**) aqueous layer with anionic detergent, (**e**) aqueous layer with NaCl, and (**f**) organic layer. Reprinted with permission from [9]. Copyright 1986 American Chemical Society

CTA^+ ions through the interface. More quantitatively, it was found that the critical (threshold) concentration C_{th} of alcohol needed to induce the oscillations (meaning the excitability in the membrane oscillatory response) depended on its hydrophobicity, expressed in terms of its partition coefficient, P, between benzene and water:

$$\log C_{th} = -a \log P + b \tag{6.5}$$

where a and b are the constant parameters of this linear dependence.

Noteworthy, a similar relationship was found for the taste reception (cf. e.g., [12]). Furthermore, a general tendency was that the frequency of the oscillations increased with an increase in the concentration of a given alcohol, particularly sharply when the concentration of alcohol exceeded a critical value. The logarithm of this frequency $R = \log f$ obeyed a dependence on the concentration similar to the mathematical form of the Langmuir adsorption isotherm:

$$\frac{C}{R} = \frac{C}{R_{th}} + \frac{1}{KR_{th}} \tag{6.6}$$

where R_{th} is the maximum response (analogous to maximum coverage for adsorption) and parameter K plays also the role analogous to the adsorption equilibrium constant. Again, a similar mathematical dependence has been reported for the concentration dependence of the magnitude of the nerve response in taste reception [13].

Yoshikawa et al. [9] have reported similar phenomena also in another system in a U-shaped glass tube. The middle liquid membrane consisted of mixture of nitrobenzene (80% v/v) and aliphatic alcohol (20% v/v), containing dissolved 2,2′-bipirydine (the latter compound was added in order to diminish the resistance of the organic phase and thus to reduce the noise in the electric response). This organic layer was placed between two aqueous solutions: 0.4 mM sodium dodecyl

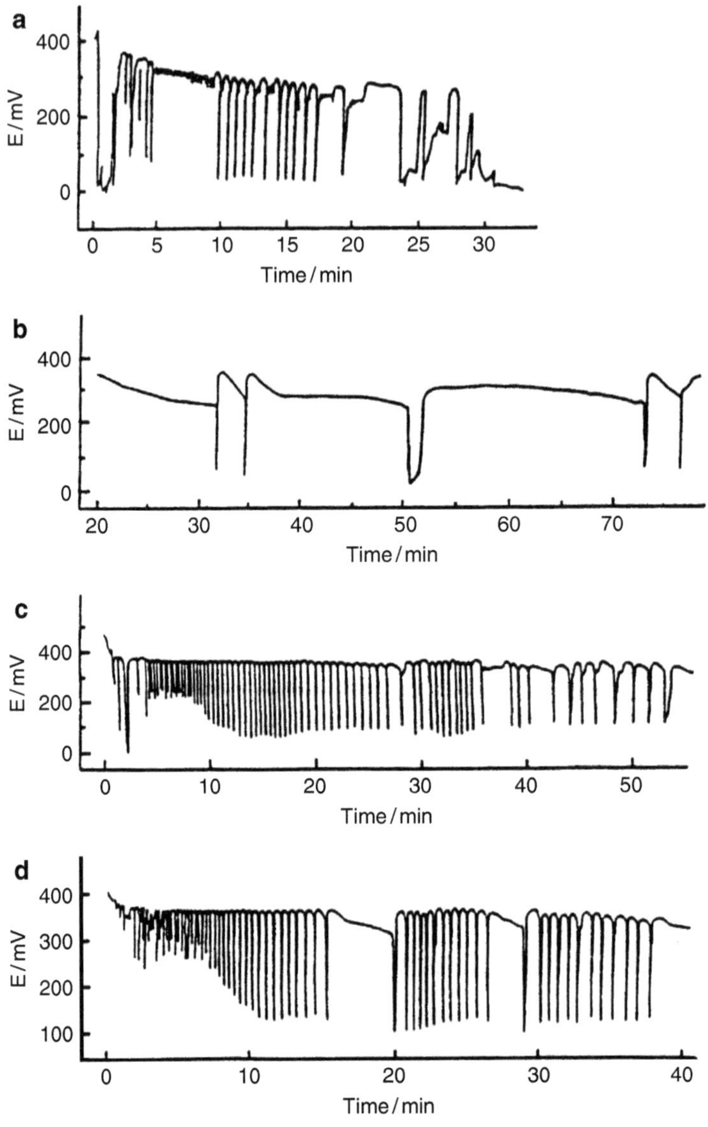

Fig. 6.4 Oscillations of electrical potential with (**a**) 0.5 M *n*-propyl alcohol, (**b**) 0.5 M isopropyl alcohol, (**c**) 0.5 M *n*-butyl alcohol, and (**d**) 0.25 M *n*-pentyl alcohol in the left aqueous phase; other experimental conditions were the same as for Fig. 6.3. Reprinted with permission from [5]. Copyright 1984 American Chemical Society

sulfate (anionic surfactant) on one side and 500 mM sodium chloride on the other. Analogously to the above mechanism, the oscillations were explained in terms of consecutive formation and destruction of monolayer structures of the dodecyl sulfate anion at the interface, with the role of alcohol similar also to the discussed above.

6.2.2 *Electrochemical Model of Taste Sensor*

In view of the analogies, described in the previous subsection, of the dynamics of the liquid membrane with taste reception, or more generally—with the response of biological chemoreceptive membranes—the same group of researchers developed further the systems with liquid membranes, mimicking the sensors of taste or smell.

As a model of the sensor of taste [6], again an apparatus with a U-shaped glass tube was used. The middle liquid membrane consisted of 5 mM solution of 2,2′-bipyridine in nitrobenzene. This organic layer was placed between two aqueous phases: the left one consisted of 0.1 mM sodium oleate (or sodium stearate) solution, containing 5% v/v aliphatic alcohol, and the right one was 0.5 M NaCl solution. In order to determine the "taste profile" of this model system, there was examined the effect of chemical substances belonging to four basic taste categories: salty (NaCl), sweet (sucrose), bitter (quinine chloride), and sour (hydrochloric acid). In relevant experiments the left liquid phase contained 0.1 mM sodium oleate and 10% v/v 1-propanol, and after ca. 1 h, when the oscillations became stable, one of the above specific chemical stimuli was added to this phase. The results of such experiments, collected in Fig. 6.5, indicate significantly different oscillatory responses of the model sensor. Of course, one can pose a question whether the system detects really the taste and not simply very different chemical species, but the aim of those experiments was to show that the system recognizes just different molecules, exemplified here as the representatives of various taste categories.

Qualitatively, these effects were explained in terms of the repetitive construction and destruction of the surfactant monolayer at the interface, affected by the presence of additional components. The mathematical model of the oscillatory responses, particularly due to the presence of irregular courses, had to be extended to three dynamical variables, which for the 2D system could not be obtained. These variables are the concentrations: X_i—of oleate near the interface, Y_i—of alcohol near the interface, and Z_i—of the aggregate of oleate and alcohol at the interface. They are involved in the network of process, involving also the bulk concentrations of both X and Y species (X_{b}, Y_{b}) (Fig. 6.6).

A very simplified kinetic model [6] led to the following set of ordinary differential equations:

$$\frac{\mathrm{d}X_i}{\mathrm{d}t} = a_1 - b_1 X_i - c_1 Z_i \tag{6.7}$$

$$\frac{\mathrm{d}Y_i}{\mathrm{d}t} = a_2 - b_2 Y_i - c_2 Z_i \tag{6.8}$$

$$\frac{\mathrm{d}Z_i}{\mathrm{d}t} = F(X_i, Y_i) - dG(Z_i) \tag{6.9}$$

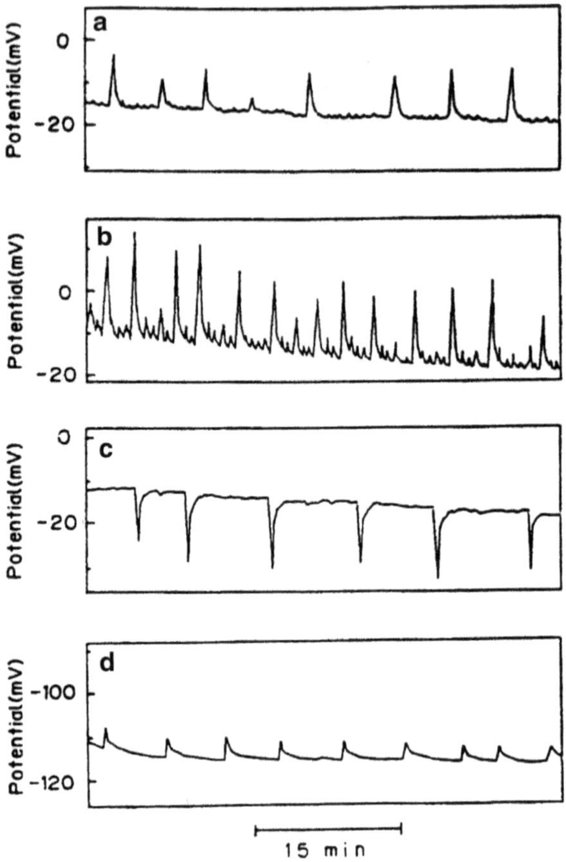

Fig. 6.5 Oscillations of electrical potential by the addition of 0.5 mL of (**a**) 0.1 M sodium chloride, (**b**) 1 M sucrose, (**c**) 0.001 M quinine chloride, and (**d**) 0.03 M hydrochloric acid to the left aqueous phase. See [6] for experimental conditions. Reprinted with permission from [6]. Copyright 1988 American Chemical Society

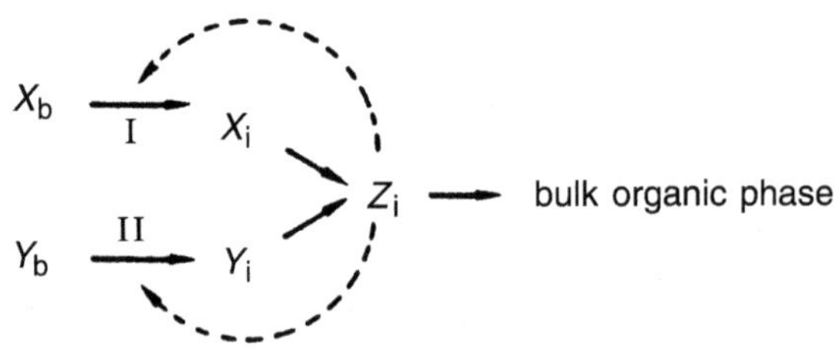

Fig. 6.6 Schematic representation of processes causing the oscillatory response in liquid membrane system mimicking the taste sensor. Reprinted with permission from [6]. Copyright 1988 American Chemical Society

in which $a_1 - b_1X_i - c_1Z_i$ and $a_2 - b_2Y_i - c_2Z_i$ describe the rates of diffusion of oleate and alcohol, respectively, from the bulk aqueous phase to the interface. In particular, the first two terms in each of these expressions denote the transfer of respective species due to the concentration gradient between the bulk phase and the interface, including the rate of change from X_i and Y_i to Z_i. The third term represents the effects of negative feedback induced by the increase in Z_i. Furthermore, in Eq. (6.9), the function $F(X_i, Y_i)$, expressing the synergetic interaction of X_i and Y_i yielding Z_i is given by the simple linear dependence $e(X_i + Y_i)$. Finally, function $G(Z_i)$ describes the rate of escape of Z_i into the bulk organic phase. Coefficients a_i, b_i, c_i, d, and e are parameters. In order to obtain oscillatory solutions, these parameters have to attain appropriate values and, last not least, the function $G(Z_i)$ has to have a N-shape nonlinearity, expressed here in a polynomial form:

$$G(Z_i) = \frac{-Z_i + Z_i^3}{3} \tag{6.10}$$

Of course, if the oscillatory courses for various added substances are very similar, the procedure of optical distinguishing between different responses can be very subjective. Thus, a more objective and quantitative analysis of the responses of model taste sensors (or rather—sensors of different species added) would be very useful. That is why, inspired by the above works, Szpakowska et al. continued studies of such model sensors [14, 15] and have proposed [16, 17] the analysis of the oscillations by means of Gábor approach [18]. The following experimental system was employed: the left aqueous phase contained a cationic surfactant—benzyldimethyltetradecylammonium chloride (BDMTAC) and ethanol, the middle organic phase (liquid membrane) was formed from the solution of picric acid in nitromethane, and the right aqueous phase contained various substances corresponding to different categories of taste: sucrose, citric acid, quinine hydrochloride, caffeine, KCl, glucose, acetic acid, lactose or fructose, in most cases as 0.1 M solutions. In the following the left aqueous phase will be called the donor phase (*d*), while the right phase will be named the acceptor phase (*a*). Thus, the corresponding interfaces with the organic membrane (*m*) will be denoted as *d*/*m* and *m*/*a*, respectively. The aim of the theoretical analysis of the electrochemical signals was to produce the kinds of "fingerprints" for every species, creating a basis for the molecular recognition realized by the oscillating liquid membrane system. The Gábor analysis is a method of treatment of non-stationary signals (e.g., formed on the drifting baseline) that are therefore not strictly suitable for the usual Fourier transform, reliable only under assumption that the wave-like signal is stationary. For the non-stationary signal the Fourier analysis yields only averaged result. In the Gábor analysis, the actual time domain is divided into a series of subdomains of constant time intervals and a Fourier transformation is applied for the oscillatory response in each subdomain (for another application of an analogous approach, see Sect. 3.8, volume I). Each time interval is viewed through a Gaussian-type window and analyzed with the Fourier transformation.

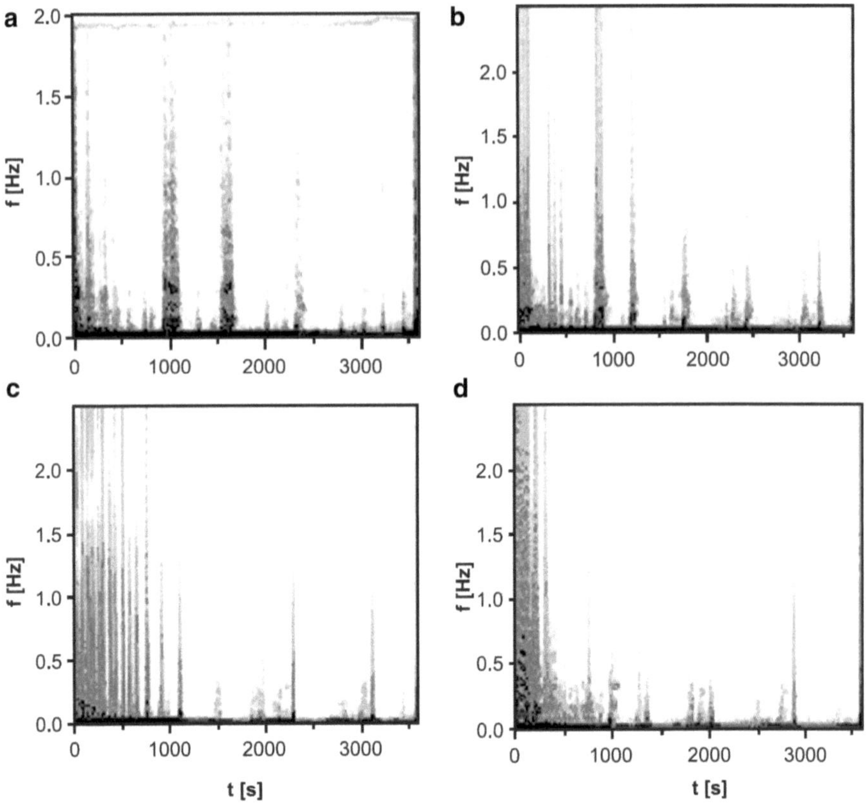

Fig. 6.7 Cross-sections of power spectra of the system with nitromethane and different sweet substances: (**a**) sucrose, (**b**) glucose, (**c**) lactose, and (**d**) fructose. Reprinted from [16]. Copyright 2005, with permission from Elsevier

The results have a form of three-dimensional plot in which the signal amplitudes are plotted as a function of frequency and time. Such plots obtained for different studied species revealed the wide spectrum of frequencies, with no dominating frequencies, which is the situation typical of chaotic behavior. Due to similar morphology of such plots, it was rather difficult to consider the desired "fingerprints" for particular substances. However, if those 3D diagrams are transformed into 2D ones, by taking cross-sections at constant signal amplitude, the images become more sensitive to particular compound. For instance, Fig. 6.7 show such cross-sections for different four carbohydrates.

Furthermore, a great advantage of such procedure of molecular recognition is that the "fingerprints" of particular species appear to be practically independent of their concentration. In later work on this subject, Szpakowska et al. [19] continued the above analysis and also formulated rather critical opinion that liquid membrane systems cannot be used in practice as taste sensor due to their instability and long time required for analysis.

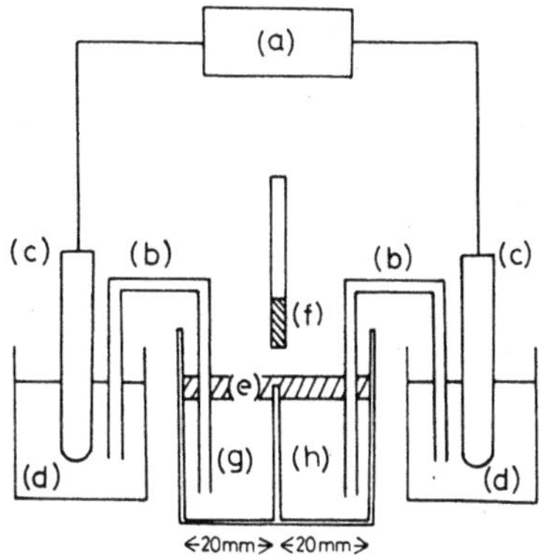

Fig. 6.8 Diagram of apparatus (**a**) millivoltmeter, (**b**) KCl salt bridge, (**c**) Ag/AgCl electrode, (**d**) 3 M KCl aqueous solution, (**e**) 90% oleic acid plus 10% 1-propanol containing 10 mM tetraphenylphosphonium chloride, (**f**) defatted cotton soaked in amine solution in a glass tube (4 mm), located ca. 10 mm above the surface of (**e**), (**g**) 0.5 M NaCl aqueous solution, and (**h**) 0.5 M KCl aqueous solution. Reprinted with permission from [8]. Copyright 1985 American Chemical Society

6.2.3 Electrochemical Model of Smell Sensor

In order to prepare the system that could detect absorbed vapors, the apparatus shown in Fig. 6.8 was constructed by Yoshikawa and Matsubara [8].

In this modification the essential part of the cell, i.e., the model chemoreceptor consists of two chambers, one of which contains 0.5 M NaCl solution, and the other one—0.5 M KCl. Both solutions are covered with a common layer of organic phase, i.e., the liquid membrane consisting of 90% oleic acid + 10% 1-propanol, with 10 mM tetraphenylphosphonium chloride dissolved for decreasing the ohmic resistance. When the liquid membrane was exposed to ammonia vapor, the electrical response remained nearly constant for about 5–20 min, then changed abruptly and this effect was accompanied with spontaneous, temporal movement of the organic phase, followed by rhythmic changes of the electric potential. This effect vanished when the exposure to ammonia was discontinued and returned when the exposure was repeated. Analogous effects were observed also for some amines. It is important to note that in all cases the reaction of the model system was observed only if the concentration of the ammonia or amine was above certain critical value, which indicates the excitable characteristics of the sensor. Typical responses on the exposure of ammonia and several amines are collected in Fig. 6.9.

6.3 Recent Developments in Liquid Membrane Oscillators

The pioneer discoveries by Dupeyrat and Nakache, followed by the group of Yoshikawa, found other continuators: Buhse, Micheau, Pimienta et al., in relation to the biphasic autocatalytic processes (as, e.g., biphasic alkaline hydrolysis of

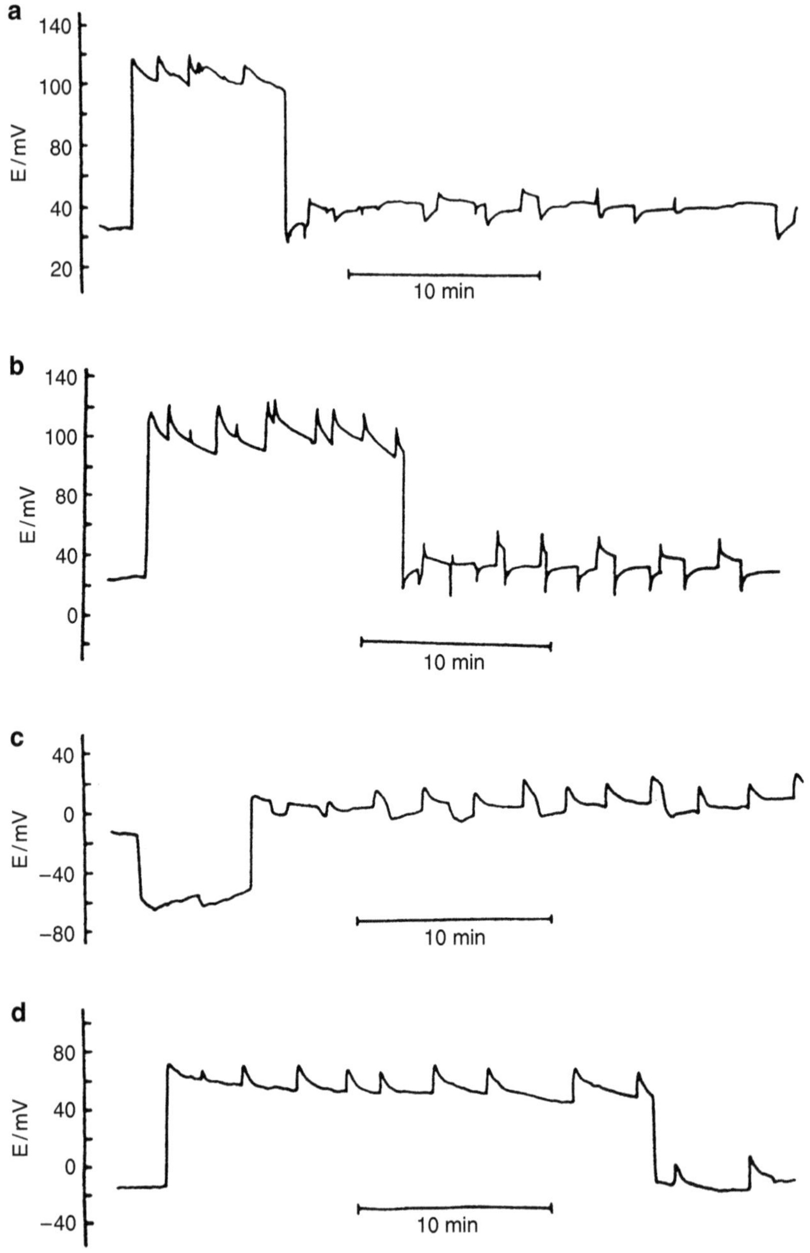

Fig. 6.9 Typical traces of changes of electrical potential on the exposure to vapor of (**a**) ammonia (6 M NH_3), (**b**) methylamine (40% aqueous solution), (**c**) pyridine, and (**d**) piperidine. The initial stage (ca. 10 min), i.e., induction period on exposure to the vapor, is omitted in these figures. Reprinted with permission from [8]. Copyright 1985 American Chemical Society

esters [20, 21]). Their works, in turn, were inspired also by earlier report by Bachmann et al. [22] who have described the ethyl caprylate hydrolysis occurring in biphasic medium to yield ethanol and surfactant ions caprylate (octanoate) which can aggregate and form micelles. The course of that reaction indicated its highly nonlinear autocatalytic kinetics, ascribed to the formation of micelles, and the problem was considered a model for prebiotic structures. As Buhse, Micheau et al. have shown later [23], this biphasic autocatalytic process turns into a bistable one if the reaction is performed in a continuous flow stirred tank reactor, in line with more general predictions on such dependence.

For the picric acid/CTAB two-phase system, introduced by Yoshikawa et al., Pimienta et al. [24] have employed UV–VIS spectroscopic studies in order to give a new insight into the possible kinetic mechanism for the oscillations reported for the U-shaped cell. In this case, the middle organic layer (liquid membrane) was formed by the 2.0 mM solution of picric acid in CH_2Cl_2. The aqueous solution in the left arm contained 5 mM CTAB and 0.4 M *n*-butanol, and the other one, in the right arm—0.1 M sucrose. Measurements of voltage between two aqueous phases have revealed additional small amplitude oscillations between the large spike, indicating the morphology typical of mixed-mode oscillations. UV/VIS measurements have revealed in turn that the key process in the system is the formation of 1:1 ion pairs between picrate and CTA^+ at the liquid/liquid interface, that desorb and move into organic phase. Simultaneously, the formation of reversed micelles, suggested by Yoshikawa et al., as a key process for the oscillations, was now questioned, based on near-IR spectroscopic measurements. The kinetic spectroscopic measurements indicated further that the presence of alcohol (essential for the occurrence of oscillations in the U-shaped system) inhibits the formation of the ion pairs. This inhibitory effect of an alcohol was suggested as key source of oscillations. The inhibition was suggested to be released presumably due to the transfer of the alcohol from the interface into the organic phase. The system's dynamics was described in terms of the S-shaped, autocatalytic-like kinetics with respect to the ion-pair formation. Accordingly, a new mechanism, based on the Langmuir–Hinshelwood kinetics was elaborated that included the competitive adsorption of CTA^+ cations, picrate anions, and the alcohol at the liquid/liquid interface. The release of inhibition which depends on the concentration of vacant interface sites and which occurs autocatalytically expresses the key process for the nonlinear behavior. This outline kinetic mechanism is presented below in more detail.

When the left aqueous phase (containing CTAB and butanol) gets in contact with the organic layer, the interface becomes rapidly covered with CTA^+ monolayer, according to the Langmuir–Hinshelwood kinetics (where symbol I stands below for the interface):

$$CTA^+ + I \rightleftarrows CTA^+ - I \tag{6.11}$$

Simultaneously, molecules of *n*-butanol (denoted below as B) move toward this interface and intercalate the CTA^+ layer in this way that they occupy free sites in between. It was assumed that already present at the interface, adsorbed CTA^+ ions accelerate the adsorption of *n*-butanol:

$$B + I \xrightleftharpoons{CTA^+-I} B - I \tag{6.12}$$

Also simultaneously, the molecules of picric acid arrive from the bulk organic phase to the interface and when get in contact with the aqueous phase, they rapidly dissociate to form respective ions; in particular, picrate anions (A^-) get trapped at the interface due to the electrostatic attraction to positive CTA^+ ions (in fact, no picrate anions were found in the left aqueous layer, in the course of the experiment). Thus, also the following step was considered to be catalyzed by CTA^+–I:

$$HA + I \xrightleftharpoons{CTA^+-I} A^- - I + H^+ \tag{6.13}$$

In summary, the above three reaction steps represent the competition between A^-, CTA^+ and *n*-butanol (B) for free sites at the interface between the left aqueous layer and the middle organic membrane, with the total number of these sites remaining constant:

$$[I]_0 = [I] + [A^- - I] + [CTA^+ - I] + [B - I] = \text{const} \tag{6.14}$$

Furthermore, in line with experimental UV–VIS observations, at the interface adsorbed CTA^+–I, and A^-–I species interact with each other to form the hydrophobic CTA–A ion pairs that desorb and move into the organic membrane. It is important to note that this step is *inhibited* by *n*-butanol, presumably due to blocking the interaction between the adsorbed CTA^+ and A^-:

$$CTA^+ - I + A^- - I \xrightleftharpoons{\text{inhibited by B-I}} CTA - A + 2I \tag{6.15}$$

The release of two adsorption sites in the latter process means that the inhibitive effect of alcohol is weakened, as its intercalation between CTA^+ and A^- ions is diminished. When the number of free sites [I] is high enough, alcohol's molecules arriving at the interface are even considered to travel directly into the organic phase in the autocatalytic (with respect to I) process:

$$B - I \xrightleftharpoons{I} B' + I \tag{6.16}$$

where B' stands for *n*-butanol in the organic phase.

With respect to the other interface, the only process considered there is the conversion of HA into A^- at this interface and the slow accumulation of A^- in the right aqueous phase; this process however is not considered to contribute to the oscillatory dynamics of the system. This also means that the role of sucrose present in this phase remains not understood.

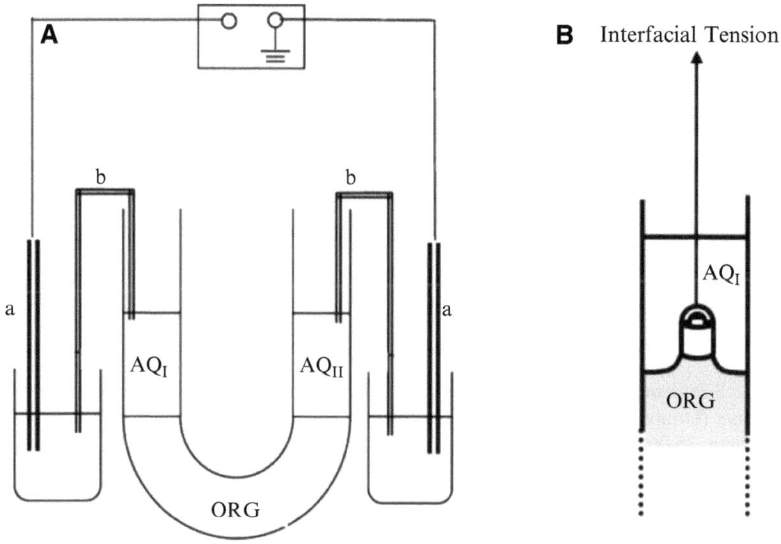

Fig. 6.10 (**A**) Device used for potential measurements in the three-phase (U-tube) geometry water/oil/water. (a) Electrodes, (b) salt bridges, AQ$_I$ 6 mL of aqueous solution of CTAB (5×10^{-3} M) and *n*-butanol (0.4 M), ORG, 13 mL of picric acid (3×10^{-3} M) in CH_2Cl_2, AQ$_{II}$, 6 ml of aqueous solution of sucrose (0.1 M). (**B**) *Left side* of the U-tube (AQ$_I$/ORG): small glass cylinder connected to a microbalance for surface tension measurements. Reprinted with permission from [25]. Copyright 2004 American Chemical Society

The reaction schemes listed above were transformed into ten kinetic differential equations describing the dynamics of every ten species involved (see [24] for details) and numerical integration yielded oscillations of the concentration of adsorbed picrate ions, i.e., A–I species, similar to those reported in the experiment.

This simplified model was however not able to reproduce either the periodic macroscopic movements of the interface or the mixed-mode oscillations. In order to understand better the mechanism of these phenomena, Pimienta et al. [25] have performed later simultaneous measurements of the variations of the voltage and the interfacial tension, in order to find correlation between them. The experimental setup employed is sketched in Fig. 6.10.

In particular, measurements of equilibrium interfacial tension at the CH_2Cl_2/water interface were performed using the stir-up technique, in which a small plastic (high-density polyethylene) cylinder, connected to the microbalance, was lowered to the liquid interface, immersed, and withdrawn until the maximum pull on the cylinder was found. Representative results of simultaneous measurements of the voltage and interfacial tension, showing their perfect correlation, are shown in Fig. 6.11. The drop in voltage corresponds to a fast decrease in the interfacial tension, indicating a fast increase of the concentration of the surfactant adsorbed on the interface.

In the mechanistic interpretation of their experimental results, Pimienta et al. go back to the primary interpretation of Dupeyrat and Nakache of the motion of the

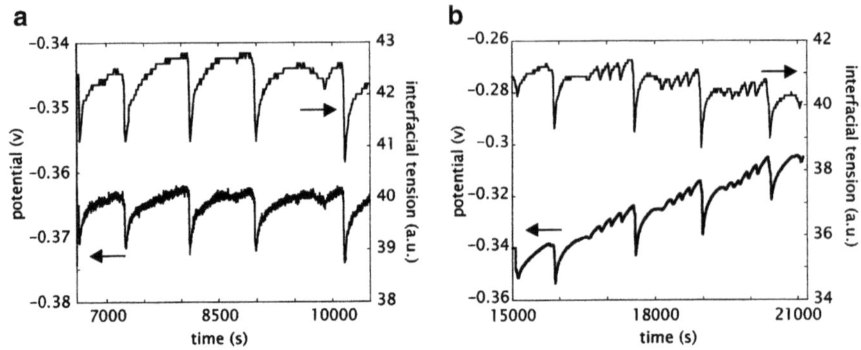

Fig. 6.11 Simultaneous measurement of the potential and the interfacial tension oscillations in the three-phase (U-tube) geometry taken from one experiment: (**a**) between 6,500 and 10,500 s, (**b**) between 15,000 and 21,000 s. *Lower curve*: electric potential (*left axis*). *Upper curve*: interfacial tension in arbitrary units (*right axis*). Reprinted with permission from [25]. Copyright 2004 American Chemical Society

interface invoking the Marangoni instability. In terms of this concept, when interfacial tension gradients reach a critical value, the system becomes unstable and switches to convective motion of the interface (i.e., Marangoni instability sets in). Convective fluxes lead to the accelerated supply of surfactants at the interface, and their adsorption: $CTA^+ + I \rightleftarrows CTA\text{–}I$, accelerated in this way, causes fast decrease in the interfacial tension. When a sufficiently packed monolayer is formed, the interfacial tension gradients vanish and the system switches to convection-free, i.e., to slower diffusion transport regime. The fast interfacial reaction: $CTA\text{–}I + A^- \rightarrow CTAP + I$ becomes then dominant and restores the interfacial tension gradients until they become critical again. Thus, the system exhibits periodic switches between the convective and diffusion regimes of reactant transports.

Further studies of such liquid membrane systems were also performed in recent years, among others by Szpakowska et al. who also has summarized recent achievements in this area in a concise review from 2009 [11]. The main point discussed in those recent works is the analysis and substantial reinterpretation of the processes responsible for the oscillations, different for the systems with cationic and anionic surfactants, as well as the representation of the oscillatory regime in a way typical of nonlinear dynamics. For the purpose of these analyses, it is again (cf. p. 383) useful to name the left aqueous phase in the U-tube as the donor phase (*d*), i.e., that phase from which the surfactant enters the middle liquid membrane (*m*) through, thus, the *d*/*m* interface. The right aqueous phase which can be eventually reached by the surfactant moving the liquid membrane, is then an acceptor phase (*a*), and the corresponding interface is denoted as *m*/*a*. In terms of this notation, Yoshikawa et al. in their early works have considered the phenomena occurring at the *d*/*m* interface as the principal source of the oscillations. The studies undertaken by Szpakowska et al. were inspired by earlier work by another Japanese group of Arai et al. [26] who, in search of the system's construction with improved reproducibility of oscillations, used the experimental arrangement involving

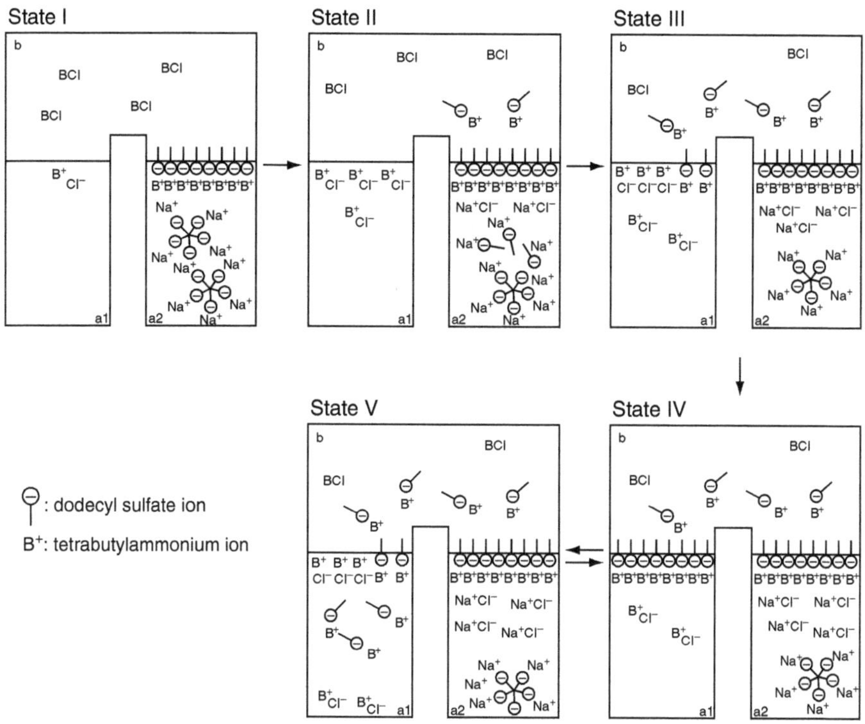

Fig. 6.12 Proposed mechanism of potential oscillation across the octanol membrane with SDS. States I, II, and III: induction period. State IV: occurrence of pulse. State V ⇄ IV: potential oscillations. Reprinted from [26]. Copyright 1995, with permission from Elsevier

the *inverted* U-shaped cell, requiring thus an organic solvent with specific gravity less than that of water. In this case the liquid membrane of octanol, with dissolved tetrabutylammonium chloride or picric acid, (denoted as b in Fig. 6.12) was placed on the top of two aqueous phases, one of which was pure water (a_1), and the other one (a_2) was the aqueous solution of sodium dodecyl sulfate (SDS) and ethanol. Analysis of this system suggested that not the *d*/*m*, (i.e. b/a_2) but the acceptor/liquid membrane (*a*/*m*, or b/a_1) interface was largely involved in the oscillatory mechanism, according to the scheme shown in Fig. 6.12.

The crucial role of processes at the *m*/*a* interface was more recently suggested also by Maeda et al. [27] for the system: aqueous solution of CTA^+Cl^- (W1)/nitrobenzene + picric acid/aqueous, salt-free phase (W2). Experimental method involved polarographic studies of the ion transfer at two (*d*/*m* and *m*/*a*) interfaces. The applicability of such approach was verified earlier by the same authors [28–30] who also have found novel oscillations of membrane potential or current accompanied by the membrane transport of H^+, Na^+, or Cs^+ ions. Based on studies of ion transfer, the novel, very detailed, mechanism was proposed in which the oscillations of electric potential were ascribed to the variation of the potential drop at the *m*/*a* (or membrane/W2) interface, occurring between the value

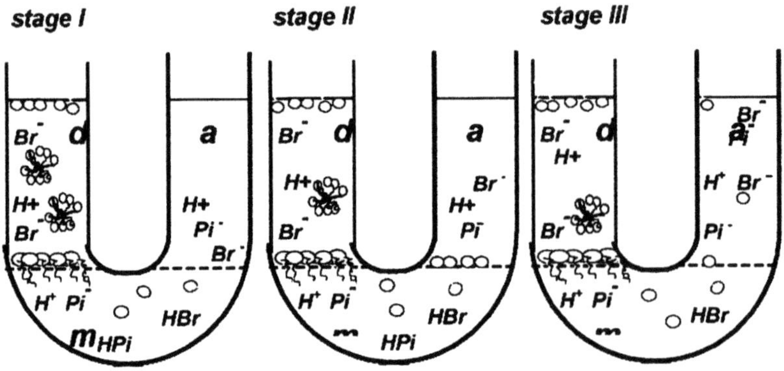

Fig. 6.13 Oscillation mechanism in the nitrobenzene oscillator containing CTAB. Reprinted from [31]. Copyright 2006, with permission from Elsevier

determined mainly by the transfer of H^+ and picrate ions from organic layer into W2 phase and the value determined by the transfer of H^+ and Cl^- to W2 phase. In other words, the transport of surfactant was considered unimportant as a source of oscillations, since it only continuously accumulated at the *a*/*m* interface.

In line with these novel achievements, Szpakowska et al. have also studied the oscillations in the CTAB-water/nitrobenzene or nitromethane/sucrose–water system [31–33]. The shape of oscillations strongly depended on the compound forming a liquid membrane and the surfactant dissolved in aqueous phase [11]. For each of these systems the detailed oscillatory mechanism was proposed, based essentially on the laws of chemical kinetics. In all cases the authors have suggested that the oscillations were generated at the *a*/*m* interface, again contrary to the original concept by the Yoshikawa group. For liquid membranes with the left aqueous phase containing initially cationic surfactant (and ethanol), the proposed mechanism of the oscillations involved sudden adsorption, followed by sudden desorption of CTA^+ cations and Br^- and picrate anions at the right *m*/*a* interface, when it is reached by the species diffusing from the opposite side of the system. The three-stage mechanism of the oscillation (I—induction period, II—first peak formation, III—creation of the first peak) was proposed, according to the sequence of processes visualized in Fig. 6.13.

For the system with anionic surfactant—sodium oleate and butanol present in the left aqueous phase, and the liquid membrane being the solution of 2,2′-bipyridine in nitromethane—the oscillations were suggested to be caused by adsorption and desorption of both alcohol and surfactant at the same, *m*/*a* interface, the whole process being controlled by the slow diffusion of these species across the liquid membrane [34]. Thus, independently of the kind of surfactant, the same *m*/*a* interface appears to be responsible for the oscillations. For all systems considered, the kinetic models were constructed and the existence of oscillatory regimes was confirmed by numerical integration of the relevant system of ordinary differential equations.

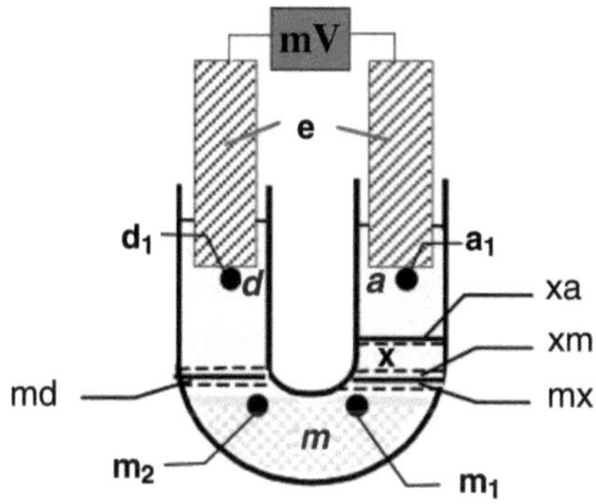

Fig. 6.14 Experimental setup of liquid membrane oscillator. d_1, a_1, m_1, and m_2: positions of electrodes in donor and acceptor aqueous phases and in liquid membrane, respectively. md, mx, xm, and xa: layer in phase i in vicinity of phase j. *e*: Ag|AgCl|Cl$^-$ electrodes. Reprinted with permission from [32]. Copyright 2006 American Chemical Society

In particular, the system with CTAB and nitromethane producing complex oscillatory patterns was described as composed of four phases, with the fourth one (denoted as phase *x* in Fig. 6.14) formed during the course of the experiment between the organic membrane and the aqueous acceptor phase. Experimentally, it was observed as formation of a new transparent and colorless zone, the thickness of which was growing in time, reaching a maximum value of ca 0.7 cm at ca. 1,200 s [32].

Accordingly, two more interfaces: *m*/*x* and *x*/*a* were formed and had to be considered. The mechanism of oscillations involved, thus, also adsorption and diffusion of the ion-pairs of the surfactant cations and picrate anions through phase *x*, and assumed, in particular, sudden adsorption and desorption of surfactant molecules at the *m*/*x* and *x*/*a* interfaces. The proposed mechanism is schematically illustrated by Fig. 6.15. The corresponding kinetic model exhibited chaotic oscillations in the course of numerical integration of nine ordinary differential equations. Interestingly, such system was considered as composed of two coupled oscillators, producing therefore appropriately complex oscillatory patterns [32]. Thus, the liquid membrane oscillator from Fig. 6.14 can be considered a supplement to Sect. 3.1 in which couplings between other types of oscillators were described.

There is no doubt that non-equilibrium systems with the liquid–liquid interface(s) remain the subject of interest and are therefore being studied using also more modern methods. For example, Ikezoe et al. [35] have described, in another experimental arrangement, the hydrodynamically induced chemical oscillations at a water/nitrobenzene interface, monitored with simultaneous measurements of the electric potential and the interfacial tension; the construction of the cell is sketched in Fig. 6.16. In order to measure a time course of interfacial tension, a time-resolved quasi-elastic laser scattering (QELS), developed by this research group (cf. [36–38]), was employed. An advantage of the QELS method is the *non-contact* and *non-perturbative* measurement of dynamic interfacial tension.

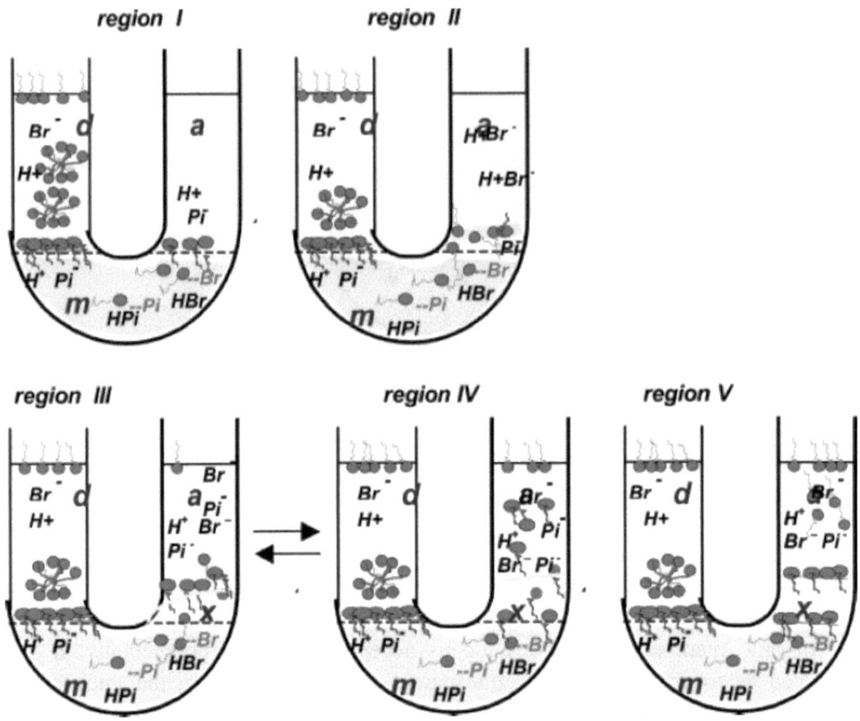

Fig. 6.15 Oscillation mechanism in the nitromethane oscillator containing CTAB (*filled circle* CTA^+ ion). Reprinted with permission from [32]. Copyright 2006 American Chemical Society

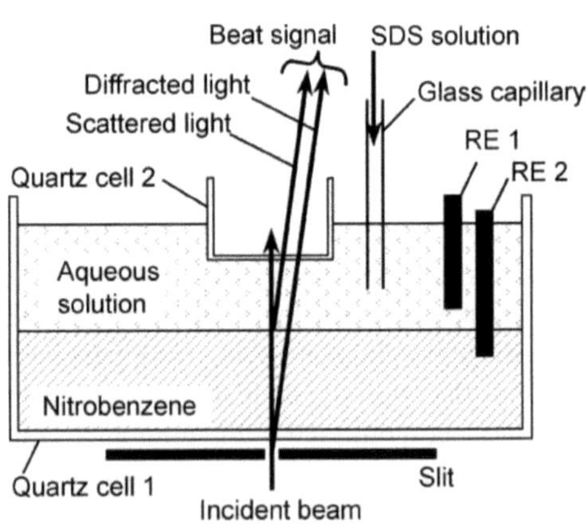

Fig. 6.16 Scheme of the experimental setup for the measurements of interfacial tension and electrical potential at the water/nitrobenzene interface. The interfacial tension was measured by QELS method. Reprinted from [35]. Copyright 2004, with permission from Elsevier

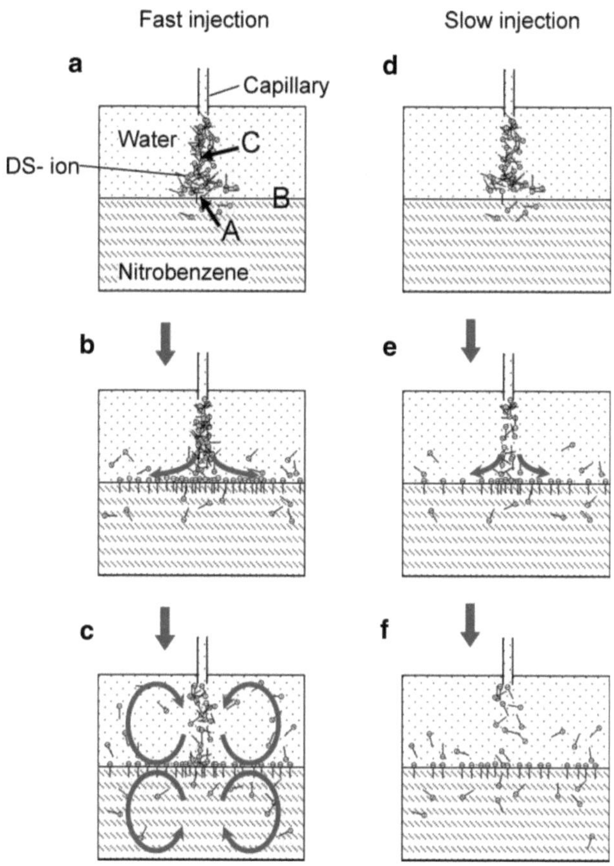

Fig. 6.17 Mechanism of generation of oscillation at the liquid/liquid interface. *Left* three figures, (**a**), (**b**), and (**c**) are for the case of one marked oscillation, *right* three figures, (**d**), (**e**), and (**f**) are for the case of periodic oscillations. Reprinted from [35]. Copyright 2004, with permission from Elsevier

The oscillations were induced by injection of sodium dodecyl sulfate (SDS) through the glass capillary. Periodic course of the oscillations occurred if the injection rate was relatively slow (e.g., $< 1\ \mu l\ min^{-1}$), while for higher flow rates only a single oscillation was reported. The mechanism of these phenomena involved alternately appearing abrupt adsorption and slow desorption processes of anionic surfactant (SDS) at the water/nitrobenzene interface. The decay of oscillations after fast injection was explained by the destructive (in this case) role of the Marangoni instability, meaning that the convection, induced by interfacial heterogeneities, enforced adsorption of DS^- ions, driving the system out of the oscillatory regime towards the equilibrium distribution of species. If the flow rate of SDS is slow, the convection is not so fast and the adsorption and desorption rates are not balanced, leading to continuous periodic oscillations of both electric potential and interfacial tension. Figure 6.17 represents schematically the course of these processes for both the fast and slow injections of SDS.

Based on the above selection of papers one can conclude that in spite of relative simplicity of construction and chemical composition of the systems with liquid membranes, the detailed mechanism of oscillations is a matter of controversy, particularly with respect to the following questions: which interface is a source of instabilities in a given system and which is the relative contribution of chemically and hydrodynamically driven instabilities to these oscillations? A more detailed analysis of these kind of phenomena is described in a recent, extensive review by Kovalchuk and Vollhardt [39]. Also, in a very recent paper, Szpakowska et al. [40] have described a latest kinetic model of the nitrobenzene-based liquid membrane oscillator (in the U-shaped cell) and have discussed the contribution of different factors to the measured oscillatory variations of the electric potential differences. It was also suggested again that hydrodynamic (Marangoni-type) instabilities cannot solely explain, but can only amplify oscillations appearing due to the decisive kinetic properties of the system. Concluding, evidently the general theory of the oscillatory systems with liquid membranes is still needed.

6.4 Outline Characteristics of Solid Membrane Oscillators

6.4.1 Biochemical Membrane Oscillators

The dynamical systems described above could in principle become the introduction to other types of membrane oscillators. Of course, the liquid membrane consisting of one-component organic liquid (with small amount of conducting species added) is a phase of simpler composition than other types of artificial membranes, not speaking about biological cell membranes. In fact, the literature on membrane oscillators, mainly due to the their relevance to biology, is huge and it is not the aim of the present work to review all or even most of these contributions. Below we shall only mention some representative examples of natural membrane oscillatory systems collected in an impressive review by Larter [41]. Since the selective permeability is typical of biological cell membranes, possessing specific ions channels (of permeability sensitive, e.g., to Ca^{2+} ions), the tremendous significance of studies of such oscillators is obvious. It is however not always clear, if the oscillatory behavior in the biological cells is a result of membrane characteristics or if it occurs inside the cell. If both subsystems are oscillatory, it can be necessary to consider coupling between them, presumably with the participation of Ca^{2+} ions. Berridge and Rapp in their review [42] have emphasized the role of calcium ions in coupling the membrane and cytoplasm oscillators. In another paper, Rapp [43] tabulated an “Atlas of cellular oscillations” in which the following true membrane-localized oscillators (known at that time) were indicated: systems exhibiting oscillations of membrane potential of secretory cells (e.g., pancreatic β cells), isolated neurons (particularly the molluscan variety), and certain cardiac cells (the sino-atrial node and the Purkinje fibers). Other known biological oscillations included oscillatory ion movements in mitochondria, oscillations in photosynthesis, glycolytic oscillations, and oscillations in protein synthesis [41]. To that list one can add the oscillations in potential (with a period of ca. 5 min), observed near a sprouding bean root [44], the phenomenon possibly related to spatial non-uniformities associated with the oscillations. Another example concerns the egg cells, at least most of which have a cellular membrane which exhibits excitability, in the sense of generating single-action potentials. Interestingly, such cells reveal a region of negative resistance, and exhibit thus bistability of membrane potential. Such properties seem to be used by the egg in the generation of a very fast change in membrane potential upon fertilization by a sperm cell; in consequence, the penetration of the egg by more than one sperm becomes electrically blocked, i.e., the polyspermy is excluded. This example shows that excitability (in this case of the cell membrane) itself can be a dynamic property of great importance for the development of the living matter [41, 45, 46]. Another discovery by Yoneda et al. indicated periodic vibrations in the surface tension of the membranes of unfertilized, but artificially activated, sea urchin eggs, from which the cell nucleus had been removed [47]. More recently, periodic oscillations in the membrane potential of cardiac cells were reported by Matsuda et al. [48]. The oscillations can also attain more complex—quasi-periodic or chaotic—nature, the latter considered important if not really crucial for the regulatory mechanisms in living organisms. For example, the electric stimulation of the nerve cells of the giant squid with sinusoidal electric signal resulted in the entrainment which could cause also chaotic response [41, 49, 50].

Finally, the oscillations in membrane systems can turn into spatial or spatiotemporal patterns [41] associated with the inhomogeneous spatial distribution of electric potential. The most important implication of such phenomena is their possible essential contribution to morphogenesis in living systems when the initially identical cells differentiate into more specialized groups of them, exhibiting appropriate shapes. Larter [41] indicates possible ambiguous interpretation of the term "morphogen" in this context. According to Turing classical work (see Sect. 1.1.4), a morphogen is species whose spatial distribution becomes inhomogeneous due to specific coupling of chemical reaction kinetics and diffusion transport. Recently, the term "morphogen" was extended to describe the species which diffuses along some sort of preexisting gradient, signaling the cells to develop in a certain way, but which may not itself be involved in the process by which the gradient arises. As an example of the morphogen, in this or another sense, one can invoke auxin, i.e., indoleacetic acid, implicated in generation of vein patterns in leaves. The above-mentioned Ca^{2+} ions are the most probable link between tissue- or organ-level development processes and appropriate cell membranes. Calcium(II) ions constitute thus important species connecting the growth and development processes to the membrane excitable properties. Of course, also the connection between the oscillations and pattern formation should be considered, as in the well-known case of concentric and spiral waves in the BZ reaction (see Fig. 1.1). Among the living systems, exhibiting such connection, the best known is the slime mold *Dictyostelium discoideum*. Recent developments in

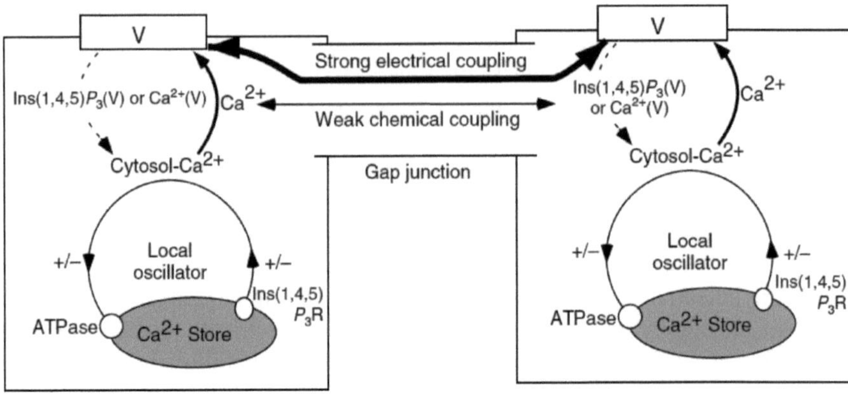

Fig. 6.18 A schematic representation of the two-cell system. Each cell is a local oscillator composed of a cytosolic store Ca^{2+}-excitable system. The cytosolic Ca^{2+} of each oscillator is transformed into membrane potential (V) oscillations by a Ca^{2+}-activated inward current. The membrane potentials of the cells are strongly linked. Each local oscillator is weakly linked to the membrane potential by a voltage-dependent feedback loop such as voltage-dependent Ins(1,4,5)P_3 synthesis or voltage-dependent Ca^{2+} influx. Ins(1,4,5)P_3R, Ins(1,4,5)P_3 receptor; ATPase, ATPase pump. Reprinted from [52] with permission of John Wiley & Sons, Inc. Copyright 2010. The Authors Journal Compilation © 2009 FEBS

general understanding of spatial pattern formation in electrochemical systems (Sect. 1.2) should allow to make the progress in studies of analogous processes in the membrane phases.

The reader interested in the studies of cardiac rhythms described in terms of competing membrane oscillators can consult a very recent review by Noble et al. [51] in which this interaction is described in terms of voltage-dependent ion channels and an intracellular calcium signal oscillator. In turn, Imtiaz et al. [52] have reviewed the synchronization of Ca^{2+} oscillations with respect to a long-range signaling mechanism in smooth muscle that results in global outcomes of local interactions; the long-range *electrochemical* coupling was found to be many orders of magnitude stronger than the coupling through diffusion of Ca^{2+} or inositol 1,4,5-triphosphate (Ins(1,4,5)P_3). The model two-cell system with electrochemical and chemical pathways is schematically depicted in Fig. 6.18.

Another example of application of nonlinear dynamics to membrane systems is the stability of electrochemical model of excitable biomembranes, described by Rex and Schulze [53]. Biochemical oscillations at the cellular level, also analyzed in terms of nonlinear dynamics, are described in the monograph by Goldbeter [54]. Furthermore, application of nonlinear dynamics in biophysical chemistry, including the effects of non-uniform fields on membrane transport, neuronal systems, calcium signaling, as well as the peroxidase–oxidase oscillator, was summarized by Larter in her later review [55]. Also, membrane oscillations and ion transport were reviewed by Kihara and Maeda [56] with the intention of extracting the fundamental information significant for understanding of oscillations at biomembranes in

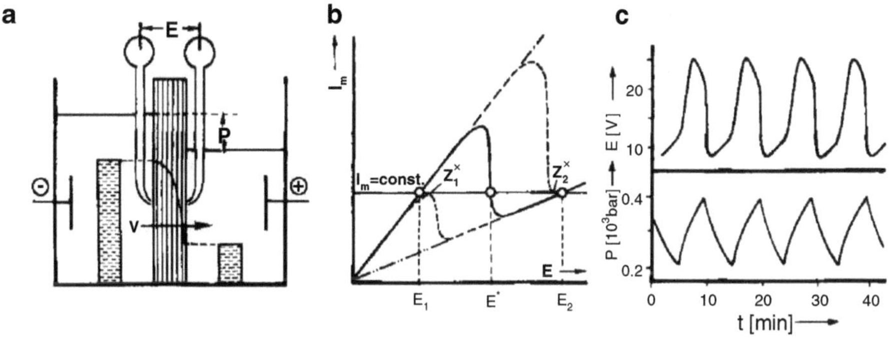

Fig. 6.19 The Teorell membrane oscillator: (**a**) schematic construction, for the case then left chamber contains higher concentration of an electrolyte, (**b**) current–potential characteristics with imposed current I_m, for which up to three steady-states can emerge, and (**c**) oscillations of the membrane potential E and the hydrostatic pressure difference P. Reproduced from [67] with permission of John Wiley & Sons, Inc. Copyright 1978 John Wiley & Sons, Inc.

living organisms. Finally, an extensive description of application of nonlinear dynamics in biological processes is given in the classical, now two-volume monograph by Murray [57, 58].

6.4.2 *Artificial Solid Membrane Oscillators*

We describe here only briefly selected systems with solid membranes, while for deeper analysis of such oscillators the interested reader is advised to consult the relevant, specialized literature, including an above-mentioned review by Larter [41]. Model membrane oscillators constitute a numerous group of oscillators in which two solutions are separated with membranes of various compositions and constructions, permeable for all or only selected components of the solutions. We shall focus on such properties of membranes that would show their relevance to other types of electrochemical oscillators. One of typical source of instabilities in such systems is the presence of negative differential resistance (NDR) in their current–potential characteristics.

As a relatively simple artificial system that exhibits basic dynamical features of nonlinear oscillators, we shall mention here the Teorell membrane oscillator [59–64]. In fact, Teorell construction aimed to mimick the neural excitation in biological systems and, noteworthy, he was able to describe the excitability in giant algae *Nitella* in terms of the theory developed for his membrane system [60, 65]. Detailed analysis of such oscillator, explaining its dynamics, including excitable characteristics, was performed, among others, by Franck [66, 67], Drouin [68], and Meares and Page [69]. The Teorell system includes either the porous glass or the ion-exchange membrane, separating two well-stirred electrolyte solutions of the same salt (e.g., NaCl), but at different concentrations. Since liquid may pass from one chamber to another via the membrane, the induced electroosmotic flow alters the level of the solutions in the chambers and in this way generates hydrostatic pressure difference, opposing this flow (Fig. 6.19a).

The current–potential characteristics of the Teorell system indicates the region of the negative resistance (Fig. 6.19b). When the flow of appropriate constant electric current, of appropriate direction, is imposed, the system sets into oscillations of the membrane potential (associated with the varying membrane resistance) and of the hydrostatic pressure (Fig. 6.19c). The oscillations, the period of which is usually 10–30 min, are generally caused by periodically changing competition between the hydrodynamic (HD) and electroosmotic (EO) flow of the fluid between the chambers of the Teorell oscillator. When the flow of current is imposed in the direction from the right to the left cell, the electroosmotic flow of water, carrying positive charge with respect to the negative fixed charges of the membrane phase, occurs in the same direction. In terms of Fig. 6.19b, the oscillations are interpreted as the periodic shift of the N-shaped *I–E* characteristics. In the bistable regime, the cross-section of the horizontal (load) line of the fixed imposed current I_m with the solid N-shaped curve determines three steady-states: an unstable middle one and two external stable. States Z_1* and Z_2* differ with the direction of the fluid flow, while the critical potential for the flow reversal is E*. The volume flux, driven in opposite directions by the voltage and the pressure difference, shifts the critical flux reversal potential in the same way as the hydrostatic pressure changes with the volume flux. Figure 6.19b shows two limiting cases, corresponding to the flux reversal, when the system in one of Z* states becomes unstable and then rather abruptly jumps to the competitive Z* state, of different membrane potential. So, the oscillations means periodic switching between states Z_1* and Z_2*.

In order to characterize the Teorell system in more detail, one should realize that the membrane potential, as related to the ohmic resistance, is determined by the concentration profile of the salt in the membrane. Figure 6.20 compares stationary and non-stationary concentration profiles (a) with the electric characteristics of membrane (b), and the direction of volume flow (c). In Fig. 6.20a, the straight line corresponds to the linear ohmic dependence $R_m = \Delta\Psi/I$ (for constant current I), where $\Delta\Psi$ denotes now the membrane potential. The nonlinear dependence in the same figure describes the dependence of the membrane resistance on potential. The intersections of those lines determine the steady-states of the system, with (note different notation, compared to Fig. 6.19) Z_1 and Z_3 being stable, and Z_2—unstable.

Further analysis involves the change of these characteristics upon the perturbation with the electric current or pressure. For example, the pressure impulse causes the horizontal shift of the nonlinear curve from Fig. 6.20b, so the respective initial steady-state can lose its stability. In terms of such approach, the oscillations of membrane potential (and hydrostatic pressure) occurring under galvanostatic conditions can be represented by the sequence of diagrams, collected in Fig. 6.21.

The physical processes underlying the oscillations can now be summarized in a following way [69]. For the sufficiently high (overcritical) electric current,

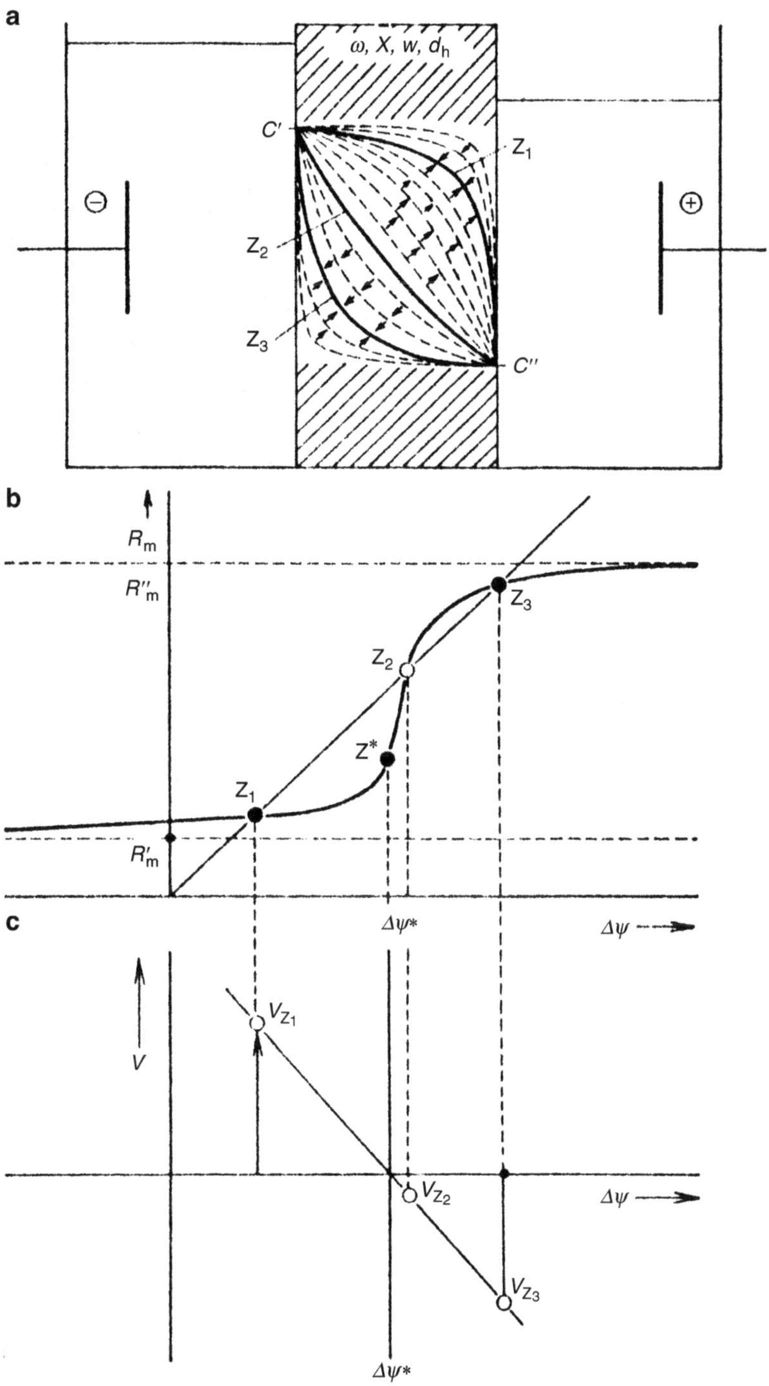

Fig. 6.20 On the stability of steady-state concentration profiles: (**a**) temporal variations of the non-stationary concentration profiles, (**b**) direction of motion of the point representing the state, on current–voltage characteristics, and (**c**) the flow v as a function of the membrane potential (denoted

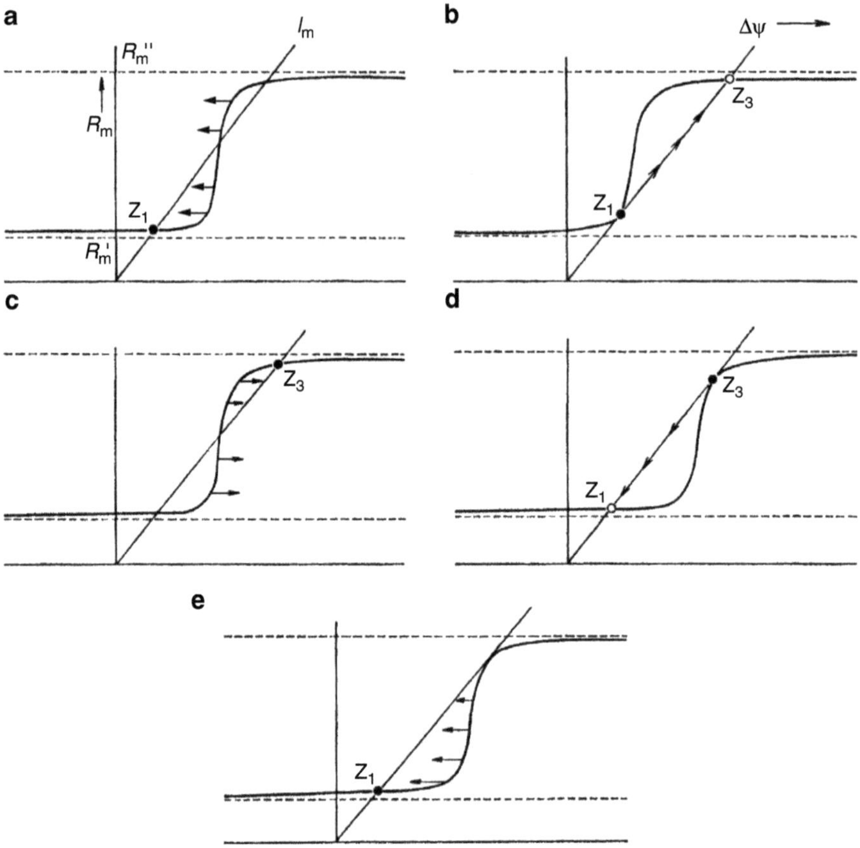

Fig. 6.21 Emergence of periodic transport phenomena in the bistable membrane system: (**a**) membrane in high-conductivity state Z_1. Upon increasing hydrostatic pressure the nonlinear curve moves to the *left*; (**b**) Z_1 loses stability and membrane state changes from Z_1 to Z_3; (**c**) membrane in low-conductivity state Z_3 with reversed direction of flow; the nonlinear characteristics moves to the *right*; (**d**) Z_3 loses stability and membrane state changes back to Z_1; and (**e**) membrane again in high-conductivity state Z_1 and the nonlinear curve moves to the *left*. Figure reprinted from [66] with kind permission of Deutsche Bunsen-Gesellschaft für Physikalische Chemie

oscillations arise because the counteracting hydrostatic pressure generated by the liquid transferred electroosmotically eventually drives concentrated solution into the membrane. The membrane resistance falls when concentrated solution enters and, at constant current, this reduces the strength of the electroosmotic flow. When the point is reached at which the direction of flow is reversed, concentrated solution flows through the membrane under the pressure difference,

---

Fig. 6.20 (continued) here as $\Delta\Psi$; both (**b**) and (**c**) diagrams have common potential axis). R_m—membrane resistance. Concentration of salt $c'' < c'$. Reprinted from [66] with kind permission of Deutsche Bunsen-Gesellschaft für Physikalische Chemie until a level is reached at which the electroosmotic pressure is sufficient to reverse the liquid flow again. The membrane now refills with dilute solution and a new cycle commences. In this cycle one can find positive feedback loop manifesting itself in the voltage dependence of the membrane resistance, and a negative feedback loop coming from the dependence between the volume flux and the hydrostatic pressure [70]. Such mechanism of oscillations in the Teorell system was also successfully modeled [69].

The oscillators of the Teorell type were studied also later, among others by Langer et al. [71] who have analyzed such a system with *fine* pore membranes, both experimentally and theoretically. This means that instead of broad pore (e.g., glass sinter as in the original version), the membranes of a hydrodynamic permeability for several orders of magnitude lower, and a fixed ion concentration for several orders of magnitude higher, were used. The assumption was that such fine pore membranes are mimicking more adequately the excitable biological tissues.

Besides the Teorell-type systems, also other membrane oscillators were intensively studied. One can divide them into several groups, depending on the membrane type. One of these groups includes systems with polyethylene membrane, often doped with iodine. Such enrichment of a polyethylene phase causes the shift of its π electrons onto the acceptor iodine molecules and this enhances the electric conductivity of the organic phase. The second group is formed by artificial lipid membranes (usually bilayer), treated as simple models of natural membrane cells. Karvaly [72] has described a system in which such membrane separated two KCl solutions, one of which (alkaline) contained also potassium iodide, and the other one (acidified)—also potassium hexacyanoferrate(III). Another experimental realization of the artificial lipid membrane involves saturation of the porous material, e.g., the Millipore filter, with appropriate organic compound, e.g., DOPH (dioleyl phosphate $[CH_3(CH_2)_7CH = CH(CH_2)_7CH_2O]_2POOH$). To the third group, one can count systems with the enzyme immobilized in the membrane. A very simple experiment shows the onset of oscillations in this type of the systems. The surface of the conventional glass electrode was covered with the layer of collodion in which the enzyme—papain—was immobilized. Papain, obtained from the fruits of the melon tree, catalyzes the hydrolysis of the ethyl ester of benzoyl-L-arginine. If the enzyme-modified glass electrode was immersed into the aqueous solution of this ester, spontaneous pH oscillations set in at the glass/membrane interface, associated with the variations of the membrane potential [41].

With respect to excitability of membranes, there were reported also some non-standard observations. For example, Mueller and Rudin [73–75] have described the lipid bilayers that were *not* excitable unless an "excitability inducing material," or EIM, was added to the bilayer. Since the EIM species was extracted from white eggs, one can suppose [41] that it was some sort of channel-forming protein. Although the detailed nature of EIM remains unclear, the significance of the discovery of Mueller and Rudin lies in the conclusion that excitability can be caused by the characteristics of the membrane (lipid bilayer) alone, so the deeper studies of the whole cell structure are, from this point of view, not necessary to explain this kind of dynamic response.

6.5 Oscillations in Conducting Polymer Systems

The use of polyethylene membranes doped with iodine in membrane oscillators may provoke a question whether oscillatory phenomena occur in processes involving conducting polymers. Apparently there are only a few reports on such processes involving conducting polymers, and two of such works are briefly described here. In 1995, Petitjean et al. [76] have described new oscillatory phenomena associated with the electropolymerization of pyrrole in $CH_3CN + N(Bu)_4PF_6$ on an iron electrode, studied by the ring-disk-electrode technique. One of surprising findings was that the frequency and amplitude of both disk and ring current oscillations were quite insensitive to hydrodynamic conditions (rotation speed of the electrode), but strongly depended on the concentration of chemical species present in the electrolytic medium, including water concentration. The following three-step mechanism of oscillations was proposed.

In *step I*, the relatively low *disk* current (0.3 mA) is due to both pyrrole and iron oxidation reactions, taking place simultaneously:

$$Fe \rightarrow Fe^{3+} + 3e \tag{6.16}$$

$$n\mathrm{Py} \rightarrow \mathrm{PPy} + (2n-2)\mathrm{H}^+ + (2n-2)\mathrm{e} \tag{6.17}$$

The polypyrrole film is formed on the electrode surface and Fe^{3+} ions are produced at the interface between the film and the electrode. Based on a very low ring current of Fe^{3+} reduction, one concludes that the polypyrrole film forms a membrane that substantially hinders the diffusion of Fe^{3+} ions towards the electrolyte. Thus, Fe^{3+} ions accumulate at the electrode/film interface or inside the film, with positive charge of accumulating Fe^{3+} ions compensated by the PF_6^- counterions, migrating through the film faster than Fe^{3+} ions. This accumulation of electrolyte at the interface weakens the polymer film. Due to its progressing swelling and appearance of internal constraints, the membrane eventually breaks down. This triggers step II.

In *step II*, the breakdown of membrane induces a sudden increase of Fe^{3+} ion emission at the disk. Entering the solution, they partly reach the area of the ring electrode, and this causes the experimentally reported dramatic increase of the *ring* cathodic current (from 8 to 30 μA), related to $Fe^{3+} + e \rightarrow Fe^{2+}$ process. Simultaneously, the considerable increase in membrane permeability triggers the large flux of pyrrole towards the disk surface. Due to high local concentration of Fe^{3+} ions, pyrrole is there catalytically oxidized to polypyrrole:

$$n\mathrm{Fe}^{3+} + n\mathrm{Py} \rightarrow n\mathrm{Fe}^{2+} + n\mathrm{Py}^+ \rightarrow \mathrm{PPy} + n\mathrm{Fe}^{2+} + (2n-2)\mathrm{H}^+ + (n-2)\mathrm{e} \tag{6.18}$$

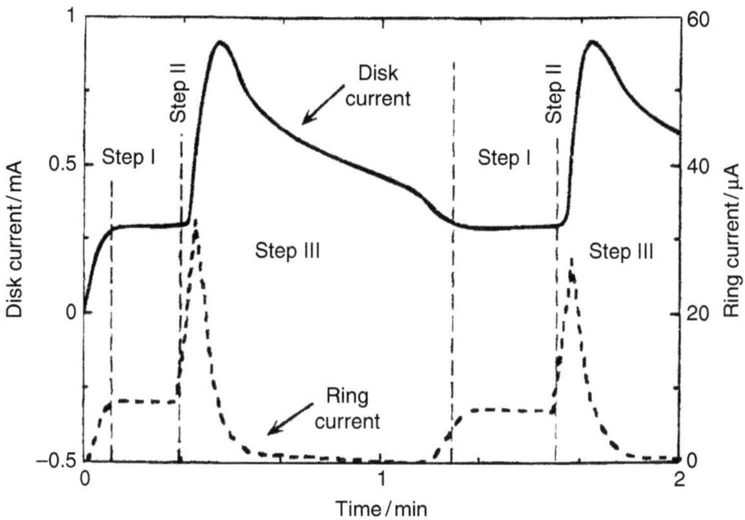

Fig. 6.22 Different steps involved in the two first periods of the current oscillations under standard conditions (MeCN + 0.005 M $N(Bu)_4PF_6$ + 0.3 M Py + 0.15 M H_2O, disk and ring electrodes polarized at 1.2 and 0.2 V, respectively). Reprinted from [76] by permission of The Electrochemical Society

with Fe^{2+} reoxidized to Fe^{3+} at the Pt disk surface. This catalytic effect of Fe^{3+} ions and the arrival of fresh pyrrole are revealed by quasi-instantaneous increase of the disk current (from 0.3 to 0.9 mA). The polymer membrane is rapidly reconstituted. Also, the local acidification of the solution enhances further dissolution of iron.

In *step III*, since PPy membrane undergoes fast reconstitution, the flux of pyrrole towards the disk surface slows down after passing a peak value. Further reconstruction and thickening of membrane eventually prevents the pyrrole molecules from penetrating into the cavity. Catalytic oxidation of pyrrole ceases and the disk current drops suddenly and stabilizes at low value, typical of step I (0.3 mA). The oscillatory cycle is completed. The relation between the disk and the ring currents is schematically depicted in Fig. 6.22, while Fig. 6.23 illustrates pictorially the relevant processes.

Another interesting case of oscillations involving conducting polymers was reported by Aoki et al. [77] who have described the periodic oxidation current of single particle made of redox latex. Such a large electroactive particle (1.3 μm in radius) was synthesized by coating the polystyrene latex with electroactive polyaniline film. The platinum disk 15 μm in diameter was used as a working microelectrode. The limiting current after eliminating the noise of the electric power source exhibited periodical oscillations with 14 Hz. The oscillations were interpreted in terms of collisions of the particle to the electrode, to cause the oxidation of the redox sites of individual particles. Thus, the observed phenomenon indicated the electrochemical event of one particle.

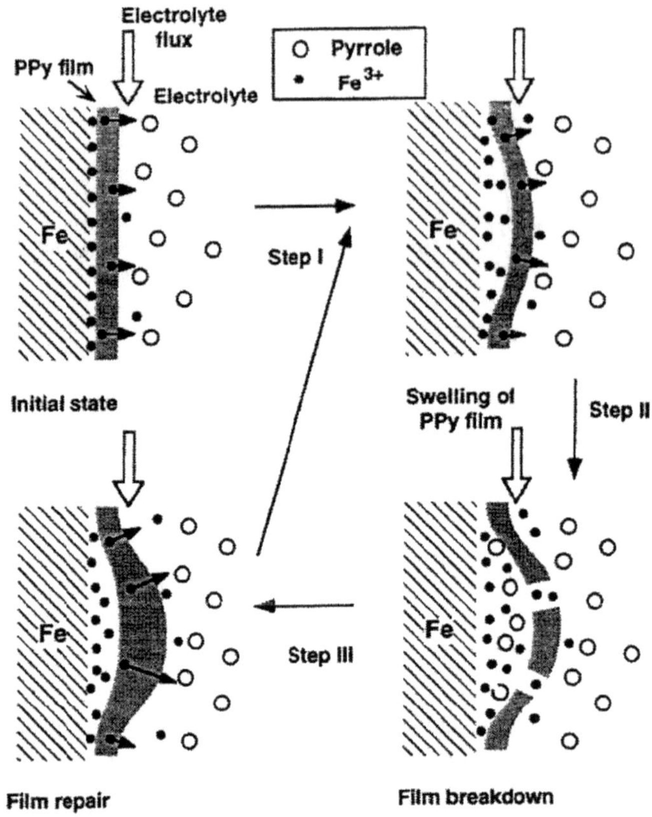

Fig. 6.23 Schematic illustration of the model proposed for the formation of the polypyrrole membrane on an iron electrode. PF_6^- counterions necessary to satisfy electroneutrality have been omitted from the figure for the sake of clarity. Reprinted from [76] by permission of The Electrochemical Society

References

1. Dupeyrat M, Nakache E (1978) Direct conversion of chemical energy into mechanical energy at an oil water interface. Bioelectrochem Bioenerg 5:134–141
2. Nakache E, Dupeyrat M (1982) Chemical reactions and oscillating potential difference variations at an oil–water interface. Bioelectrochem Bioenerg 9:583–590
3. Yoshikawa K, Matsubara Y (1983) Spontaneous oscillation of pH and electrical potential in an oil–water system. J Am Chem Soc 105:5967–5969
4. Toko K, Yoshikawa K, Tsukiji M, Nosaka M, Yamafuji K (1985) On the oscillatory phenomenon in an oil/water interface. Biophys Chem 22:151–158
5. Yoshikawa K, Matsubara Y (1984) Chemoreception by an excitable liquid membrane: characteristic effects of alcohol on the frequency of electrical oscillations. J Am Chem Soc 106:4423–4427
6. Yoshikawa K, Shoji M, Nakata S, Maeda S, Kawakami H (1988) An excitable liquid membrane possibly mimicking the sensing mechanism of taste. Langmuir 4:759–762

7. Yoshikawa K, Omochi T, Matsubara Y, Kourai H (1986) A possibility to recognize chirality by an excitable artificial liquid membrane. Biophys Chem 24:111–119
8. Yoshikawa K, Matsubara Y (1985) Oscillation of electrical potential across a liquid membrane induced by amine vapor. Langmuir 1:230–232
9. Yoshikawa K, Nakata S, Omochi T, Colacicco G (1986) Novel liquid membrane oscillator with anionic surfactant. Langmuir 2:715–717
10. Noble RD, Way JD (1987) Liquid membrane theory and application, ACS symp Ser, vol 374. ACS, Washington, DC
11. Szpakowska M (2009) Liquid membrane oscillators. Desalination 241:349–356
12. Deutsch EW, Hansch C (1966) Dependence of relative sweetness on hydrophobic bonding. Nature (London) 211:75
13. Beidler LM (1954) A theory of taste simulation. J Gen Physiol 38:133–139
14. Szpakowska M, Czaplicka I, Szwacki J, Nagy OB (2002) Oscillatory phenomena in systems with bulk liquid membranes. Chem Pap 56:20–23
15. Szpakowska M, Szwacki J, Lisowska-Oleksiak A (2004) Investigation of some taste substances using a set of electrodes with lipid-modified membranes. Desalination 163:55–59
16. Płocharska-Jankowska E, Mátéfi-Tempfli S, Nagy OB (2005) On the possibility of molecular recognition of taste substances studied by Gábor analysis of oscillations. Biophys Chem 114:85–93
17. Płocharska-Jankowska E, Szpakowska M, Mátéfi-Tempfli S, Nagy OB (2006) A new approach to the spectra analysis of liquid membrane oscillators by Gábor transformation. J Phys Chem B 110:289–294
18. Gábor D (1946) Theory of communication. J Inst Electr Eng (London) 93:429–457
19. Szpakowska M, Magnuszewska A, Szwacki J (2006) On the possibility of using liquid or lipid, lipid like-polimer membrane systems as taste sensor. J Membr Sci 273:116–123
20. Buhse T, Nagarajan R, Lavabre D, Micheau JC (1997) Phase-transfer model for the dynamics of "micellar autocatalysis". J Phys Chem A 101:3910–3917
21. Tixier J, Pimienta V, Buhse T, Lavabre D, Nagarajan R, Micheau JC (2000) (2000) Ester containing aggregates in the autocatalytic biphasic hydrolysis of ethyl alkanoate colloids surf. Coll Surf A 167:131–142
22. Bachmann PA, Luisi PL, Lang J (1992) Autocatalytic self-replicating micelles as models for prebiotic structures. Nature 357:57–59
23. Buhse T, Pimienta V, Lavabre D, Micheau JC (1997) Experimental evidence of kinetic bistability in a biphasic surfactant system. J Phys Chem A 101:5215–5217
24. Pimienta V, Etchenique R, Buhse T (2001) On the origin of electrochemical oscillations in the picric acid/CTAB two-phase system. J Phys Chem A 105:10037–10044
25. Pimienta V, Lavabre D, Buhse T, Micheau JC (2004) Correlation between electric potential and interfacial tension oscillations in a water-oil-water system. J Phys Chem B 108:7331–7336
26. Arai K, Fukuyama S, Kusu F, Takamura K (1995) Role of surfactant in the electrical potential oscillation across a liquid membrane. Electrochim Acta 40:2913–2920
27. Maeda K, Nagami S, Yoshida Y, Ohde H, Kihara S (2001) Voltammetric elucidation of the process of self-sustained potential oscillation observed with a liquid membrane system composed of water containing cetyltrimethylammonium chloride–nitrobenzene containing picric acid–pure water. J Electroanal Chem 496:124–130
28. Maeda K, Kihara S, Suzuki M, Matsui M (1990) Voltammetric study on the oscillation of the potential difference at a liquid/liquid or liquid/membrane interface accompanied by ion transfer. J Electroanal Chem 295:183–201
29. Kihara S, Maeda K, Shirai O, Yoshida Y, Matsui M (1993) In: Pungor E (ed) Bioelectroanalysis, vol 2. Akadémiai Kiadó, Budapest, p 331
30. Shirai O, Kihara S, Yoshida Y, Matsui M (1995) Ion transfer through a liquid membrane or a bilayer lipid membrane in the presence of sufficient electrolytes. J Electroanal Chem 389:61–70

31. Szpakowska M, Czaplicka I, Nagy OB (2006) On the mechanism of nitrobenzene liquid membrane oscillators containing hexadecyltrimethylammonium bromide. Biophys Chem 120:148–153
32. Szpakowska M, Czaplicka I, Nagy OB (2006) Mechanism of a four-phase liquid membrane oscillator containing hexadecyltrimethylammonium bromide. J Phys Chem A 110:7286–7292
33. Szpakowska M, Czaplicka I, Płocharska-Jankowska E, Nagy OB (2003) Contribution to the mechanism of liquid membrane oscillators involving cationic surfactant. J Colloid Interface Sci 261:451–455
34. Szpakowska M, Magnuszewska A, Nagy OB (2008) Mechanism of nitromethane liquid membrane oscillator containing sodium oleate. J Colloid Interface Sci 325:494–499
35. Ikezoe Y, Ishizaki S, Takahashi T, Yui H, Fujinami M, Sawada T (2004) Hydrodynamically induced chemical oscillation at a water/nitrobenzene interface. J Colloid Interface Sci 275:298–304
36. Takahashi S, Harata A, Kitamori T, Sawada T (1994) Quasi-elastic laser scattering method for monitoring capillary wave frequency at a water/nitrobenzene interface. Anal Sci 10:305–308
37. Zhang Z, Tsuyumoto I, Kitamori T, Sawada T (1998) Observation of the dynamic and collective behavior of surfactant molecules at a water/nitrobenzene interface by a time-resolved quasi-elastic laser-scattering method. J Phys Chem B 102:10284–10287
38. Uchiyama Y, Kitamori T, Sawada T, Tsuyumoto I (2000) Role of the liquid/liquid interface in a phase-transfer catalytic reaction as investigated by in situ measurements using the quasi-elastic laser scattering method. Langmuir 16:6597–6600
39. Kovalchuk NM, Vollhardt D (2006) Marangoni instability and spontaneous nonlinear oscillations produced at liquid interfaces by surfactant transfer. Adv Colloid Interface Sci 120:1–31
40. Szpakowska M, Płocharska-Jankowska E, Nagy OB (2009) Molecular mechanism and chemical kinetic description of nitrobenzene liquid membrane oscillator containing benzyldimethyltetradecylammonium chloride surfactant. J Phys Chem B 113:15503–15512
41. Larter R (1990) Oscillations and spatial nonuniformities in membranes. Chem Rev 90:355–381
42. Berridge MJ, Rapp PE (1979) A comparative survey of the function, mechanism and control of cellular oscillators. J Exp Biol 81:217–279
43. Rapp PE (1979) An atlas of cellular oscillators. J Exp Biol 81:281–306
44. Scott BIH (1957) Aust J Biol Sci 10:164
45. Dale B, de Santis A (1981) Maturation and fertilization of the sea urchin oocyte: an electrophysiological study. Dev Biol 85:474–484
46. Whitaker M, Steinhardt RA (1982) Ionic regulation of egg activation. Q Rev Biophys 15:593–666
47. Yoneda M, Ikeda M, Washitani S (1978) Periodic change in the tension at the surface of activated non-nucleate fragments of sea-urchin eggs. Dev Growth Differ 20:329–336
48. Matsuda H, Noma A, Kurachi Y, Irisawa H (1982) Transient depolarization and spontaneous voltage fluctuations in isolated single cells from guinea pig ventrices. Calcium-mediated membrane potential fluctuations. Circ Res 51:142–151
49. Guttman R, Feldman L, Jakobsson E (1980) Frequency entrainment of squid axon membrane. J Membr Biol 56:9–18
50. Matsumoto G, Aihara K, Ichikawa M, Tasaki A (1984) Periodic and nonperodic responses of membrane potentials in squid giant axons during sinusoidal current stimulations. J Theor Neurobiol 3:1–14
51. Noble D, Noble PJ, Fink M (2010) Competing oscillators in cardiac pacemaking. Historical background. Circ Res 106:1791–1797
52. Imtiaz MS, von der Weid PY, van Helden DF (2010) Synchronization of Ca^{2+} oscillations: a coupled oscillator-based mechanism in smooth muscle. FEBS J 277:278–285
53. Rex S, Schulze KD (1994) Stability analysis of an electrochemical model for excitable biomembranes. Z phys Chem 186:65–76

54. Goldbeter A (1997) Biochemical oscillations and cellular rhythms: the molecular bases of periodic and chaotic behavior. Cambridge University Press, Cambridge, MA
55. Larter R (2003) Understanding complexity in biophysical chemistry. J Phys Chem B 107:415–429
56. Kihara S, Maeda K (1994) Membrane oscillations and ion transport. Prog Surf Sci 47:1–54
57. Murray JD (2002) Mathematical biology. I. An introduction, 3rd edn. Springer, New York, NY
58. Murray JD (2003) Mathematical biology. II. Spatial models and biomedical applications, 3rd edn. Springer, New York. NY
59. Teorell T (1951) Zur quantitativen Behandlung der Membranpermeabilität. Z Elektrochem 55:46–469
60. Teorell T (1958) Exp Cell Res Suppl 5:83
61. Teorell T (1959) Electrokinetic membrane processes in relation to properties of excitable tissues. I. Experiments on oscillatory transport phenomena in artificial membranes. J Gen Physiol 42:831–846
62. Teorell T (1959) Electrokinetic membrane processes in relation to properties of excitable tissues. II. Some theoretical considerations. J Gen Physiol 42:847–863
63. Teorell T (1961) Oscillatory electrophoresis in ion exchange membranes. Ark Kemi 18:401–408
64. Teorell T (1962) Excitability phenomena in artificial membranes. Biophys J 2(suppl):27–52
65. Teorell T (1961) An analysis of the current–voltage relationship in excitable *nitella* cells. Acta Physiol Scand 53:1–6
66. Franck UF (1963) Über das elektrochemische Verhalten von porösen Ionenaustauschmembranen. Ber Bunsenges Phys Chem 67:657–671
67. Franck UF (1978) Chemische Oszillationen. Angew Chem 90:1–16
68. Drouin H (1969) Experimente mit dem Teorellschen Membranoszillator. Ber Bunsenges Phys Chem 73:223–229
69. Meares P, Page KR (1974) Oscillatory fluxes in highly porous membranes. Proc R Soc Lond A 339:513–532
70. Franck UF (1980) The Teorell membrane oscillator—a complete nerve model. Upsala J Med Sci 85:265–282
71. Langer P, Page KR, Wiedner G (1981) A Teorell oscillator system with fine pore membranes. Biophys J 36:93–107
72. Karvaly B (1973) Oscillatory electrochemical reactions at bimolecular lipid membrane-redox electrolyte interfaces. Nature 244:25–26
73. Mueller P, Rudin DO (1967) Action potential phenomena in experimental bimolecular lipid membranes. Nature (London) 213:603–604
74. Mueller P, Rudin DO (1968) Action potentials induced in biomolecular lipid membranes. Nature (London) 217:713–719
75. Mueller P, Rudin DO (1968) Resting and action potentials in experimental bimolecular lipid membranes. J Theor Biol 18:222–224
76. Petitjean J, Aeiyach S, Ferreira CA, Lacaze PC, Takenouti H (1995) A new oscillatory electrochemical phenomenon observed in the electropolymerization of pyrrole in MeCN + $N(Bu)_4PF_6$ on an iron electrode studied by the ring-disk-electrode technique. J Electrochem Soc 142:136–142
77. Lei T, Aoki K, Fujita K (2000) Periodical oxidation current of single particle made of redox latex. Electrochem Commun 2:290–294

Chapter 7
Control of Electrochemical Chaos and Unstable Steady-States

7.1 Application of Map-Based Control Algorithms

Modern research in the area of deterministic chaotic systems [1] is oriented on acquiring control of this apparently uncontrollable regime. An outline characteristics of the control of chaos was given in Sect. 1.10.3 of volume I. The reader particularly interested in this area can consult, e.g., a recent specialized monograph on that subject [2]. Two landmark papers on the chaos control appeared in 1990. In one of them Ott et al. [3] have described the (OGY) algorithm of converting a chaotic (strange) attractor to any one of a large number of possible attracting time-periodic motions by making only small time-dependent perturbations of an available system parameter. In other words, by applying the appropriate feedback-based strategy one can suppress chaos by maintaining the system's dynamics on a selected, desired periodic phase trajectory. In the second paper, written by Pecora and Carroll [4], it was shown that certain subsystems of nonlinear, chaotic systems can be made to synchronize by linking them with common signals. These theoretical achievements triggered increasing interest also in achieving the stabilization of any unstable states, both in theoretical models and real systems, including electrochemical processes. Several particular algorithms were employed for such strategies based on appropriate feedbacks.

For a *highly* dissipative system (when the strange attractor has a dimension slightly higher than 2), it is convenient to reduce the original OGY approach to a simple map-based (SPF) control algorithm, first introduced by Peng et al. [5, 6]. The principles of this algorithm, described in Sect. 1.10.3 of volume I, for clarity of the present notation [7] are here briefly repeated. The SPF approach uses the 1-D return maps (i.e., x_{n+1} vs. x_n dependence) which in turn are constructed from the Poincaré section of the chaotic trajectories in the phase space. According to SPF principles, the map is expressed as $x_{n+1} = \mathrm{f}(x_n, p)$ functional dependence, where p is a control parameter that determines the position of the map. This map is further linearized near the unstable fixed point, denoted here, according to notation in [7], as x_f instead of previously used symbol x^*. This fixed point is determined by the intersection of the map with the straight line of unit slope (corresponding to $x_{n+1} = x_n$).

M. Orlik, *Self-Organization in Electrochemical Systems II*, Monographs in Electrochemistry, DOI 10.1007/978-3-642-27627-9_7,

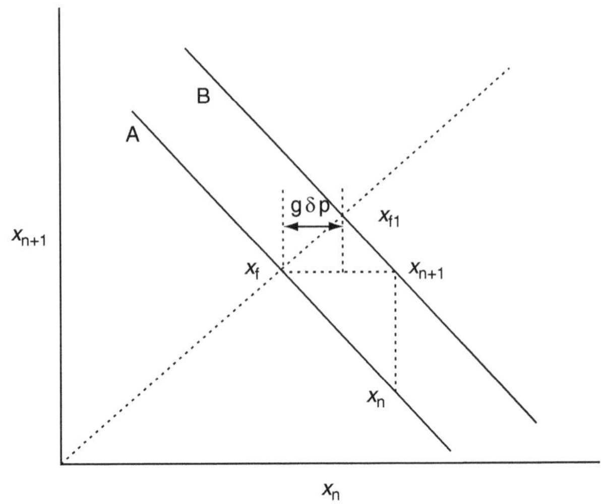

Fig. 7.1 A and B are return maps for parameter values p and $p + \delta p$, respectively. x_f and x_{f1} are the fixed points for the two return maps, and δp is chosen so that x_{n+1}, the next iterate of the return map, is placed on x_f, the fixed point for A. Reprinted from [7]. Copyright 1996, with permission from Elsevier

Thus, achieving the control of chaos means that every time when the phase trajectory intersects the Poincaré section in the vicinity of the fixed point x_f, the map should be appropriately shifted so that the next x value is equal to the original steady-state: $x_{n+1} = x_f$. Based on this principle, one derives the expression defining the required perturbation δp of parameter defining the map position [7]:

$$\delta p = \frac{\mu}{(\mu - 1)g}(x_n - x_f) = K(x_n - x_f), \tag{7.1}$$

where μ is the slope of the return map $x_{n+1} = f(x_n, p)$ at the fixed point x_f and g is the quantity defined as: $g = (x_{f1} - x_f)/\delta p$ (see Fig. 7.1).

Parmananda has used this approach [7] to suppress chaos in the three-variable model of In(III)-SCN-polarographic oscillator, elaborated by Koper and Gaspard [8, 9] (cf. Sect. 4.3.2, volume I). The external voltage U was chosen as the control parameter p. Switching the appropriate feedback stabilized either period-1 or period-2 oscillations. This is illustrated in Fig. 7.2 which shows the current *minima* plotted as a function of time (or iteration number in the sequence of numerical integration steps).

The SPF algorithm was further successfully applied by Kiss et al. [10] to controlling electrochemical chaos in the copper–phosphoric acid system, both in terms of theoretical calculations (involving the three-variable model by Koper and Gaspard [9]) and experimental results. In particular, the aim of the work was to control chaotic current oscillations at such Cu disk rotation rates where chaos developed through simple period-doubling bifurcations (see Fig. 7.3 for the experimental bifurcation diagram).

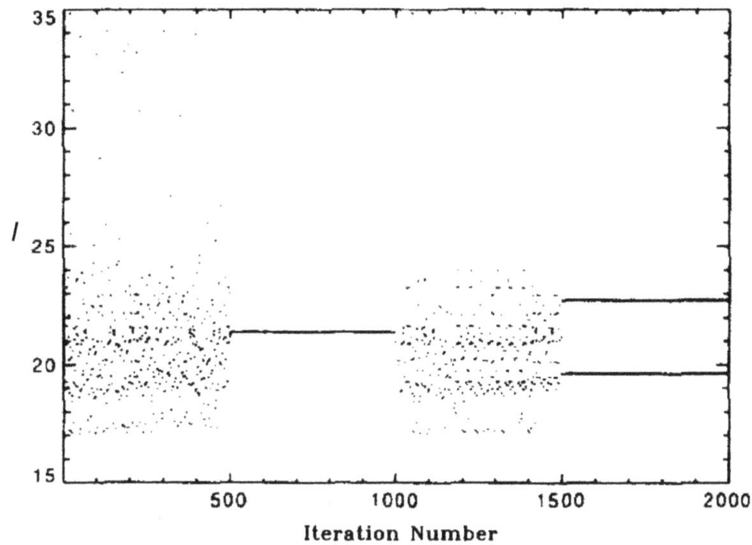

Fig. 7.2 Sequence of minima in current I plotted over a time period during which the control for period one is turned ON and OFF and then the control of period two is turned on. Reprinted from [7]. Copyright 1996, with permission from Elsevier

Application of small, appropriate perturbations δV of anodic potential of Cu electrode led to stabilization of selected periodic behaviors. Figure 7.4 illustrates the successful control of period-1 oscillations, while Fig. 7.5—in terms of the phase portrait—shows the stabilization of period-2 orbit. In addition, the power law relating the average chaotic transient time to the size of maximum perturbation was tested and found to be in good agreement with the theoretical predictions by Ott et al. [3].

Since the SPF algorithm may not always work [11], Rollins et al. [12], based on approach by Dressler and Nitsche [13], have proposed a simple recursive proportional-feedback (RPF) approach. The latter strategy was applied by the same group [14] to the control of chaos in experimental electrochemical system, consisting of electrodissolution of a rotating copper disk anode in an acetate buffer. In this way the chaotic response of the system was stabilized on an unstable period-1 orbit (Fig. 7.6). Later, Kiss and Gáspár have used SPF and RPF algorithms in their implementation of artificial neural networks (ANN) to the experimental chaos control in oscillatory electrodissolution of copper in phosphoric acid [15].

7.2 Application of Derivative Control Strategy

The *derivative control strategy* was introduced by Bielawski et al. [16] in order to stabilize spontaneous instability (ascribed to unstable focus) in Nd-doped optical-fiber laser. Generally, this approach means the feedback designed in this way that

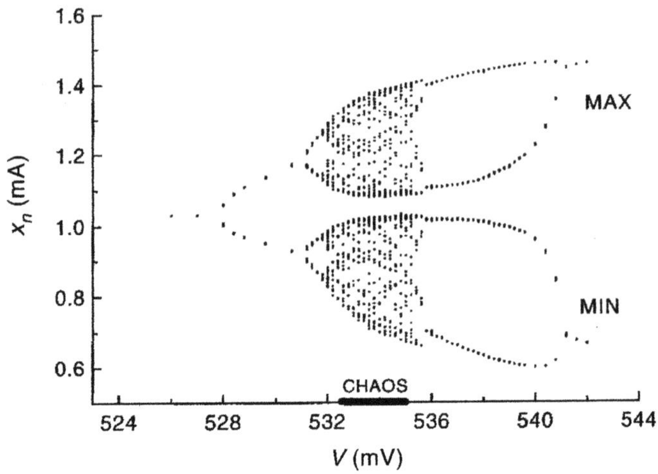

Fig. 7.3 Bifurcation diagram of the copper RDE–phosphoric acid system showing both maxima and minima of current oscillations as a function of the applied anodic potential. Total ohmic resistance of the circuit is 202 Ω and rotation rate is 1,850 rpm. Reprinted with permission from [10]. Copyright 1997 American Chemical Society

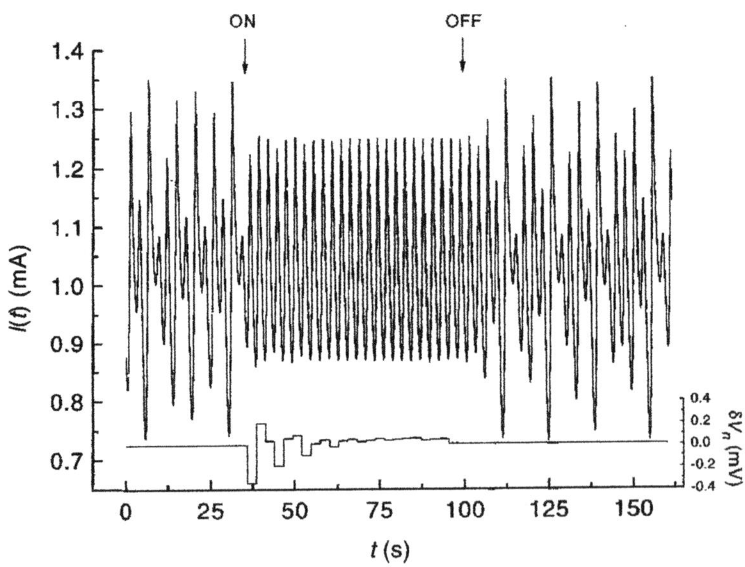

Fig. 7.4 Current (*left axis*) vs. time for an interval when control for stabilizing period-one oscillations has been switched on at 35.2 s and then switched off at 99.3 s. The control constant K in the $\delta V_n = K(I_n - I_f)$ control formula, for stabilizing period-one oscillations is -17 ± 4 mV/mA, while δV_{max} was set to 0.5 mV. The potential perturbations δV_n (*right axis*) applied at the successive returns to the Poincaré section are shown in the insert. See [10] for experimental conditions and other details. Reprinted with permission from [10]. Copyright 1997 American Chemical Society

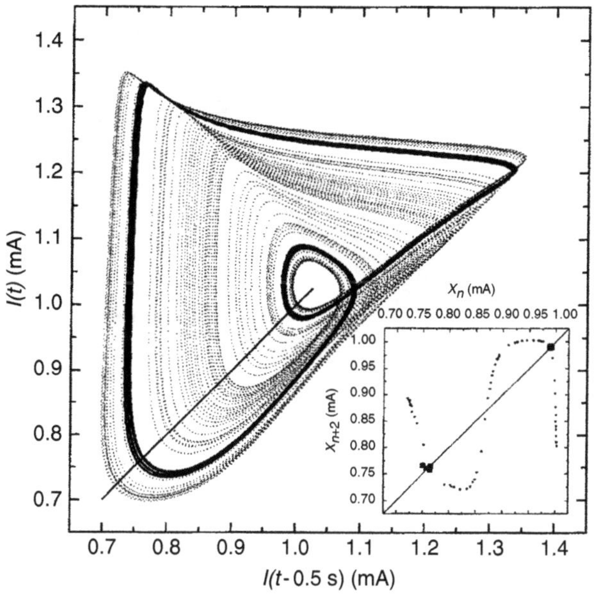

Fig. 7.5 Stabilized period-two orbit ($x_f = 0.7726$ mA) embedded in the chaotic attractor (*dots*). The next-return map x_{n+2} vs. x_n has been generated by using successive current values on the Poincaré section (*thick line*) (cf. also Fig. 5 in [10]). The control parameter K for period-two control is -11 mV/mA and δV_{max} is 0.5 mV. Superimposed on the map are the next-return values (*filled squares*) while control is being implemented. The applied anodic potential in the uncontrolled system is 527 mV, total ohmic resistance is 220 Ω, and rotation rate is 1,900 rpm [10]. Reprinted with permission from [10]. Copyright 1997 American Chemical Society

the perturbation signal is proportional to the time derivative of the appropriate dynamic variable or control parameter.

Parmananda et al. [17] have applied this strategy to the potentiostatic oscillatory electrodissolution of copper disk in a sodium acetate–acetic acid buffer, in terms of the kinetic, three-variable mechanism of the electrochemical corrosion, elaborated earlier [18, 19]:

$$\frac{\mathrm{d}y}{\mathrm{d}\tau} = p(1 - \theta_{\mathrm{OH}} - \theta_{\mathrm{O}}) - qy \tag{7.2}$$

$$\frac{\mathrm{d}\theta_{\mathrm{OH}}}{\mathrm{d}\tau} = y(1 - \theta_{\mathrm{OH}} - \theta_{\mathrm{O}}) - [\exp(-\beta\theta_{\mathrm{OH}}) + r]\theta_{\mathrm{OH}} + 2s\theta_{\mathrm{O}}(1 - \theta_{\mathrm{OH}} - \theta_{\mathrm{O}}) \tag{7.3}$$

$$\frac{\mathrm{d}\theta_{\mathrm{O}}}{\mathrm{d}\tau} = r\theta_{\mathrm{OH}} - s\theta_{\mathrm{O}}(1 - \theta_{\mathrm{OH}} - \theta_{\mathrm{O}}) \tag{7.4}$$

where y—concentration of metal ions in the electrolyte solution, θ_{O}—electrode coverage of oxygen species, θ_{OH}—electrode coverage of hydroxide species, and the parameters p, q, r, s, β are determined by chemical reaction rates in the model.

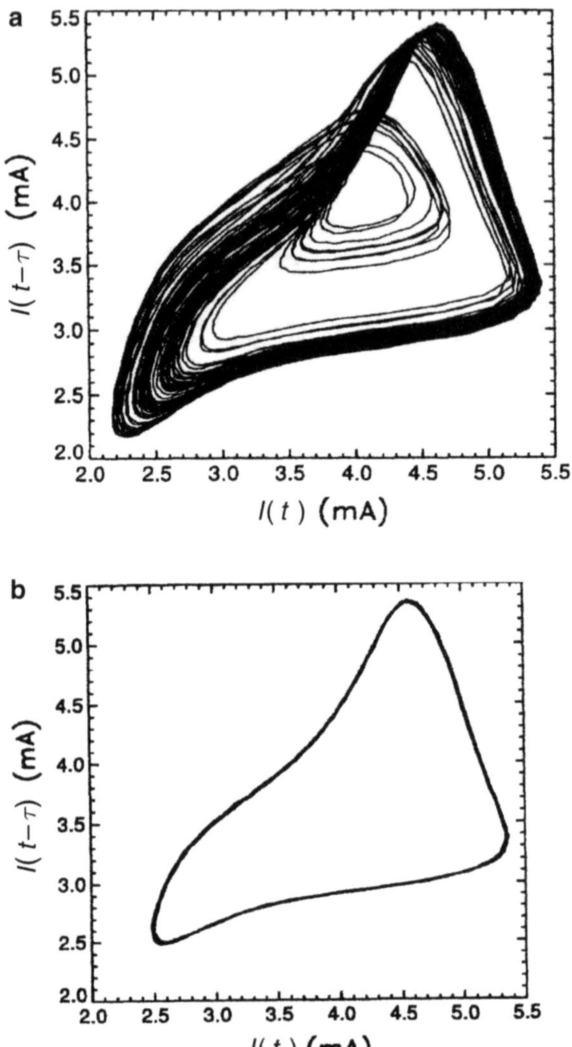

Fig. 7.6 (**a**) The reconstructed chaotic attractor in the embedding phase space from the experimental time series using a delay time of 120 ms; (**b**) shows the trajectory while the control is being implemented. Reproduced from with permission [14]. http://link.aps.org/abstract/PRE/v47/pR3003. Copyright 1993 by the American Physical Society

In this case, practical application of derivative control means adding, to the potential of the copper anode, a feedback term proportional to the time derivative of the electrical current passing through the cell:

$$\delta V = \alpha \frac{\mathrm{d}I}{\mathrm{d}t} \tag{7.5}$$

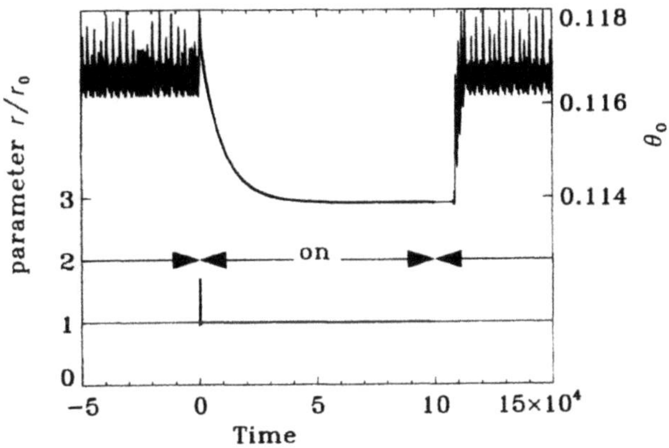

Fig. 7.7 The model: at $t = 0$, control is initiated (*thick line*). The control parameter r undergoes a sharp peak and then returns immediately back near its original value r_0. The system represented by θ_0 (*upper graph*) quits its chaotic oscillations and reaches the steady-state. At $t = 10 \times 10^4$ control is released. After a delay, the system returns to its original chaotic state. Reproduced with permission from [17]. http://link.aps.org/abstract/PRE/v49/p5007. Copyright 1994 by the American Physical Society

In the model, such parameter values were chosen that they generated chaotic current response. Then, the switching of the feedback caused leaving by the system of the chaotic attractor and trip to the steady-state, always the same for given control parameters, i.e., independently of initial conditions. In the absence of this feedback control, this steady-state is an unstable focus, lying inside the stable limit cycle. This effect is clearly illustrated by Fig. 7.7, as a decay of the oscillations in the time window, when the feedback was on.

In the experiment, a continuous feedback signal, proportional to the time derivative of the current was generated by an analog differentiator circuit. This feedback signal was used to continuously perturb the potential of rotating Cu disk anode in contact with buffer acetate solution. In this way it was possible to suppress the (then periodic) oscillations and stabilize the steady-state (Fig. 7.8). It should be emphasized that, as in the case of other feedback-based approaches, the control strategy oriented on the stabilization of the given state does not require any prior knowledge of the mechanism of the process and therefore can be easily implemented to the systems of any complex dynamics.

Another example of application of derivative control strategy to (electro)chemical systems was described by Parmananda and Eiswirth [20]. This time the strategy was employed to stabilization of unstable fixed points (saddles) in three models exhibiting bistability: (i) in a strictly one-dimensional bistable system, (ii) in a two-variable model with one fast-relaxing variable, effectively being one-dimensional, and (iii) in a two-variable model, exhibiting also oscillations, but controlled in the bistable regime. Generally, implementation of derivative control strategy in bistable systems requires first applying an initial perturbation (parameter spike) of the system, in order to drive the system away of one of the stable nodes.

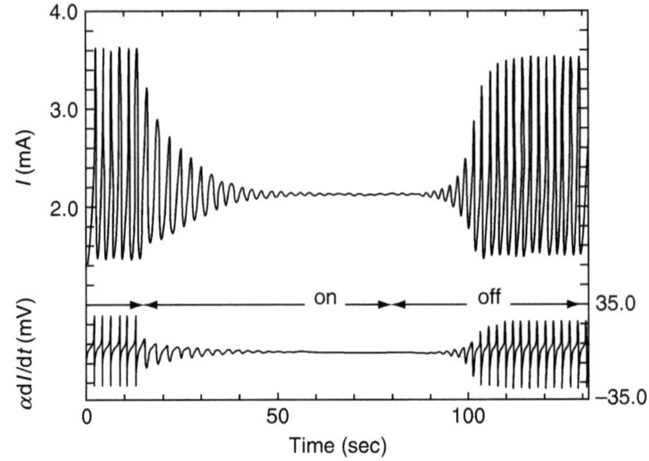

Fig. 7.8 Experiment: the time series of anodic current plotted continuously over a time during which the control is switched on and then turned off. The derivative signal used to achieve control is superimposed on the bottom graph. The rotation rate = 1,400 rpm, the anodic potential with control off is $V_0 = 0.715$ V. Reproduced with permission from [17]. http://link.aps.org/abstract/PRE/v49/p5007. Copyright 1994 by the American Physical Society

In case (i) it was shown analytically that implementation of the derivative control strategy would stabilize the saddle of a bistable system in the vicinity of a cusp bifurcation. The cusp is present, e.g., in the dynamical system $\mathrm{d}y/\mathrm{d}t = (\mu y - y^3)$, with $\mu > 0$. Of three steady-states 0 and $\pm\sqrt{\mu}$, the first one is a saddle which was stabilized in numerical computations. In case (ii) as an example there was chosen a numerical model corresponding to the Langmuir–Hinshelwood mechanism of reactions occurring in an adsorbed state on catalytic surfaces. If x and y denote the surface coverages of two reactants, respectively, their dynamics is described by the following equations, consisting of terms describing the adsorption, desorption, and two-molecular reaction between X and Y species (with Y needing two adsorption sites):

$$\frac{\mathrm{d}x}{\mathrm{d}t} = k_1(1 - x - y) - k_2x - k_3xy \tag{7.6}$$

$$\frac{\mathrm{d}y}{\mathrm{d}t} = k_4(1 - x - y)^2 - k_5y^2 - k_3xy \tag{7.7}$$

In the bistable regime there are two stable states with predominant x and y coverages, respectively, separated by a saddle point where appreciable amounts of x and y coexist on the surface. For example, for the parameter set [k_1, k_2, k_3, k_4, k_5] = [8, 1, 60, 15, 0, 2], the two stable fixed points are $(x,y) = (0.82, 9.1 \times 10^{-3})$ and $(x,y) = (0.25, 0.247)$, while the saddle point, which acquires stability under derivative control, is characterized with $(x,y) = (0.335, 0.177)$.

Finally, in case (iii), the model of electrochemical corrosion suggested by Talbot and Oriani [21] was used (see also Sect. 6.1.2.2, volume I) which in dimensionless form is defined by equations:

$$\frac{dy}{d\tau} = p(1 - \theta_{OH}) - qy \tag{7.8}$$

$$\frac{d\theta_{OH}}{d\tau} = y(1 - \theta_{OH}) - \exp(-\beta\theta_{OH})\theta_{OH} \tag{7.9}$$

with y meaning the concentration of metals ions dissolved into the solution. The stabilization of the saddle fixed point for two set of control parameters was achieved.

Before this work, the derivative control strategy was used only to suppress the oscillatory (chaotic or periodic) behavior. Thus, the novelty of the above results lies in showing that derivative control in combination with an initial perturbation can be used to stabilize also the saddle points in the bistable regime. Noteworthy, earlier successful approach to stabilize saddle in the bistable regime, but using different feedback scheme, was published by Hjelmfelt and Ross [22], who, by appropriate regulation of the inflow of I^- ions to the stirred reactor (CSTR) containing the components of the homogeneous ClO_2^-–I^- oscillator, were able to suppress the oscillations and stabilize otherwise directly unobservable unstable steady-state.

Finally, the derivative feedback control was applied by Strasser et al. [23] to selected electrochemical oscillators under galvanostatic and potentiostatic conditions. Then, the time derivative of a measured quantity—the current or the concentration of an electroactive species, was fed back to the externally applied voltage or current, respectively. For that purpose the following experimental systems and their theoretical models were employed: H_2 oxidation on polycrystalline Pt, in the presence of Cu^{2+} and Br^- or Cl^- poisons, and formic acid oxidation on the polycrystalline Pt electrode. Two variants of derivative feedbacks were tested: $(dI/dt)/U$ control and the $(d\theta/dt)/I$ control. The latter type was implemented by means of the Au ring electrode, which served as a probe for the instantaneous values of $d\theta_{Br}/dt$ or $d\theta_{Cu}/dt$ during galvanostatic H_2 oxidation. The detection of these species was based on the oxidation of Br^- to Br_2 or the reduction of Cu^{2+} to Cu, at respective potentials. Theoretical treatment involved linear stability analysis, with the results presented in the form of the Jacobian sign patterns. Special emphasis was for the subclassification of HN-NDR oscillators, with respect to the nature of the current providing species (cf. Sect. 3.4, volume I).

7.3 Application of Delayed-Feedback Control

Stabilization of unstable steady-states and periodic responses can also be achieved using the delayed-feedback control strategy introduced by Pyragas [24] and further studied by Just et al. [25] in terms of linear stability analysis. Parmananda et al. [26]

have applied this approach to both the model and experimental electrochemical systems, and obtained suppression of periodic and chaotic oscillations to the steady-state, as well as conversion of chaotic into periodic dynamics. Let x_1 denote, e.g., the faradaic current (measurable system variable) and p_1—the accessible control parameter (e.g., the electrode potential) which can be continuously perturbed in the following way:

$$p_1(t) = p_1(0) + \gamma[x_1(t) - x_1(t - \tau)] \tag{7.10}$$

where $p_1(0)$ is the initial value of the parameter and the term $\gamma[x_1(t) - x_1(t - \tau)]$ defines the superimposed delayed feedback, which changes only the stability of the target dynamics without altering the location of fixed points and/or creating new periodic orbits in the phase space [16]. If $\tau \to 0$, the delayed-feedback technique reduces to the derivative control described above, used for stabilizing steady-states, as discussed above. If $\tau \to \tau_{\mathrm{UPO}}$, where τ_{UPO} is the period of an unstable periodic orbit (UPO) embedded in a strange attractor, the method reduces to the original Pyragas method [24] of controlling chaos. Thus, if τ is varied within the limits $0 < \tau \leq \tau_{\mathrm{UPO}}$, the method described here can be used to stabilize both unstable steady-state and periodic solutions. Numerical studies were performed for the model of the electrochemical dissolution of metals used also earlier [Eqs. (7.2)–(7.4)]. A value of parameter p was changed in order to introduce the control, according to the following particular form of Eq. (7.10):

$$p_1(t) = p_1(0) + \gamma[\theta_{\mathrm{OH}}(t) - \theta_{\mathrm{OH}}(t - \tau)] \tag{7.11}$$

Depending on the parameters, various initial states and effects of control could in fact be achieved in model calculations: stabilization of an unstable steady-state in the periodic oscillatory regime and the stabilization of an unstable steady-state in the chaotic oscillatory regime. Note that in both these cases, when the target state is reached, i.e., when $\theta_{\mathrm{OH}}(t) = \theta_{\mathrm{OH}}(t - \tau)$, the time-dependent perturbation vanishes. This is an important feature of the delayed-feedback control approach, since it warrants that no new steady-state solution is being created, but only the stability of the original fixed point is reversed, i.e., in this case the unstable steady-state transforms into unstable one. For condition $\tau = \tau_{\mathrm{UPO}}$, meaning the Pyragas method of controlling chaos, it was transformed into periodic regime (Fig. 7.9).

As an experimental system, the anodic dissolution of rotating Cu disk in acetate buffer medium was studied, in a three-electrode arrangement. In order to create a delayed-feedback control, the potential of Cu anode (measured with respect to SCE), and denoted below as V, was continuously perturbed according to the formula:

$$V(t) = V(0) + \gamma[I(t) - I(t - \tau)] \tag{7.12}$$

where I is the anodic current. In this way it was possible to stabilize the unstable state for the case of periodic oscillations in an unperturbed system.

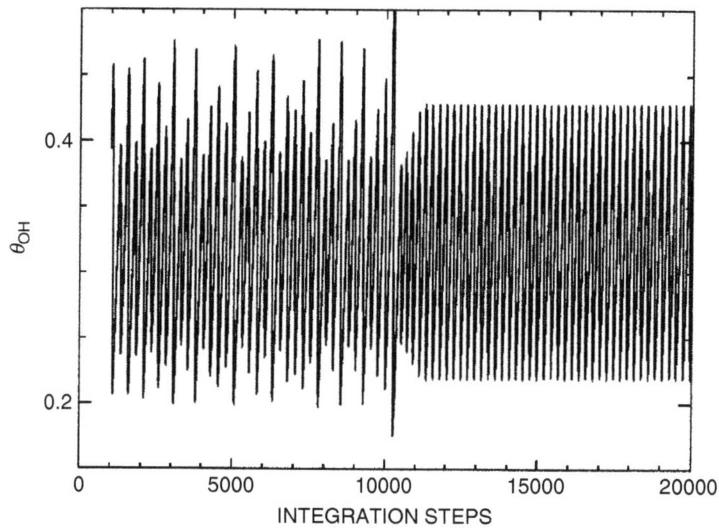

Fig. 7.9 Controlling chaos in the model Eqs. (7.2)–(7.4) using the Pyragas method. Parameters $\gamma = -0.11$ and $\tau = \tau_{UPO} = 24$ integration steps. See [26] for other parameters. Reproduced with permission from [26]. http://link.aps.org/abstract/PRE/v59/p5266. Copyright 1999 by the American Physical Society

Less perfectly, but still successfully it appeared possible to suppress the chaotic regime and stabilize the unstable steady-state.

Finally, using Pyragas method, the UPO embedded in the strange attractor was successfully stabilized (Fig. 7.10).

More recently, Kiss et al. [27] have used delayed-feedback control to tracking unstable steady-states and periodic orbits of oscillatory and chaotic electrochemical systems. Two experimental systems were used for the application of the control procedure. For tracking the unstable steady-states, the anodic Ni dissolution in sulfuric acid was chosen (cf. Sect. 6.1.4, volume I), while for tracking of UPOs the chaotic electrodissolution of copper in phosphoric acid (cf. Sect. 6.1.3.3, volume I) was employed. For programming the perturbation on the system's parameter V, either a simple feedback formula:

$$\delta V(t) = K[i(t) - i(t - \tau)] \tag{7.13}$$

or an extended time-delay autosynchronization (ETDAS) method was employed:

$$\delta V(t) = K\left[i(t) - (1 - r)\sum_{k=1}^{\infty} r^{k-1} i(t - k\tau)\right] \tag{7.14}$$

where $i(t)$ is a state variable at time t, or at time with a delay $k\tau$, respectively, K is the control gain. In turn, $-1 \leq r < 1$ is a control parameter which regulates the weight of information from the past, with $r \to 0$ meaning the simple delayed feedback, and $r \to 1$ required for stabilization of highly unstable orbits. Tracking

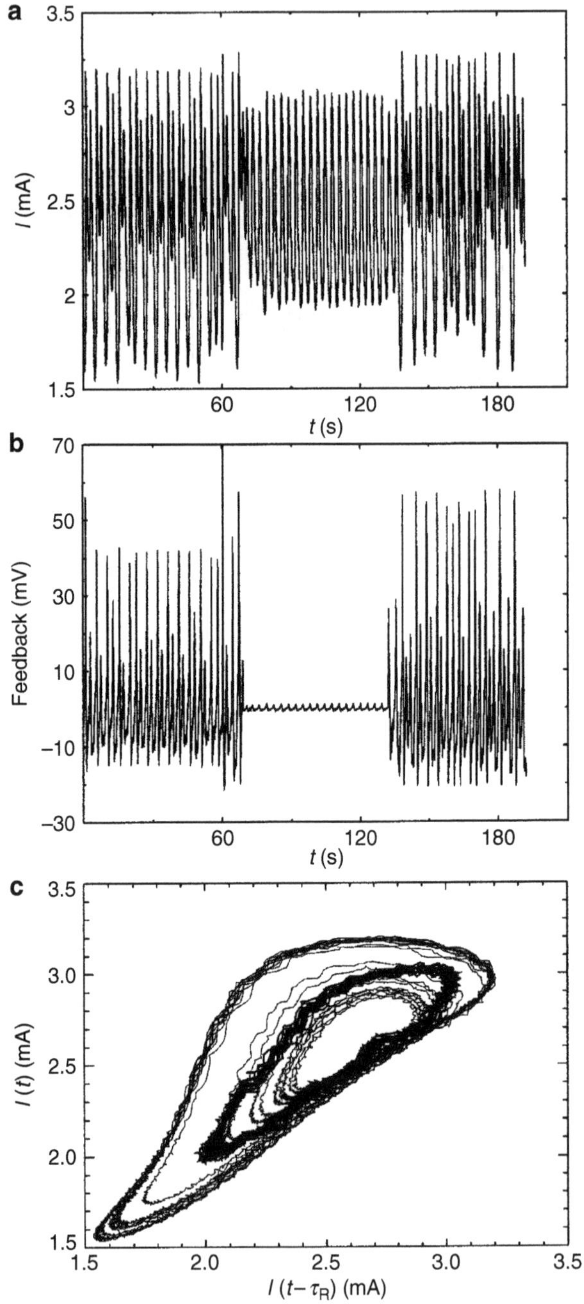

Fig. 7.10 Controlling chaos in the electrochemical system using the Pyragas method. The rotation rate is 2,900 rpm, while the anodic potential $V(0)$ is 0.715 V. The control parameters are $\gamma = -0.19$ mV/mA and $\tau = 65$ sampling steps. (**a**) Anodic current plotted over a period time during which the control is switched off, on, and off again. (**b**) Calculated feedback signal plotted over the same period but applied during the control session only. (**c**) Phase-space reconstruction for a segment of time series from Fig. 7.10a. The value of τ_R used for reconstruction is the period of 30 sampling steps. Reproduced with permission from [26]. http://link.aps.org/abstract/PRE/v59/p5266. Copyright 1999 by the American Physical Society

of unstable orbits is achieved by continuous updating K, r and τ, with the last parameter being of greatest importance. In this way the unstable steady-state in the Ni electrodissolution, as well as unstable orbit belonging to the strange attractor of the copper electrodissolution in phosphoric acid, were stabilized.

Furthermore, these tracking algorithms were used for constructing the experimental bifurcation diagrams of both systems, indicating the location of Hopf, saddle-node, saddle-loop, and period-doubling bifurcations. One should note that typical bifurcation diagrams are usually determined based on observable (stabilized) states, while the present method allowed to complement them with unstable states (not observable without control techniques).

7.4 Application of Sinusoidal Forcing

Besides specially programmed system's perturbation, the simple external sinusoidal modulation imposed on oscillating electrochemical system can induce control of its dynamics. Other examples of the role of such perturbations are described in Sect. 3.2, while here we concentrate on the stabilization effects. Such method of control is of *non-feedback* type, since the perturbation does not depend on the actual system's state and thus it is an approach based on principle alternative to OGY and analogous algorithms of controlling chaos. In electrochemistry, the application of sinusoidal forcing was described in terms of both numerical model and experimental studies by Parmananda et al. [28, 29]. The modulation of the control parameter p_1, e.g., the potential of metallic anode, is described with the following dependence:

$$p_1 = p_1(0) + \gamma \sin(\omega t) \tag{7.15}$$

Theoretical analysis of the transformation of system's dynamics subject to such perturbation was analyzed in terms of three-dimensional model of oscillatory electrochemical dissolution [Eqs. (7.2)–(7.4)]. The following phenomena were obtained: (i) period-1 oscillations emerging from chaotic regime (chaos → P1 transition), (ii) period-3 oscillations emerging from chaotic regime (chaos → P3), (iii) period-2 oscillations stabilized in the region of period-1 oscillations (P1 → P2), and (iv) period-3 oscillations stabilized in the region of period-1 oscillations (P1 → P3). The latter two cases mean transformation to oscillatory dynamics of higher periodicity. Cases (i) and (iv) are illustrated by exemplary Figs. 7.11 and 7.12.

While the fact that suppression of chaotic behavior to periodic states via the parameter entrainment was discussed in physical literature [30–33], the increase of dynamics complexity, induced in this way, is a new result.

For verification of the above theoretical predictions, the experimental system involving again electrodissolution of rotating Cu disk electrode in acetate buffer media, with the SCE and Pt foil serving as the reference and counter electrodes, respectively, was employed. The potentiostatically controlled potential of the copper electrode was continuously perturbed by the external sinusoidal voltage signal: $V(t) = V(0) + \gamma[\sin(2\pi\nu t)]$. Experimental studies exhibited all theoretically

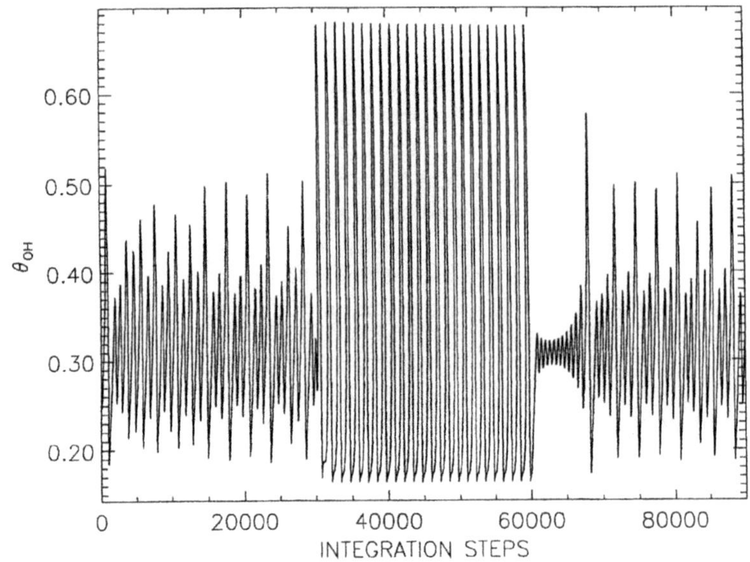

Fig. 7.11 Stabilization of a period-1 (P1) state in the model Eqs. (7.2)–(7.4) exhibiting chaotic oscillations. Parameters: $p = 2.0\times10^{-4}$; $q = 1.0\times10^{-3}$; $r = 2.0\times10^{-5}$; $s = 9.7\times10^{-5}$; $\beta = 5.0$. control parameters in Eq. (7.15): $\gamma = 5\times10^{-5}$; $\omega = 0.005$ rad s^{-1}. Reprinted from [28]. Copyright 1999, with permission from Elsevier

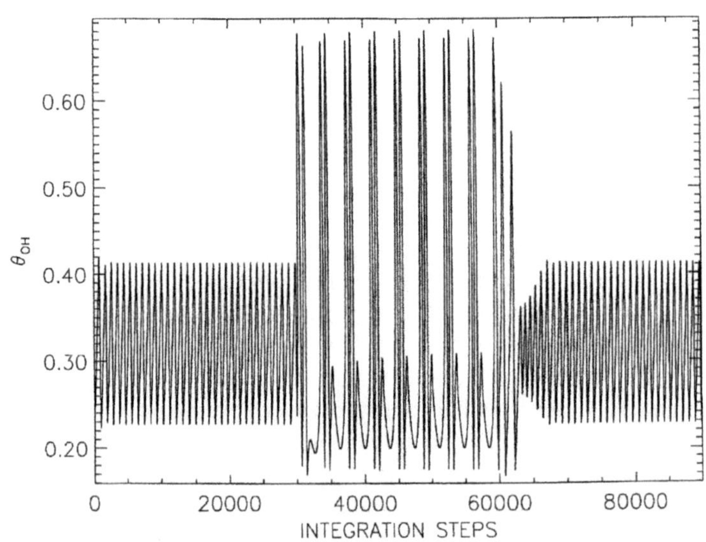

Fig. 7.12 Stabilization of a period-3 (P3) state in the model Eqs. (7.2)–(7.4) exhibiting period-1 oscillations. Parameters: p, q, r, β - as for Fig. 7.11, $s = 9.61\times10^{-5}$. Control parameters in Eq. (7.15): $\gamma = 5\times10^{-5}$; $\omega = 0.0017$ rad s^{-1}. Reprinted from [28]. Copyright 1999, with permission from Elsevier

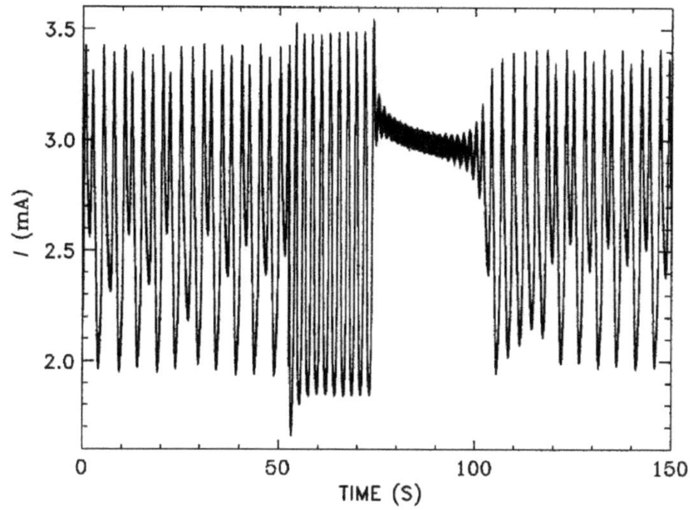

Fig. 7.13 Stabilization of a period-1 (P1) state in the electrochemical system Cu/acetate buffer exhibiting chaotic oscillations. The rotation rate is 2,500 rpm, while the anodic potential $V(0)$ is 0.770 V. Control parameters: $\gamma = 39.1$ mV; $\nu = \omega/2\pi = 0.7$ Hz. Anodic current is plotted over a period during which the external forcing control is switched off, on, and off again. Reprinted from [28]. Copyright 1999, with permission from Elsevier

predicted transitions: (chaos → P1), (chaos → P3), (P1 → P2), and (P1 → P3) [28, 29]. One should emphasize that, as other discussed here strategies, acquiring the control of the system's dynamics did not require prior knowledge of the reaction mechanism. The concordance between the (idealized) theoretical results and real experimental system indicates also that the applied control strategy appears to be robust to experimental noise. Exemplary Figs. 7.13 and 7.14 show experimental results showing the same types of transitions as Figs. 7.11 and 7.12.

7.5 Application of Sinusoidal Forcing to Control of Spatio-Temporal Behavior

In previous section the control of the temporal instabilities was described. Parmananda et al. [29] have summarized these approaches and presented their extension for the control of complex *spatio-temporal* behavior. Theoretical approach was based on *partial* differential equations and the electrochemical cell involving multiple anodes was used for corresponding experiments.

Generally, if the system consists of coupled non-linear oscillatory units, its behavior depends on the nature of coupling and the dynamical evolution of a local oscillator. If the system's parameters can be tuned, various dynamical regimes can emerge, including strongly disordered (chaotic) behavior. Then, control of such spatio-temporal complexity can be achieved by applying either global or local forcing.

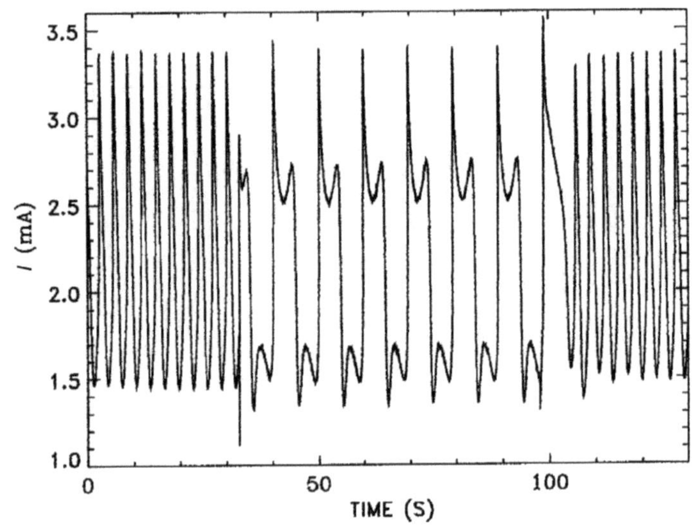

Fig. 7.14 Stabilization of a period-3 (P3) state in the electrochemical system exhibiting period-1 oscillations. The rotation rate is 1,700 rpm, while the anodic potential $V(0)$ is 0.680 V. Control parameters: $\gamma = 70.2$ mV; $\nu = \omega/2\pi = 3$ Hz. Anodic current is plotted over a period during which the external forcing control is switched off, on, and off again. Reprinted from [28], Copyright 1999, with permission from Elsevier

For studies of the effect of global forcing, the *model* for CO oxidation on a Pt (110) single crystal under UHV conditions, which proceeds via the Langmuir–Hinshelwood mechanism and associated surface structural changes, was employed. The spatial coupling of adjacent regions on the catalyst is provided by surface diffusion of CO [34, 35]. The model may exhibit traveling pulse behavior, amplitude turbulence, and phase turbulence. Now, in order to implement control, the entire array of oscillators, exhibiting amplitude turbulence, was perturbed with periodic sinusoidal forcing [$\gamma\sin(\omega t)$]. In consequence, the dynamics were stabilized on a spatially homogeneous and a temporally periodic state. For comparison with the *experiment*, the $Fe/(H_2SO_4 + Na_2SO_4)$ electrochemical system was used, in which the anode was an array of iron electrodes shrouded by epoxy, as described before by Fei and Hudson [36] (see Fig. 7.15).

The individual currents of the electrodes in the array were measured. In the absence of external forcing, different, *chaotic* current responses of individual electrodes were reported. When periodic forcing of the electrode potential was imposed to *all* the electrodes, the dynamical behavior of *each* electrode was converted to *regular* (periodic) dynamics (see representative Figs. 7.16 and 7.17).

In turn, the application of *local* forcing was inspired by earlier reports [37, 38] that high-dimensional chaotic dynamics (typical of spatio-temporal systems) could be converted to a well-defined ordered state without having perturbed the entire system, i.e., only locally. Accordingly, in the numerical model, only one of the sites of the spatially extended system was perturbed. For the appropriate choice of the control parameters, the model dynamics exhibited the induction of the order and its

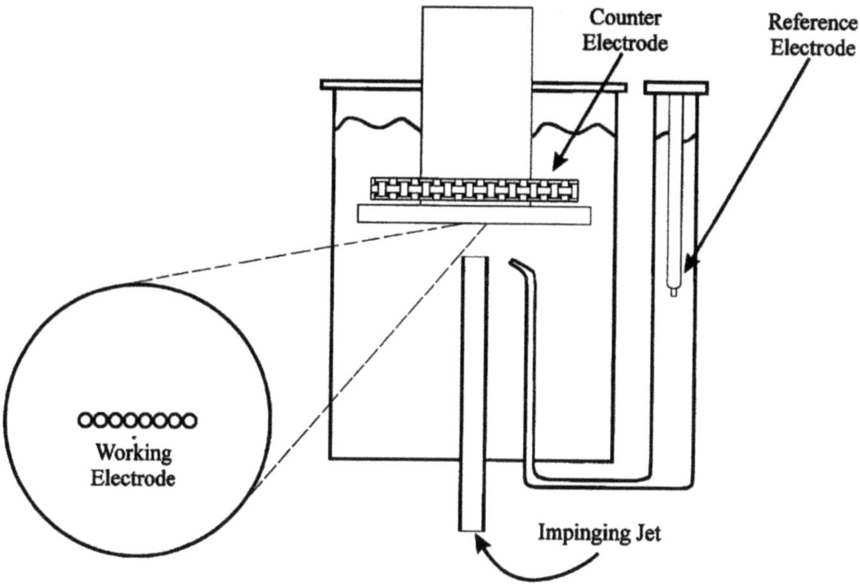

Fig. 7.15 Experimental setup used to observe and subsequently control the spatio-temporal complexity observed during potentiostatic dissolution of iron in sulfuric acid. Reprinted from [29]. Copyright 2005, with permission from Elsevier

subsequent propagation up until complete suppression of spatio-temporal complexity was achieved. The controlled state comprised of stable wave trains, propagating in one spatial dimension. These coherent waves were initiated at the point of stimulation and traversed the entire extended system [29]. For the experimental verification, the same experimental setup, as described above for global forcing, was used, with eight iron electrodes shrouded by epoxy. In the absence of forcing, irregular current responses of each electrodes were again observed. When now only one of the electrodes (No. 1) was perturbed sinusoidally in the galvanostatic mode, this chaotic dynamics have been converted to *period-1 oscillations for all the electrodes*. Thus, locally applied perturbation was able to suppress the global spatio-temporal complexity. The evolution of this phenomenon, as in the model, started from the perturbed electrode and spread over the entire set, but the propagation of the wave was not smooth (the second and the third electrode almost instantaneously switched to periodic dynamics, while the fourth electrode joined this regime with a little delay, the fifth and next electrodes did it with larger delay). However, one should not expect too close similarities between the model, in which coupling between the sites was realized through surface diffusion only, while in the electrochemical system also coupling through migration and convection operates.

The above results confirm the advantage of periodic forcing strategies in control of dynamical systems, both with respect to temporal and (particularly impressive) dynamics. The authors of the work [29] indicate also possible references of their results with an information theoretical context in biology. For example, local temporal information of biological relevance could be in this way encoded in a spatial pattern that creates corresponding temporal signals at distinct sites [37]. Furthermore, the pattern formation inside a cell could act as a biologically relevant encoding mechanism to transfer extracellular signals to targeted sites of biochemical action. Such situation was found experimentally by Camacho and Lechlieter

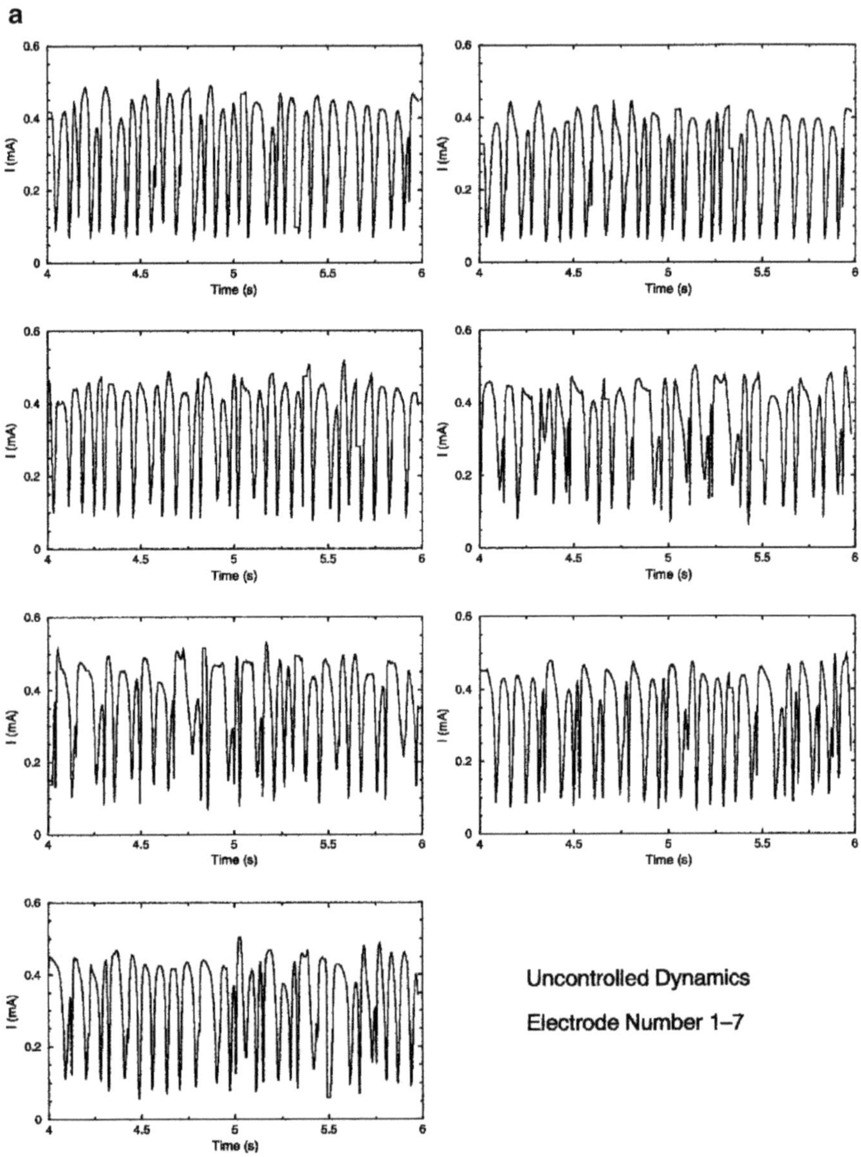

Fig. 7.16 (continued)

[39] who have reported experimental formation of a regular spatial calcium wave pattern following a local receptor activation by applying an external concentration level of bombesin to *Xenopus* oocytes.

The reader interested in control and synchronization of chaos in high-dimensional systems can also consult, e.g., the review paper by Ding et al. [40].

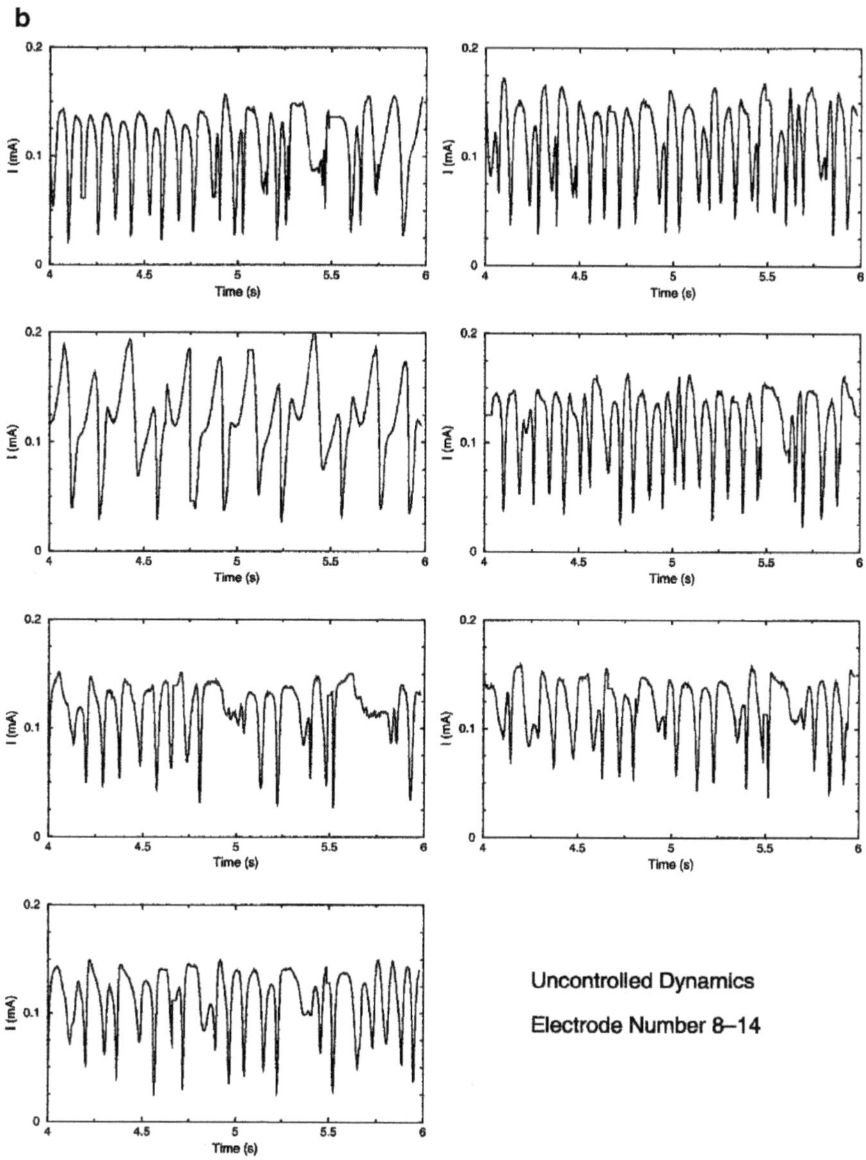

Fig. 7.16 (continued)

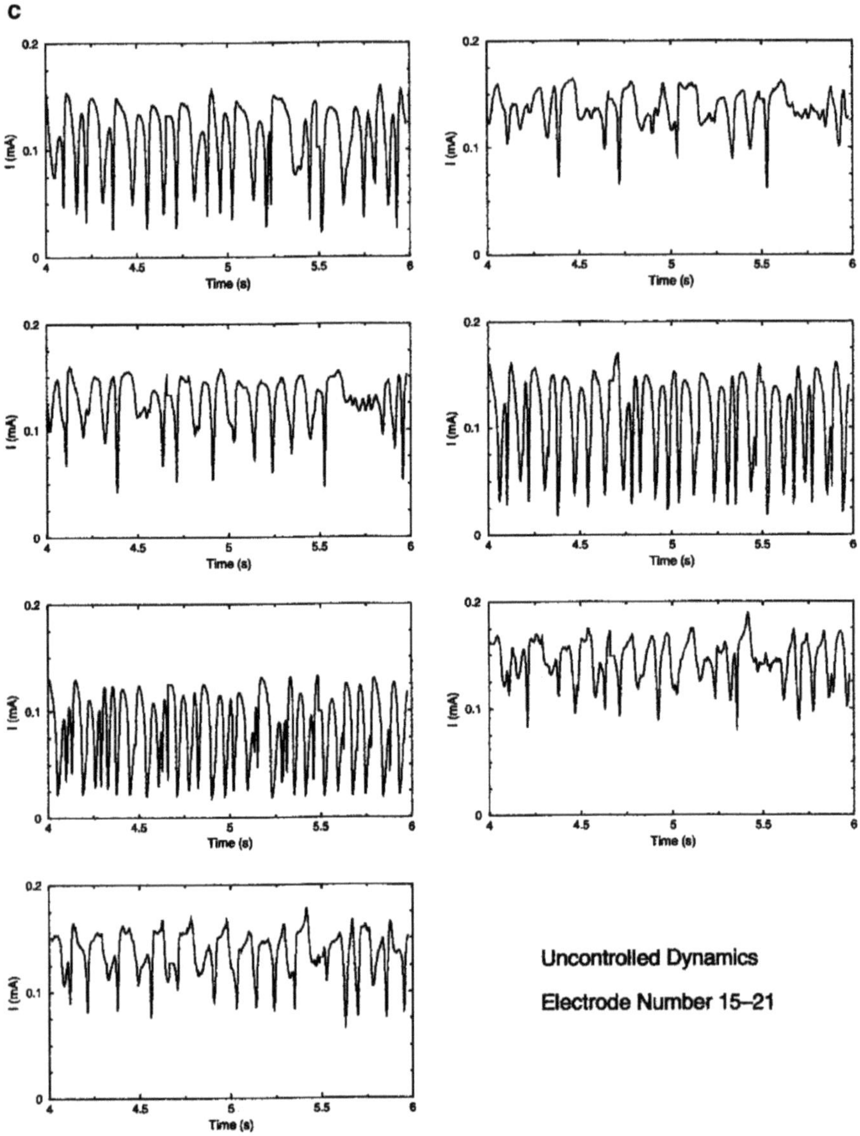

Fig. 7.16 Experimental chaotic uncontrolled time series of the 21 electrodes embedded in epoxy in a ring geometry [29]. Reprinted from [29]. Copyright 2005, with permission from Elsevier

7.6 Noise-Induced Order in Electrochemical Systems

This section of this chapter is devoted to external perturbation in the form of noise imposed on electrochemical systems. It was found earlier for various dynamical systems that the effect of noise can be counterintuitive, i.e., the noise can generate ordered behavior in chaotic systems [41]. In the year 1983, Matsumoto, Tsuda et al. [42–44] have described the *noise-induced order* obtained for the model of the chaotic course of the Belousov–Zhabotinsky process, in which the bifurcation

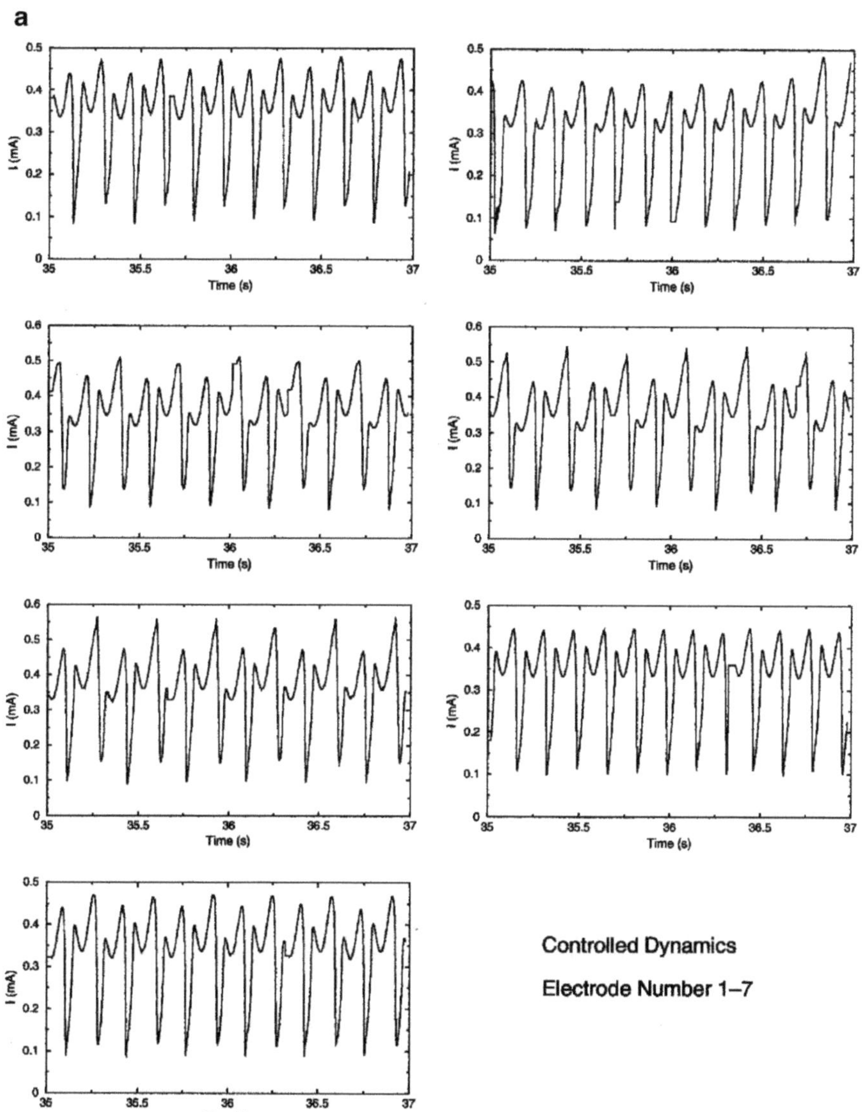

Fig. 7.17 (continued)

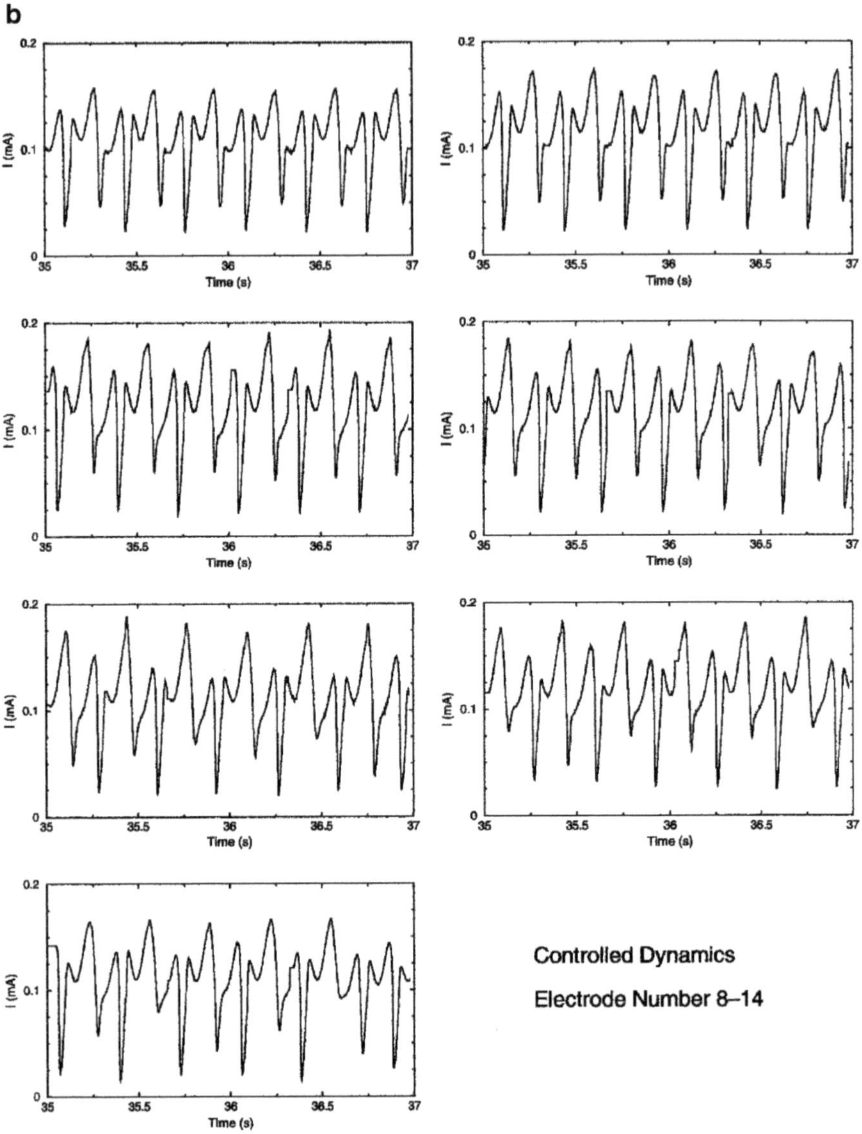

Fig. 7.17 (continued)

parameter was randomly perturbed. In consequence, the power spectrum indicated amplification of the oscillation amplitudes for certain region of frequencies or, in other words, the rather flat power spectrum for chaotic system exhibited a distinct peak for medium range of frequencies. Numerous examples of such *stochastic resonance* (SR) were reported for real physical and chemical systems.

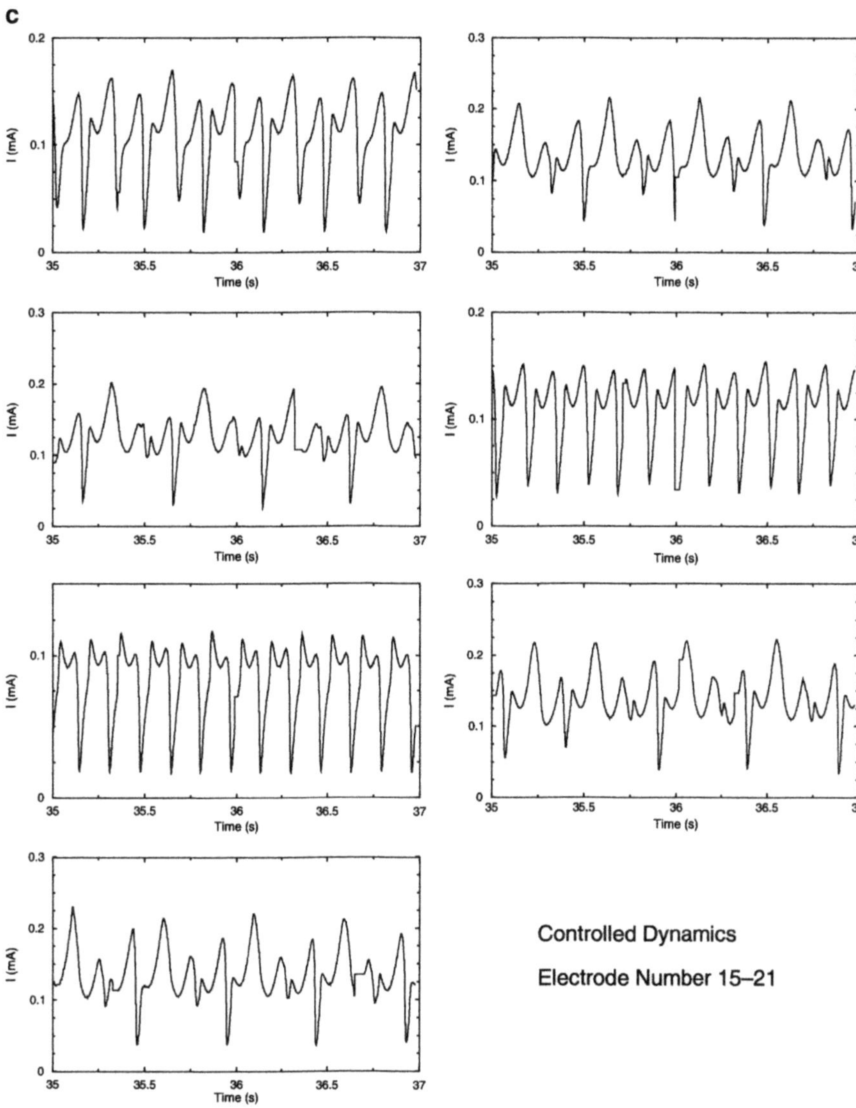

Fig. 7.17 Experimental controlled time series of the 21 electrodes when subjected to global forcing. Reprinted from [29]. Copyright 2005, with permission from Elsevier

Here we shall describe the specific effect of external noise in electrochemical systems, reported by Kiss et al. [45] who have found the experimental and numerical evidence for the *coherence resonance* (CR) in the anodic nickel dissolution in sulfuric acid and in its mathematical model (cf. Sect. 6.1.4, volume I). The occurrence of this resonance means the emergence of coherence in *noise-induced oscillations*[1]. External noise with a Gaussian distribution was superimposed on the system when the anodic current exhibited steady-state dynamics just below a supercritical Hopf bifurcation. When the amplitude of the stochastic perturbations was monotonically increased, the regularity of the noise-induced oscillations reached a maximum value for an optimum noise-level. Quantitatively, the periodicity of the noise-provoked oscillations was expressed in terms of the coherence factor β:

$$\beta = H/W \tag{7.16}$$

where H is the height and W is relative width of the dominant peak of the power spectrum. Figures 7.18 and 7.19 illustrate the CR for the experimental Ni/H_2SO_4 system. Analogous results were obtained for the corresponding two-variable, dimensionless Haim et al. model of Ni electrodissolution.

In the explanation of this phenomenon, the idea of noisy precursors to a Hopf bifurcation was invoked, meaning that the frequencies of the induced oscillations were close to that of the autonomous oscillations above a neighboring bifurcation point. One should note the more general importance of the phenomenon of CR, as this concept was used in the FitzHugh–Nagumo model of neurons [46] and for other physical dynamical systems.

Let us also note that all real systems exhibit inherent *internal* noise. The electrochemical noise (EN) is used to analyze mainly corrosion, and also deposition processes (cf. e.g., [47–49]). In the latter case it is called electrocrystallization EN (cf. e.g., [50–53]). Recently, Zhang et al. [54] have studied the zinc electroplating process using EN technique, in conjunction with the scanning electron microscopy (SEM). It was found that the EN generated during the galvanostatic electroplating of dendritic (cf. Sect. 2.11) or large conglomerate zinc deposit had large potential oscillation amplitude and positive potential drift, while the compact zinc deposit possessed small noise amplitude and little potential drift. These dependences were attributed to the change of the rate determining step from diffusion-control, through the mixed-control to the activation (or charge-transfer) control, associated with the corresponding changes of the deposit structure from dendritic to compact. Note that similar correlations, although referring to regular oscillations rather than noisy dynamics, were outlined in Sect. 2.11. In conclusion, Zhang et al. [54] claim that EN technique could give more information about the electrodeposit structure and electroplating mechanism than other normal electrochemical measurements, such as linear potential sweep (including cyclic) methods.

But in view of the content of the present section it is of primary importance to find out whether such internal noise can play an analogous, ordering role as an external

[1] Coherence resonance (CR), as meaning the emergence of coherence in noise-induced oscillations, is thus essentially different from stochastic resonance (SR) - amplification of deterministic (periodic) signal upon addition of noise.

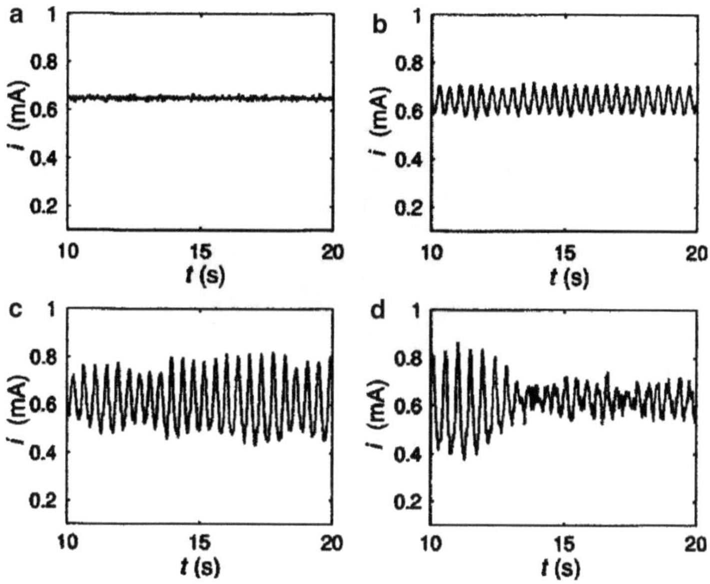

Fig. 7.18 Experiments. Time series of current at different noise strengths ($V_0 = 1.153$ V, $R = 250\ \Omega$). (**a**) without noise, noise amplitude $D = 0$ mV; (**b**) small added noise, $D = 3$ mV; (**c**) optimal coherence at intermediate noise strength, $D = 12$ mV; and (**d**) large noise amplitude, $D = 18$ mV. Reproduced with permission from [45] http://link.aps.org/abstract/PRE/v67/p035201. Copyright 2003 by American Physical Society

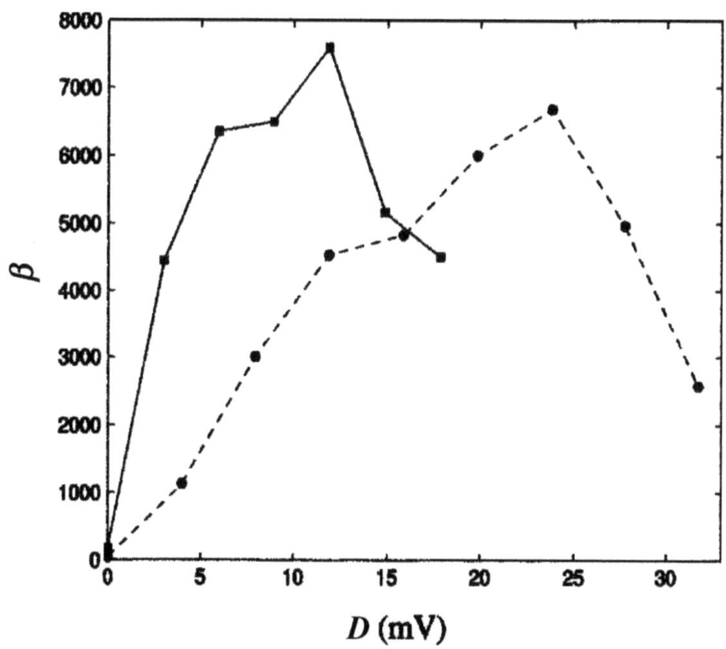

Fig. 7.19 Experiments. The coherence factor β as a function of noise amplitude D for just below ($V_0 = 1.153$ V, *squares*) and further away ($V_0 = 1.148$ V, *circles*) from the Hopf bifurcation point. Reproduced with permission from [45] http://link.aps.org/abstract/PRE/v67/p035201. Copyright 2003 by American Physical Society

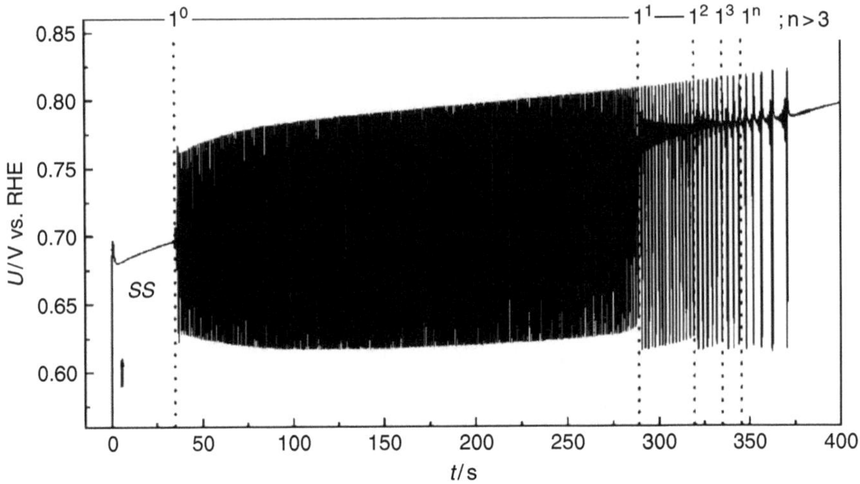

Fig. 7.20 Potential time series during galvanostatic (j = 1.00 mA cm^{-2}) electro-oxidation of methanol. The following sequence spontaneously evolves: the birth of harmonic oscillations from a stationary state (ss), the transition to mixed mode oscillations, and the death of oscillations at 371 s. Reprinted with permission from [61]. Copyright 2010 American Chemical Society

one, causing the CR. Schmid et al. [55] have shown, in terms of the Hodgkin–Huxley model of the excitable cell membrane, that *internal* noise caused by the fluctuations of individual ion channels in their assembly can induce the *intrinsic coherence resonance* (ICR). This term means that the internal noise caused a most regular spike activity, for an optimal size of the membrane patch. An analogous ICR phenomenon was reported in an electrochemical system by Rivera et al. [56]. The iron anode was polarized in solution of sulfuric acid and potassium sulfate, and as a source of internal noise, an appropriate amount of potassium chloride was added, causing pitting corrosion by Cl^- ions [57–60]. It was found that the dependence of the coherence factor (a quantitative measure of the extent of coherence) vs. the concentration of added Cl^- ions had a unimodal shape indicating thus the emergence of ICR. One can expect further development of studies of various noise effects in dynamical systems, including the electrochemical ones.

7.7 Stabilization of Oscillations in the Systems with Spontaneous Drift

To conclude with this section the content of this book, one should emphasize that controlling chemical instabilities is evidently one of the most important and promising subjects of contemporary studies in non-linear dynamics. In electrochemistry, besides various methods derived from physical studies, as the ones described above, also approaches more specifically oriented on the properties of the given system, are developed. Very recently, Nagao et al. [61] have reported

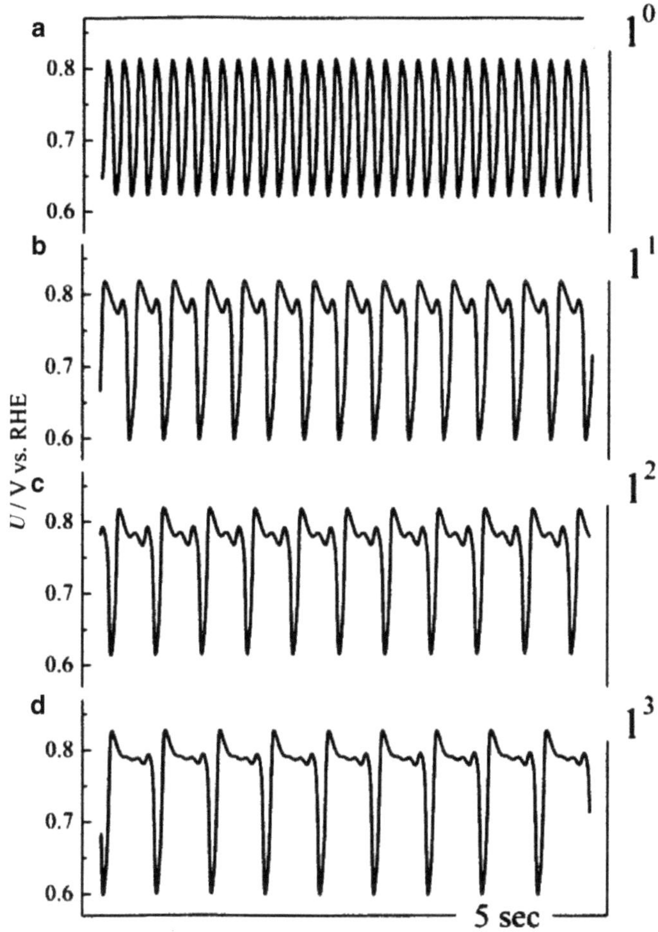

Fig. 7.21 Period-one and mixed mode oscillations stabilized by a *negative galvanodynamic sweep* of different rates for an initial current density of 1.00 mA cm^{-2}: (**a**) −2.39 μA s^{-1} cm^{-2}; (**b**) −1.00 μA s^{-1} cm^{-2}; (**c**) −1.33 μA s^{-1} cm^{-2}; (**d**) −0.58 μA s^{-1} cm^{-2}. Reprinted with permission from [61]. Copyright 2010 American Chemical Society

the stabilization of non-stationary time series in the electrocatalytic oxidation of methanol on platinum, in $HClO_4$ medium. The non-stationarity was observed under galvanostatic conditions when the spontaneously born, period-1 oscillations of the electrode potential underwent further a homoclinic transition to mixed-mode oscillations, along with slow drift of the average electrode potential towards more anodic values, leading eventually to the death of oscillations (see Fig. 7.20).

It was thus logical to stabilize the given oscillatory pattern, attained at initially imposed current density, by applying further the negative galvanostatic sweep (NGS), preventing the increase of the electrode potential. In this way, among others, simple period-one and mixed more L^S states were stabilized (see

Fig. 7.21), meaning that the number of given type of oscillatory cycles increased from 5 to 20 times.

The obtained experimental results were discussed in terms of the electrochemical reaction mechanism of the methanol oxidation. Finally, one should emphasize practical importance of such stabilization related to methanol-based fuel cells, the efficiency of which was found to increase under oscillatory regime, in dependence on their period and amplitude (cf. e.g., [62–65] and Sec. 3.2.1.2). This, in turn, appears to be another, perhaps of even more general significance, observation of efficiency increase (or dissipation decrease) under oscillatory conditions, as Richter and Ross have suggested earlier for glycolysis [66].

References

1. Schuster HG, Just W (2005) Deterministic chaos. An introduction, 4th edn. Wiley-VCH, Weinheim
2. Schöll E, Schuster HG (eds) (2008) Handbook of chaos control, 2nd edn. Wiley-VCH, Weinheim
3. Ott E, Grebogi C, Yorke JA (1990) Controlling chaos. Phys Rev Lett 64:1196–1199
4. Pecora LM, Tl C (1990) Synchronization in chaotic systems. Phys Rev Lett 64:821–824
5. Peng B, Petrov V, Showalter K (1991) Controlling chemical chaos. J Phys Chem 95:4957–4959
6. Peng B, Petrov V, Showalter K (1992) Controlling low-dimensional chaos by proportional feedback. Physica A 199:210–216
7. Parmananda P (1996) Controlling chaos in a model for an electrochemical oscillator using a map-based algorithm. Electrochim Acta 41:377–380
8. Koper MTM, Gaspard P (1991) Mixed-mode and chaotic oscillations in a simple model of an electrochemical oscillator. J Phys Chem 95:4945–4947
9. Koper MTM, Gaspard P (1992) The modeling of mixed-mode and chaotic oscillations in electrochemical systems. J Chem Phys 96:7797–7813
10. Kiss IZ, Gáspár V, Nyikos L, Parmananda P (1997) Controlling electrochemical chaos in the copper-phosphoric acid system. J Phys Chem A 101:8668–8674
11. Petrov V, Peng B, Showalter K (1992) A map-based algorithm for controlling low-dimensional chaos. J Chem Phys 96:7506–7513
12. Rollins RW, Parmananda P, Sherard P (1993) Controlling chaos in highly dissipative systems: a simple recursive algorithm. Phys Rev E 47:R780–R783
13. Dressler U, Nitsche G (1992) Controlling chaos using time delay coordinates. Phys Rev Lett 68:1–4
14. Parmananda P, Sherard P, Rollins RW, Dewald HD (1993) Control of chaos in an electrochemical cell. Phys Rev E 47:R3003–R3006
15. Kiss IZ, Gáspár V (2000) Controlling chaos with artificial neural network: numerical studies and experiments. J Phys Chem A 104:8033–8037
16. Bielawski S, Bouazaoui M, Derozier D, Glorieux P (1993) Stabilization and characterization of unstable steady states in a laser. Phys Rev A 47:3276–3279
17. Parmananda P, Rhode MA, Johnson GA, Rollins RW, Dewald HD, Markworth AJ (1994) Stabilization of unstable steady states in an electrochemical system using derivative control. Phys Rev E 49:5007–5011
18. Markworth AJ, McCoy JK, Rollins RW, Parmananda P (1992) In: Stringer J, Kim J (eds) Applied chaos. Wiley, New York, NY, p 227

19. McCoy JK, Parmananda P, Rollins RW, Markworth AJ (1993) Chaotic dynamics in a model of metal passivation. J Mater Res 8:1858–1865
20. Parmananda P, Eiswirth M (1996) Stabilizing unstable fixed points using derivative control. J Phys Chem 100:16568–16570
21. Talbot JB, Oriani RA (1985) Steady state multiplicity and oscillations in passive film formation. Electrochim Acta 1985:1277–1284
22. Hjelmfelt A, Ross J (1994) Experimental stabilization of unstable steady states in oscillatory and excitable reaction systems. J Phys Chem 98:1176–1179
23. Strasser P, Lübke M, Parmananda P, Eiswirth M, Ertl G (1998) Mechanistic analysis of electrochemical oscillators using derivative feedback control techniques. J Phys Chem B 102:3227–3237
24. Pyragas K (1992) Continuous control of chaos by self-controlling feedback. Phys Lett A 170:421–428
25. Just W, Bernard T, Ostheimer M, Reibold E, Benner H (1997) Mechanism of time-delayed feedback control. Phys Rev Lett 78:203–206
26. Parmananda P, Madrigal R, Rivera M, Nyikos L, Kiss IZ, Gáspár V (1999) Stabilization of unstable steady-states and periodic orbits in an electrochemical system using delayed-feedback control. Phys Rev E 59:5266–5271
27. Kiss IZ, Kazsu Z, Gáspár V (2006) Tracking unstable steady states and periodic orbits of oscillatory and chaotic electrochemical systems using delayed feedback control. Chaos 16:033109-1–0033109-7
28. Parmananda P, Rivera M, Madrigal R (1999) Altering oscillatory dynamics of an electrochemical system using external forcing. Electrochim Acta 44:4677–4683
29. Parmananda P, Rivera M, Green BJ, Hudson JL (2005) Controlling complexity using forcing: simulations and experiments. Appl Math Comput 164:467–491
30. Hubler A (1989) Adaptive control of chaotic systems. Helv Phys Acta 62:343–346
31. Jackson EA (1990) The entrainment and migration controls of multiple-attractor systems. Phys Lett A 151:478–484
32. Mettin R, Kurz T (1995) Optimized periodic control of chaotic systems. Phys Lett A 206:331–339
33. Mettin R, Hubler A, Scheeline A, Lauterborn W (1995) Parametric entrainment control of chaotic systems. Phys Rev E 51:4065–4075
34. Bär M, Gottschalk N, Eiswirth M, Ertl G (1994) Spiral waves in a surface reaction: model calculations. J Chem Phys 100:1202–1214
35. Bär M, Eiswirth M (1993) Turbulence due to spiral breakup in a continuous excitable medium. Phys Rev E 48:R1635–R1637
36. Fei Z, Hudson JL (1998) Chaotic oscillations on arrays of iron electrodes. Ind Eng Chem Res 37:2172–2179
37. Baier G, Sven S, Chen JP, Hoff AA (1999) Local stimulation induces long-range order in spatio-temporal disorder. J Chem Phys 110:3251–3255
38. Parmananda P, Eiswirth M (1999) Suppression of chemical turbulence using feedbacks and forcing. J Phys Chem A 103:5510–5514
39. Camacho P, Lechleiter JD (1995) In: Brock GR, Ackrill K (eds) Calcium waves, gradients and oscillations. Wiley, Chichester
40. Ding M, Ding EJ, Ditto WL, Gluckman B, In V, Peng JH, Spano ML, Yang W (1997) Control and synchronization of chaos in high dimensional systems: review of some recent results. Chaos 7:644–652
41. Serra R, Andretta M, Compiani M, Zanarini G (1986) Introduction to the physics of complex systems. Pergamon Press, Oxford
42. Matsumodo K, Tsuda I (1983) Noise-induced order. J Stat Phys 31:87–106
43. Matsumoto K (1984) Noise-induced order II. J Stat Phys 34:111–127
44. Doi S (1989) A chaotic map with a flat segment can produce a noise-induced order. J Stat Phys 55:941–964

45. Kiss IZ, Hudson JL, Escalera Santos GJ, Parmananda P (2003) Experiments on coherence resonance: noisy precursors to Hopf bifurcation. Phys Rev E 67:035201-1–035201-3
46. Lindner B, Schimansky-Geier L (2000) Coherence and stochastic resonance in a two-state system. Phys Rev E 61:6103–6110
47. Cottis RA (2001) Interpretation of electrochemical noise data. Corrosion 57:265–285
48. Bertocci U, Huet F, Jaoul B, Rousseau P (2000) Frequency analysis of transients in electrochemical noise: mathematical relationships and computer simulations. Corrosion 56:675–683
49. Darowicki K, Zieliński A (2004) The analysis of stationary electrochemical noise. Pol J Chem 78:1261–1268
50. Hudson JL, Tsotsis TT (1994) Electrochemical reaction dynamics: a review. Chem Eng Sci 49:1493–1572
51. Blanc G, Gabrielli C, Keddam M (1975) Measurement of the electrochemical noise by a cross correlation method. Electrochim Acta 20:687–689
52. Switzer JA, Huang CJ, Huang LY, Switzer ER, Kammler DR, Golden TD, Bohannan EW (1998) Electrochemical self-assembly of copper/cuprous oxide layered nanostructures. J Am Chem Soc 120:3530–3531
53. Bonnefont A, Kostecki R, McLarnon F, Arrayet JC, Servant L, Argoul F (1999) *In situ* atomic force microscopy imaging of electrodeposition of mixed layers of copper/cuprous oxide. J Electrochem Soc 149:4101–4104
54. Zhang Z, Leng WH, Cai QY, Cao FH, Zhang JQ (2005) Study of the zinc electroplating process using electrochemical noise technique. J Electroanal Chem 578:357–367
55. Schmid G, Goychuk I, Hänggi P (2001) Stochastic resonance as a collective property of ion channel assemblies. Europhys Lett 56:22–28
56. Rivera M, Escalera Santos GJ, Uruchurtu-Chavarín J, Parmananda P (2005) Intrinsic coherence resonance in an electrochemical cell. Phys Rev E 72:030102-1–030102-4
57. Pagitsas M, Diamantopoulou A, Sazou D (2002) General and pitting corrosion deduced from current oscillations in the passive–active transition state of the Fe|H_2SO_4 electrochemical system. Electrochim Acta 47:4163–4179
58. Sazou D, Diamantopoulou A, Pagitsas M (2000) Chemical perturbation of the passive–active transition state of Fe in a sulfuric acid solution by adding halide ions. Current oscillations and stability of the iron oxide film. Electrochim Acta 45:2753–2769
59. Pagitsas M, Diamantopoulou SD (2003) A point defect model for the general and pitting corrosion of iron|oxide|electrolyte interface deduced from current oscillations. Chaos, Solitons Fract 17:263–275
60. Sazou D, Pagitsas M (2003) Non-linear dynamics of the passivity breakdown of iron in acidic solutions. Chaos, Solitons Fract 17:505–522
61. Nagao R, Sitta E, Varela H (2010) Stabilizing nonstationary electrochemical time series. J Phys Chem 114:22262–22268
62. Zhang J, Datta R (2005) Electrochemical preferential oxidation of CO in reformate. J Electrochem Soc 152:A1180–A1187
63. Lu H, Rihko-Struckman L, Hanke-Rauschenbach R (2009) Improved electrochemical CO removal via potential oscillations in serially connected PEM fuel cells with PtRu anodes. Electrochim Acta 54:1184–1191
64. Mota A, Lopes PP, Ticianelli EA, Gonzalez ER, Varela H (2010) Complex oscillatory response of a PEM fuel cell fed with H_2/CO and oxygen. J Electrochem Soc 157: B1301–B1304
65. Lopes PP, Ticianelli EA, Varela H (2011) Potential oscillations in a proton exchange membrane fuel cell with a Pd–Pt/C anode. J Power Sources 119:84–89
66. Richter PH, Ross J (1981) Concentration oscillations and efficiency: glycolysis. Science 211:715–717

About the Author

Marek Orlik studied at the Faculty of Chemistry of the University of Warsaw, where he in 1990 received his PhD for the studies of electrode mechanisms of cyanide complexes of transition metals, under the supervision of Zbigniew Galus. At the same University in 2000, he completed habilitation for the studies of self-organization phenomena in electrochemical systems. In 2004, he became an Associate Professor at the home Faculty, and in 2010, he was nominated a Titular (full) Professor of Chemistry. In the years 1996–1998, as a scholarship holder of the *Alexander von Humboldt Foundation*, he worked with Karl Doblhofer and Gerhard Ertl in the Fritz Haber Institute of the Max Planck Society in Berlin, studying self-organizing electro-convective phenomena. At the University of Linz (Austria), he cooperated with Gerhard Gritzner in the area of electrode kinetics in nonaqueous solvents and computer modeling of electrode processes. His current research activities, including both experimental and theoretical studies, focus on self-organization phenomena in electroreduction processes of complex compounds, in electrooxidation of metal electrodes, and in homogeneous redox processes involving hydrogen peroxide. He has published several dozen papers in the area of both classical electrode kinetics and dynamic instabilities in electrochemical and chemical processes. He is also an author of a book "Oscillating reactions. Order and Chaos," published in Poland in 1996 and a coauthor (with Zbigniew Galus) of three chapters, devoted to electrochemistry of gold, silver, and mercury in a recent edition of "Encyclopedia of Electrochemistry" (Wiley, 2006). Besides scientific and educational activity at the University of Warsaw, he also significantly contributes to the chemistry education at the secondary school level, through the long-year involvement in the Polish and International Chemistry Olympiads, as well as is the editor of the Polish journal for school teachers.

M. Orlik, *Self-Organization in Electrochemical Systems II*,
Monographs in Electrochemistry, DOI 10.1007/978-3-642-27627-9,

About the Editor

Fritz Scholz is Professor at the University of Greifswald, Germany. Following studies of chemistry at Humboldt University, Berlin, he obtained a Dr. rer. nat. and a Dr. sc. nat. (habilitation) from that University. In 1987 and 1989, he worked with Alan Bond in Australia. His main interest is in electrochemistry and electroanalysis. He has published more than 280 scientific papers, and he is editor and coauthor of the book "Electroanalytical Methods" (Springer, 2002, 2005, 2010, and Russian Edition: BINOM, 2006), coauthor of the book "Electrochemistry of Immobilized Particles and Droplets" (Springer 2005), coeditor of the "Electrochemical Dictionary" (Springer, 2008), and coeditor of volumes 7a and 7b of the "Encyclopedia of Electrochemistry" (Wiley-VCH 2006). In 1997, he has founded the *Journal of Solid State Electrochemistry* (Springer) and serves as Editor-in-Chief since that time. He is the editor of the series "Monographs in Electrochemistry" (Springer) in which modern topics of electrochemistry are presented. Scholz introduced the technique "Voltammetry of Immobilized Microparticles" for studying the electrochemistry of solid compounds and materials, he introduced three-phase electrodes to determine the Gibbs energies of ion transfer between immiscible liquids, and currently he is studying the interaction of free oxygen radicals with metal surfaces, as well as the interaction of liposomes with the surface of mercury electrodes in order to assess membrane properties.

M. Orlik, *Self-Organization in Electrochemical Systems II*,
Monographs in Electrochemistry, DOI 10.1007/978-3-642-27627-9,

Index

M. Orlik, *Self-Organization in Electrochemical Systems II*,
Monographs in Electrochemistry, DOI 10.1007/978-3-642-27627-9,

MIX
Papier aus verantwortungsvollen Quellen
Paper from responsible sources
FSC® C105338

If you have any concerns about our products,
you can contact us on
ProductSafety@springernature.com

In case Publisher is established outside the EU,
the EU authorized representative is:
Springer Nature Customer Service Center GmbH
Europaplatz 3, 69115 Heidelberg, Germany

Printed by Libri Plureos GmbH
in Hamburg, Germany